Quantum Waveguides

Theoretical and Mathematical Physics

The series founded in 1975 and formerly (until 2005) entitled *Texts and Monographs in Physics* (TMP) publishes high-level monographs in theoretical and mathematical physics. The change of title to *Theoretical and Mathematical Physics* (TMP) signals that the series is a suitable publication platform for both the mathematical and the theoretical physicist. The wider scope of the series is reflected by the composition of the editorial board, comprising both physicists and mathematicians.
The books, written in a didactic style and containing a certain amount of elementary background material, bridge the gap between advanced textbooks and research monographs. They can thus serve as basis for advanced studies, not only for lectures and seminars at graduate level, but also for scientists entering a field of research.

More information about this series at http://www.springer.com/series/720

Pavel Exner · Hynek Kovařík

Quantum Waveguides

Springer

Pavel Exner
Faculty of Nuclear Sciences and Physical Engineering
Czech Technical University
Prague
Czech Republic

and

Department of Theoretical Physics
Nuclear Physics Institute ASCR
Řež
Czech Republic

Hynek Kovařík
DICATAM, Sezione di Matematica
University of Brescia
Brescia
Italy

ISSN 1864-5879 ISSN 1864-5887 (electronic)
Theoretical and Mathematical Physics
ISBN 978-3-319-18575-0 ISBN 978-3-319-18576-7 (eBook)
DOI 10.1007/978-3-319-18576-7

Library of Congress Control Number: 2015938752

Mathematics Subject Classification: 81Q37, 58J50, 81Q35, 35P15, 35P25, 35J05, 35J10

Springer Cham Heidelberg New York Dordrecht London

Printed on acid-free paper

Springer International Publishing AG Switzerland is part of Springer Science+Business Media
(www.springer.com)

To Jana and Riccarda

Preface

The title of this book is short and one cannot resist thinking of Milan Kundera's observation that one quality we have lost is slowness. At times when books were not so numerous and readers were patient, one might have preferred to speak about quantum mechanics of particles confined to regions of tubular form, in particular, relations between their spectral and scattering properties and the geometry of confinement, *et cetera*. But habits are different nowadays, hence *quantum waveguides*, even if the guided objects are not exactly waves, and not a small part of what we are going to discuss concerns states in which the particles do not move. However, although the term we have coined may not be fully fitting, it has the advantage of linking the subject of the book to related problems in areas of classical physics such as acoustics and electromagnetism.

Guided quantum dynamics, as discussed in this book, attracted attention in the second half of the 1980s. The motivation came from two sources. On the one hand, new developments in solid-state physics called for a theoretical analysis of such effects, and on the other hand, from the mathematical point of view these questions opened new and unexplored areas in spectral geometry. The older one of the authors had been lucky to participate in those studies from the beginning, the younger one joined this effort a decade later. The subject proved to be rich and looking back at those years we see many interesting results obtained by numerous people; we feel that the time may be right to summarize the understanding achieved as well as to identify new challenges.

The questions we address in the book are physical, or at least they come from physics, and the instruments we use are mathematical. This means, in particular, that the claims are made with full rigor, the proofs being either given completely or sketched to a degree allowing the reader to fill in the details. Some of these exercises are delegated to problems accompanying each chapter. The level of those vary, some boil down to simple if tedious computations or extensions of the results derived in the main text, while others represent more complicated questions which may constitute the contents of a research paper.

Since mathematics is a tool we employ, not the goal, our theorems are formulated with a reasonable degree of generality, however, we do not strive for the weakest possible assumptions and a mathematically minded reader will find a lot of room for improvements. Technically speaking, our arguments come mostly from applied functionals analysis, but we also need results from differential geometry, probability, and other areas. We decided not to burden the book with appendices summarizing this material; we assume the reader is acquainted with the basic concepts and we provide references whenever we find it necessary.

Most problems discussed in the book involve various simple geometric considerations, and consequently, it would be easy to accompany the text with numerous drawings. We resist this temptation, believing the reader will profit from working these things out while going through the text. Old textbooks used to come with a parenthetical encouragement—(Draw a picture!)—but we are sure he or she would know when such a visual support is needed. In addition, many original papers we cite, including some of our own, are full of illustrations.

Dealing with problems of different kinds, we also have to think about the notation. We try to be consistent but not pedantic. For instance, we use vector notation at places where it is convenient due to a frequent use of components but drop the arrows elsewhere. Similarly, tensor notation is employed only when needed to work with objects like curved surfaces, layers, or networks, etc.

Since our goal is to provide a summary of the research activities of numerous people over a quarter of a century, we had to augment the exposition with a reasonable representative, if not exhaustive, bibliography which will allow the reader to understand the history and pursue the further development of each topic discussed here. We strived to keep it up to date during the writing, being aware, of course, that the field is full of life and new interesting papers will surely keep appearing after the book is published.

Working on quantum waveguide problems over the years we benefited from the opinions of many colleagues whom we want to thank for the pleasure of fruitful discussions and common work. They were numerous and we have to do it in part anonymously, mentioning only some names. In the first place our thanks go to Petr Šeba and the late Pierre Duclos who understood importance of quantum waveguides and made weighty contributions to the field at its early stages. We are also grateful to our other coauthors, especially to F. Bentosela, D. Borisov, T. Cheon, T. Ekholm, M. Fraas, R. Frank, E. Harrell, T. Ichinose, A. Joye, S. Kondej, D. Krejčiřík, J. Lipovský, M. Loss, K. Němcová (Ožanová), O. Post, G. Raikov, P. Šťovíček, M. Tater, O. Turek, S. Vugalter, T. Weidl, K. Yoshitomi, as well as to J. Avron, C. Cacciapuoti, J.-M. Combes, E.B. Davies, G.F. Dell'Antonio, P. Freitas, F. Gesztesy, A. Laptev, E. Lieb, H. Neidhardt, K. Pankrashkin, A. Sadreev, E. Soccorsi, V. Zagrebnov, and many, many others. Last but not least, we are deeply obliged to our wives and our families for their understanding and support which made the writing of this book possible.

Prague
Brescia

Pavel Exner
Hynek Kovařík

Contents

Symbols

$B_r, B_r(x)$	Ball of radius r and center x
$\mathcal{B}(\mathcal{H})$	Bounded operators on Hilbert space $\mathcal{H}$
$D'(\mathbb{R})$	Space of distributions on $C_0^\infty(\mathbb{R})$
$\mathbb{C}^{n,m}$	Complex matrices with k rows and l columns
Dom	Domain of an operator or a form
e	Electron charge
$h, \hbar$	Planck's constant
H^k	Sobolev space $W^{k,2}$
$H_0^1(\Omega)$	Functions of $H^1(\Omega)$ vanishing at $\partial\Omega$
$H_{\alpha,\vec{a}}$	Point-interaction Hamiltonian
$K, \mathcal{K}$	Gauss curvature, local and total
L^p_{loc}	Space of functions which are locally L^p
L^∞_ε	L^∞ Functions vanishing at infinity
$M, \mathcal{M}$	Mean curvature, local and total
K_0	Macdonald function
$\mathbb{N}_0$	Set of non-negative integers
$\mathbb{N}$	Set of positive integers
$Q(A)$	Form domain of operator A
S^1	Unit sphere in $\mathbb{R}^d$
$W^{k,p}$	Sobolev space with indices k, p
$W_0^{1,p}(\Omega)$	Functions of $W^{1,p}(\Omega)$ vanishing at $\partial\Omega$
$\gamma, \gamma(s)$	Signed curvature
γ_{E}	Euler's constant
ϵ_d	Transverse energy $\left(\frac{\pi}{d}\right)^2$
κ_n	Square roots of threshold energies
$\sigma(A)$	Spectrum of operator A
σ_{ac}	Absolutely continuous spectrum
σ_{disc}	Discrete spectrum
σ_{ess}	Essential spectrum

σ_{p}	Point spectrum
σ_{sc}	Singularly continuous spectrum
$\tau, \tau(s)$	Torsion
∇_g	Vector of covariant derivatives on (Σ, g)
$-\Delta_D^{\Omega}$	Dirichlet Laplacian in Ω
Δ_g	Laplace-Beltrami operator on (Σ, g)
$\partial\Omega$	Boundary of Ω
$[\cdot]$	Integer part
$\|\cdot\|_g$	Norm in $L^2(\Omega_0, g^{1/2}\mathrm{d}s\mathrm{d}u)$
$\|\cdot\|_{\mathrm{HS}}$	Hilbert-Schmidt norm
$\|\cdot\|_p$	L^p norm
$\|\cdot\|_\infty$	Supremum norm
$\sharp M$	Cardinality of set M

Introduction

The worst of all is to fear something that has no shape

Karel Schulz, Stone and Pain

Introductions are here to give the reader a feeling what to expect in the pages that would follow, a rough map of the territory he or she is entering. They may be skipped and often they are. The usefulness of such an opening guide depends, of course, on the subject the book is going to treat. In the present case, the main motivation comes from the fact that the text which follows uses methods of mathematical physics and as such it could be perceived as highly technical, at least in some places. Before plunging into it, we want therefore to describe in simple words what this book contains. We will do it without a single formula; there will be more than enough of them in the chapters to follow.

To start on a general note, let us first recall the well-known fact that the birth of quantum mechanics three generations ago marked one of the big leaps in our understanding of Nature. For the first time we had a theory capable of explaining the structure of the matter at the atomic level, and it was only natural that most attention had been paid in the opening period to such fundamental questions. However, this quest led simultaneously to the discovery of various quantum effects which proved to be of practical importance and influenced our daily lives substantially—looking around, one has to admit that the mechanics of the numerous appliances we depend upon is of a quantum nature.

This broad use required extensive experimental and manufacturing explorations, and those in their turn had a reverse influence on the theory. While at the time of the founding fathers one considered electrons either flying as a beam of particles or moving in a crystal regarded, according to a good theoretical tradition, as an infinite homogeneous environment, the newly acquired manufacturing skills posed questions concerning the solution of the known equations of motion in more complicated settings, for instance, in regions of nontrivial shapes. A good example is represented by *quantum dots*, tiny crystals of a semiconductor material, the size of which is typically in tens of nanometers, studied intensely in the last three decades.

They can be regarded as artificial atoms held together by the boundary confinement at the material interface instead of the electrostatic attraction of a nucleus.

A motivation for this book can be derived from another type of object studied in solid-state physics, often called a *quantum wire*. In contrast to quantum dots, quantum wires represent a mixed-dimensionality sort of confinement, in which the electron motion is restricted to a very small size in one direction but it is extended in the other direction(s) so that a transport is possible. A comment is due on the used adjective, because not every thin thread of a conductive material is a quantum wire. The difference is in the type of the transport. Most materials are "dirty" and charge carriers move in them in a diffusive way. The situation changes if the concentration of impurities, measured by the mean free path, is large enough. In high-quality semiconductor materials, this quantity can be at least of the order of tens of micrometers, being thus comparable with typical lengths of the "wires" studied in such experiments. The main consequence is that the quantum nature of transport becomes dominant; we can observe this, in particular, from the fact that the features of the conductivity we know well from the macroscopic experience, such as Ohm's law, cease to be valid.

Even if we neglect possible impurities, the description of an electron motion would still be a formidable task. What makes it accessible is the crystalline structure of the material, which allows us to pass to a one-body problem with the material properties being encoded in the *effective mass* of the particle. Moreover, in most parts of this treatise the actual value of the effective mass will be unimportant and we can get rid of it by an appropriate choice of units.

While we have used semiconductor quantum wires as a motivation to study a guided quantum motion, there are other systems that can serve this purpose. Another class of objects which has recently attracted a lot of attention are *carbon nanotubes* obtained by folding the hexagonal lattice of a graphene sheet into a cylindrical form (in reality, of course, the nanotubes were found before the existence of graphene as a two dimensional carbon crystal was established experimentally). In other settings, the particle guiding may involve true "ducts" as is the case with atoms confined within a hollow glass fiber; if the cavity cross section is small enough, quantum effects again become important.

So far, we have been vague concerning the way the particle is forced to stay in a prescribed part of the configuration space. There are different physical situations to which different models suit. In semiconductor quantum wires the boundary is in reality usually an interface between two semiconductor materials, hence it could naturally be described as a potential step. Often the latter is large on the scale of the effects one investigates, so it is possible to simplify the description by regarding the boundary as a hard wall at which the wave functions vanish. This is the assumption we shall use for most of the book, however, there will be notable exceptions. At places we shall consider systems in which the Dirichlet condition is replaced by a Neumann or a mixed condition. For instance, the main results of Chap. 8 concern systems of thin tubes the boundary of which is Neumann (or is absent). Furthermore, the closing chapter is devoted to situations where the confinement is even "softer", being realized by appropriate singular potentials.

Speaking of boundaries, we should mention some other classes of systems to which the techniques discussed in this book can be applied. While we are mainly concerned with quantum systems in which the dynamics is governed by the appropriate Schrödinger equation, similar behavior can be encountered elsewhere, at least as long as stationary situations are considered. In Chap. 3, for instance, we shall briefly discuss some properties of acoustic waveguides. Equally important, one is also able to say something about electromagnetic waveguides. True, the dynamics there is governed by Maxwell's equations, but in particular situations the description of some field components can be reduced to the appropriate Helmholtz equation—we shall mention this in the notes to Chap. 1—which, in particular, introduces a simple way to experimentally verify some of the mathematical results derived here. To amuse the reader, we add that this does not exhaust the list of possible "classical" applications; you may think of soap bubbles on extended frames of nontrivial shapes and other exotic objects.

The central theme of the book is the relationship between, on one hand, the geometry of the confinement, and, on the other hand, the spectral and scattering properties of the confined particle, expressed in terms of the relevant observables, in the first place the Hamiltonian. The term we use to describe such systems refers to a guided motion, hence it would seem natural to start the investigation with transport in tubes. We choose a different departing point, however, and look first at states which remain localized as the time runs. Apart from methodical reasons, this will allow us to better appreciate the effects that a nontrivial geometry can induce.

The simplest of them is *binding by bending*. A straight hard-wall tube in the form of an infinite cylinder has a purely continuous spectrum. If we bend it in a way which keeps it straight at both ends, at least asymptotically, the corresponding Hamiltonian appears to have isolated eigenvalues referring to bound states localized around the bent part of the tube, their number and positions depending on the geometry of the problem. In fact, this simple and surprising result gave the initial impetus to the investigation of quantum waveguides.

To understand why this effect is so intriguing, one should notice first that it has no classical analogue. There are, of course, closed trajectories in such a tube but their family has measure zero in the phase space, and therefore it cannot give rise to a discrete spectrum according to the conventional quantization rules. Nevertheless, some classical physics considerations may help us to grasp what is happening in such a system. Everybody has probably seen a bobsleigh race and will be able to tell, even without solving the Newton equations, what occurs in a curved part of the track: the sleigh "climbs" the wall there from the bottom of the channel, being slowed down due to energy conservation. A quantum "sleigh" could not do it continuously, however, for the simple reason that its motion perpendicular to the channel is quantized. A "climb" then means jumping into a higher transverse state, and if the longitudinal part of the energy is not sufficient, such an object would be reflected from the bend; if it rests in the bend, being in the lowest transverse state, it has nowhere to go.

This is, of course, a hand-waving argument, and is far from truly explaining the effect. First of all, it obviously fails if the channel has a rectangular form with flat bottom, which is the case we shall be mostly interested in. Second, it does not allow us to understand why the binding effect is robust—we shall see that it is produced by tube bends of any shape, not necessarily smooth ones, and also occurs in higher dimensions, even for bent tubes in $\mathbb{R}^d$ which are mostly of mathematical interest and our original motivation no longer applies to them.

On the other hand, there are higher dimensional systems of strong physical interest, in the first place geometrically nontrivial layers in $\mathbb{R}^3$. While from the experimental point of view it may sometimes be easier to prepare such a curved layer than a bent tube, for instance, fabrication of a quantum wire is more involved than preparation of a thin semiconductor film on a nonflat substrate, mathematically the layer dynamics represents a more difficult problem. We shall discuss this in Chap. 4, paying most attention to curved layers of a constant width built over nonplanar surfaces. Although in this case we do not have a general result which would ensure the existence of localized states in *any* such layer unless it is flat, we shall be able to identify wide classes of layers which exhibit such geometrically induced bound states.

Bending is not the only hard-wall tube deformation which can produce bound states. The same is true, for instance, if a straight tube is locally protruded. In this case, there can be a family of classically closed trajectories of nonzero measure. In fact, the phase-space picture is more complicated having in general a mixed structure with tori referring to integrable motion and a chaotic component. Nevertheless, a binding occurs here even for arbitrarily small protrusions, which again is not in accord with the usual quantization rules.

Bending is also not the only instance when a quantum waveguide exhibits bound states despite the absence a non-negligible phase-space component referring to restricted classical motion. Another example is provided by a lateral coupling. Having two adjacent planar hard-wall strips, we can connect them by opening a window in the common boundary; this again produces a discrete spectrum which is present even if the window width is very small. In a sense, this localization is caused by reflections from the window edges. For a classical particle such an event can occur, of course, with zero probability, however, one can imagine that the reflection becomes more likely once the point particle is replaced by a wave packet spread over some nonzero distance.

While the last claim may again only be of heuristic importance, it does help us to understand some other purely quantum effects. In this connection, one may recall the so-called *Šeba billiard*, that is, a quantum particle confined to a rectangular region with a point interaction in its interior. Such a system is reminiscent of the classical Sinai example of chaotic motion in which a circular obstacle is placed in the center of a rectangular cavity. If the obstacle size shrinks to zero in this example, however, the motion becomes integrable again as the probability of hitting the point is zero. In the quantum case, on the other hand, its presence is felt and shows up in the spectral statistics in a way which is regarded as a chaoticity manifestation. We shall encounter another example of this type in Chap. 7 where we show how an array of point interactions can give rise to a magnetic transport having no classical counterpart.

From the mathematical point of view, the bound states in tubes with local geometrical modifications come from the coupling between the transverse modes that such a perturbation causes; in a straight tube they are decoupled. The trouble is that the effect of such a coupling is not *a priori* clear. While bending or protruding a tube leads to an effective attractive interaction, for other perturbations the effect might be the opposite. A simple example is a local squeeze of a straight tube, a less trivial one we encounter in three dimensions. If a straight tube of noncircular cross section is locally twisted, it has no bound states, and adding an attractive interaction of some sort we can produce such states only if the strength of the latter exceeds a certain critical value.

The mentioned mode coupling depends, of course, on particular geometric properties. In some situations it is weak, one notable example being thin smoothly bent tubes. If one considers a family of such tubes built over the same nonstraight curve and with their diameters shrinking, the transverse and longitudinal motion are found to become adiabatically decoupled in the limit. Consequently, we get an asymptotic expansion for the bound-state energies with the next-to-leading term containing the geometric information expressed in terms of a one-dimensional Schrödinger operator. We have to add a *caveat*, though. It is vital for the last claim that all the tubes have the same axis. If a shrinking tube family tends to a non-smooth curve, the limiting spectral properties can be completely different—we shall see an example of this in Chap. 8.

There are many other situations in which the mode coupling is weak and one can derive asymptotic estimates and expansions of various types. We devote Chap. 6 to this problem presenting there several methods which one can use for that purpose. We focus mainly on the examples previously discussed and analyzed, in particular, the spectral behavior of slightly bent tubes, mildly curved layers, or waveguides coupled through a narrow window.

In addition to weak coupling, a wide variety of questions can be asked concerning the relationship between the confinement geometry and spectral properties of the corresponding operators such as bounds on individual eigenvalues, their number and moments of the discrete spectrum, as well as the influence of waveguide boundaries expressed through the appropriate boundary conditions. We are going to provide at least some answers in Chap. 3, and we will also discuss there what happens if a bent tube contains a finite number of particles interacting through an electrostatic repulsion.

However, the emphasis made so far on the discrete spectrum must not overshadow the importance of the transport phenomena which we shall discuss beginning in Chap. 2. Here again, the transverse-mode coupling is a source of nontrivial effects. In a straight tube each of those modes propagate independently behaving as a free particle in the longitudinal direction. Once we introduce a perturbation, a nontrivial scattering may appear meaning that the particle can get reflected, and also, that it can leave the deformed region in a transverse mode different from the one in which it entered it, naturally provided the energy conservation does not prevent it.

An effect that particularly deserves attention is that of *resonances* in waveguides. From the mathematical point of view, one can treat them as in other similar situations, namely as perturbations of embedded eigenvalues, however, there are different ways in which this mechanism applies. One possibility is that the system has indeed an eigenvalue embedded in the continuum due to a particular symmetry; a resonance than appears once this symmetry is violated, for instance by a potential perturbation or by a magnetic field. On the other hand, there are situations where the embedded eigenvalues do not correspond to any actual system. Prominent examples are thin bent tubes with the adiabatic decoupling of the transverse and longitudinal motion mentioned above. If we neglect the mode-coupling terms in the Hamiltonian, there will be curvature-induced eigenvalues below higher transverse thresholds analogous to the bound states we have described. The coupling does not vanish for any finite tube diameter, though, turning these eigenvalues into resonances, exponentially narrow with respect to the tube width. This behavior can be understood as a manifestation of the fact that the interaction includes tunneling between transverse modes the energetic distances of which grow as the tube gets thinner.

We have indicated above that quantum effects in transport can be manifested, for instance, through an uncommon conductance behavior. This requires a comment; to explain why something like that is possible, we have to make a link between one-body transmission probabilities and macroscopic effects such as the electric current flowing through the quantum wire when we attach it to a "battery". Fortunately, they are related in a simple way, found by R. Landauer and M. Büttiker, which we will mention in the notes to Chap. 2. In particular, if the voltage bias between the reservoirs is tiny and the temperature of the environment is very low, the conductance becomes just a multiple of the transmission probability which makes the resonances observable.

As we have said, purity of the material is crucial for the one-body quantum-mechanical model to describe actual semiconductor wires. On the other hand, impurities typically consisting of alien atoms will, of course, influence the electron motion. A natural way to describe them is to add suitable local potential perturbations to the Hamiltonian. The resulting problem, however, may be mathematically complicated requiring the solution of an appropriate partial differential equation. If the characteristic size of the impurities is much smaller that the waveguide diameter, we can simplify this task by using point perturbations as we will do in Chap. 5. This approach can indeed produce a family of solvable models reducing the spectral and scattering analysis in essence to an algebraic problem. Using it we have to be aware, however, of the peculiar nature those pointlike interactions have in dimension two and three—recall that Fermi used in this connection the term *pseudopotential*—demonstrated by the fact that approximating them by narrow regular potential wells one must suppose that the latter have a zero-energy resonance and employ an involved coupling-constant renormalization when passing to the zero-radius limit.

A natural way to control the behavior of a charged particle in the waveguide is to apply an external electric or magnetic field. There is a variety of such situations and in Chap. 7 we are going to analyze some of them demonstrating, for instance, how such fields can destroy bound states in a waveguide or turn them into resonances. Effects of a particular importance concern the *magnetic transport*. It is well known that a homogeneous magnetic field localizes a charged particle in the plane perpendicular to the field direction, both classically and quantum mechanically. The localization can be removed by an "infinitely long" obstacle which forces such a particle to move along it, thus creating a particular guide for the so-called *edge states*. The obstacle in question may be of different types, a hard wall, a potential, or a variation of the magnetic field itself. In the quantum case the magnetic transport is of a more universal character, in the sense that the obstacle removes the localization fully, and that it can also occur in situations which have no classical analogue.

In most parts of this book, we consider quantum motion in regions which are topologically simple, typically a single tube with nontrivial geometric or potential perturbations. From the engineering point of view, however, such "wires" are only a construction material from which more complicated objects of network form can be built. There is no doubt that a theoretical analysis of such systems may be rather complicated, and looking for ways to simplify the task, one naturally focuses on situations when the network constituents are thin and the motion in them can be treated as essentially one-dimensional.

This brings us naturally to the subject of *quantum graphs*, the models in which the motion of a quantum particle is confined to a metric graph. They are a good illustration of the role that physical motivation can play in the development of a mathematical theory. The concept was proposed originally in the early days of quantum chemistry, but it attracted little attention and had the status of a slightly obscure example until the end of the 1980s when the progress of fabrication techniques mentioned earlier suddenly brought a variety of tiny artificial objects for which the graph description was a useful model. Properties of quantum graphs represent a vast topic, and in this book we are going to deal only with a particular question concerning approximations of quantum graphs by families of thin-tube networks shrinking to the graph "skeleton".

This problem is of importance for quantum-graph theory itself. This is connected with the fact that to construct a graph Hamiltonian one has to fix a way in which wave functions are coupled at the graph vertices. One naturally requires self-adjointness—or in physical terms, conservation of the probability current at each vertex—but this leaves a lot of freedom and tells us nothing about the physical nature of such a vertex coupling. Approximation of a graph by an appropriate family of "fat graphs" appears to be an obvious way to resolve this problem, but as is often the case with apparent ideas, its implementation proved to be mathematically rather hard. First of all, the answer depends on the type of tube boundary used in the approximation. If the tubes are of Neumann type, the limit leads to the simplest coupling usually called *Kirchhoff*. It appears, however, that adding suitably scaled potentials and changing the graph topology locally, one is able to approximate *any* admissible vertex coupling; in Chap. 8 we are going to describe a complete solution to this problem. We shall also show how a nontrivial limit can be achieved in the case of a Dirichlet tube boundary, which is completely different.

Apart from the crystalline character of the material which appeared in our considerations only through the effective mass, there are other ways in which waveguides can acquire a periodic character, such as periodically arranged shape modulations or point perturbations. From the physical point of view, the possibility of producing structures of this type is a tool to control the band spectrum, and via that the transport properties. This is a key element in the production of various *metamaterials* of which the most popular examples at present are *photonic crystals* but many others will surely follow. Several periodic systems will be analyzed in Chap. 9. We also discuss there random perturbations of waveguides and the associated localization effects which are again of practical interest, in particular, because real waveguides have never ideal shapes.

Finally, in Chap. 10 we describe one more way in which guided quantum dynamics can be treated. While in the other parts of the book we have assumed that the motion is confined to a tube, layer, graph, or another fixed subset of Euclidean space, from the physical viewpoint it is often an idealization. We have mentioned that a boundary of a semiconductor quantum wire is in fact a potential jump, hence if two such wires are placed close to each other, the particle can tunnel between them, which would be impossible if the guide had a hard-wall boundary. To take the tunneling effect into account, we analyze a class of models in which particles are confined by a potential "ditch" or a system of ditches; for simplicity we shall assume that the potentials are singular, being supported by curves, graphs, surfaces, etc. We are going to show that, despite the different configuration space, such models have a lot in common with those discussed in the previous chapters, for instance, they exhibit curvature-induced bound states, and in the strong-coupling limit the dynamics is effectively one-dimensional, being reminiscent of the behavior of particles in thin tubes.

With the hope that the previous pages have given the reader an idea of what to expect in the following chapters, let us stop and turn now to a discussion of the subject using the proper tools.

Chapter 1
Geometrically Induced Bound States

Ad methodum philosophicam perdiscendam multum valet mathematica, et imprimis arithmetica et geometria.
Jan Amos Komenský, Orbis Pictus

The object of our interest in the first three chapters is a spinless quantum particle confined to a spatial region $\Omega \subset \mathbb{R}^d$ of a tubular form, which results from various local perturbations of a straight tube, $\Omega_0 = \mathbb{R} \times M$ with some precompact cross section $M \subset \mathbb{R}^{d-1}$. We shall be mostly concerned with the situation when the tube boundary is a hard wall. In the absence of external fields the particle Hamiltonian is then a multiple of the appropriate Dirichlet Laplacian,

$$H = -\frac{\hbar^2}{2m^*}\,\Delta_D^\Omega\,. \tag{1.1}$$

For the sake of simplicity in the following we mostly employ rationalized units, putting $\hbar^2/2m^* = 1$. The ***Dirichlet Laplacian*** is defined for any open region, in general not connected, as the unique self-adjoint operator on $L^2(\Omega)$ associated with the sesquilinear form which is the closure of $q\,:\, q(\phi,\psi) = \int_\Omega \overline{\nabla\phi}\cdot\nabla\psi\,\mathrm{d}x$ on $C_0^\infty(\Omega)$—cf. [RS, Sect. XIII.15]. However, we shall deal with regions Ω having a "nice" boundary for which this operator can be alternatively defined in the classical way,

$$-\Delta_D^\Omega\psi = \sum_{j=1}^{d} \frac{\partial^2\psi}{\partial x_j^2}$$

with the domain consisting of all ψ from the local Sobolev space $H_0^1(\Omega)$ such that $-\Delta\psi$, understood in the sense of distributions, belongs to L^2—cf. [Da, Theorem 1.2.7]. The subset consisting of $\psi : \bar{\Omega} \to \mathbb{C}$ which are C^∞ on Ω with $-\Delta\psi \in L^2$ and satisfy the ***Dirichlet condition***,

P. Exner and H. Kovařík, *Quantum Waveguides*,
Theoretical and Mathematical Physics, DOI 10.1007/978-3-319-18576-7_1

$$\psi(x) = 0 \quad \text{for} \quad x \in \partial\Omega\,, \tag{1.2}$$

at its boundary forms an operator core for $-\Delta_D^\Omega$. The differentiability requirement can be modified, of course, for instance to C^k with some $k \geq 2$.

Spectral properties of the Hamiltonian (1.1) referring to a straight tube $\Omega_0 = \mathbb{R} \times M$ are easy to find, since in this case we can separate variables. The spectrum of $-\Delta_D^{\Omega_0}$ is thus absolutely continuous and equal to $[\nu_1, \infty)$, where ν_1 is the lowest eigenvalue of the cross-section Dirichlet Laplacian $-\Delta_D^M$. In addition, the purely discrete spectrum of the last named operator determines the points where the multiplicity of $\sigma\big(-\Delta_D^{\Omega_0}\big)$ changes. The eigenfunctions χ_n and eigenvalues ν_n of $-\Delta_D^M$ will be used often in the following; we shall usually refer to them as ***transverse modes*** and ***thresholds***, respectively.

The separation of variables means at the same time that different transverse modes are not coupled, i.e. that $-\Delta_D^{\Omega_0}$ is reduced by the projections onto the subspaces spanned by $L^2(\mathbb{R})\otimes\{\chi_n\}$. A perturbation of Ω_0 such as bending, a local deformation, or a local change of boundary conditions, generally results in a coupling between the transverse modes, which may be manifested in the spectral properties of $-\Delta_D^\Omega$. We are going to discuss different aspects of this problem in detail; in the present chapter we will consider the discrete spectrum which such a perturbation can induce.

1.1 Smoothly Bent Strips

The simplest, and at the same time, practically important case is that of ***planar waveguides***, where $d = 2$ and the set M is a segment of the real axis. Our first aim is to show that the bending of a planar strip pushes the spectral threshold down. In particular, if such an Ω is in a suitable sense asymptotically straight, it will follow that $-\Delta_D^\Omega$ has at least one isolated eigenvalue.

Suppose thus that $\Omega \subset \mathbb{R}^2$ is a smoothly bent strip of a fixed width $d = 2a$. The geometry of Ω is conveniently described by means of its axis, which is by assumption a curve Γ of infinite length in $\mathbb{R}^2$ without angles and self-intersections. At each point of it we take the segment of the normal of length $2a$ centered at the curve; the strip is then the union of these segments. If necessary we shall employ labels referring to strip axis and halfwidth such as $\Omega_{\Gamma,a}$. It is also possible to use an off-center curve, i.e. to replace the interval $(-a, a)$ with another one of length $2a$. At times, for instance, it will be useful to choose one of the boundaries of Ω as the generating curve.

The points of Ω can be written parametrically using the natural curvilinear coordinates in the strip. Let s be the arc length of Γ and u the normal distance of a strip point from the curve, then its Cartesian coordinates are

$$x(s,u) \;=\; \xi(s) - u\dot\eta(s)\,, \quad y(s,u) \;=\; \eta(s) + u\dot\xi(s)\,, \tag{1.3}$$

where the functions ξ, η represent a parametric expression of Γ. For brevity, we shall often drop their arguments writing ξ instead of $\xi(s)$, etc. By definition, $\dot\xi^2 + \dot\eta^2 \;=\; 1$;

a dot will always denote the derivative with respect to the arc length. One may say alternatively that Ω is the image of $\Omega_0 = \mathbb{R}\times(-a,a)$ by the map $x : \Omega_0 \to \mathbb{R}^2$ given by (1.3). The strip axis is characterized by the ***signed curvature*** γ of Γ defined by

$$\gamma = \dot{\eta}\ddot{\xi} - \dot{\xi}\ddot{\eta}.$$

Up to the sign, γ coincides with the curvature understood as the inverse radius of the osculation circle, $|\gamma| = (\ddot{\xi}^2 + \ddot{\eta}^2)^{1/2}$. It is clear that if Γ is a C^k–class curve, the function γ is C^{k-2}. It is also useful to introduce the ***bending***,

$$\beta(s_2, s_1) := \int_{s_1}^{s_2} \gamma(s)\,\mathrm{d}s\,, \tag{1.4}$$

which is interpreted as the angle between tangent vectors at the respective points of Γ. In particular, we write $\beta(s) \equiv \beta(s,0)$ and introduce

$$\beta_\Gamma := \lim_{s\to\infty} [\beta(s) - \beta(-s)] = \int_{\mathbb{R}} \gamma(s)\,\mathrm{d}s$$

as the ***total bending*** of the curve Γ provided the right-hand side makes sense.

The signed curvature is important particularly because it determines the curve, uniquely up to Euclidean transformations, by the relations

$$\xi(s) = \xi(s_0) + \int_{s_0}^{s} \cos\beta(s_1, s_0)\,\mathrm{d}s_1\,,\;\; \eta(s) = \eta(s_0) - \int_{s_0}^{s} \sin\beta(s_1, s_0)\,\mathrm{d}s_1\,. \tag{1.5}$$

Unless stated otherwise, we shall always set $s_0 = 0$ and $\xi(0) = \eta(0) = 0$ when using these formulæ, which means that Γ is tangent to the x axis at the origin of coordinates in the chosen frame. Other useful identities are

$$\sin\beta = -\dot{\eta}\,, \quad \cos\beta = \dot{\xi}$$

with $\beta \equiv \beta(s)$ and

$$\ddot{\xi} = \gamma\dot{\eta}\,, \quad \ddot{\eta} = -\gamma\dot{\xi}\,.$$

Through the generating curve the function γ determines the geometry of the strip. If it is zero identically, Γ is a line and Ω is a straight strip. Apart from this trivial case, a strip is called ***simply bent*** if γ is not sign-changing, and ***multiply bent*** otherwise. If the strip is simply bent and the transverse coordinate u runs through the asymmetric interval $(0,d)$, then Γ corresponding to $u = 0$ represents the "inner" boundary of Ω if $\gamma(s) \geq 0$ and vice versa.

The curvilinear coordinates $s,\ u$ are locally orthogonal so the metric properties of Ω express through a diagonal metric tensor, $dx^2 + dy^2 = g_{ss}ds^2 + g_{uu}du^2$, where the transverse component $g_{uu} = 1$ and the longitudinal one is

$$g_{ss} \equiv g = (1 + u\gamma)^2 .$$

It is also easy to compute the Jacobi matrix of the coordinate transformation and its determinant, which is equal to $\sqrt{g} = 1 + u\gamma$.

Since our goal is to find relations between the strip geometry and spectral properties of $-\Delta_D^\Omega$ it is useful to list various assumptions one can impose on Ω. As indicated we consider here smoothly bent strips and employ the curvilinear coordinates. Consequently, the latter must be well defined: this is true if

(i) the map (1.3) is *injective*.

This requirement means, in particular, a restriction to the strip halfwidth since $u\gamma(s) < 1$ must hold everywhere as is seen from the Jacobian of the map (1.3) given above. This ***local injectivity*** is ensured if $a\|\gamma\|_\infty < 1$, which we shall assume throughout (see, however, Remark 1.1.4 below).

Remark 1.1.1 In addition, the injectivity requirement has global consequences. For instance, a strip with a total bending exceeding π necessarily has a self-intersection. Since it is sometimes useful to consider a strongly coiled Ω, e.g., as a model of a flat three-dimensional spiral, it is possible to bypass this restriction replacing the plane by a multi-sheeted Riemannian surface.

Another regularity assumption concerns smoothness properties of the generating curve. With the correspondence (1.5) in mind we can express them through those of the function γ; we shall usually suppose that

(ii)$_k$ γ is C^k*-smooth* for $k = 1, 2$,

which is true if Γ is of the class C^{k+2}. Since the curve is infinite it can exhibit a singular behavior even if it is smooth. To prevent this, we assume

(iii)$_k$ *regularity at infinity*: the function γ together with its derivatives up to the k-th order is bounded in $\mathbb{R}$.

Taken together, the above assumptions ensure that the relations (1.3) define a (global) C^{k+1}-diffeomorphism between the two strips, Ω_0 and Ω. This map has a transparent geometrical meaning; we shall refer to it as to the ***straightening*** transformation.

It provides us with a tool to study the operator $-\Delta_D^\Omega$. The substitution (1.3) defines a unitary operator $\tilde{U}$ from $L^2(\Omega)$ to $L^2(\Omega_0, g^{1/2}\mathrm{d}s\,\mathrm{d}u)$ by $(\tilde{U}\psi)(s,u) := \psi(x,y)$, which transforms the Hamiltonian into

$$\tilde{H} := \tilde{U}\left(-\Delta_D^\Omega\right)\tilde{U}^{-1} = -g^{-1/2}\partial_s g^{-1/2}\partial_s - g^{-1/2}\partial_u g^{1/2}\partial_u . \qquad (1.6)$$

The left-hand side of this relation always makes sense; to ensure the existence of the ***Laplace-Beltrami operator*** appearing on the right-hand side it is sufficient to put $k = 1$ in the above assumptions. We employ here the usual shorthands, $\partial_s \equiv \partial/\partial_s$, etc.

Furthermore, it is often useful to work on a Hilbert space without the weight in the inner product. To this end, we have to replace $\tilde{U}$ by the unitary operator U from $L^2(\Omega)$ to $L^2(\Omega_0)$ acting as $U\psi := g^{1/4}\tilde{U}\psi$. This requires a stronger smoothness hypothesis, namely to take $k = 2$ in the above assumptions. The Dirichlet Laplacian is now transformed to

$$H := U\left(-\Delta_D^\Omega\right)U^{-1} = -\partial_s(1+u\gamma)^{-2}\partial_s - \partial_u^2 + V(s,u) \tag{1.7}$$

with the curvature-induced ***effective potential***

$$V(s,u) := -\frac{\gamma^2}{4(1+u\gamma)^2} + \frac{u\ddot{\gamma}}{2(1+u\gamma)^3} - \frac{5}{4}\frac{u^2\dot{\gamma}^2}{(1+u\gamma)^4} \tag{1.8}$$

(Problem 1). Abusing the notation we use for the transformed operator (1.7) the same symbol as for the original Hamiltonian.

Remarks 1.1.2 (a) If $a\|\gamma\|_\infty < 1$ the factors $g^{\pm 1/4}$ are bounded, so the domain of the operators $\tilde{H}$, H given implicitly in (1.6) and (1.7), respectively, consists of $\psi \in H_0^1(\Omega_0)$ with $\Delta\psi \in L^2$ in the sense of distributions. As usual with unbounded operators, we shall need various cores of these operators. One is analogous to that described in the opening—cf. (1.2). To find a still smaller core, notice that (1.7) can be regarded under the assumptions *(ii)*$_2$ and *(iii)*$_2$ as a result of a relatively bounded perturbation of $-(1-a\|\gamma\|_\infty)^{-2}\partial_s^2 - \partial_u^2$, with the relative weight less than one. Hence any core of the last named operator is by [Ka, Sect. V.4] also a core of H, for instance, the family of all finite sums $\sum_j f_j(s)\chi_j(u)$, where χ_j are the transverse modes and $f_j \in C_0^\infty(\mathbb{R})$—see [BEH, Theorem 5.7.2]. In the same way one can construct cores for $\tilde{H}$.
(b) One way to weaken the above smoothness requirements is to investigate the operator (1.6) through its quadratic form. Then instead of *(ii)*$_1$ and *(iii)*$_1$ we just need $\gamma \in L^\infty(\mathbb{R})$.
(c) By an easy modification of the above argument one gets the "straightened" operators for an annular strip Ω built over a closed loop, and for a finite or a semi-infinite strip with given boundary conditions at the "cuts".

In this way we have replaced the original operator $-\Delta_D^\Omega$ by unitarily equivalent operator which acts on the straight strip Ω_0. The price we pay for this is a more complicated form of the new Hamiltonian. However, since the geometrical information is now contained in coefficients of the operator, it is easier to treat it using standard functional-analytic methods.

Let us first look at the problem from a heuristic point of view. A natural scale to measure the strip thickness is given by the maximum curvature of Γ, in particular, the strip is ***thin*** if $a\|\gamma\|_\infty \ll 1$. In such a case the factor $g^{1/2}$ does not differ much from one. Furthermore, for small enough a the effective potential (1.8) is dominated by its first term. In other words, the Hamiltonian (1.7) effectively decouples,

$$H = -\partial_s^2 - \frac{1}{4}\gamma(s)^2 - \partial_u^2 + \mathcal{O}(a)\,, \tag{1.9}$$

in the limit $a \to 0+$. The transverse part is specified by the Dirichlet condition at $u = \pm a$. Its spectrum is discrete and simple, with the eigenvalues $\{\kappa_n^2\}_{n=1}^\infty$, where $\kappa_n := \pi n/d$, corresponding to the eigenfunctions

$$\chi_{2j+1}(u) = \sqrt{\frac{2}{d}}\,\cos\kappa_{2j+1}u\,, \quad \chi_{2j}(u) = \sqrt{\frac{2}{d}}\,\sin\kappa_{2j}u\,. \tag{1.10}$$

The longitudinal part is a Schrödinger operator the spectrum of which is determined by the potential $-\frac{1}{4}\gamma(s)^2$. Specifically, if the curvature vanishes at large distances, the essential spectrum covers the positive halfline. Furthermore, since the operator is one-dimensional and the potential is purely attractive, there is at least one isolated eigenvalue unless $\gamma = 0$. The mode-coupling terms appear as a perturbation here, hence one may expect that the spectral picture will not change provided the strip halfwidth is small enough.

This is indeed the case. Let us first notice that the essential spectrum of H is not affected by the bending as long as the strip is in a suitable sense ***asymptotically straight***, irrespective of its halfwidth a. We formalize the last requirement in the following assumption,

(iv) $\gamma, \dot\gamma, \ddot\gamma \in L_\varepsilon^\infty(\mathbb{R})$;

recall that $L_\varepsilon^\infty(\mathbb{R})$ is the set of L^∞ functions f on $\mathbb{R}$ with the property that to any $\varepsilon > 0$ there is a compact K_ε such that $\|f \upharpoonright (\mathbb{R}\setminus K_\varepsilon)\|_\infty < \varepsilon$.

Remark 1.1.3 This notion of asymptotic straightness is rather weak. Recall that the existence of asymptotes to Γ is ensured if γ decays sufficiently fast at large distances (Problem 2). Assumption *(iv)* covers curves which behave asymptotically as a parabola, $|\gamma(s)| \sim |s|^{-3/2}$, Archimedes spiral, $|\gamma(s)| \sim |s|^{-1}$, etc. In the last example γ is not integrable. Should assumption *(i)* be satisfied, however, a total bending not exceeding π must still exist in the principal-value sense—as an illustration consider a U-shaped strip coiled into a spiral.

Proposition 1.1.1 *Let assumptions (i), (ii)$_2$, (iv) be satisfied and $a\|\gamma\|_\infty < 1$. Then we have $\sigma_{\mathrm{ess}}(-\Delta_D^\Omega) = [\kappa_1^2, \infty)$.*

Proof Since *(ii)*$_2$ and *(iv)* imply *(iii)*$_2$, we may check the claim for the operator (1.7). To prove that

$$\inf \sigma_{\mathrm{ess}}(-\Delta_D^\Omega) \geq \kappa_1^2$$

we employ Neumann bracketing [RS, Sect. XIII.15]: we impose an additional Neumann condition at the normal cuts of Ω placed at $s = \pm s_0$. Then, by the variational principle, we get the operator inequality $-\Delta_D^\Omega \geq H_N := H_l \oplus H_0 \oplus H_r$, and consequently, $\inf \sigma_{\mathrm{ess}}(-\Delta_D^\Omega) \geq \inf \sigma_{\mathrm{ess}}(H_N)$. The middle operator has a purely discrete spectrum by the minimax principle, since it is bounded below by $a(-\partial_s^2)_N + (-\partial_u^2)_D - b$ with suitable constants. Let J be any of the intervals $(-\infty, -s_0)$ and (s_0, ∞). By Remark 1.1.2c the corresponding operator on $L^2(J)$ is bounded from below by

$$-(1 + a\|\gamma_J\|_\infty)^{-2}\partial_s^2 + (-\partial_u^2)_D + V_-^{(J)}(s)$$

with the Neumann condition at the endpoint of J, where we set $V_-^{(J)}(s) := \inf\{V(s, \cdot) : |u| < a\}$ and $\gamma_J := \gamma \restriction J$. By assumption both the norms $\|V_-^{(J)}\|_\infty$ and $a\|\gamma_J\|_\infty$ tend to zero as $s_0 \to \infty$, so using the minimax principle again we infer that $\inf \sigma_{\mathrm{ess}}(-\Delta_D^\Omega) > \kappa_1^2 - \varepsilon$ holds for any $\varepsilon > 0$. In this way we get $\sigma_{\mathrm{ess}}(-\Delta_D^\Omega) \subset [\kappa_1^2, \infty)$; the opposite inclusion is verified using appropriate Weyl sequences (cf. [We, Theorem 7.24] and Problem 3). ∎

Using a simple variational estimate, we can then confirm the above heuristic conclusion if we adopt another decay assumption,

(v) $\gamma, \dot\gamma, |\ddot\gamma|^{1/2} \in L^2(\mathbb{R}, |s|\,ds)$.

While in general *(v)* does not imply *(iv)*, both assumptions are satisfied, for instance, if $\gamma(s), \dot\gamma(s), |\ddot\gamma(s)|^{1/2} = \mathcal{O}(|s|^{-1-\varepsilon})$ as $|s| \to \infty$.

Proposition 1.1.2 *Assume (i)–(v) with $k = 2$. If Ω is not straight, the operator $-\Delta_D^\Omega$ has at least one eigenvalue below $-\kappa_1^2$ for small enough a.*

Proof By the minimax principle it is enough to estimate H from above by an operator with the same essential-spectrum threshold which has a nonempty discrete spectrum. If $a\|\gamma\|_\infty < \frac{1}{2}$, we may use

$$H \leq -\partial_u^2 - 4\,\partial_s^2 - \frac{1}{9}\,\gamma(s)^2 + \frac{a}{16}\,|\ddot\gamma(s)|,$$

and the claim is valid if the Schrödinger operator on $L^2(\mathbb{R})$ with the potential $-\frac{1}{36}\,\gamma(s)^2 + \frac{a}{64}\,|\ddot\gamma(s)|$ has a negative eigenvalue. This is true, of course, for $a = 0$ when the potential is purely attractive, and by analytic perturbation theory the eigenvalue persists for all sufficiently small a. ∎

It appears, however, that a much stronger claim can be made, namely that curvature leads to an effective attraction irrespective of the strip width. This again follows from a variational estimate, as the next theorem shows.

Theorem 1.1 *Let (i) and (ii)$_1$ be satisfied together with* $a\|\gamma\|_\infty < 1$. *Then* $\inf\sigma(-\Delta_D^\Omega) < \kappa_1^2$ *holds unless* $\gamma = 0$ *identically.*

Proof It is clearly sufficient to find a trial function ψ from $Q(\tilde{H})$, the form domain of $\tilde{H}$, which makes the quadratic form

$$q[\psi] := \|\tilde{H}^{1/2}\psi\|_g^2 - \kappa_1^2\|\psi\|_g^2 = \|g^{-1/4}\partial_s\psi\|^2 + \|g^{1/4}\partial_u\psi\|^2 - \kappa_1^2\|g^{1/4}\psi\|^2$$

negative; here $\|\cdot\|_g$ means the norm in $L^2(\Omega_0, g^{1/2}\mathrm{d}s\,du)$. We shall seek it in the form $\psi = \phi_\lambda\chi_1 + \varepsilon f$, where χ_1 is the lowest transverse mode (1.10) and the functions ϕ_λ and f have to be properly chosen.

The first of them serves to control the tails of the trial function. Let $K := [-s_0, s_0]$ for some $s_0 > 0$ and choose a function $\phi \in C_0^\infty(\mathbb{R})$ such that $\phi(s) = 1$ on K. A suitable family of ϕ_λ is then obtained by a scaling exterior to K,

$$\phi_\lambda(s) := \begin{cases} \phi(s) & \dots\ |s| \le s_0 \\ \phi(s_0\,\mathrm{sgn}\,s + \lambda(s - s_0\,\mathrm{sgn}\,s)) & \dots\ |s| > s_0 \end{cases} \tag{1.11}$$

This allows us to make the positive contribution to the energy coming from the trial function tails small. Indeed, we find easily

$$q[\phi_\lambda\chi_1] := \int_{\mathbb{R}} \langle g^{-1/2}\rangle(s)\,|\dot{\phi}_\lambda(s)|^2\,\mathrm{d}s \le \frac{\lambda}{1 - a\|\gamma\|_\infty}\,\|\dot{\phi}\|^2,$$

where $\langle\cdot\rangle$ denotes the transverse average w.r.t. χ_1^2. In the next step we deform the trial function in the central part of Ω_0. This can be done in various ways. For instance, we can choose $f := \jmath^2(\tilde{H}-\kappa_1^2)\phi_\lambda\chi_1$, where $\jmath$ is a function from $C_0^\infty(K \times (-a, a))$, or more explicitly

$$f(s,u) = -\left(\jmath^2\frac{\gamma}{1+u\gamma}\chi_1'\right)(s,u) = \jmath^2(s,u)\sqrt{\frac{2}{d}}\,\frac{\kappa_1\gamma(s)}{1+u\gamma(s)}\,\sin\kappa_1 u\,.$$

In view of *(ii)*$_1$ and the choice of $\jmath$, this function belongs to $Q(\tilde{H})$, and a straightforward computation yields

$$q[\phi_\lambda\chi_1 + \varepsilon f] = q[\phi_\lambda\chi_1] + 2\varepsilon\|\jmath(\tilde{H}-\kappa_1^2)\phi_\lambda\chi_1\|_g^2 + \varepsilon^2(f, (\tilde{H}-\kappa_1^2)f)_g\,.$$

Importantly, the coefficient of the linear term on the right-hand side is independent of λ because the scaling acts only out of the support of $\jmath$. If γ is nonzero in K, we can always choose $\jmath$ in such a way that the coefficient is nonzero, in which case the sum of the last two terms is negative for all sufficiently small ε. We fix such an ε. The above estimate then shows that for λ small enough we have $q[\phi_\lambda\chi_1 + \varepsilon f] < 0$, which is what we set out to prove. ∎

Corollary 1.1.1 *Let the assumptions of the preceding theorem be satisfied and let* $\inf \sigma_{\mathrm{ess}}(-\Delta_D^\Omega) = \kappa_1^2$ *hold, e.g. by Proposition 1.1.1. Then* $-\Delta_D^\Omega$ *has at least one isolated eigenvalue of finite multiplicity.*

Remark 1.1.4 In the above proof we have used the quadratic form of the operator (1.6) only, hence by Remark 1.1.2b we may assume $\gamma \in L^\infty(\mathbb{R})$ instead of *(ii)*$_1$. The claim of *Theorem* 1.1 then remains valid, e.g., if there exists a set $M \subset \mathbb{R}$ of nonzero Lebesgue measure and $c > 0$ such that γ does not change sign in M and $|\gamma(s)| \ge c$ holds for all $s \in M$. In addition, the condition $a\|\gamma\|_\infty < 1$ was only used outside $[-c, c]$, hence one can allow $a\gamma(s) = \pm 1$ for some s from a bounded set $M' \subset \mathbb{R}$. This includes a possibility that M' has an open subset in which case the strip boundaries may have angles. The assumptions of the above corollary can also be weakened; notice that hypotheses *(ii)*$_2$ and *(iv)* in *Proposition* 1.1.1 are again used outside a compact region only, and moreover, in assumption *(iv)* only the decay of γ itself matters. The proofs require just small modifications of the arguments used here (Problems 4–6 and 3.11).

Notice that the existence of bound states in smoothly bent strips is *a purely quantum effect*. Since there is no external field, any classical particle trajectory consists of line segments with “geometric-optics” reflections at the walls. Of course, if such a particle moves perpendicularly to the axis of Ω, it follows a closed trajectory bouncing between the walls, however, the corresponding set of initial conditions has measure zero in the phase space.

Proposition 1.1.3 *Let the map (1.3) be injective with* γ *piecewise continuous and such that* $a\|\gamma\|_\infty < 1$. *Then apart from the trivial set mentioned above there are no closed trajectories.*

Proof We employ parametrization (1.3) together with (1.5). Since a segment between subsequent reflections is given by $x\,\cos\varphi + y\,\sin\varphi + c = 0$ for some c, φ, in the curvilinear coordinates it is expressed as

$$u(s) = \frac{c - \int_0^s \cos\big(\beta(s')+\varphi\big)\,ds'}{\sin\,(\beta(s)+\varphi)}\,,$$

where we may put $s_0 = 0$ without loss of generality. This expression makes sense if its denominator is nonzero; it is clear that the choice $\sin\,(\beta(\tilde{s})+\varphi) = 0$ and $c = \int_0^{\tilde{s}} \cos\big(\beta(s')+\varphi\big)\,ds'$ corresponds to the mentioned trivial set. Suppose that a nontrivial closed trajectory exists. Mapped in the described way to the straightened strip, it should have turning points; since it is smooth, there must be a point where $\dot{u}(s)$ diverges. We may again assume that this happens at $s = 0$. If $\sin\varphi \ne 0$ we have

$$u(0) = \frac{c}{\sin\varphi}\,, \qquad \dot{u}(0) = -(1+u(0)\gamma(0))\cot\varphi\,,$$

hence a turning point requires $\sin\varphi = 0$. In that case we can compute

$$\lim_{s\to 0} u(s) = \lim_{s\to 0} \frac{\pm\int_0^s \cos\beta(s')\, ds'}{\sin\beta(s)} = \pm\left(\lim_{s\to 0}\gamma(s)\right)^{-1},$$

with the last limit one-sided if γ has a jump at $s = 0$. However, this is a contradiction, since $|u(0)| \le a$, while $|\gamma(0\pm)|^{-1} > a$ holds by assumption. ∎

1.2 Polygonal Ducts

The smooth strips discussed above are not the only systems in which a local change of geometry induces the existence of bound states. Another often considered class of two-dimensional waveguides consists of ***polygonal ducts*** assembled from pieces of a straight strip. Naturally, they do not have a fixed width in the bends and the technique of the previous section using locally orthogonal coordinates cannot be used directly. On the other hand, they sometimes yield solvable examples since they decompose into a union of simple regions and the spectral problem can be handled by solution-matching techniques.

Let $\mathcal{P} \equiv \mathcal{P}_{\ell,\beta}$ be a polygonal path in $\mathbb{R}^2$ consisting of a pair of halflines joined by a piecewise linear curve characterized by the segment lengths $\ell = \{\ell_1, \dots, \ell_n\}$ and the angle family $\beta = \{\beta_1, \dots, \beta_{n+1}\}$, where $\beta_j \in [0, \pi)$ is the (signed) angle between the $(j-1)$-th and j-th segment; the two halflines are naturally labeled with $j = 0,\ n+1$, respectively. Without loss of generality we may assume that $\beta_j \ne 0$ for $j = 1, \dots, n$, since a pair of adjacent segments with $\beta_j = 0$ can be replaced by a single segment of the combined length $\ell_{j-1}+\ell_j$.

It is clear that $\mathcal{P}$ is determined by the families ℓ and β uniquely up to Euclidean transformations; a mirror image is obtained by the change of sign convention for the angles. In the language of the previous section $\mathcal{P}$ is a curve corresponding to the signed curvature

$$\gamma(s) := \pm\sum_{j=1}^{n} \beta_j \delta(s-s_j), \quad s_j := \sum_{i=1}^{j-1} \ell_i .$$

We suppose that $\mathcal{P}$ does not intersect itself. We can write it as union of its segments, $\mathcal{P} = \cup_{j=0}^{n+1}\Gamma_j$. With each segment Γ_j one can associate two sets S_j^d and W_j. The former is the straight open strip of width $d = 2a$ such that Γ_j is a subset of its axis, the latter is the closed set delineated by the axes of the angles between Γ_j and the neighboring segments; for $1 \le j \le n$ it is a wedge unless Γ_{j+1} and Γ_{j-1} are parallel in which case it may be a strip, for $j = 0,\ n+1$ it is a halfplane. Now we may define $\Omega_{j,d} := S_j^d \cap W_j$ and the corresponding ***finitely bent polygonal duct*** of width d as

$$\Omega_{\mathcal{P},d} := \bigcup_{j=0}^{n+1} \Omega_{j,d} .$$

We restrict ourselves to the situation when each Γ_j contains at least one point such that a perpendicular open segment of the width d centered at Γ_j is contained in $\Omega_{j,d}$. It is easy to see that this is true if the conditions

$$a\left(\tan\frac{|\beta_j|}{2} + \tan\frac{|\beta_{j+1}|}{2}\right) \le \ell_j\,, \quad j = 1, \dots, n\,, \tag{1.12}$$

are satisfied; the larger the angles, the stronger is the restriction.

Theorem 1.2 *Let $\Omega \equiv \Omega_{\mathcal{P},d}$ be a finitely bent polygonal duct which satisfies conditions (1.12) and does not intersect itself, in other words $\Omega_{i,d} \cap \Omega_{j,d} = \emptyset$ for any $i, j = 0, \dots, n+1$ with $|i-j| \ge 2$. Then $\sigma_{\mathrm{ess}}(-\Delta_D^\Omega) = [\kappa_1^2, \infty)$, where $\kappa_1 := \pi/d$. Furthermore, $-\Delta_D^\Omega$ has at least one isolated eigenvalue of finite multiplicity provided Ω is not a straight strip, i.e. $n = 1$ and $\beta_1 = 0$.*

Proof The essential spectrum is localized as in *Proposition* 1.1.1 by combination of a Neumann bracketing with Weyl's criterion. The fact that the discrete spectrum is non-empty unless Ω is straight is obtained by Dirichlet bracketing. We employ the estimate $-\Delta_D^\Omega \le H_D$ where H_D is the Laplacian with additional Dirichlet conditions imposed at the circular segments S_j, $j = 1, \dots, n$, defined as follows: S_j has its center at the "inner" vertex of $\partial\Omega$ between $\Omega_{j-1,d}$ and $\Omega_{j,d}$ and radius d, it connects the two tangential points at which the circle touches the opposite boundary. In view of conditions (1.12) such segments exist and determine a decomposition of the strip,

$$\Omega = \Omega_r \cup \bigcup_{j=1}^{n+1} C_j\,,$$

where Ω_r is the strip with the rounded outer corners and C_j are the appropriate "caps". It is clear that Ω_r is a strip of the class considered in the previous section, apart from the fact that the curvature of its axis equals a^{-1} in the bends. Since this concerns a finite part of Ω_r, we have $\sigma_{\mathrm{disc}}(-\Delta_D^{\Omega_r}) \ne \emptyset$ by Remark 1.1.4, and by the minimax principle the same is true for $-\Delta_D^\Omega$. ∎

Remark 1.2.1 The Dirichlet bracketing can be used even if the conditions (1.12) are not valid—cf. Problem 7 for an example. If the strip built over a polygonal path is too thick, individual segments may lose meaning. For instance, consider $\mathcal{P}$ with $n = 2$ corresponding to $\{\ell, \ell\}$ and $\{\beta, -2\beta, \beta\}$ with some $\ell > 0$ and $0 < \beta < \pi/2$. If $a\tan\beta > \ell$, the respective $\Omega_{\mathcal{P},d}$ is a strip with a triangular protrusion which fits into the class discussed in Sect. 1.4 below.

In the above proof we have ignored the contribution from the caps split off by the Dirichlet bracketing. The corner regions become important if some of the bends of the generating path $\mathcal{P}$ are sharp.

Proposition 1.2.1 *Let Ω be as in the above theorem, then the number of eigenvalues counted with multiplicity satisfies the bound*

$$\sharp\left\{\sigma_{\mathrm{disc}}(-\Delta_D^\Omega)\right\} \geq \sum_{j=1}^{n} \left[\frac{2c}{\pi-\beta_j}\right],$$

where $[\cdot]$ *means the integer part and* $c := \left(1-2^{-2/3}\right)^{3/2}$.

Proof Let us define $\alpha_j := (\pi-\beta_j)/2$. We again employ Dirichlet bracketing, this time inserting into the j-th triangular corner a rectangle parallel to the corner axis of halfwidth $b_j \in \left(\frac{d}{2}, d(\cos\alpha_j)^{-1}\right)$ and length $(d-b_j\cos\alpha_j)/\sin\alpha_j$. Eigenvalues of the corresponding Dirichlet Laplacian are easily found. The number of them below κ_1^2 equals $[\nu]$, where

$$\nu := \frac{\sqrt{4\mu_j^2-1}}{2\mu_j}\,\frac{1-\mu_j\cos\alpha_j}{\sin\alpha_j}$$

with $\mu_j := b_j/d$. The right-hand side reaches its maximum at $\mu_j = (4\cos\alpha)^{-1/3}$, so the claim follows by an easy estimate. ∎

The simplest nontrivial example of a polygonal duct is a ***broken strip***, $n = 1$, built over the path $\mathcal{P}$ which consists of two halflines joined at the vertex with an angle $\pi-\beta$. Since a change of the strip width, $d \to d'$, amounts in this case to a scaling transformation which modifies the spectrum by the multiplicative factor $(d/d')^2$, we can put without loss of generality $d = \pi$. Let us denote such a broken strip by Ω_β.

Proposition 1.2.2 $\sigma_{\mathrm{disc}}\left(-\Delta_D^\Omega\right)$ *for the broken strip* $\Omega \equiv \Omega_\beta$ *with* $\beta > 0$ *is non-empty and consists of* $N \equiv N(\beta)$ *eigenvalues* $\{\epsilon_n(\beta)\}_{n=1}^N \subset \left(\frac{1}{4}, 1\right)$ *arranged conventionally in the ascending order. To any integer m there is a* β *such that* $N(\beta) \geq m$. *Each eigenfunction is symmetric with respect to the symmetry axis of* Ω. *The functions* $\epsilon_n(\cdot)$ *are continuously decreasing in* $[\beta^{(n)}, \pi)$, *where* $\beta^{(1)} = 0$ *and* $\beta^{(n)}$ *for* $n \geq 2$ *is the critical angle at which the n-th eigenvalue emerges.*

Proof Again let $\alpha := (\pi-\beta)/2$. The discrete spectrum is a non-void subset of the interval $(0, 1)$ by *Theorem* 1.2, and the previous proposition implies that $N(\beta)$ can exceed any integer provided β is chosen close enough to π. Since Ω_β has mirror symmetry, $-\Delta_D^\Omega$ decomposes into a direct sum of the symmetric and antisymmetric part w.r.t. the axis of symmetry.

This enables us to reduce the task to the spectral problem for the Laplacian on the skewed halfstrip $\Sigma_\alpha := \{\vec{x} \in \mathbb{R}^2 : x > 0,\ 0 < y < \min\{\pi, x\tan\alpha\}\}$ with the Dirichlet condition at $y = 0, \pi$. The antisymmetric part of the original operator corresponds to the Dirichlet condition at the remaining piece of the boundary, $\mathcal{C}_\alpha := \{\vec{x} \in \partial\Sigma_\alpha : y = x\tan\alpha\}$. It obviously does not contribute to the discrete spectrum, and therefore we can restrict our attention to the symmetric part leading to the Neumann condition at $\mathcal{C}_\alpha$.

The halfstrips Σ_α can be compared by means of the unitary operators $U : L^2(\Sigma_\alpha) \to L^2(\Sigma'_\alpha)$ defined by $(U\phi)(x, y) := \sigma^{1/2}\phi(\sigma x, y)$, where $\sigma := \tan\alpha'/\tan\alpha$. In particular, the Laplacian on $L^2(\Sigma'_\alpha)$ is mapped to the operator $U^{-1}(-\Delta)U = -\sigma^2\partial_x^2 - \partial_y^2$ on $L^2(\Sigma_\alpha)$ so the eigenvalue monotonicity follows by the minimax principle. Moreover, the difference $(1-\sigma^2)\partial_x^2$ is relatively bounded w.r.t. $-\Delta$, hence each $\epsilon_n(\cdot)$ is continuous by [Ka, Sect. VII.6.5]. ∎

Remark 1.2.2 It should be noted that since the half-lines defining the boundary of Ω_β are connected in a non-smooth way, the operator domain of $-\Delta_D^{\Omega_\beta}$ does not coincide with $H^2(\Omega_\beta) \cap H_0^1(\Omega_\beta)$, being instead given by

$$\mathrm{Dom}\,(-\Delta_D^{\Omega_\beta}) = H^2(\Omega_\beta) \cap H_0^1(\Omega_\beta) \oplus \mathbb{C}[u_{\mathrm{sing}}], \tag{1.13}$$

where $\mathbb{C}[u_{\mathrm{sing}}]$ denotes the span of a function u_{sing} singular at the inner bend where the boundary angle of the broken strip exceeds π making the region non-convex. In the polar coordinates (r, θ) centered at such a vertex this function takes the form

$$u_{\mathrm{sing}}(r, \theta) = \chi(r)\, r^{\frac{\pi\theta}{\pi+\beta}} \sin\left(\frac{\pi\,\theta}{\pi+\beta}\right)$$

with $\chi(\cdot)$ being a smooth cut-off function equal to one in a neighborhood of the vertex. A simple calculation then shows that $\Delta\, u_{\mathrm{sing}} = 0$ in the vicinity of the vertex while $u_{\mathrm{sing}} \notin H^2(\Omega_\beta)$. The same remark applies, of course, to more complicated ducts with the boundary having angles.

Apart from the case of small α the above results give no quantitative information about the discrete spectrum. A straightforward approach to this problem is to solve the corresponding Helmholtz equation. One way to do that, well suited for polygonal ducts, is known as the ***mode-matching method***. It is based on decomposing the duct into a union of regions such that in each of them we can write an Ansatz for the solution using the transverse-mode basis. To belong to the Hamiltonian domain, of course, the Ansätze must match smoothly at the common boundary of the regions; this yields conditions on coefficients of the expansions which determine the spectrum. Let us illustrate how this technique works using a particular case of the above example known as the ***L-shaped strip***, see Fig. 1.1.

Proposition 1.2.3 *We have $N(\pi/2) = 1$ and $\epsilon_1(\pi/2) = 0.9291\ldots$. The corresponding eigenfunction is exponentially decaying, $|\psi_1(\vec{x})| \le c \exp[-q_1 s(\vec{x})]$ for some $c > 0$, where $q_1 = 0.2663\ldots$ and $s(\vec{x}) := \max\{x, y\}$.*

Proof As in the previous proposition it is sufficient to look for a half of the symmetric solution in the cut strip $\Sigma_{\pi/4}$ which we write as $\Sigma_{\mathrm{I}} \cup \Sigma_{\mathrm{II}}$, where $\Sigma_{\mathrm{I}} := \{\vec{x} : x \ge \pi,\ 0 < y < \pi\}$ and $\Sigma_{\mathrm{II}} := \{\vec{x} : 0 < x \le \pi,\ 0 < y < \pi\}$. We shall seek a solution of the equation $-\Delta_D^{\Omega}\psi = \epsilon\psi$ in the form

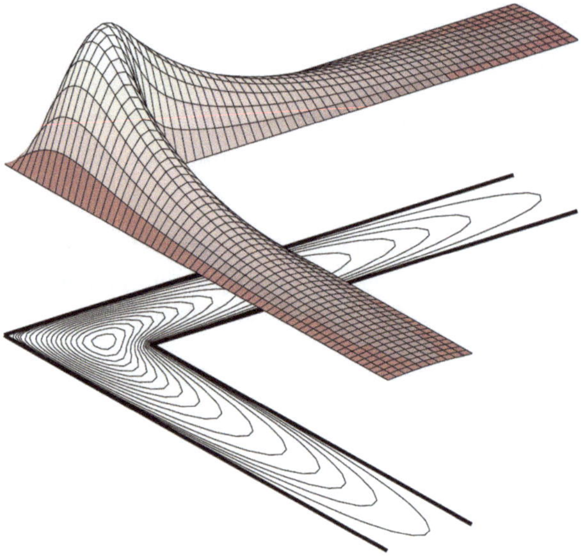

Fig. 1.1 Eigenfunction of an L-shaped waveguide

$$\psi(\vec{x}) = \begin{cases} \sum_{j=1}^{\infty}(-1)^{j+1}a_j\, e^{q_j(\pi-x)}\chi_j(y) & \ldots\ \vec{x}\in\Sigma_{\mathrm{I}} \\ \sum_{j=1}^{\infty}(-1)^{j+1}a_j\left[s_j(x)\chi_j(y)+s_j(y)\chi_j(x)\right] & \ldots\ \vec{x}\in\Sigma_{\mathrm{II}} \end{cases} \tag{1.14}$$

where $q_j \equiv q_j(\epsilon) := \sqrt{j^2-\epsilon}$, the functions $\chi_j : \chi_j(y) = (2/\pi)^{1/2}\sin(jy)$ constitute a transverse basis, and $s_j(x) := \sinh(q_j x)/\sinh(q_j\pi)$. This choice ensures that the boundary conditions are satisfied and $\psi_{\mathrm{I}}(\pi, y) = \psi_{\mathrm{II}}(\pi, y)$ holds for $y \in (0,\pi)$ provided the series converges; this and other properties of ψ can be expressed in terms of the sequence $\{a_j\}$ (Problem 8).

To match the two parts of the Ansatz smoothly, we also have to require $\partial_x\psi_{\mathrm{I}}(\pi+, y) = \partial_x\psi_{\mathrm{II}}(\pi-, y)$. This yields the condition $Ca = a$, where $C = \{C_{jk}\}$ is the operator on $\ell^2(j)$ defined through its matrix elements,

$$C_{jk} := \frac{1-e^{-2\pi q_j}}{\pi q_j}\,\frac{jk}{j^2+k^2-\epsilon}\,.$$

A straightforward way to solve this condition is to consider a sequence of truncated matrices and to find the sought $\{a_j\}$ as a limit when the cut-off is removed. However, it is not *a priori* clear whether such an approximation would converge (cf. Problem 9), hence we take another route and look for solution in the class of sequences $a_j = j^{-s}r_j$ with $\{r_j\} \in \ell^\infty$, where $s > 0$ will be determined later. Since the system $Ca = a$ to be solved is linear, we may put $a_1 = r_1 = 1$ without loss of generality. The remaining part $r = \{r_j\}_{j=2}^{\infty}$ of the sequence then satisfies the equation $r = t + Kr$, where

$$t_j := \frac{1-\mathrm{e}^{-2\pi q_j}}{\pi q_j} \, \frac{j^{s+1}}{j^2+1-\epsilon}\,, \quad K_{jk} := \frac{1-\mathrm{e}^{-2\pi q_j}}{\pi q_j} \, \frac{j^{s+1}k^{s-1}}{j^2+k^2-\epsilon}$$

for $j, k = 2, 3, \ldots$. It has a unique solution, namely $r = (I - K)^{-1}t$, provided $\|K\|_\infty < 1$. We have

$$|K\xi_j| \le \|\xi\|_\infty \frac{1}{\pi}\left(1 - \frac{\epsilon}{j^2}\right)^{-(s+1)/2} \int_{q_j}^\infty \frac{\mathrm{d}\eta}{(1+\eta)^2\eta^{s-1}}$$

for $s \in [1, 2)$, and since $\epsilon \in (0, 1)$ and $j \ge 2$, it follows that

$$\|K\|_\infty \le \frac{1}{\pi}\left(\frac{4}{3}\right)^{(s+1)/2} \int_0^\infty \frac{\mathrm{d}\eta}{(1+\eta)^2\eta^{s-1}} =: N(s)\,,$$

provided $\|K\|_\infty < 1$, where $\|K\|_\infty$ denotes the norm of the operator K on ℓ^∞. We have

$$|(K\xi)_j| \le \|\xi\|_\infty \frac{1}{\pi}\left(1 - \frac{\epsilon}{j^2}\right)^{-(s+1)/2} \int_1^\infty \frac{\eta^{s-1}}{1+\eta^2}\,\mathrm{d}\eta$$

for $s \in [1, 2)$, and since $\epsilon \in (0, 1)$ and $j \ge 2$, it follows that

$$\|K\|_\infty \le \frac{1}{\pi}\left(\frac{4}{3}\right)^{(s+1)/2} \int_0^\infty \frac{\eta^{s-1}}{1+\eta^2}\,\mathrm{d}\eta =: N(s)\,.$$

We have $N(1) = 2/3$ and the right-hand side is a continuous function of s, hence there is a $\delta > 0$ such that $N(1+\delta) < 1$. In other words, there is a unique sequence $r \in \ell^\infty$ which satisfies the equation $r = t + Kr$; the corresponding sequence a decays as $\mathcal{O}(j^{-(1+\delta)})$ for $j \to \infty$, and thus by Problem 8 it gives rise to a vector $\psi \in \mathrm{Dom}\,(-\Delta_D^\Omega)$.

Finally, to find the value of ϵ one has to solve the first equation of the original system, $\sum_j C_{1j}a_j = 1$ with $a_1 = 1$ and $a_j = j^{-s}r_j\,,\ j \ge 2$, obtained above. This yields the condition

$$F(\epsilon) := \frac{1-\mathrm{e}^{-2\pi q_1}}{\pi q_1} \left[\frac{1}{2-\epsilon} + \sum_{j=2}^\infty \frac{r_j}{j^2+1-\epsilon}\right] = 1 \qquad (1.15)$$

which has in $(0, 1)$ a unique solution equal to $\epsilon = 0.9291...$ (Problem 10). Notice that this does not imply uniqueness of the solution to our spectral problem because the class of sequences we have employed does not coincide with $\ell^2(j)$. However, it is easy to check the relation $N(\pi/2) = 1$ directly (Problem 11). The exponential decay is obvious from (1.14) because at large distances the term corresponding to the first transverse mode dominates. ∎

In a similar way one can treat Ω_β with $\beta \neq \pi/2$ and more complicated polygonal ducts—see the notes. The mode-matching method is often used in the physical literature; the convergence of the truncation procedure is rarely checked, typically being guessed from the numerical stability of the algorithm.

1.3 Bent Tubes in $\mathbb{R}^3$

The mechanism which leads to the existence of localized states is not restricted to planar waveguides. Let us now consider a three-dimensional Dirichlet tube. As before it will be built around a sufficiently smooth curve Γ with which we can conventionally associate the Frenet triad frame (t, n, b). We assume that the latter exists at least piecewise globally, i.e. that there is an increasing sequence $\{s_j\} \subset \mathbb{R}$, possibly empty and without finite accumulation points in case it is infinite, such that the triad is well defined in each interval (s_j, s_{j+1}); recall that this is the case when $\ddot{\Gamma}$ vanishes nowhere in the interval.

Tubes in $\mathbb{R}^3$ can have a variety of cross sections. We suppose that $M \subset \mathbb{R}^2$ is an open precompact set which contains zero, and put $a := \sup_{x\in M} |x|$. Given a piecewise smooth function $\alpha : \mathbb{R} \to \mathbb{R}$ we define a map $f : \mathbb{R}\times M \to \mathbb{R}^3$ by

$$f(s, r, \theta) := \Gamma(s) - r\,[\,n(s)\, \cos(\theta - \alpha) + b(s)\, \sin(\theta - \alpha)\,]\ , \tag{1.16}$$

where r, θ are polar coordinates in $\mathbb{R}^2$. In this way we associate with Γ and M a tube $\Omega \equiv \Omega_{\Gamma,M}$ which is the range of this map, $\Omega := f(\mathbb{R}\times M)$. While f depends on α we do not use it as a label for Ω because we restrict ourselves in the following to tubes which satisfy condition (1.18) below.

In analogy with (1.5) the map f introduces curvilinear coordinates in Ω which will be useful in analyzing the spectrum of $-\Delta_D^\Omega$. Recall first that the vectors t, n, b are related by the ***Frenet formula***

$$\begin{pmatrix} \dot t \\ \dot n \\ \dot b \end{pmatrix} = \begin{pmatrix} 0 & \gamma & 0 \\ -\gamma & 0 & \tau \\ 0 & -\tau & 0 \end{pmatrix} \begin{pmatrix} t \\ n \\ b \end{pmatrix},$$

where γ, τ are the curvature and torsion of Γ, respectively, and the dot means again differentiation w.r.t. the arc length s. Using it we find the metric tensor

$$(g_{ij}) = \begin{pmatrix} (1 + r\gamma \cos(\theta - \alpha))^2 + r^2(\tau - \dot\alpha)^2 & 0 & r^2(\tau - \dot\alpha) \\ 0 & 1 & 0 \\ r^2(\tau - \dot\alpha) & 0 & r^2 \end{pmatrix} \tag{1.17}$$

of this coordinate system (Problem 12a). In particular, we have

$$g := \det(g_{ij}) = r^2\,(1 + r\gamma \cos(\theta - \alpha))^2$$

and the Jacobian of f equals $g^{1/2} = r(1+r\gamma\cos(\theta-\alpha))$. It is clear that to make the described coordinates locally orthogonal one has to require

$$\dot{\alpha} = \tau\,, \tag{1.18}$$

i.e. the system has to rotate around t with respect to the Frenet triad with the angular velocity equal to the torsion when we move along Γ.

Remarks 1.3.1 (a) This assumption, sometimes called ***Tang's condition***, which we shall adopt in the following, can always be satisfied for a circular tube, $M = B_a$ with $B_a := \{x \in \mathbb{R}^2 : |x| < a\}$, which has a cross section invariant with respect to rotations. In all the other cases (1.18) represents a nontrivial requirement on the tube geometry.
(b) Without loss of generality we may also assume that M is connected because in the opposite case the problem reduces to a spectral analysis of the family of disjoint tubes corresponding to connected components of the cross section.

As in the planar case we want the curvilinear coordinate system to be defined globally. Hence we assume that Ω does not intersect itself, i.e. that

(i) the map f is injective.

In three-dimensional space this does not imply restrictions on the bending angle as in Remark 1.1.1, and in addition, the tube can be knotted in various ways. The smoothness requirement can be expressed as follows

(ii)$_k$ $\Gamma \in C^{k+2}(\mathbb{R}, \mathbb{R}^3)$, $k \geq 0$, and the functions γ, τ together with their derivatives up to the k-th order are bounded in $\mathbb{R}$.

In combination these assumptions ensure that f is a C^k-diffeomorphism between $\Omega_0 := \mathbb{R} \times M$ and Ω—cf. Problem 12b—which provides the straightening transformation in the present case. This claim is obvious if the Frenet frame exists globally, however, since α may have isolated jumps and condition (1.18) determines it only up to a constant, such a coordinate transformation can be constructed even if Γ satisfies only the weaker hypothesis made above. It is sufficient that the one-sided limits of $\alpha(s)$ exist at the exceptional points which is, in view of the condition (1.18), true if the torsion is locally bounded (Problem 12c).

Using $\tilde{U} : L^2(\Omega) \to L^2(\Omega_0, g^{1/2}\mathrm{d}s\,\mathrm{d}r\,\mathrm{d}\theta)$ defined through the variable change given by f we can pass from $-\Delta_D^\Omega$ to the unitarily equivalent operator

$$\tilde{H} := -g^{-1/2}\partial_s r^2 g^{-1/2}\partial_s - g^{-1/2}\partial_r g^{1/2}\partial_r - g^{-1/2}\partial_\theta r^{-2} g^{1/2}\partial_\theta\,,$$

which makes sense with $k = 1$ in the above assumption, and with $k = 0$ or an even weaker assumption if we define it through its quadratic form,

$$\|\tilde{H}^{1/2}\phi\|_g^2 = \|rg^{-1/4}\partial_s\phi\|^2 + \|g^{1/4}\partial_r\phi\|^2 + \|g^{1/4}r^{-1}\partial_\theta\phi\|^2 \tag{1.19}$$

for $\phi \in Q(\tilde{H})$, where the subscript labels the norm in $L^2(\Omega_0, g^{1/2}\mathrm{d}s\,\mathrm{d}r\,\mathrm{d}\theta)$.

Under the stronger assumption *(ii)*$_2$ one can remove the Jacobian passing to another representation of $-\Delta_D^\Omega$ by means of $U : L^2(\Omega) \to L^2(\Omega_0)$ defined as $U\phi = g^{1/4}\phi \circ f$. After a short calculation we get

$$H := U(-\Delta_D^\Omega)U^{-1} = -\partial_s h^{-2}\partial_s - \Delta_D^M + V(s, r, \theta)\,, \tag{1.20}$$

where $-\Delta_D^M$ is the Dirichlet Laplacian on the tube cross section and the ***effective potential*** is given by

$$V(s, r, \theta) = -\frac{\gamma^2}{4h^2} + \frac{1}{2}\frac{h_{ss}}{h^3} - \frac{5}{4}\frac{h_s^2}{h^4} \tag{1.21}$$

with

$$\begin{aligned} h &:= g^{1/2}r^{-1} = 1 + r\gamma\cos(\theta-\alpha)\,, \\ h_s &:= r\gamma\tau\sin(\theta-\alpha) + r\dot{\gamma}\cos(\theta-\alpha)\,, \\ h_{ss} &:= r(\ddot{\gamma} - \gamma\tau^2)\cos(\theta-\alpha) + r(2\dot{\gamma}\tau + \gamma\dot{\tau})\sin(\theta-\alpha) \end{aligned}$$

(Problem 13). The form of the straightened operator is similar to that of (1.7) and the claim about its domain and cores from Remark 1.1.2 adapts easily to the three-dimensional situation.

The same is true for the essential spectrum. The transverse part $-\Delta_D^M$ of (1.20) has a purely discrete spectrum $\{\nu_n\}_{n=1}^\infty$, arranged conventionally in the ascending order, corresponding to the eigenfunction family $\{\chi_n\}_{n=1}^\infty \subset L^2(M)$. The ground-state eigenvalue ν_1 is simple and positive, in fact, it is bounded below by the ***Faber-Krahn inequality***,

$$\nu_1 \geq \pi\, j_{0,1}^2\, |M|^{-1}\,, \tag{1.22}$$

where $|M|$ is the area of M and $j_{0,1} \approx 2.40$ is the first zero of the Bessel function J_0, so the minimal value is for a fixed $|M|$ attained by the disc. Let us again assume that the tube is asymptotically straight in the sense that

(iii) $\gamma, \dot{\gamma}, \ddot{\gamma} \in L_\varepsilon^\infty(\mathbb{R})$.

This allows us to localize the essential spectrum.

Proposition 1.3.1 *Let Γ have a piecewise global Frenet frame. Assume (i), (ii)$_2$, and (iii) together with $a\|\gamma\|_\infty < 1$. Then we have $\sigma_{\mathrm{ess}}(-\Delta_D^\Omega) = [\nu_1, \infty)$.*

Proof Similar to that of *Proposition* 1.1.1 (Problem 14). ∎

Notice that we imposed no restriction on the torsion apart from the boundedness contained in *(ii)*$_2$. Moreover, the stated assumptions may be weakened allowing local deformations or a weaker curvature decay (cf. Problems 14 and 3.11).

The heuristic argument of Sect. 1.1 applies again: for thin tubes the effective potential (1.21) is dominated by the curvature-induced attractive potential, so one can establish the existence of bound states using a simple perturbation-theory argument (Problem 15). Furthermore, using the GJ-trick one can prove that such a result remains valid for tubes not necessarily thin.

Theorem 1.3 *Assume that Γ has a piecewise global Frenet frame, (i) is satisfied and $a|\gamma(s)| \le c < 1$ holds outside a finite part of Ω. Moreover, let $\gamma, \tau \in L^\infty(\mathbb{R})$ be such that (1.18) is satisfied. Then $\inf \sigma(-\Delta_D^\Omega) < \nu_1$ unless $\gamma = 0$ identically. In particular, $-\Delta_D^\Omega$ has at least one eigenvalue below ν_1 if the assumptions of Proposition 1.3.1 are valid.*

Proof To construct a trial function one starts from a generalized eigenfunction at energy ν_1. It is straightforward to check that

$$\int_M \left[|\partial_r \chi_1|^2 + r^{-2} |\partial_\theta \chi_1|^2 - \nu_1 |\chi_1|^2 \right] \mathrm{d}r\, \mathrm{d}\theta = 0\,,$$

and therefore one can adapt the argument from the two-dimensional case taking Problem 4 into account (Problem 16). ∎

Notice that we made no use here of the particular form of the transverse eigenfunctions. As a result the argument also works when the Hamiltonian contains a potential provided the latter depends on the transverse variables only and conforms with condition (1.18)—cf. Problem 17. The same is true even if the additional interaction is strongly singular as is the case for a tube threaded by a magnetic flux line—see the notes. On the other hand, we shall see later that condition (1.18) is necessary.

1.4 Local Perturbations of Straight Tubes

Of course, bending is not the only way to create localized states. In this section we shall briefly mention three mechanisms which give rise to a discrete spectrum in a straight tube. We shall not restrict the dimension here, i.e. the unperturbed tube will be $\Omega_0 = \mathbb{R} \times M$, where $M \subset \mathbb{R}^{d-1}$ is an open precompact set; we suppose that M is pathwise connected and that ∂M is piecewise smooth. The free Hamiltonian is the corresponding Dirichlet Laplacian, $H_0 = -\Delta_D^{\Omega_0}$ with form domain $H_0^1(\Omega_0)$. The variables $\vec{x} = (x, y)$ with $y \in M$ can be separated in the free Hamiltonian, so we have

$$H_0 = \overline{-\partial_x^2 \otimes I + I \otimes (-\Delta_D^M)}\,, \tag{1.23}$$

where $-\Delta_D^M$ is the Dirichlet Laplacian on $L^2(M)$. Due to the compactness of $\overline{M}$ this operator has a purely discrete spectrum; we denote by $\chi_n,\ \nu_n$ with $n = 1, 2, \ldots$ its eigenfuctions and eigenvalues, respectively. The eigenfunctions can be chosen to be real-valued, which we shall assume throughout.

First we perturb this H_0 with a potential. In most situations we shall consider below the potential will be bounded, however, in general a much weaker regularity is required. For instance, using the corresponding quadratic form,

$$t_V[\psi] = \int_{\Omega_0} |\nabla\psi(\vec{x})|^2\, d\vec{x} + \int_{\Omega_0} V(\vec{x})\, |\psi(\vec{x})|^2\, d\vec{x}\,, \tag{1.24}$$

one can define $H = H_0 \dot{+} V$ for any $V \in L^1_{\mathrm{loc}}(\Omega_0)$. To decide which potentials can exhibit bound states we first have to localize the essential spectrum. One naturally expects that the latter will be preserved if V vanishes along the tube at large distances in both directions. In fact, it is enough if the potential decay is controlled in the integral sense.

Proposition 1.4.1 *$\sigma_{\mathrm{ess}}(H) = \sigma_{\mathrm{ess}}(H_0) = [\nu_1, \infty)$ holds for any V belonging to $(L^p + L^\infty_\varepsilon)(\Omega_0)$, where $p \geq \max\left\{2, \frac{d}{2}\right\}$ for $d \neq 4$ and $p > 2$ for $d = 4$.*

Proof By Weyl's theorem it is sufficient to check that $V(H_0 - z)^{-1}$ is compact for some $z < 0$. Since the Green function of H_0 is majorized by that of the Laplacian in $\mathbb{R}^d$ (Problem 18), the result follows from the usual Schrödinger operator theory – cf. [RS, Sect. XIII.4]. ∎

For the existence of bound states in such a tube it is then decisive whether the potential V is attractive in a suitable sense.

Proposition 1.4.2 *$\sigma_{\mathrm{disc}}(H) \neq \emptyset$ holds provided the assumptions of Proposition 1.4.1 are satisfied and*

$$\int_{\mathbb{R}}\int_M V(x, y)\, \chi_1(y)^2\, d\vec{x} < 0\,. \tag{1.25}$$

Proof In view of the previous proposition we just have to find a trial function which makes the shifted energy form, $\psi \mapsto t_V[\psi] - \nu_1 \|\psi\|^2$, negative. Using assumption (1.25) one finds that this can be achieved, e.g., by choosing $\psi(x, y) = \phi_\lambda(x)\chi_1(y)$, where ϕ_λ is the function (1.11) with λ small enough. ∎

Naturally, these bound states are not of a geometric origin but it is useful to include them in the discussion; it will help us later, for instance in estimating eigenvalue moments in Sect. 3.1.2 or in analyzing the weak coupling behavior in Sect. 6.2. The essentially one-dimensional character of the problem is seen from the fact that the bound state existence is not affected by the potential strength. Furthermore, weakly coupled states may exist even if the above integral is zero, as we shall show in Sect. 6.1 below. On the other hand, notice that condition (1.25) expressing attractiveness in the mean does not concern the potential itself but rather its projection onto the lowest transverse mode.

Binding may result not only from a potential but also from *a modification of the kinetic term*. While the former is typically associated with an external field, the latter is a natural model for a local change of the waveguide material parameters. The Hamiltonian can then be defined through the quadratic form

$$t_\rho[\psi] = \int_{\Omega_0} \rho(\vec{x})|\nabla\psi(\vec{x})|^2\,\mathrm{d}\vec{x}\,, \tag{1.26}$$

where ρ is a measurable function such that $c_1 \le \rho(\vec{x}) \le c_2$ holds for some positive c_1, c_2 and all $\vec{x} \in \Omega_0$. For simplicity we restrict ourselves to the situation when the perturbation is local, i.e. the function $\rho(\cdot)-1$ has a compact support; then the operator H associated with (1.26) has the following spectral properties.

Proposition 1.4.3 *Under the stated assumptions about the function* ρ, *we have* $\sigma_{\mathrm{ess}}(H) = [\nu_1, \infty)$, *and moreover,* $\sigma_{\mathrm{disc}}(H)$ *is nonempty provided*

$$\int_{\Omega_0} (\rho(\vec{x})-1)\,\chi_1(y)^2\,\mathrm{d}\vec{x} < 0\,. \tag{1.27}$$

Proof In view of the hypotheses the essential spectrum is easily localized by means of Neumann bracketing, minimax estimates, and constructing suitable Weyl sequences. Next we consider the shifted energy form, $\psi \mapsto t_\rho[\psi] - \nu_1\|\psi\|^2$ and again choose $\psi(x,y) = \phi_\lambda(x)\chi_1(y)$ as a trial function with $\phi_\lambda(x) = 1$ on the support of $\rho(\cdot)-1$. Then the contribution from the integral in (1.27) is negative and independent of λ, hence it prevails over $\lambda\|\dot{\phi}\|^2_{L^2(\mathbb{R})}$ and thus determines the sign of the form for λ small enough. ∎

Of course, the two effects can appear in various combinations, and the functions V, ρ can either model actual physical quantities or they may arise from a transformation of the Hamiltonian as was the case in Sects. 1.1 and 1.3.

Finally, the third binding mechanism we will mention here is again of a geometrical nature, being connected with *a local variation of the tube shape*. Consider a set-valued function $x \mapsto M_x$ which assigns to each $x \in \mathbb{R}$ a bounded set $M_x \subset \mathbb{R}^{d-1}$ with the properties described at the beginning of this section. We shall consider tubes of a varying cross section defined as

$$\Omega := \bigcup_{x\in\mathbb{R}} M_x\,. \tag{1.28}$$

Suppose that the variation has a local character, i.e. that M_x equals a fixed set M outside a compact set, and that the cross section of the tube (1.28) varies in a piecewise continuous manner. More specifically, we assume that apart from a discrete set of points, to each $x \in \mathbb{R}$ and $\varepsilon > 0$ there is an open $O \ni x$ such that for any $x' \in O$ the symmetric difference $M_x \Delta M_{x'}$ is contained in the ε-neighborhood of ∂M_x. Moreover, the deformation is supposed to satisfy a global bound, i.e. there is a precompact

$N \subset \mathbb{R}^{d-1}$ such that $M_x \subset N$ holds for all $x \in \mathbb{R}$. With these assumptions, we have the following result.

Theorem 1.4 *$\sigma_{\rm ess}(-\Delta_D^\Omega) = [\nu_1, \infty)$, where ν_1 is the lowest eigenvalue of $-\Delta_D^M$. The discrete spectrum is empty if $M_x \subset M$ for all $x \in \mathbb{R}$. On the other hand, if $M_x \supset M$ for each $x \in \mathbb{R}$ and there is an interval where $M_x \setminus M$ has a nonzero measure, then $-\Delta_D^\Omega$ has at least one eigenvalue in $(0, \nu_1)$.*

Proof The essential spectrum is determined as in *Proposition* 1.1.1 (Problem 19). For a squeezed tube we have $\Omega \subset \Omega_0 := \mathbb{R} \times M$ and by Dirichlet bracketing $-\Delta_D^\Omega \geq -\Delta_D^{\Omega_0}$ holds on $L^2(\Omega)$ so the discrete spectrum is void. For a protruded Ω we have to find $\psi \in Q(-\Delta_D^\Omega)$ such that $q[\psi] := \|\nabla\psi\|^2 - \nu_1\|\psi\|^2 < 0$.

In view of the assumption about the x-dependence of the cross section we may suppose without loss of generality that Ω is a smooth volume-expanding deformation of Ω_0, because otherwise we may choose $\Omega' \subset \Omega$ with this property and arrive at the conclusion for Ω using Dirichlet bracketing. The transverse spectrum is purely discrete for any $x \in \mathbb{R}$; we denote by $\chi_{j,x}$ and $\nu_{j,x}$, respectively, the positive eigenfunctions (with unit norm in $L^2(M_x)$) and eigenvalues of $-\Delta_D^{M_x}$. Moreover, we extend the function χ_1 defined originally on $\mathbb{R} \times M$ by zero to the whole Ω. Now we choose $\psi(x, y) := \phi_\lambda(x)\chi_1(y) + \varepsilon f(x)\chi_{1,x}(y)$ in analogy with *Theorem* 1.1, where ϕ_λ is again the function (1.11) with the parameter s_0 picked in such a way that $M_x = M$ for $|x| \geq s_0$ and f is a non-negative function from $C_0^\infty(-s_0, s_0)$. If the cross section varies smoothly with x so does $\chi_{1,x}$ and it is easy to check that ψ belongs to $H_0^1(\Omega)$. A short calculation using integration by parts with respect to y then gives

$$q[\phi_\lambda\chi_1 + \varepsilon f\chi_{1,x}] = \lambda\|\dot\phi\|^2_{L^2(\mathbb{R})} + \varepsilon^2 \int_{\mathbb{R}} |\partial_x(f(x)\chi_{1,x})|^2\, \mathrm{d}x$$
$$+ 2\varepsilon \int_{\mathbb{R}} f(x)\,(\nu_{1,x} - \nu_1) \int_M \chi_1(y)\,\chi_{1,x}(y)\,\mathrm{d}y\,\mathrm{d}x \;+ \varepsilon^2 \int_{\mathbb{R}} f^2(x)\,(\nu_{1,x} - \nu_1)\,\mathrm{d}x\,.$$

The function $x \mapsto \nu_{1,x}$ is continuous so the integrals make sense. Furthermore, the ground state eigenvalue is strictly monotonous with respect to a (nonzero capacity) expansion of the cross section by [GZ94]. Since $\chi_{1,x}$ and χ_1 are positive on M, it follows that the term proportional to ε is negative, hence it is sufficient to choose λ and ε small enough. ∎

If the variation of Ω_0 does not have a local character and vanishes slowly enough at infinity, it may produce infinitely many eigenvalues below the threshold of the essential spectrum. We shall illustrate this effect in two-dimensional waveguides. Let Ω be given by

$$\Omega = \Big\{(x, y) \in \mathbb{R}^2 \,:\, x \in \mathbb{R}\,,\; 0 < y < 1 + f(x)\Big\}\,, \tag{1.29}$$

where f is a suitable function vanishing at infinity. Although the statements of the following proposition hold under much weaker assumptions, for the sake of simplicity we consider only a particular class of functions f.

Proposition 1.4.4 *Let $f \in C^2(\mathbb{R})$ be positive and such that*

$$f(x) = |x|^{-\alpha}, \qquad |x| \geq x_0 \tag{1.30}$$

for some $0 < \alpha < 2$ and $x_0 > 0$. Then $\sigma_{\text{ess}}(-\Delta_D^\Omega) = [\pi^2, \infty)$ and $-\Delta_D^\Omega$ has infinitely many eigenvalues in $(0, \pi^2)$.

Proof We start with the essential spectrum. Let $\varepsilon > 0$, then there is an $x_\varepsilon > x_0$ such that $|f(x)| < \varepsilon$ provided $|x| \geq |x_\varepsilon|$. We introduce additional Neumann boundary conditions at $\{x = \pm x_\varepsilon\}$. In this way we obtain the operators H_l, H_c and H_r which correspond to the restriction of $-\Delta_D^\Omega$ on $x < -x_\varepsilon$, $|x| \leq |x_\varepsilon|$, and $x > x_\varepsilon$, respectively. By the Neumann bracketing we get the operator inequality

$$-\Delta_D^\Omega \geq H_l \oplus H_c \oplus H_r\,. \tag{1.31}$$

It follows easily that

$$\inf \sigma_{\text{ess}}(-\Delta_D^\Omega) \geq \inf \sigma_{\text{ess}}(H_l \oplus H_c \oplus H_r) = \min\{\inf \sigma_{\text{ess}}(H_l),\ \inf \sigma_{\text{ess}}(H_r)\}\,.$$

On the other hand, extending the test functions from the form domain of H_r by zero to $\Omega^r = (x_\varepsilon, \infty) \times (0, 1 + f(x_\varepsilon))$, we obtain the inequality

$$H_l \geq -\Delta_{D,N}^{\Omega^r},$$

where the subscript N indicates the Neumann boundary condition at $x = x_\varepsilon$. Hence

$$\inf \sigma_{\text{ess}}(H_r) \geq \inf \sigma_{\text{ess}}(-\Delta_{D,N}^{\Omega^r}) \geq \frac{\pi^2}{(1+\varepsilon)^2}\,.$$

The same reasoning shows that the latter inequality also holds for $\inf \sigma_{\text{ess}}(H_l)$. Since ε can be chosen arbitrary, it follows that

$$\inf \sigma_{\text{ess}}(-\Delta_D^\Omega) \geq \pi^2.$$

The inclusion $[\pi^2, \infty) \subset \sigma_{\text{ess}}(-\Delta_D^\Omega)$ can again be proved by constructing a suitable Weyl sequence in the standard way.

In order to show that $-\Delta_D^\Omega$ has infinitely many eigenvalues in $(0, \pi^2)$, we look at the quadratic form associated to $-\Delta_D^\Omega$ restricted to the functions

$$u(x, u) = u_0(x, y)\, v(x), \quad u_0(x, y) = \sin\left(\frac{\pi y}{1 + f(x)}\right).$$

A straightforward but lengthy calculation then shows that

$$\int_\Omega |\nabla u|^2\, dxdy - \pi^2 \|u\|_2^2 = \int_{\mathbb{R}} \frac{1+f(x)}{2}\, |v'(x)|^2\, dx - \int_{\mathbb{R}} W_f(x)\, v^2(x)\, dx, \tag{1.32}$$

where

$$W_f(x) = \frac{\pi^2(2f(x)+f^2(x))}{2(1+f(x))} - \frac{\pi^2 f''(x)}{4} + \frac{f'(x)^2}{2(1+f(x))}\left(\pi - \frac{\pi^2}{6} - \frac{1}{4}\right).$$

From the assumptions on f it follows that we can find some $R > 0$ large enough such that $W_f(x) \geq c\, x^{-\alpha}$ holds for some $c > 0$ and all $x > R$. Next we put an additional Dirichlet boundary condition at $x = R$ and note that by the variational principle and by (1.32) the number of eigenvalues of $-\Delta_D^\Omega$ below π^2 is greater than or equal to the number of negative eigenvalues of the operator H_f in $L^2(R, \infty)$ generated by the form

$$Q[v] = \int_R^\infty \frac{1+f(x)}{2}\, |v'(x)|^2\, dx - c \int_R^\infty x^{-\alpha}\, v^2(x)\, dx, \quad v \in H_0^1(R, \infty). \tag{1.33}$$

Now we construct a sequence $\{v_n\}$ of test functions given by

$$v_n(x) = \begin{cases} 2^{-n}(x - 2^n) & \ldots\ 2^n \leq x < 2^{n+1} \\ 1 & \ldots\ 2^{n+1} \leq x < 2^{n+2} \\ 1 - 2^{-n-2}(x - 2^{n+2}) & \ldots\ 2^{n+2} \leq x < 2^{n+3} \end{cases}$$

and $v_n(x) = 0$ otherwise, which satisfy $v_n \in H_0^1(R, \infty)$ as long as $2^v > R$. By a direct calculation we find that

$$Q[v_n] < 0$$

holds for all n large enough (depending on α). Since the v_n are linearly independent, it follows that H_f has infinitely many negative eigenvalues, and therefore the same is true for the operator $-\Delta_D^\Omega - \pi^2$. ∎

If Ω is a deformation of Ω_0 which is neither a protrusion nor a squeeze, there is no general rule for the existence of bound states. In particular, it is not important whether the deformation is protruding or squeezing in the mean.

Example 1.4.1 Consider a pair of functions $\varphi_\pm \in C_0^\infty((0,1))$ satisfying the inequalities $0 < \varphi_+(x) < \varphi_-(x) < 1$ for all $x \in (0,1)$. Given $L > 0$ we set $f_L(x) := \pi[1 + \varphi_+(x-L) - \varphi_-(L-x)]$ and consider $-\Delta_D^{\Omega_L}$ corresponding to the strip

$$\Omega_L := \{\vec{x} : 0 < y < f_L(x)\,,\ x \in \mathbb{R}\}$$

having a protrusion and a dent spaced $2L$ apart; by assumption the total volume added by the deformation is *negative.* We get an upper bound to $-\Delta_D^{\Omega_L}$ adding a Dirichlet boundary at the perpendicular strip cut at $x = 0$. Since both operators obviously have the same essential spectrum $[1, \infty)$, a bound state in Ω_L will exist if one of the operators $-\Delta_D^{\Omega_L^\pm}$ corresponding to the halfstrips

$$\Omega_L^\pm := \{\vec{x} : 0 < y < \pi[1 \pm \varphi_\pm(x)],\ -L < x < \infty\}$$

with Dirichlet condition at the cut has an eigenvalue in $(0, 1)$. Since the family $\{\Omega_L^+\}$ is increasing with L to $\Omega_\infty^+ := \bigcup_{L>0} \Omega_L^+$ and the boundary has the segment property, the operators $-\Delta_D^{\Omega_L^+} P_L$, where P_L is the projection onto $L^2(\Omega_L^+)$ in $L^2(\Omega_\infty^+)$, converge by [RT75, Lemma I.1] to $-\Delta_D^{\Omega_\infty^+}$ in the norm resolvent sense. Since the latter has an eigenvalue in $(0, 1)$ by *Theorem* 1.4, it follows that the same is true for $-\Delta_D^{\Omega_L}$ for all L large enough.

Example 1.4.2 Consider another locally deformed strip. This time it will be $\Omega_{d,L}$ with $d \in \left(\frac{1}{2}, 1\right)$ and $L > 1$, defined as $\Omega_{d,L} := \{\vec{x} : |y| < f_{d,L}(x),\ x \in \mathbb{R}\}$, where

$$f_{d,L}(x) := \begin{cases} \pi L & \dots\ 2|x| < \pi d \\ \pi d/2 & \dots\ \pi d \geq 2|x| < 2\pi L \\ \pi/2 & \dots\ |x| \geq \pi L \end{cases}$$

Such a deformation of the unperturbed strip $\Omega_{1,1/2}$ means a *positive* volume change if $d > 2L[1 - \sqrt{1-(2L)^{-1}}]$ which, for a fixed d, can be achieved by choosing L large enough. The operator $-\Delta_D^{\Omega_{d,L}}$ can be estimated from below by imposing Neumann condition at the cuts $\{\pm\pi L\} \times (-\pi d/2, \pi d/2)$, and this estimate will not be spoiled if we replace the Dirichlet condition by the Neumann one at the segments $(-\pi d/2, \pi d/2) \times \{\pm\pi L\}$. The tail part spectrum fills the interval $[1, \infty)$, so an isolated eigenvalue of $-\Delta_D^{\Omega_{d,L}}$ would give rise to an eigenvalue $\nu \in (0, 1)$ of $-\Delta_D^{\tilde{\Omega}_{d,L}}$, where $\tilde{\Omega}_{d,L}$ is the middle part, a rectangular cross with Neumann "lids". By Problem 20 the ground state of such an operator tends to $\tilde{\nu} d^{-2}$ as $L \to \infty$, where $\tilde{\nu} \approx 0.66$. Consequently, the discrete spectrum of $-\Delta_D^{\Omega_{d,L}}$ is void, e.g., for $d = \frac{3}{4}$ and L large enough.

On the other hand, the sign of the added volume plays a decisive role in the situation where the deformation is gentle, as we shall discuss in Chap. 6.

1.5 Coupled Two-Dimensional Waveguides

Let us return now to the two-dimensional situation and consider more complicated structures composed of several strips. There is naturally a large number of possible combinations and we restrict ourselves to discussion of a few simple examples.

Notice, however, that one can often draw conclusions about the existence of bound states using the results we have already proved. For instance, if a system of tubes contains a subset of the form of a bent guide (and no components pushing the essential spectrum threshold down such as outgoing asymptotically straight ducts of a larger width) then an isolated eigenvalue exists in accordance with Problem 6, and the same is true if we can insert a locally protruded tube, etc.

1.5.1 A Lateral Window Coupling

The first example concerns a pair of parallel strips of widths d_1, d_2 with a common boundary containing a window of width $\ell = 2a$. In other words, we consider the set $\Omega \equiv \Omega_{d_1,d_2,\ell} := \{\vec{x} : y \in (-d_2, d_1)\,,\ x \in \mathbb{R}\}\setminus B_a$, where $B_a := \{0\}\times((-\infty, -a]\cup [a, \infty))$. We regard the operator $-\Delta_D^\Omega$ as the Hamiltonian of such ***laterally coupled waveguides***. Let us introduce some notation. We set $d := \max\{d_1, d_2\}$, $D := d_1+d_2$, and $\varrho := d^{-1}\min\{d_1, d_2\}$. Furthermore, we employ the shorthands $\epsilon_d := \left(\frac{\pi}{d}\right)^2$, and $\epsilon_D = \epsilon_d(1+\varrho)^{-2}$, ϵ_ℓ corresponding to D and ℓ, respectively, in the same way. Then the following claim is valid.

Theorem 1.5 *We have $\sigma_{\mathrm{ess}}(-\Delta_D^\Omega) = [\epsilon_d, \infty)$ and $\sigma_{\mathrm{disc}}(-\Delta_D^\Omega) \neq \emptyset$ for any $a > 0$. The eigenvalues $\epsilon_m = \epsilon_m(a)$, $m = 1, \dots, N$, are contained in (ϵ_D, ϵ_d) and decrease continuously as functions of a; their number satisfies the bounds $N_a \le N \le N_a+1$, where $N_a := \min\left\{1, \left[\frac{2a}{d}\sqrt{1-(1+\varrho)^{-2}}\right]\right\}$ and $[\cdot]$ denotes the integer part. The spectrum is simple and the eigenvalues are bounded by*

$$\left(\frac{d}{2a}\right)^2 (m-1)^2 \le \frac{\epsilon_m}{\epsilon_d} - (1+\varrho)^{-2} < \left(\frac{d}{2a}\right)^2 m^2\,. \tag{1.34}$$

Similarly, for the critical values a_m, $m = 2, \dots,$ at which the m-th eigenvalue emerges from the continuum we have

$$\frac{d(m-1)}{\sqrt{1-(1+\varrho)^{-2}}} \le a_m < \frac{dm}{\sqrt{1-(1+\varrho)^{-2}}}\,. \tag{1.35}$$

Proof The essential spectrum can be localized in the same way as in *Proposition* 1.1.1. The existence of eigenvalues is obtained by a GJ-type argument; the eigenvalue estimates follow from the Dirichlet-Neumann bracketing by comparison with the spectrum of the Laplacian in the box $(-a, a) \times (-d_2, d_1)$ with appropriate boundary conditions. The sharp inequalities in (1.34) and (1.35) follow from the strict monotonicity of the estimating eigenvalues w.r.t. a smooth outward deformation of the Dirichlet box which can be checked in the same way as in *Theorem* 1.4 (Problem 20). Finally, continuity of $\epsilon_m(\cdot)$ can be checked in a similar way using a scaling in the x direction. ∎

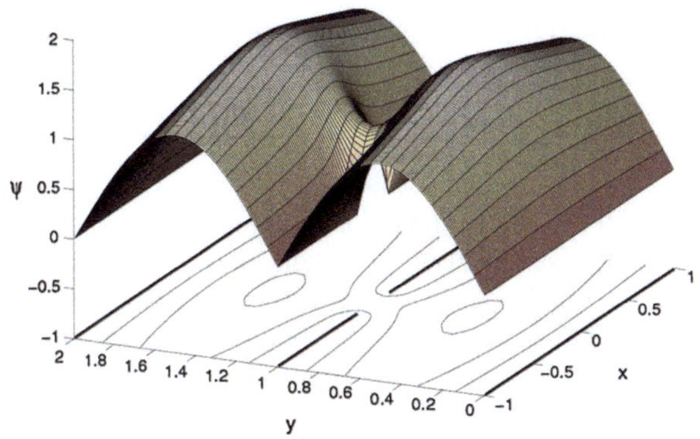

Fig. 1.2 Ground-state eigenfunction in a waveguide with lateral coupling

In fact, the lower bounds in (1.34) and (1.35) are sharp too. Before proceeding further, let us recall that the bound states described in the theorem are again of a purely quantum nature. The reason is the same as in *Proposition* 1.1.3: the set of closed classical trajectories has zero measure, which is here even more obvious. Of course, the phase-space topology is now more complicated, because a particle whose trajectory is not perpendicular to the strip axes may end up in any of the ducts after passing the window region, depending on the initial conditions, but it cannot be turned back (unless it hits a window edge, i.e. the tip of the barrier, which is an event of probability zero).

More about eigenvalues and eigenfunctions can be learned from the numerical solution which is found by the mode-matching method. Since Ω is mirror-symmetric with respect to the line $x = 0$, the operator $-\Delta_D^\Omega$ decomposes into a direct sum of two parts with definite parities, and one can therefore consider the halfstrip problems with Neumann and Dirichlet conditions, respectively, at the cut. Moreover, in the case $d_1 = d_2 = d$ there is another mirror symmetry, which allows us to study one (half)duct only. The antisymmetric part is trivial in this case and the symmetric one is equivalent to the strip with Neumann condition in the window. The matching procedure is straightforward and we leave the task of working out the details to the reader (Problem 21). An eigenfunction example in such a waveguide system with a narrow window is shown in Fig. 1.2.

We shall describe the asymmetric case in more detail, because it reveals another feature of the mode matching which is useful to keep in mind. Without loss of generality we may assume $d_2 < d_1 = d$. As indicated above, we consider the right-halfplane part of Ω with Neumann and Dirichlet conditions on the segment $\mathcal{C} := (-d_2, d_1)$ of the y-axis. We expand the sought solution in terms of the transverse bases. In the window part, $0 < x \le a$, we use

$$\eta_k(y) := \sqrt{\frac{2}{D}}\, \sin(K_k(d_1 - y))\,, \quad k = 1, 2, \dots\,,$$

where $K_k := kK_1 = k\kappa_1(1+\varrho)^{-1}$ with $\kappa_1 = \sqrt{\epsilon_d}$, while in the ducts we take

$$\chi_j^{(\pm)}(y) := \sqrt{\frac{2}{d}}\,\varrho^{-(1\mp 1)/4}\,\sin\left(\kappa_j y \varrho^{-(1\mp 1)/2}\right)\, i_\pm(y)\,, \quad j = 1, 2, \dots,$$

where $\kappa_j := j\kappa_1$ and $i_\pm$ are the indicator functions of the intervals $\mathcal{C}_+ := (0, d_1)$ and $\mathcal{C}_- := (-d_2, 0)$, respectively. The union of the two bases is, of course, an orthonormal basis in $L^2(\mathcal{C})$. Since numerical computations involve a truncation, however, a proper ordering is needed. For that purpose we arrange the numbers $j,\ k\varrho^{-1}$ with $j, k = 1, 2, \dots$ into a nondecreasing sequence (if ϱ is rational and there is a coincidence, any order can be chosen in such a pair); we denote its elements by θ_m, i.e.

$$\theta_1 := 1\,, \quad \theta_2 := \min\{2, \varrho^{-1}\}\,, \quad \text{etc.} \tag{1.36}$$

The corresponding ordered basis in $L^2(\mathcal{C})$ is

$$\xi_m : \xi_m(y) = \chi_j^{(\pm)}(y) \quad \text{if} \quad \theta_m = j\varrho^{-(1\mp 1)/2}\,.$$

The even solutions of energy $\epsilon\kappa_1^2$ with $(1+\varrho)^{-2} < \epsilon < 1$, which correspond to the Neumann condition at $x = 0$, are sought using the Ansatz

$$\psi(x,y) := \begin{cases} \sum_{k=1}^{\infty} a_k \frac{\cosh(p_k x)}{\cosh(p_k a)}\, \eta_k(y) & \dots 0 < x \le a \\ \sum_{j=1}^{\infty} b_j^{(\pm)}\, e^{q_j^{(\pm)}(a-x)}\, \chi_j^{(\pm)}(y) & \dots x \ge a\,,\ y \in \mathcal{C}_\pm \end{cases} \tag{1.37}$$

where $p_j := \kappa_1\sqrt{j^2(1+\varrho)^{-2} - \epsilon}$ and $q_j^{(\pm)} := \kappa_1\sqrt{j^2\varrho^{-(1\mp 1)} - \epsilon}$. The duct part of (1.37) can be written in a unified way as

$$\psi(x,y) = \sum_{m=1}^{\infty} c_m\, \mathrm{e}^{r_m(a-x)}\, \xi_m(y)\,,$$

where we have set $c_m := b_j^{(\pm)}$ and $r_m := q_j^{(\pm)}$ for $\theta_m = j\varrho^{-(1\mp 1)/2}$. Using the continuity of ψ and its normal derivative at $x = a$ together with the orthonormality of $\{\chi_j^{(\pm)}\}$, we find conditions which the coefficient sequences must satisfy; they can be concisely written as

$$c_m = \sum_{k=1}^{\infty} a_k\,(\xi_m, \eta_k)\,, \quad r_m c_m + \sum_{k=1}^{\infty} a_k\, p_k\, \tanh(p_k a)\,(\xi_m, \eta_k) = 0\,;$$

substituting from the first equation to the second one, we obtain the spectral condition $Ca = 0$, where C is the operator on $\ell^2(j)$ with the matrix elements

$$C_{mk} := (r_m + p_k \tanh(p_k a))\,(\xi_m, \eta_k)\,, \tag{1.38}$$

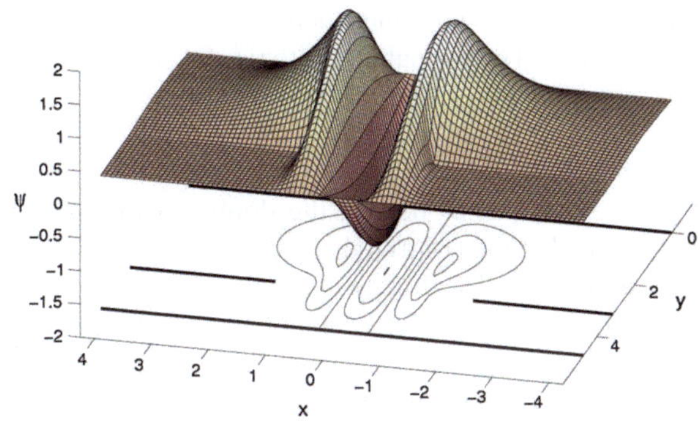

Fig. 1.3 The third eigenfunction in an asymmetric waveguide system

where the overlap integrals are given by

$$(\chi_j^{(\pm)}, \eta_k) = \frac{2j}{\pi}\, \frac{\varrho^{(1\mp 1)/4}}{\sqrt{1+\varrho}}\, \frac{\sin\left(\frac{\pi k}{1+\varrho}\right)}{j^2 - \left(\frac{k}{1+\varrho}\right)^2}\,.$$

In the odd case, which corresponds to the Dirichlet condition at $x = 0$, the only change consists of replacing tanh by coth in (1.38). The eigenvalues and eigenfunctions are obtained by solving the equation $Ca = 0$ numerically (Problem 21). Convergence of the truncated approximations is checked similarly as in *Proposition* 1.2.3: since an eigenfunction belongs to the domain of any power of $-\Delta_D^{\Omega}$, we may seek the solution in $\ell^2(j^s)$ for s large enough such that the operator C is compact. Let us stress, however, that it is crucial for the convergence to use the proper ordering (1.36) of the transverse basis. An eigenfunction example in an asymmetric waveguide system with a lateral coupling is shown in Fig. 1.3.

Remark 1.5.1 The mode matching also shows that all eigenfunctions are exponentially decaying, more specifically that $|\psi_m(\vec{x})| \le c\, \exp\big[-q_{1,m}^{(+)}|x|\big]$ holds for some $c > 0$, where $q_{1,m}^{(+)} := \sqrt{1-\epsilon_m}$, because at large distances the term corresponding to the lowest transverse mode dominates the series. Similar exponential bounds hold for bent tubes with a compactly supported curvature as well as for tubes with local deformations considered in the previous section (Problem 22).

1.5.2 A Leaky Interface

The waveguides considered so far have been ideal in the sense that their walls are impenetrable. Since in real systems they are rather potential steps and tunneling cannot be a priori excluded, we are going to discuss now a modification of the previous example in which two adjacent ducts have a common boundary through which the particle can leak. The formal Hamiltonian of such a system has the form

$$H_\alpha = -\Delta_D^\Omega + \alpha(x)\delta(y)\,,$$

where $\Omega := \mathbb{R} \times (-d_2, d_1)$ is, as before, the double strip and α is the coupling strength. We are interested in the nontrivial situation when the $\alpha(x)$ varies along the strip; our aim is to show that a local increase of the tunneling probability provides another binding mechanism. The outer edges of Ω are supposed to be ideal, with Dirichlet boundary conditions.

Hamiltonians of the above type will be discussed in more detail later, particularly in Chaps. 6 and 10. Here we just recall that they can be defined in different ways. A general method is to employ quadratic forms: for $\psi \in Q(-\Delta_D^\Omega)$ we put

$$t_\alpha[\psi] = \int_\Omega |\nabla\psi(\vec{x})|^2\,\mathrm{d}\vec{x} + \int_\mathbb{R} \alpha(x)|\psi(x,0)|^2\,\mathrm{d}x\,, \tag{1.39}$$

where the last integral makes sense by standard Sobolev imbedding for any Borel measurable function $\alpha : \mathbb{R} \to \mathbb{R}$; the self-adjoint operator associated with the form t_α is identified with H_α. Since the singular interaction support is a smooth curve, one can also say that for a regular enough α, say piecewise smooth, the operator acts as $H_\alpha\psi = -\Delta\psi$ away from the line $y = 0$ and its domain consists of all $\psi \in H^1_{\mathrm{loc}}(\Omega)$ which satisfy the Dirichlet condition at $\partial\Omega$ and

$$\psi(x,0+) = \psi(x,0-)\,, \quad \partial_y\psi(x,0+) - \partial_y\psi(x,0-) = \alpha(x)\psi(x,0) \tag{1.40}$$

holds for any $x \in \mathbb{R}$, and $-\Delta\psi$ in the sense of distributions belongs to $L^2(\Omega)$.

As elsewhere in this chapter we consider local perturbations of a system with separating variables, i.e. we assume $\lim_{|x|\to\infty} \alpha(x) = \alpha_0$ with the same limit in both directions. While we speak for definiteness about a barrier dividing two strips, which would mean choosing an $\alpha_0 > 0$, the sign of α_0 is not important in the rest of this section. To determine the spectral properties of H_α, one first has to find the transverse modes, i.e. to solve the equation

$$-\chi''(y) + \alpha\delta(y)\chi(y) = \nu\chi(y) \tag{1.41}$$

in $(-d_2, d_1)$ with Dirichlet conditions at the endpoints, where the δ-interaction is understood in the sense of the boundary conditions (1.40).

Lemma 1.5.1 *The problem (1.41) has a complete system of eigenfunctions,*

$$\chi_j(y;\alpha) = \mp N_j \,\sin\left(\sqrt{\nu_j(\alpha)}\, d_\varrho^{(1\pm 1)/2}\right)\, \sin\left(\sqrt{\nu_j(\alpha)}\left(y \mp d_\varrho^{(1\mp 1)/2}\right)\right)$$

for $y \in C_\pm$, where $N_j > 0$ is chosen so that $\|\chi_j\| = 1$. The corresponding eigenvalues $\{\nu_j(\alpha)\}_{j=1}^\infty$ solve the condition $-\alpha = \sqrt{\nu}\left(\cot\sqrt{\nu}d_1 + \cot\sqrt{\nu}d_2\right)$. The function

$\alpha \mapsto \nu_j(\alpha)$ is continuous and strictly increasing for any $j = 1, 2, \dots$. Furthermore, let $\{\theta_j\}_{j=1}^{\infty}$ be the sequence (1.36) obtained by natural ordering of the set $\mathbb{N} \cup \varrho^{-1}\mathbb{N}$, then for any $j = 2, 3, \dots$ we have

$$\frac{\pi}{2d}(j-1) \le \frac{\pi}{d}\theta_{j-1} < \sqrt{\nu_j(\alpha)} < \frac{\pi}{d}\theta_j \le \frac{\pi j}{d}\,.$$

Proof is left to the reader (Problem 23).

Remarks 1.5.2 (a) If the δ-interaction is attractive enough, $\alpha < -(d_1^{-1}+d_2^{-1})^{-1}$, the lowest eigenvalue is negative and the corresponding eigenfuction is a combination of hyperbolic sines analogous to the above expression.
(b) If the interval widths are rationally related, $\varrho = \frac{p}{q}$, the spectrum also includes the points with $\sqrt{\nu} = \frac{\pi p}{d_1} n = \frac{\pi q}{d_2} n$, $n = 1, 2, \dots$, where the above eigenvalue condition is not well defined. From the point of view of our original problem it represents a trivial part which can be left out without loss of generality. The prime example is a symmetric waveguide pair, $d_1 = d_2$, where this concerns all the antisymmetric solutions. The eigenvalue inequalities remain valid if we take the corresponding ordered set $\mathbb{N} \cup \varrho^{-1}\mathbb{N}$ without repetitions.

It is now easy to localize the essential spectrum through the transverse ground state. The existence of bound states depends on the shape of the function α, because the leaky barrier is similar to the systems considered in Sect. 1.4; here also the interaction can be both attractive and repulsive.

Proposition 1.5.1 *Suppose that $\alpha \in L^1_{\mathrm{loc}}(\mathbb{R})$ and there is a number α_0 such that $\alpha(x)-\alpha_0 = \mathcal{O}(|x|^{-1-\varepsilon})$ for some $\varepsilon > 0$ as $|x| \to \infty$; then $\sigma_{\mathrm{ess}}(H_\alpha) = [\nu_1(\alpha_0), \infty)$ and the discrete spectrum is nonempty provided*

$$\int_{\mathbb{R}} (\alpha(x)-\alpha_0)\,\mathrm{d}x < 0\,. \tag{1.42}$$

Proof The essential spectrum is found as before: adding Neumann cuts at $\pm y$ with y large enough we check that $\inf \sigma_{\mathrm{ess}}(H_\alpha) > \nu_1(\alpha_0) - \eta$ for any $\eta > 0$, then one has to repeat the argument of Problem 3 with $\chi_1(y;\alpha_0)\,\mathrm{e}^{ips}$ multiplied by a suitable family of mollifiers. Next we employ functions (1.11) again and choose $\psi(x, y) = \phi_\lambda(x)\chi_1(y;\alpha_0)$ as a trial function, obtaining

$$t_\alpha[\psi] - \nu_1(\alpha_0)\|\psi\|^2 = \lambda\|\dot{\phi}\|^2 + \chi_1(0;\alpha_0)^2 \int_{\mathbb{R}} (\alpha(x)-\alpha_0)|\phi_\lambda(x)|^2\,\mathrm{d}x\,.$$

The value $\chi_1(0;\alpha_0) = N_1 \sin\big(\sqrt{\nu_1(\alpha_0)}d\big) \sin\big(\sqrt{\nu_1(\alpha_0)}d\varrho\big)$ is nonzero, which is true even if $\nu_1(\alpha_0) = 0$, because the normalization factor explodes in the limit $\alpha_0 \to -(d_1^{-1} + d_2^{-1})^{-1}$ (cf. Problem 23). In view of the assumption the right-hand side is then negative for λ small enough. ∎

Since the δ-interaction can be approximated by a family of scaled potentials (see the notes) there is an analogy between the present example and the potential binding of Sect. 1.4. However, the problem has also a geometric aspect. This can be seen when we look at eigenvalues and eigenfunctions of the operator H_α in the situation when a constant barrier of the "height" α_0 has an opening of a fixed length and compare the results with those of Sect. 1.5.1, in particular, in the situation when α_0 is large so that the tunneling is small and there is a little difference from the ideal (Dirichlet) barrier (Problem 24).

1.5.3 Crossed Strips

In the previous two examples the strips were parallel. Suppose now instead that they cross at an angle $\theta \in (0, \frac{1}{2}\pi]$, i.e. take the scissor-shaped region $\Omega \equiv \Omega_{D,\theta}$ which consists of all $(x, y) \in \mathbb{R}^2$ satisfying the inequalities

$$\max\left\{0, \frac{|x|\sin\frac{\theta}{2} - \frac{d}{2}}{\cos\frac{\theta}{2}}\right\} < |y| < \frac{|x|\sin\frac{\theta}{2} + \frac{d}{2}}{\cos\frac{\theta}{2}};$$

for simplicity we assume that all the arms have the same width d.

Proposition 1.5.2 *$\sigma_{\mathrm{ess}}(-\Delta_D^\Omega) = [\epsilon_d, \infty)$ and $\sigma_{\mathrm{disc}}(-\Delta_D^\Omega) \neq \emptyset$ holds for any angle $\theta \in (0, \frac{1}{2}\pi]$. The eigenvalues $\epsilon_m = \epsilon_m(\theta)$, $m = 1, 2, \dots$, are contained in $(\epsilon_{2d}, \epsilon_d)$ and increase continuously as functions of θ in $(0, \theta^{(m)})$, where $\theta^{(m)}$ is the critical angle at which the m-th eigenvalue disappears in the continuum, $\theta^{(m)} < \frac{1}{2}\pi$ for $m \geq 2$. Their number satisfies the bound*

$$\sharp\left\{\sigma_{\mathrm{disc}}(-\Delta_D^\Omega)\right\} \geq 2\left[\frac{2c}{\theta}\right],$$

where c is the constant from Proposition 1.2.1. The ground state eigenvalue $\epsilon_1(\theta) \leq \epsilon_1(\pi/2) \approx 0.66\epsilon_d$. The eigenfunctions ϕ_m corresponding to ϵ_m are even w.r.t. the x-axis for any $\theta \in (0, \frac{1}{2}\pi]$ and $m = 1, 2, \dots$.

Proof The argument is analogous to that of *Propositions* 1.2.1 and 1.2.2. The single eigenvalue for the symmetric cross-shaped region, $\theta = \frac{1}{2}\pi$, can be found by mode matching (Problem 26). ∎

1.6 Thin Bent Tubes

We have seen that in specific examples one can learn a lot more about the discrete spectrum than just the fact that it is nonempty. Returning to a single strip or tube one can ask what general properties of the eigenvalues can be deduced. In this section we shall demonstrate one result of this type; the discussion of the discrete spectrum will continue in Chap. 3.

A powerful tool for spectral analysis is the perturbation theory. We have already used relation (1.9) to establish the existence of bound states by comparison with a suitable one-dimensional Schrödinger operator. Now we want to study thin bent strips and tubes in more detail. For the sake of simplicity we shall speak here mostly of planar strips; we encourage the reader to work out the generalization to tubes in $\mathbb{R}^3$ (Problem 27).

Consider thus a curved strip Ω_a of halfwidth a generated by a curve Γ which is described by the transformed Hamiltonian (1.7). To reveal how spectral properties change when a varies, it is useful to pass to the unitarily equivalent operator acting on $L^2(\Omega_0)$ with $\Omega_0 := \mathbb{R} \times (-1, 1)$ obtained by transverse scaling

$$H := -a^{-2}\partial_u^2 + T\,, \qquad T \equiv T(a) := -\partial_s(1+au\gamma(s))^{-2}\partial_s + V(s, au)\,,$$

where $V(s, au)$ is the rescaled effective potential (1.8). We shall use the decomposition

$$H = -\partial_s^2 - \frac{1}{a^2}\,\partial_u^2 - \frac{\gamma^2}{4} + a\,\beta(s, u, \partial_s)\,, \tag{1.43}$$

where $\beta(s, u, \partial_s)$ is the shorthand for the differential expression

$$\partial_s\, \frac{2u\gamma + au^2\gamma^2}{(1+au\gamma)^2}\,\partial_s + \frac{au^2\gamma^4 + 2\,u\gamma^3}{4(1+au\gamma)^2} + \frac{u\ddot{\gamma}}{2(1+au\gamma)^3} - \frac{5}{4}\,\frac{au^2\dot{\gamma}^2}{(1+au\gamma)^4}\,.$$

Let χ_j be the eigenfunctions (1.10) corresponding to the eigenvalues $\kappa_j^2 = (\pi j/2)^2$. We also consider the one-dimensional Schrödinger operator

$$T_0 := -\partial_s^2 - \frac{1}{4}\gamma^2\,. \tag{1.44}$$

We have a perturbation expansion in terms of the strip halfwidth.

Theorem 1.6 *Adopt the assumption (i)–(iv) of Sect. 1.1 with $k = 2$. Then for each negative eigenvalue λ of T_0 there exists an $a_0 > 0$ such that for all $a \in (0, a_0)$ there is a unique simple eigenvalue $\epsilon(a)$ of $-\Delta_D^{\Omega_a}$ given by*

$$\epsilon(a) = \kappa_1^2 a^{-2} + \lambda(a)\,,$$

where $\lambda(a) \in C^\infty[0, a_0)$, in particular, there exist coefficients $d_{\lambda,m}$ such that

$$\lambda(a) = \lambda + \sum_{m=1}^{\infty} d_{\lambda,m}\, a^m\,. \tag{1.45}$$

Proof We just sketch the main ideas of the argument referring for details to the literature mentioned in the notes. We start from the equation

$$a^2(H - \lambda)\, g = f\,. \tag{1.46}$$

Let P be the projector on the subspace $L^2(\mathbb{R}, \mathrm{d}s) \otimes \{\chi_1\}$ of $L^2(\Omega_0)$ and let $Q := I - P$. Accordingly, we have the decomposition

$$g(s,u) = w_1(s)\chi_1(u) + g_2(s,u)\,, \quad w_1(s) := (\chi_1, g(s,\cdot)\chi_1)_{L^2(-1,1)}\,, \quad g_2 := Qg\,,$$

and similarly the function f can be expressed as $f(s,u) = f_1(s)\chi_1(u) + f_2(s,u)$. If we put $\mu = \epsilon - \kappa_1^2 a^{-2}$ and apply the projectors P and Q to the Eq. (1.46), we obtain the system

$$\begin{aligned} -\partial_s^2 w_1 + a w_1 - \mu w_1 + a\beta_{11}(s,\partial_s)\, w_1 + a\beta_{12}(s,\partial_s) g_2 &= f_1(s)\,,\\ -a^2\partial_s^2 g_2 - (\partial_u^2 + \kappa_1^2 + a^2\mu) g_2 + a\beta_{21}(s,\partial_s) w_1 + a\beta_{22}(s,\partial_s) g_2 &= f_2(s,u)\,, \end{aligned}$$

where we have denoted by

$$\beta_{11} := (\chi_1, \beta\chi_1)_{L^2(-1,1)}\,, \quad \beta_{12} := (\beta, \chi_1)_{L^2(-1,1)}\,, \quad \beta_{21} := Q\beta\chi_1\,, \quad \beta_{22} := Q\beta$$

the differential operators in the variable s, dropping the arguments in the first two formulæ. Let now ϕ be the eigenfunction of T_0 associated with the eigenvalue λ and consider the following problem: given the triple $(f_1(s), f_2(s,u), r)$, find $(w_1(s), g_2(s,u), t)$ such that

$$\begin{aligned} -\partial_s^2 w_1 + a w_1 - \mu w_1 + t\,\phi + a\beta_{11}(s,\partial_s)\, w_1 + a\beta_{12}(s,\partial_s) g_2 &= f_1(s)\,,\\ -a^2\,\partial_s^2 g_2 - (\partial_u^2 + \kappa_1^2 + a^2\mu) g_2 + a\,\beta_{21}(s,\partial_s) w_1 + a\beta_{22}(s,\partial_s) g_2 &= f_2(s,u)\,,\\ (w_1, \phi)_{L^2(-1,1)} &= r\,, \end{aligned}$$

where r and $t = t(\mu, a, f_1, f_2, r)$ are real numbers. The central point of the proof is to observe, with the help of the two systems, that ϵ is an eigenvalue of H *iff*

$$t(\mu, a, 0, 0, 1) = 0\,, \qquad \mu = \epsilon - \kappa_1^2 a^{-2}\,.$$

Moreover, one can check that $t(\mu, a, 0, 0, 1)$ is infinitely differentiable as a function of (μ, a) on $(\lambda - \delta, \lambda + \delta) \times [0, a_0)$ with some $\delta > 0$. Finally, passing to the limit $a \to 0$, a direct calculation shows that $t(\mu, 0, 0, 0, 1) = \mu - \lambda$. Hence $\partial_\mu t(\mu, 0, 0, 0, 1) = 1$ and from the implicit function theorem we conclude that $\mu(a) \in C^\infty[0, a_0)$; expansion (1.45) then follows by Taylor's formula. ∎

1.7 Twisted Tubes

But how curiously it twists! It's more like a corkscrew than a path!
Lewis Carrol, Through the Looking Glass

The above discussed geometric deformations of waveguides are, of course, not the only ones possible. To conclude this chapter let us mention another important case. It concerns three-dimensional straight tubes which are *twisted* in the sense defined properly below. We shall see that a local twisting can have an effect opposite to the deformations discussed up to now, namely that it can preclude the existence of bound states in such a tube. The reason for this is a Hardy-type inequality induced by a twisting of the waveguide.

1.7.1 A Hardy Inequality for Twisted Tubes

Consider first a straight three-dimensional tube $\Omega_0 = \mathbb{R} \times M$, where $M \subset \mathbb{R}^2$ is an open connected subset of $\mathbb{R}^2$. For $\vec{x} = (x_1, x_2, x_3) \in \Omega_0$ we write $x = (\vec{x}_t, x_3)$ with $\vec{x}_t = (x_1, x_2)$. Given a function $\alpha \in C^2(\mathbb{R})$ with the first and second derivatives bounded on $\mathbb{R}$, we define the twisted domain by

$$\Omega_\alpha := \{r_\alpha(x_3)\vec{x} \,:\, x \in \mathbb{R} \times M\},$$

where $\vec{x}$ is taken as a column vector and

$$r_\alpha(x_3) = \begin{pmatrix} \cos\alpha(x_3) & \sin\alpha(x_3) & 0 \\ -\sin\alpha(x_3) & \cos\alpha(x_3) & 0 \\ 0 & 0 & 1 \end{pmatrix}.$$

We have encountered such a rotation along the tube already: Ω_α is nothing else than the bent tube defined at the beginning of Sect. 1.3 in the particular case $\gamma = \tau = 0$. As before we consider the Dirichlet Laplacian on $L^2(\Omega_\alpha)$, i.e. the self-adjoint operator on $L^2(\Omega_\alpha)$ associated with the closed quadratic form

$$\tilde{Q}_\alpha[\varphi] = \int_{\Omega_\alpha} |\nabla\varphi(\vec{x})|^2 \,\mathrm{d}\vec{x}\,, \quad \varphi \in \mathrm{Dom}(\tilde{Q}_\alpha) = H_0^1(\Omega_\alpha)\,.$$

We introduce the "straightening" transformation

$$(U\varphi)(\vec{x}) = \varphi\left(r_\alpha(x_3)\vec{x}\right)\,, \quad \vec{x} \in \Omega_0,\; f \in L^2(\Omega_\alpha)\,. \tag{1.47}$$

It is easy to see that U is a unitary operator from $L^2(\Omega_\alpha)$ onto $L^2(\Omega_0)$; note also that $U(H_0^1(\Omega_\alpha)) = H_0^1(\mathbb{R} \times M)$. Set

$$\nabla_t := (\partial_1, \partial_2)\,, \qquad \partial_\varphi := x_1\partial_2 - x_2\partial_1\,.$$

The derivative $\dot\alpha$ exists by assumption and we define the operator $H_{\dot\alpha}$ as the self-adjoint operator on $L^2(\Omega_0)$ associated with the closed quadratic form

$$Q_{\dot\alpha}[f] := \tilde{Q}_\alpha[U^{-1}f] = \int_{\Omega_0} (|\nabla_t f|^2 + |\dot\alpha(x_3)\partial_\varphi f + \partial_3 f|^2)(\vec{x})\,\mathrm{d}\vec{x} \tag{1.48}$$

with $f \in H_0^1(\mathbb{R}\times M)$. Evidently, $H_{\dot\alpha} = U(-\Delta_D^{\Omega_{\dot\alpha}})\,U^{-1}$. By a straightforward computation we find that $H_{\dot\alpha}$ acts on its domain as

$$H_{\dot\alpha} = -\Delta_D^M - (\dot\alpha(x_3)\partial_\varphi + \partial_3)^2.$$

Denote by ν_1 the lowest eigenvalue of the Dirichlet Laplacian $-\Delta_D^M$. The indicated Hardy-type inequality for twisted tubes then reads as follows.

Theorem 1.7 *Let $M \subset \mathbb{R}^2$ be a bounded open connected set containing the origin and with C^2-regular boundary. Suppose that M is not a disc centered at the origin. Assume, moreover, that $\dot\alpha$ is a compactly supported continuous function with bounded derivative and that $\dot\alpha$ is not identically zero. Then for all $f \in H_0^1(\mathbb{R}\times M)$ and any s_0 such that $\dot\alpha(s_0) \neq 0$ we have*

$$Q_{\dot\alpha}[f] - \nu_1\,\|f\|_2^2 \;\geq\; c_h \int_{\mathbb{R}\times M} \frac{|f(\vec{x}_t, x_3)|^2}{1+(x_3-s_0)^2}\,\mathrm{d}\vec{x}_t\,\mathrm{d}x_3\,, \tag{1.49}$$

where c_h is a positive constant independent of f but depending on s_0, $\dot\alpha$ and M.

Remarks 1.7.1 (a) If M is a disc, then $c_h = 0$ as expected, since in this case Ω_α coincides with Ω_0 as a set and $H_{\dot\alpha}$ is unitarily equivalent to H_0.
(b) For a compactly supported $\dot\alpha$ the decay rate of the weight on the right-hand side of (1.49) cannot be improved—see the notes for further details.

To prove *Theorem* 1.7 we need some preliminary results. First we define

$$\eta(M) = \inf_{v\in H_0^1(M)} \frac{\int_M (|\nabla_t v|^2 + |\partial_\varphi v|^2 - \nu_1 v^2)(\vec{x}_t)\,\mathrm{d}x_t}{\|v\|^2_{L^2(M)}}, \tag{1.50}$$

which depends on M only. Obviously $\eta(M) \geq 0$, and moreover, we have

Lemma 1.7.1 *Let M satisfy the assumptions of Theorem 1.7. Then $\eta(M) > 0$.*

Proof Since the Sobolev space $H_0^1(M)$ is compactly embedded into $L^2(M)$, the operator $-\Delta_D^M - \partial_\varphi^2$ associated with the quadratic form

$$\int_M (|\nabla_t v|^2 + |\partial_\varphi v|^2)\,\mathrm{d}x_t\,, \quad v \in H_0^1(M)\,,$$

has a purely discrete spectrum. Denote by λ_1 its principal eigenvalue and by v_1 the corresponding normalized eigenfunction, then $\eta(M) = \lambda_1 - \nu_1 \geq 0$ clearly holds. Assume that $\eta(M) = 0$; this would imply that

$$\|\partial_\varphi v_1\|_{L^2(M)} = 0\,,$$

and consequently, $\partial_\varphi v_1 = 0$ identically in M since $v_1 \in C^\infty(M)$; in other words, v_1 would be radial and would satisfy the equation

$$(\Delta_D^M + \lambda_1)v_1 = 0 \tag{1.51}$$

in M. Pick $\epsilon > 0$ such that the open disc $B_\epsilon := \{x_t \in \mathbb{R}^2 \,:\, |x_t| < \epsilon\}$ is contained in M. Since v_1 is radial, regular, and satisfies (1.51) in B_ϵ we find that

$$v_1(x_t) = \tilde{v}(x_t)\,, \quad x_t \in B_\epsilon \tag{1.52}$$

with $\tilde{v}(x_t) = cJ_0(\lambda_1^{1/2}|x_t|)$, $x_t \in \mathbb{R}^2$, where J_0 is the zero-order Bessel function and $c \neq 0$; if $c = 0$, the unique continuation principle would imply $v_1 = 0$ identically in M which contradicts the fact that v_1 is an eigenfunction. Note that $(\Delta_D^M + \lambda_1)\tilde{v} = 0$ holds in the whole $\mathbb{R}^2$. Comparing the last equation with (1.51) and bearing in mind the unique continuation principle, we find that (1.52) holds for all $x_t \in M$. Let now $\{\mathcal{C}_\omega\}$ be the set of the connected components of ∂M. Fix ω and introduce the function ϱ_ω by $\mathcal{C}_\omega \ni x_t \mapsto \varrho_\omega(x_t) := |x_t| \in (0, \infty)$. Set $\mathcal{I}_\omega := \varrho_\omega(\mathcal{C}_\omega)$. Since $\mathcal{C}_\omega$ is connected and ϱ_ω is continuous, $\mathcal{I}_\omega$ should be connected too, i.e. $\mathcal{I}_\omega$ is either a one-point set or a bounded interval of positive length. Due to the Dirichlet boundary conditions, we have $v_1(x_t) = 0$ for all $x_t \in \mathcal{C}_\omega$, i.e. $J_0(\lambda_1^{1/2} r) = 0$ for all $r \in \mathcal{I}_\omega$. However, J_0 has at most a finite number of zeros on any bounded interval, so all $\mathcal{I}_\omega$ are one-point sets, in other words, all $\mathcal{C}_\omega$'s are arcs of circles centered at the origin. Since $\partial M \in C^2$, all $\mathcal{C}_\omega$'s are circles. Since M is connected and contains the origin, it is a disc centered at the origin which contradicts, however, the assumptions of *Theorem* 1.7. ∎

To estimate the form (1.48) it is useful to have the transverse and longitudinal variables separated. Let us consider the terms entering the expression separately. By assumption on $\dot{\alpha}$ there is an interval $I = (a, b) \subset \mathbb{R}$ such that $\operatorname{supp}\dot{\alpha} \subset \bar{I} = [a, b]$. Pick $f \in H_0^1(\mathbb{R} \times M)$ and define

$$\begin{aligned} T_1[f] &:= \|\nabla_t f\|^2 - \nu_1\,\|f\|^2\,, & T_3[f] &:= \|\dot{\alpha}\,\partial_\varphi f\|^2\,,\\ T_2[f] &:= \|\partial_3 f\|^2\,, & T_{2,3}[f] &:= -2\,\mathrm{Re}\,(\partial_3 f,\, \dot{\alpha}\,\partial_\varphi f)\,, \end{aligned}$$

where the norms and the inner product refer to $L^2(\mathbb{R} \times M)$. The following lemma allows us to control the mixed term $T_{2,3}[f]$.

Lemma 1.7.2 *Let the assumptions of Theorem 1.7 be satisfied. Then for each $\mu > 0$ and $\nu > 0$ there exists a constant $\gamma(\mu, \nu)$ such that*

$$|T_{2,3}[f]| \le \gamma(\mu, \nu)\, T_1[f] + \mu\, T_2[f] + \nu\, T_3[f]$$

holds for any $f \in H_0^1(\mathbb{R} \times M)$.

Proof Here and in the sequel we denote by c numerical constants, which may depend on $\dot\alpha$ and M, but not on f and whose values may change from line to line. It suffices to check the claim for $f \in C_0^\infty(\mathbb{R} \times M)$ which can always be written as $f(x) = \psi_1(x_t)\, u(x)$, where ψ_1 is the unique normalized positive eigenfunction of the Dirichlet Laplacian $-\Delta_D^M$ corresponding to ν_1 and $u \in C_0^\infty(\mathbb{R} \times M)$. By a direct calculation we get values of the terms $T_j[f]$ for this f, namely

$$\begin{aligned} T_1[f] &= \|\psi_1 \nabla_t u\|^2 , \qquad T_3[f] = \|\dot\alpha\, (\psi_1 \partial_\varphi u + u \partial_\varphi \psi_1\|^2 , \\ T_2[f] &= \|\psi_1 \partial_3 u\|^2 , \qquad T_{2,3}[f] = -2\,\mathrm{Re}\,(\psi_1 \partial_3 u, \chi_I\, \dot\alpha\, (\psi_1 \partial_\varphi u + u \partial_\varphi \psi_1) , \end{aligned}$$

where, in order to establish the identity for $T_1[f]$, we have integrated by parts as follows

$$2 \int_{\Omega_0} \psi_1\, u\, \nabla_t u \cdot \nabla_t \psi_1 \,\mathrm{d}\vec{x} = - \int_{\Omega_0} u^2\, |\nabla_t \psi_1|^2 \,\mathrm{d}\vec{x} + \nu_1 \int_{\Omega_0} \psi_1^2\, u^2 \,\mathrm{d}\vec{x} .$$

Using

$$|\dot\alpha\, \partial_\varphi u|^2 \le c\, |\nabla_t u|^2$$

and applying the Cauchy-Schwarz inequality, the first term in the sum of $T_{2,3}[f]$ can be estimated as follows,

$$\left| 2\left(\psi_1 \partial_3 u,\, \dot\alpha\, \psi_1\, \partial_\varphi u\right)\right| \le 2\sqrt{c}\, \sqrt{T_1[f]}\, \sqrt{T_2[f]} \le \frac{2\,c}{\mu}\, T_1[f] + \frac{\mu}{2}\, T_2[f] . \quad (1.53)$$

To estimate the second term, we first combine integrations by parts to get

$$2\left(\psi_1 \partial_3 u,\, \dot\alpha\, u\, \partial_\varphi \psi_1\right) = -\left(u\, \psi_1,\, \ddot\alpha\, u\, \partial_\varphi \psi_1\right) = \left(u,\, \ddot\alpha\, \psi_1^2\, \partial_\varphi u\right) .$$

From $|\ddot\alpha\, \partial_\varphi u|^2 \le c\, |\nabla_t u|^2$ and from the Cauchy-Schwarz inequality we get

$$\left|\left(u,\, \ddot\alpha\, \psi_1^2\, \partial_\varphi u\right)\right|^2 \le c\, T_1[f]\, \|\chi_I\, \psi_1 u\|^2 ,$$

where χ_I is the shorthand for the characteristic function of the set $I \times M$. Obviously, we can find an open interval $J \subset \operatorname{supp} \dot\alpha \subset \bar{I}$ such that there exists a certain positive number δ, for which $\dot\alpha(x_3) \ge \delta$ holds for all $x_3 \in J$. On the other hand, there exists a constant c, which depends on I and J, such that for any $\phi \in H^1(I)$, the following inequality holds:

$$\|\phi\|^2_{L^2(I)} \le c\left(\|\phi\|^2_{L^2(J)} + \|\phi'\|^2_{L^2(I)}\right). \tag{1.54}$$

The proof of this inequality is left to the reader as an exercise (Problem 29). In this way we get

$$\|\chi_I\,\psi_1 u\|^2 \le c\left(T_2[f] + \|\chi_J\,\psi_1 u\|^2\right) \le c\left(T_2[f] + \delta^{-2}\|\chi_J\,\dot\alpha\psi_1 u\|^2\right), \tag{1.55}$$

where χ_J denotes the characteristic function of the set $J \times M$. Moreover, for each fixed value $x_3 \in \mathbb{R}$ we have $\dot\alpha(x_3)\,\psi_1\,u(\cdot, x_3) \in H^1_0(M)$, and therefore we can apply *Lemma* 1.7.1 to obtain

$$\|\chi_J\,\dot\alpha\psi_1 u\|^2 \le \|\dot\alpha\psi_1 u\|^2 \le \eta(M)^{-1}\left(T_3[f] + \|\dot\alpha\|^2_\infty\,T_1[f]\right). \tag{1.56}$$

We then easily conclude that

$$\begin{aligned}\left|\left(u, \chi_I\,\ddot\alpha\psi_1^2\,\partial_\varphi u\right)\right|^2 &\le c\,T_1[f]\left(\|\dot\alpha\|^2_\infty\,T_1[f] + \eta(M)\,\delta^2\,T_2[f] + T_3[f]\right)\\ &\le \left(c(\mu,\nu)\,T_1[f] + \frac{\mu}{2}\,T_2[f] + \nu\,T_3[f]\right)^2\end{aligned} \tag{1.57}$$

for any $\nu > 0$ and $c(\mu, \nu)$ large enough depending also on $\eta(M)$ and δ. The desired estimate then follows by combining (1.53) with (1.57). ∎

We note that a stronger version of *Lemma* 1.7.2, and consequently of *Proposition* 1.7.1 below can be proved—see the notes. As the next step we shall use the above results to demonstrate a version of the Hardy-type inequality with the integral weight proportional to $\dot\alpha^2$.

Proposition 1.7.1 *Let the assumptions of Theorem 1.7 hold and let further $I \subset \mathbb{R}$ be the open interval considered above. Then there exists a positive constant c_I such that for all $f \in H^1_0(\mathbb{R}\times M)$ we have*

$$\int_{\mathbb{R}\times M}\left(|\nabla_t f|^2 + |\partial_3 f - \dot\alpha\,\partial_\varphi f|^2 - \nu_1 f^2\right)(\vec x)\,\mathrm{d}\vec x \;\ge\; c_I\,\eta(M)\int_{\mathbb{R}\times M}|\dot\alpha f|^2(\vec x)\,\mathrm{d}\vec x\,.$$

Proof If we choose $\mu = 1$ and $\nu < 1$, the previous lemma yields the estimate

$$\begin{aligned}\int_{\Omega_0}\left(|\nabla_t f|^2 + |\partial_3 f - \dot\alpha\,\partial_\varphi f|^2 - \nu_1 f^2\right)(\vec x)\,\mathrm{d}\vec x &= T_1[f] + T_2[f] + T_3[f] + T_{2,3}[f]\\ &\ge \frac{1}{2}\,T_1[f] + \left(1 - \frac{1}{2\gamma(1,\nu)}\right)\left(T_2[f] + T_3[f] - |T_{2,3}[f]|\right) + \frac{1-\nu}{2\gamma(1,\nu)}\,T_3[f];\end{aligned}$$

without loss of generality we may suppose that $2\gamma(1, \nu) > 1$. Since we have at the same time $T_2[f] + T_3[f] - |T_{2,3}[f]| \ge 0$, it follows from *Lemma* 1.7.1 that the expression in question is estimated from below by $c_I\left(\|\dot\alpha\|^2_\infty\,T_1[f] + T_3[f]\right)$ with

$$c_I := \frac{1}{2} \min\left\{ \frac{1}{\|\dot{\alpha}\|_\infty^2}, \frac{1-\nu}{\gamma(1,\nu)} \right\};$$

from here the claim follows easily. ∎

Proof of Theorem 1.7 We may suppose that $s_0 = 0$. We choose an interval $J = (-a, a) \subset \operatorname{supp} \dot{\alpha} \subset \bar{I}$ with some $a > 0$, such that $|\dot{\alpha}| > 0$ holds on $\bar{J}$, and define $\tilde{u} : \mathbb{R} \to \mathbb{R}$ by

$$\tilde{u}(s) := 1 \quad \text{if} \quad |s| \geq a, \qquad \tilde{u}(s) := |s|/a \quad \text{if} \quad |s| < a.$$

Let us write $f = \tilde{u} f + (1 - \tilde{u}) f$. Applying the one-dimensional Hardy inequality

$$\int_{\mathbb{R}} \frac{|v(x)|^2}{x^2}\, dx \leq 4 \int_{\mathbb{R}} |v'(x)|^2\, dx, \quad v \in H^1(\mathbb{R}), \quad v(0) = 0,$$

to the function $x_3 \mapsto \tilde{u}(x_3) f(\vec{x})$ with x_1, x_2 fixed, we arrive at

$$\begin{aligned} \int_{\mathbb{R}\times M} \frac{f(\vec{x})^2}{1+x_3^2}\, d\vec{x} &\leq 2 \int_{\mathbb{R}\times M} \frac{|\tilde{u}(x_3) f(x)|^2}{x_3^2}\, dx + 2 \int_{J\times M} |(1-\tilde{u}) f|^2\, dx \\ &\leq 16\, \|(\partial_3 \tilde{u}) f\|^2 + 16\, \|\tilde{u}\, \partial_3 f\|^2 + 2\, \|\chi_J (1-\tilde{u}) f\|^2 \\ &\leq \left(\frac{16}{a^2} + 2\right) \|\chi_J f\|^2 + 16\, \|\partial_3 f\|^2, \end{aligned} \tag{1.58}$$

where χ_J is again the shorthand for the characteristic function of $J \times M$. *Proposition* 1.7.1 implies the existence of a constant $c > 0$ depending on $\dot{\alpha}$ such that

$$\|\chi_J f\|^2 \leq \left(c\, \eta(M) \min_J |\dot{\alpha}|\right)^{-1} \left(Q_{\dot{\alpha}}[f] - \nu_1 \|f\|^2\right).$$

To assess the second term on the right-hand side of (1.58) we rewrite the inequality of *Lemma* 1.7.2 for $\nu = 1$ as $\gamma_\mu^{-1}\, |T_{2,3}[f]| \leq T_1[f] + \mu\, \gamma_\mu^{-1}\, T_2[f] + \gamma_\mu^{-1}\, T_3[f]$, where $\gamma_\mu := \max\{1, \gamma(\mu, 1)\}$ and $\mu \in (0, 1)$. Substituting this inequality into

$$Q_{\dot{\alpha}}[f] - \nu_1 \|f\|^2 = T_1[f] + T_2[f] + T_3[f] + T_{2,3}[f],$$

writing $T_{2,3}[f] = \gamma_\mu^{-1}\, T_{2,3}[f] + (1 - \gamma_\mu^{-1})\, T_{2,3}[f]$ and using the positivity of the form $T_2[f] + T_3[f] + T_{2,3}[f]$, we obtain

$$T_2[f] = \|\partial_3 f\|^2 \leq \gamma_\mu (1-\mu)^{-1} \left(Q_{\dot{\alpha}}[f] - \nu_1 \|f\|^2\right).$$

Summing up all the contributions and using (1.58), we arrive at the inequality (1.49) with $s_0 = 0$. ∎

1.7.2 Stability of the Spectrum

The Hardy-type inequality (1.49) demonstrated above has several important consequences. One immediate corollary is the stability of the discrete spectrum against sufficiently weak potential perturbations. Specifically, we have

Proposition 1.7.2 *Under the assumptions of Theorem 1.7, let $V : \Omega_\alpha \to \mathbb{R}$ be a bounded function which satisfies $V(\vec{x}) = \mathcal{O}(|x_3|^{-2})$ as $|x_3| \to \infty$; then the operator*

$$-\Delta_D^{\Omega_\alpha} + \lambda V, \quad \lambda \in \mathbb{R},$$

has empty discrete spectrum provided λ is small enough.

The proof of this statement is elementary and we leave it to the reader. It is worth noticing that for compactly supported potentials the critical value of λ in *Proposition* 1.7.2 decays with the square of the distance between the supports of V and $\dot{\alpha}$, see also Problem 30.

Another consequence of *Theorem* 1.7 is the stability of the discrete spectrum in twisted bent tubes introduced in Sect. 1.3. Recall also that the existence of a discrete spectrum in bent tubes was demonstrated under the condition that the cross-section M was rotated in an appropriate way, related to the torsion τ of the reference curve Γ by condition (1.18). It appears that this requirement is necessary at the same time as long as the tube is only mildly bent.

Theorem 1.8 *Adopt the notation of Sect. 1.3 and suppose that $\dot{\alpha}$ and M satisfy the assumptions of Theorem 1.7. If $\dot{\alpha} \neq \tau$, there exists a positive ε, depending on τ, $\dot{\alpha}$ and M, such that*

$$\|\gamma\|_\infty + \|\dot{\gamma}\|_\infty \leq \varepsilon \;\text{ implies }\; \sigma(-\Delta_D^{\Omega_0}) = [\nu_1, \infty)\,. \tag{1.59}$$

We refer to the notes for the proof and also for further comments and references. We have thus found that if the cross-section M is not a disc, and at the same time it is not rotated in a particular way described by Tang's condition (1.18), then a sufficiently mild bending of the tube does not produce any discrete spectrum. This contrasts, of course, with bent strips in dimension two, where any (sufficiently regular) nontrivial bending leads to the appearance of bound states in accordance with *Theorem* 1.1.

1.7.3 Periodically Twisted Tubes and Their Perturbations

One of the essential features in the above analysis of twisted tubes was the assumption that the twisting was local, in other words the function $\dot{\alpha}$ had a compact support. Now we turn to situations in which the twisting has a global character. We begin with a discussion of a periodically twisted tube which corresponds to $\dot{\alpha}$ being a nonzero constant, $\dot{\alpha}(x_3) = \beta_0$.

Since such a tube exhibits a translational invariance, we can employ a trick described in more detail in Chap. 7, namely a partial Fourier transform in the x_3 direction, by which the operator H_{β_0} is unitarily equivalent to a direct integral, symbolically $H_{\beta_0} \simeq \int_{\mathbb{R}}^{\oplus} h(p)\,\mathrm{d}p$ with the fiber operator

$$h(p) = -\Delta_D^M + \big(p - i\,\beta_0(x_1\partial_2 - x_2\partial_1)\big)^2$$

in $L^2(M)$ subject to Dirichlet boundary conditions at ∂M. In view of the compactness of M the spectrum of $h(p)$ is purely discrete; we denote by

$$E := \inf \sigma(h(0))$$

the first eigenvalue of the operator $h(0)$. Of course, $E \in \sigma(H_{\beta_0})$ holds in view of the direct-integral decomposition; it appears that it also determines the spectral threshold of the original operator H_{β_0}.

Proposition 1.7.3 *In the described situation we have $\sigma(H_{\beta_0}) = [E, \infty)$.*

Proof Let $\psi \in L^2(M)$ be a real-valued eigenfunction of $h(0)$ corresponding to the eigenvalue E. Since it is the ground state, we can choose it to be positive; this allows us to write any given $u \in C_0^\infty(\Omega_0)$ as the product

$$u(\vec{x}_t, x_3) = \psi(\vec{x}_t)v(\vec{x}_t, x_3)$$

with the appropriate v. Integration by parts now gives

$$\begin{aligned} Q_{\beta_0}[u] - E\|u\|^2 = \int_{\Omega_0} \Big(&\psi^2\,|\nabla_t v|^2 - (\Delta_D^M\psi)\,\psi\,|v|^2 + \psi^2\,|\partial_3 v|^2 \\ &+\beta_0\,\psi\,\partial_\varphi\psi\,(v\,\partial_3\bar{v} + \bar{v}\,\partial_3 v) + \beta_0\,\psi^2(\partial_3\bar{v}\,\partial_\varphi v + \partial_\varphi\bar{v}\,\partial_3 v) \\ &+\beta_0^2\,\psi^2\,|\partial_\varphi v|^2 - \beta_0^2\,(\partial_\varphi^2\psi)\,\psi\,|v|^2 - E\psi^2\,|v|^2\Big)\,\mathrm{d}x_3\,\mathrm{d}\vec{x}_t\,. \end{aligned}$$

Since $u \in C_0^\infty(\Omega_0)$ by assumption, we have $\int_{\mathbb{R}}(v\,\partial_3\bar{v} + \bar{v}\,\partial_3 v)\mathrm{d}x_3 = 0$. Combining this with $-\Delta_D^M\psi - \beta_0^2\partial_\varphi^2\psi = h(0)\psi = E\psi$, we obtain

$$Q_{\beta_0}[u] - E\,\|u\|^2 = \int_{\Omega_0} \psi^2\left(|\nabla_t v|^2 + |\partial_3 v + \beta_0\,\partial_\varphi v|^2\right)\mathrm{d}x_3\,\mathrm{d}\vec{x}_t \geq 0\,,$$

which proves $\inf\sigma(H_{\beta_0}) = E$. It remains to check that the spectrum is purely essential covering the halfline. The easiest way to do this is to construct a suitable Weyl sequence for every point in $[E, \infty)$; we leave this to the reader. ∎

We thus see that a periodic twisting changes the essential spectrum in a way depending on β_0. It inspires the question of what a local variation of the twist would cause, in particular, what happens if the twist is locally slowed down. We thus consider rotation angles α of the form

$$\dot{\alpha}(x_3) = \beta_0 - \beta(x_3)\,, \tag{1.60}$$

where $\beta : \mathbb{R} \to \mathbb{R}$ is a bounded function supported in an interval $[-a, a]$ for some $a > 0$. From the compactness of its support in combination with *Proposition* 1.7.3 it follows by standard perturbation arguments that

$$\inf(\sigma_{\text{ess}}(H_{\dot{\alpha}})) = \inf(\sigma_{\text{ess}}(H_{\beta_0})) = E\,.$$

Let us now look for conditions under which a variation of the twist can give rise to a nonempty discrete spectrum.

Theorem 1.9 *Suppose that M satisfies the assumptions of Theorem 1.7 and $\dot{\alpha}$ is given by (1.60). If*

$$\int_{\mathbb{R}} (\dot{\alpha}^2(x_3) - \beta_0^2)\, \mathrm{d}x_3 < 0\,, \tag{1.61}$$

we have $\inf(\sigma(H_{\dot{\alpha}})) < E$, *and consequently,* $\sigma_{\text{disc}}(H_{\dot{\alpha}}) \neq \emptyset$.

Proof We construct a family of trial functions depending on a parameter $\delta > 0$. Let $u_\delta(\vec{x}_t, x_3) = \psi(\vec{x}_t)v(x_3)$, where ψ is the ground state of the operator $h(0)$ and

$$v(x_3) = \begin{cases} \mathrm{e}^{\delta(x_3+a)} & \text{if} \quad x_3 \leq -a\,, \\ 1 & \text{if} \quad -a \leq x_3 \leq a\,, \\ \mathrm{e}^{-\delta(x_3-a)} & \text{if} \quad x_3 \geq a\,. \end{cases} \tag{1.62}$$

It is easy to see that $u_\delta \in \text{Dom}\,(Q_{\dot{\alpha}})$. A straightforward calculation then gives

$$Q_{\dot{\alpha}}[u_\delta] - E\|u_\delta\|^2 = \delta\, \|\psi\|^2_{L^2(M)} - \|\partial_\varphi \psi\|^2_{L^2(M)} \int_{-a}^{a} \left(\dot{\alpha}^2(x_3) - \beta_0^2\right) \mathrm{d}x_3$$

and $\|u_\delta\|^2 = (\delta^{-1} + 2a)\|\psi\|^2_{L^2(M)}$, and in the limit $\delta \to 0$ we get

$$\frac{Q_{\dot{\alpha}}[u_\delta] - E\,\|u_\delta\|^2}{\|u_\delta\|^2} = \delta\, \frac{\|\partial_\varphi \psi\|^2_{L^2(M)}}{\|\psi\|^2_{L^2(M)}} \int_{-a}^{a} \left(\dot{\alpha}^2(x_3) - \beta_0^2\right) \mathrm{d}x_3 + \mathcal{O}(\delta^2)\,.$$

From the proof of *Lemma* 1.7.1 and the assumptions on the cross section M we conclude that $\|\partial_\varphi \psi\|^2_{L^2(M)} > 0$; it is thus enough to choose δ small enough to ensure that

$$\frac{Q_{\dot{\alpha}}[u_\delta] - E\,\|u_\delta\|^2}{\|u_\delta\|^2} < 0$$

and the claim of the theorem follows. ∎

Under additional assumptions one can also extend the above result to the critical case when the integral (1.61) vanishes.

Theorem 1.10 *In addition to the hypotheses adopted in Theorem 1.9, assume that* $\dot\alpha(x_3) + \beta_0 > 0$ *holds whenever* $|x_3| \le a$*, and that* $\ddot\alpha \in L^2(-a,a)$*. If*

$$\int_{\mathbb{R}} \left(\dot\alpha^2(x_3) - \beta_0^2\right) \mathrm{d}x_3 = 0\,, \tag{1.63}$$

we have $\inf(\sigma(H_{\dot\alpha})) < E$*, and consequently,* $\sigma_{\mathrm{disc}}(H_{\dot\alpha}) \neq \emptyset$*.*

Proof We use a GJ-type argument modifying the trial functions from the previous proof by a slight deformation in the central region, $u_{\delta,\varepsilon}(\vec x_t) := \psi(\vec x_t) v_\varepsilon(x_3)$, where

$$v_\varepsilon(x_3) = \begin{cases} \mathrm{e}^{\delta(a+x_3)} & \text{if } \; x_3 \le -a\,, \\ 1 + \varepsilon\,(\beta_0 - \dot\alpha(x_3)) & \text{if } \; -a \le x_3 \le a\,, \\ \mathrm{e}^{-\delta(x_3-a)} & \text{if } \; x_3 \ge a\,, \end{cases} \tag{1.64}$$

with $\varepsilon > 0$. By a direct calculation we find the shifted energy form,

$$Q_{\dot\alpha}[u_{\delta,\varepsilon}] - E\|u_{\delta,\varepsilon}\|^2 = \int_{\Omega_0} \left[v_\varepsilon^2 (\partial_\varphi \psi)^2 \left(\dot\alpha^2 - \beta_0^2\right) + \psi^2 \dot v_\varepsilon^2 \right] \mathrm{d}x_3\, \mathrm{d}\vec x_t\,.$$

Using the assumptions of the theorem one can check that in the limit $\varepsilon, \delta \to 0$ the integrals appearing in the last expression behave as

$$\int_{-a}^{a} v_\varepsilon^2(x_3) \left(\dot\alpha^2(x_3) - \beta_0^2\right) \mathrm{d}x_3 = -2\varepsilon \int_{-a}^{a} (\dot\alpha(x_3) - \beta_0)^2\, (\dot\alpha(x_3) + \beta_0)\, \mathrm{d}x_3 + \mathcal{O}(\varepsilon^2)$$

and

$$\int_{\mathbb{R}} \dot v_\varepsilon(x_3)^2\, \mathrm{d}x_3 = \delta + \varepsilon^2 \int_{-a}^{a} \ddot\alpha(x_3)^2\, \mathrm{d}x_3 = \mathcal{O}(\delta) + \mathcal{O}(\varepsilon^2)\,.$$

The last two relations imply

$$\frac{Q_{\dot\alpha}[u_{\delta,\varepsilon}] - E\|u_{\delta,\varepsilon}\|^2}{\|u_{\delta,\varepsilon}\|^2} = -2\varepsilon\delta\, \frac{\|\partial_\varphi\psi\|^2_{L^2(M)}}{\|\psi\|^2_{L^2(M)}} \int_{-a}^{a} (\dot\alpha(x_3) - \beta_0)^2\, (\dot\alpha(x_3) + \beta_0)\, \mathrm{d}x_3$$
$$+\delta\, \mathcal{O}(\varepsilon^2) + \mathcal{O}(\delta^2)\,;$$

setting then $\varepsilon = \sqrt{\delta}$ and using again the fact that $\|\partial_\varphi\psi\|^2_{L^2(M)} > 0$, we conclude that for δ small enough we have

$$\frac{Q_{\dot\alpha}[u_{\delta,\varepsilon}] - E\, \|u_{\delta,\varepsilon}\|^2}{\|u_{\delta,\varepsilon}\|^2} < 0\,,$$

which concludes the proof. ∎

1.8 Notes

Section 1.1 The definition of the Dirichlet Laplacian is standard, for Sobolev spaces see [Ad]. When the arc length serves as parameter, the term *unit-speed curve* is employed in differential geometry. The curvilinear coordinates we use here are sometimes called *Fermi coordinates*—cf. [Fe22, Gra]—although they were known already to Gauss. In the present context they appeared for the first time in [KJ71], where the existence of bound states in the formal limit $a \to 0$ was also noticed. Recall that some authors define the curvature γ with the opposite sign (in analogy with the Weingarten tensor for surfaces—see Sect. 4.1.1 below); the sign convention for planar curves is much less important than in higher dimensions. The effective potential (1.8) was computed first by J. Tolar in 1977, the result being published with a decade delay in [To88].

The assumption *(ii)*$_k$ does not cover various interesting cases such as strips with γ being a (nontrivial) step function. A notorious example is the *bookcover* waveguide consisting of two semi-infinite strips connected by an annular segment [SM90]. If we nevertheless require a global smoothness, it is because this leads to a simplification of the analysis. However, in appropriate places—such as Remark 1.1.4 here—we indicate what can be done if the boundary satisfies weaker regularity hypotheses. Sometimes a modified basis of transverse eigenfunctions is used. If one of the boundaries of Ω is used as the reference curve, one conventionally chooses $\chi_n(u) = (2/d)^{1/2} \sin \kappa_n u$ which differs by sign from (1.10) for the even index values.

The existence result contained in *Propositions* 1.1.1 and 1.1.2 was first demonstrated in [EŠ89a]. Three years later J. Goldstone and R.L. Jaffe [GJ92] came with a deep insight which is the essence of the argument used to prove *Theorem* 1.1, namely that the resonance at the bottom of the continuous spectrum is not a stationary point of the energy form unless the strip is straight. Consequently, one can get a negative contribution to the energy, dominating over the energy surplus from slowly decaying tails, by deforming the trial function properly in the "central" curved part. This idea, which we shall refer to as the *GJ-argument*, will be useful when dealing with various waveguide systems below. The authors of [GJ92] did not strive to find a class of strips for which the argument works—they just remarked correctly that the curvature need not be continuous—and did not construct an appropriate deformation of the trial function. This gap was filled in a little later in the thesis of W. Renger—see its summary in [RB95] where a generalization of *Proposition* 1.1.1 was also given—and independently in [DE95]; the proof presented here follows these papers. Curved quantum waveguides with a discontinuous curvature were studied recently in [KŠ12].

Proposition 1.1.3 and its analogy for the double waveguide discussed in Sect. 1.5 can be interpreted as a manifestation that the binding effect is of a purely quantum nature. The relation between classical and quantum motion in tubular neighborhoods of manifolds—especially in the limit of zero transverse width motivated by the need to understand motion with constraints—is more involved, though. The heuristic bobsleigh example mentioned in the introduction suggests that one might try to replace the "hard" confinement by a soft one through a potential of, say, oscillator shape perpendicular to the manifold. In this case, indeed, the shrinking limit can yield the classical motion governed by the effective curvature-induced potential provided the energy of the classical particle is assumed to change in the same way as that of the quantum transverse confinement. We refer to [FH01] where the general situation of an n-dimensional (compact) manifold embedded in $\mathbb{R}^{n+m}$ is treated. Somewhat similar limits allowing again for a semiclassical interpretation will be discussed in Chap. 10.

Section 1.2 The idea of the estimate which we have used in *Proposition* 1.2.1 is taken from [ABGM91]; the asymptotic behavior of the eigenvalues with respect to the angle was worked out in [DR12]. As for Eq. (1.13) we refer to [Kon67] for more information on Laplacian domains in non-convex regions with non-smooth boundaries. The existence of a bound state in L-shaped strips was probably noticed first in [He65], where the inequalities $0.870 < \epsilon_1(\pi/2) < 0.951$ were derived. It was rediscovered in [LLM86]; investigating meson scattering these authors mentioned a caricature model which involved hard confinement at a fixed distance. The mode-matching argument used in *Proposition* 1.2.3 comes from [EŠŠ89], the same eigenvalue $\epsilon = 0.929\ldots$ was obtained by a direct truncation scheme for crossed strips in [SRW89], later in [MKSW91], and experimentally checked in [CLM92] with a 1 % accuracy in a different physical setting which needs a separate comment.

While the main objects of this book are guided quantum particles, some properties discussed here can be tested in appropriate microwave devices. It was noticed repeatedly, e.g., in [EŠ90], [SM90] and [GJ92], that in a rectangular waveguide built over a planar strip the z-component of the electric field for TE_{0m} modes satisfies the same Helmholtz equation with Dirichlet boundary conditions as the one we consider here. Moreover, making the resonator sufficiently flat one can achieve that different components of the spectrum can be clearly distinguished and the TM and TE_{nm} modes, $n \neq 0$, are effectively suppressed. The mentioned first measurement which confirmed the L-shaped strip eigenvalue was followed by [CLM93] where multiple bound states in sharply broken strips were demonstrated, and by [CLMT97] where Z-shaped ducts such as those considered in Problem 7 were investigated. More about these experiments and geometric effects in electromagnetic waveguides can be found in the book [LCM], see especially Chaps. 4 and 5, and also in [BDM13].

Speaking of observations of geometrically induced bound states one has to keep in mind that in any actual measurement the channel is of a finite length, coupled to a source and a drain, and the bound states investigated here give rise to resonances the width of which decreases with the length of the leads. They must not be confused with the resonances in the continuous spectrum of $-\Delta_D^\Omega$ which will be discussed in the next chapter, although from the practical point of view the latter are also manifested in finite-length channels. A discussion of this effect for a Z-shaped quantum wire including a comparison with experiment is given in [CLM97], see also [LCM, Sect. 5.4]. At a model level the emergence of resonances coming from perturbations of infinite-channel bound states was analyzed in [Ex90a].

The mode-matching method is particularly convenient if the channel in question consists only of rectangular parts as for the L-shaped strip discussed here. On the other hand, if this is not the case one has to look for other "solvable" components. For instance, a broken strip with $\beta \neq \pi/2$ can be treated by matching the lead Ansatz $\psi_{\rm I}$ of (1.14) with a general wedge solution in the corner area [WSM92, CLM93, DR12]. For channels with a larger number of components it is useful to establish a procedure for matching the solutions along the channel. One possibility is the *transfer-matrix method* described in Chap. 2 of the book [LCM] together with other methods of numerical solutions of such Helmholtz equations and the corresponding bibliography. On the other hand, for discussion of the discrete spectrum of broken and branched waveguides by variational methods refer, e.g., to [Na11b] or [ART12].

Section 1.3 The rotating coordinate frame which allows us to avoid mixed derivatives between the longitudinal and transverse coordinates of the "straightened" operator is known in the classical waveguide theory as the *Tang system* [TG89]. It is not unique, locally there is a family of such systems with the corresponding functions α differing by a constant. This property makes it possible to construct the Tang system even if Γ has no global Frenet frame; the construction suggested in Problem 12c can be used even in cases when points of non-uniqueness may accumulate at some $s \in \mathbb{R}$, cf. [EK04]. In the present context the rotating frame was first noticed in [KJ71], see also [dC81]. The rotation is also the reason why the polar coordinates in the cross-section plane are suitable. It is not difficult, however, to express $-\Delta_D^\Omega$ in the Cartesian coordinates $x = r\cos\theta,\ y = r\sin\theta$. For $\dot\alpha = 0$ this gives the result of [Pi82]; disputing this choice R. da Costa [dC83] pointed out a relation between the rotating frame and the orientation of the force due to a smooth confining potential around the curve, see also an illustrative example suggested by M. Berry described in [DJ93].

For the inequality (1.22) giving a lower bound to the transverse ground state see [Fa23, Kra25]. The existence of bound states in thin bent tubes (Problem 15) was demonstrated in [Ex90b]; the general variational proof is again due to [GJ92], see also [DE95]. Adding a transverse potential is easy as long as it is relatively bounded w.r.t. the operator $-\Delta_D^\Omega$ (Problem 17). A more complicated situation arises if the interaction is strongly singular and one has to choose a self-adjoint extension of $-\Delta_D^\Omega + V$. An example of a bent tube threaded by a magnetic flux line, where the transverse part of operator is the Aharonov-Bohm Hamiltonian $(-i\nabla - A)^2$ in $L^2(M)$ with the vector potential A giving a magnetic field which vanishes everywhere outside the tube axis, cf. [Ru83], was discussed in [DJ93].

Section 1.4 The useful Green's function inequality of Problem 18 which we employed in *Proposition* 1.4.1 can be derived in alternative ways. For instance, an analogous inequality between the heat kernels—see [Da, Sect. 2.1] or [DvC, Appendix D]—gives the sought result through the Laplace transform. A classical way to demonstrate this property is through Hadamard's formula—see [Ha08] or [Ga, Chap. 15].

The ground state of $-\Delta_D^M$ is strictly decreasing with respect the domain expansion by the result of [GZ94] provided the added volume $M_x \setminus M$ has a nonzero capacity, which is the case for *Theorem* 1.4 in view of the inequality $m(O) \le \mathrm{cap}(O)$, where m is Lebesgue measure, valid for any open O—cf. [Fu].

Section 1.5 The spectrum of the double waveguide with a coupling through the boundary window was found in [EŠTV96], the existence of bound states for $d_1 = d_2$, which is equivalent to the Dirichlet-to-Neumann boundary condition switch in a single duct, was independently noticed in [BGRS97]. This system is an idealization of quantum-wire couplers—cf. [AEu90, HTW93, Ku93]. In the electromagnetic waveguide context, the bracketing argument was used earlier in [Pop86] to establish the existence of trapped modes in a wide enough window. The example of waveguides coupled through a leaky interface is taken from [EKr99]. For the approximation of a δ-interaction by a family of scaled potentials see [AGHH, Sect. I.3]; in a similar vein we shall in Chap. 10 study approximations of δ-interactions supported by curves of a more general family by scaled potential "ditches". A related problem of an attractive singular interaction supported by two parallel lines in the plane, again related to the subject of Chap. 10, is analyzed thoroughly in [KonK13].

As we have noticed, the bound states in waveguides with such a window-type coupling, both for the Dirichlet wall or for a local increase of barrier transparency, are of a purely quantum nature. Such a non-classical binding can nevertheless be strong as one can see from the eigenfunctions which we suggest the reader to compute in Problems 21 and 24, in particular, their nodal lines. It is true that apart from the central one the latter cannot be straight (Problem 25), but the numerical results of [EŠTV96] and [EKr99] show that they differ little from line segments, so the wave packets feel the barrier "spikes" almost as hard walls.

The bound state in a cross-shaped region found in [SRW89] was one of the first examples of a geometrically induced discrete spectrum in a non-compact Ω. A scissor-shaped Ω with a general angle θ has been discussed in [BEPS02]. In addition to properties stated in *Proposition* 1.5.2, a numerical analysis suggests that the eigenfunction ϕ_m has parity $(-1)^{m-1}$ with respect to the y-axis and that $(\epsilon_{2m} - \epsilon_{2m-1})(\theta) \to 0$ holds with an exponentially fast convergence for $\theta \to 0$ as can be expected.

Section 1.6 Thin tube behavior of curvature induced bound states was first analyzed in [Ex90b]. *Theorem* 1.6 and its extension to tubes in $\mathbb{R}^3$ given in Problem 27 is due to Grushin, see [Gr09] for details of the proof, and also Problem 28. As mentioned in the notes to Sect. 1.1, the thin tube limit can use potentials instead of Dirichlet conditions; one can regard the problem alternatively as an adiabatic approximation [WT10].

Section 1.7 The repulsive effect of torsion was first pointed out in [CBr96]. *Theorem* 1.7 is due to [EKK08], where local versions of *Lemma* 1.7.2 and *Proposition* 1.7.1 can also be found. The proof of *Lemma* 1.7.1 is taken from [BKRS09]. Spectral stability of the Laplace operator in bent three-dimensional tubes as a consequence of the Hardy inequality is discussed in [EKK08] and later in more detail in [Kr08]. A different approach to the repulsive effect of twisting, without appeal to a Hardy-type inequality, was used in [Gr04,Gr05], see also [BMT07]. Other consequences of twisting on the behavior of the discrete spectrum, apart from the absence of bound states in mildly bent tubes, will be discussed in Chap. 3.

Another observation which follows from inequality (1.49) is the fact that twisting removes the Green's function singularity at the threshold of the essential spectrum.

This, in turn, influences the long time behavior of solutions to the heat equation in twisted waveguides. Indeed, in [KrZu10] it was shown that the L^2 norm of the solutions to the equation in twisted waveguides decay faster than in the straight waveguides provided the initial data has a fast enough decay at infinity. The corresponding pointwise estimates on the heat kernel were established later in [GKP14]. The effects of a repulsive nature similar to twisting have also been observed in two-dimensional strips in the presence of a local magnetic field, see *Theorem* 7.3, and in strips with "twisted" Dirichlet-Neumann boundary conditions, see [EkKo05] and [KoKr08].

Theorems 1.9 and 1.10 are taken from [EKo05]. For a compactly supported perturbation the induced discrete spectrum is finite. The situation changes when the variation of the twist extends beyond any compact and decays slowly enough at large distances; then the operator $H_{\dot\alpha}$ has infinitely many discrete eigenvalues which accumulate at the threshold E, see [BKRS09]. On the other hand, slowing down the twist is not the only geometric deformation which induces a discrete spectrum in a periodically twisted tube; it was shown in [EFr07] that a similar effect also appears in the situation when the twisting is constant, but the cross-section is radially scaled with the scaling parameter depending on x_3 in a suitable way.

1.9 Problems

1. Prove relations (1.6)–(1.8). Check that the first term of the effective potential equals $-\frac{1}{4}\gamma_u(s)^2$ where γ_u is the signed curvature of the parallel curve Γ_u defined by relations (1.3) with a fixed value of the transverse variable u.
2. Prove that an infinite planar curve possesses asymptotes if its signed curvature satisfies $\gamma(s) = \mathcal{O}(|s|^{-2-\varepsilon})$ as $s \to \pm\infty$ for some $\varepsilon > 0$.
Hint: $|\sin\beta(s,\infty)| \le |\beta(s,\infty)|$
3. Define $\psi_n(s,u) := \frac{1}{\sqrt{n}}\,\phi\left(\frac{s-s_n}{n}\right)\, \mathrm{e}^{ips}\chi_1(u)$ for a fixed $p \in \mathbb{R}$, where $\phi \in C_0^\infty(\mathbb{R})$ and $\{s_n\}$ is a sequence such that $|s_n|/n \to \infty$. Check that under the assumptions of *Proposition* 1.1.1, $\psi_n \to 0$ weakly and $\|(H - \kappa_1^2 - p^2)\psi_n\| \to 0$ as $n \to \infty$.
4. Prove the first claim in Remark 1.1.4.
Hint: Replace f in the proof by $f(s,u) := j(s)u\chi_1(u)$ with a suitable $j \in C_0^\infty(K)$.
5. Let Ω satisfy the assumptions of *Proposition* 1.1.1 and suppose that Ω_e is a set with C^4 smooth boundary which coincides with Ω outside a compact set in $\mathbb{R}^2$. Moreover, the boundary part $\partial\Omega_\mathrm{e} \setminus \partial\Omega$ may not be smooth. Then $\sigma_\mathrm{ess}(-\Delta_D^{\Omega_\mathrm{e}}) = [\kappa_1^2, \infty)$.
6. If $\Omega_\mathrm{e} \supset \Omega$ holds in the previous problem, the respective eigenvalues of $-\Delta_D^{\Omega_\mathrm{e}}$ and $-\Delta_D^{\Omega}$ satisfy the inequalities $\epsilon_j^\mathrm{e} \le \epsilon_j,\ j = 1, 2, \dots$. In particular, $\sigma_\mathrm{disc}(-\Delta_D^{\Omega_\mathrm{e}}) \ne \emptyset$ holds when Ω obeys the assumptions of *Theorem* 1.1.
7. Let $\mathcal{P}$ be the polygonal path with two right-angle bends, which is the union of the halflines $\{\pm x \ge 0,\ y = \pm\ell/2\}$ with the segment $\{x = 0,\ |y| \le \ell/2\}$. Prove that the operator $-\Delta_D^\Omega$ corresponding to the polygonal duct $\Omega \equiv \Omega_{\mathcal{P},d}$ built over $\mathcal{P}$ has a bound state even in the case $d > \ell$ which is not covered by *Theorem* 1.2.

Hint: Replace a part of the duct axis by the graph of $p(x) = \frac{\ell}{2}\sin\frac{\pi x}{2a}$, $|x| \le a$.
8. The pointwise convergence in (1.14) is ensured if $\{a_j\} \subset \ell^1$. The conditions $\{a_j\} \in \ell^2(j^{\mp 1})$ are equivalent to $\psi \in L^2(\Omega)$ and $\psi \in \mathrm{Dom}\,(-\Delta_D^\Omega)$, respectively, where ψ is the symmetric extension of the function (1.14) to the L-shaped strip $\Omega \equiv \Omega_{\pi/2}$.
9. The operator $A = (A_{jk})$ on ℓ^2 with $A_{jk} = \sqrt{jk}/(j^2+k^2)$ is non-compact.
Hint: A is compact if and only if $\|A - P_n A\| \to 0$ as $n \to \infty$, where P_n is the projection spanned by the first n basis vectors. Consider $\{j^{-(1+\mu)/2}\}$ with a suitable μ.
10. $\epsilon = 0.9291...$ is the unique solution of spectral condition (1.15).
Hint: The maps $\epsilon \mapsto t,\ K$ are ℓ^∞-continuous and monotonous, so F is continuous and increasing; check that $F(0) \le (2\pi)^{-1}\left[1 + (12/5\pi)\sum_{j=2}^\infty j^{-2}\right] < 1$ while $F(1) > 2$. For the numerical solution see [EŠŠ89].
11. $-\Delta_D^\Omega$ on the L-shaped strip $\Omega \equiv \Omega_{\pi/2}$ has a single bound state.
Hint: Use Neumann bracketing at $x = \pi$ and $y = \pi$.
12. (a) Prove relation (1.17).
(b) The map (1.16) is a local C^k diffeomorphism if $\Gamma \in C^{k+2}$, $\alpha \in C^k$, and $g \ne 0$ holds in $\mathbb{R} \times M$; it becomes global under the assumption *(i)* of Sect. 1.3. In particular, this is true if $a\|\gamma\|_\infty < 1$, condition (1.18) holds, and Γ has a global Frenet frame.
(c) Find examples of C^∞ curves without global Frenet frame. Construct a global diffeomorphism f satisfying the condition (1.18) in case of a piecewise global triad.
Hint: Consider curves with $\ddot\Gamma$ vanishing at a point or in an interval.
13. Check relations (1.19)–(1.21).
14. Prove *Proposition* 1.3.1. Extend the claim made in Problem 5 to the three-dimensional situation.
15. In addition to hypotheses of *Proposition* 1.3.1, suppose that γ, $\dot\gamma$, $|\ddot\gamma|^{1/2}$ and τ, $\dot\tau$ belong to $L^2(\mathbb{R}, |s|\,ds)$. Using the minimax principle, show that $\sigma_{\rm disc}(-\Delta_D^\Omega) \ne \emptyset$ holds for all a small enough. Find an alternative set of assumptions which does not involve a requirement of torsion integrability.
16. Fill in the details of the proof of *Theorem* 1.3.
17. Let Ω be a smoothly bent tube in $\mathbb{R}^d$, $d = 2, 3$. Let a function $v \in L^2(M)$ define the potential which depends on transverse variables only, $V(x,y) := v(u)$ if $d = 2$ and $V : (V \circ f)(s,r,\theta) = v(r,\theta)$ for $d = 3$. Show that the conclusions of *Theorems* 1.1 and 1.3 extend to the operator $H_v := -\Delta_D^\Omega + V$ if $\nu_1 = \kappa_1^2$ is understood as the lowest eigenvalue of $-\Delta_D^M + v$.
18. Let $-\Delta_D^\Omega$ and $-\Delta_D^{\Omega'}$ be Dirichlet Laplacians in the regions $\Omega \subset \Omega' \subset \mathbb{R}^d$ with piecewise smooth boundaries, and let further $G_\Omega(\cdot,\cdot;z)$, $G_{\Omega'}(\cdot,\cdot;z)$, respectively, be the corresponding Green's functions. Then the inequalities $0 < G_\Omega(x,x';z) \le G_{\Omega'}(x,x';z)$ hold for any $x, x' \in \Omega$, $x \ne x'$, and $z < 0$.
Hint: To check the monotonicity use the fact that $-\Delta_D^\Omega$ is approximated in the strong resolvent sense by the projection of $-\Delta_D^\Omega + \mu\chi_{\Omega'\setminus\Omega}$ to $L^2(\Omega)$ as $\mu \to +\infty$, cf. [BD79], together with with the identity

$$(H_0 + \hat U U - z)^{-1} = (H_0 - z)^{-1} - \hat U (H_0 - z)^{-1}(I + U(H_0 - z)^{-1}\hat U)^{-1}(H_0 - z)^{-1}U.$$

For positivity see [RS, Appendix 1 to Sect. XIII.12].

19. Fill in the details of the proof of *Theorem* 1.4. Show that the result remains valid if M_x is only asymptotically constant in the following sense: there is an M such that the symmetric difference $M_x \Delta M$ is for each $x \in \mathbb{R}$ contained in some ε_x-neighborhood of ∂M, where $\varepsilon_x \to 0$ holds as $|x| \to \infty$.
Hint: Use once more the domain continuity of Dirichlet eigenvalues [RT75].
20. Prove *Theorem* 1.5.
Hint: Use a GJ-argument and bracketing as in [EŠTV96]; to check the sharp inequalities, use an outward deformation of $B := (-a, a) \times (-d_2, d_1)$ which can be mapped onto B by a smooth coordinate change and proceed as in [Ka, Sect. VII.6.5].
21. Consider the operator $-\Delta_D^\Omega$ of the previous problem in the symmetric case, $d_1 = d_2$. Find the eigenvalues and eigenfunctions by mode matching. Do the same for $d_1 > d_2$ using operator (1.38) and its antisymmetric counterpart. Show that the condition $Ca = 0$ can be replaced by $c + Kc = 0$, where

$$K_{jm} := \frac{1}{r_j} \sum_{k=1}^{\infty} (\xi_j, \eta_k)\, p_k \tanh(p_k a)\, (\eta_k, \xi_m)\,.$$

Compare the convergence of the truncated approximants in the two cases.
Hint: Cf. [EŠTV96].
22. Let Ω be a bent tube in $\mathbb{R}^d$, $d = 2, 3$, which is straight outside a compact region. Suppose that $-\Delta_D^\Omega$ has an eigenvalue $\epsilon < \kappa_1^2$, then there is a $c > 0$ such that the corresponding eigenfunction ψ satisfies the inequality $|\psi(s, u)| \le c \exp\left(-|s|\sqrt{\kappa_1^2 - \epsilon}\right)$ for all $s \in \mathbb{R}$. Similar exponential bounds hold for local perturbations of Sect. 1.4.
Hint: We have $\psi(s, u) = \sum_{j=1}^\infty c_j \exp\left(\pm s\sqrt{\kappa_j^2 - \epsilon}\right) \chi_j(u)$ in any straight semi-inifinite part of the tube Ω.
23. Prove *Lemma* 1.5.1. Find the normalization factors N_j and show that the sequence $\{\chi_j(0; \alpha)\}$ is bounded for any fixed $\alpha \in \mathbb{R}$.
Hint: Cf. [EKr99] and [EKr01b].
24. Find the spectrum of the operator H_α corresponding to (1.39) for the function $\alpha : \alpha(x) = \alpha_0(1 - \chi_{(-a,a)}(x))$ with fixed $\alpha_0, a > 0$. Compare the results with those for the similar problem of Sect. 1.5.1, in particular, in the case when $\alpha_0 d \gg 1$.
Hint: Use mode matching—cf. [EKr99].
25. Consider the eigenfuctions ϕ_m corresponding to the eigenvalues ϵ_m appearing in *Theorem* 1.5 with $m = 3, 4, \ldots$ and their analogues from the previous problem. Show that none of their nodal lines, with the exception of the central one for even m, is a line segment perpendicular to the strip axis.
Hint: If so ϕ_m should be mirror symmetric w.r.t. the nodal line.
26. Let $\Omega_L := \{\vec{x} : |y| < \frac{\pi}{2}, |x| < \pi L\} \cap \{\vec{x} : |x| < \frac{\pi}{2}, |y| < \pi L\}$ and denote by Ω the infinite cross structure, $\Omega := \cup_{L>0} \Omega_L$. Let $H_D := -\Delta_D^{\Omega_L}$ and H_N be the analogous operator with the Dirichlet boundary condition switched to the Neumann condition at the cross arm cuts. Show that $-\Delta_D^\Omega$ has a single eigenvalue $\nu_1 \approx 0.66$ in $(0, 1)$, and that the ground state eigenvalues ν_1^D, ν_1^N of H_D and H_N, respectively, tend to ν_1 from above and from below as $L \to \infty$.

Hint: By symmetry the problem is analogous to that of *Proposition* 1.2.3 with Neumann condition at the outer boundary. For a finite cross the exponentials in (1.14) are replaced by hyperbolic functions.

27. Prove that for a bent tube in $\mathbb{R}^3$ with a fixed cross section of radius a one has the result analogous to *Theorem* 1.6, in particular, one has

$$\epsilon(a) = j_{0,1}^2 a^{-2} + \lambda(a)\,,$$

where $j_{0,1} \approx 2.40$ is the first zero of the Bessel function J_0 when the cross section is circular. In this relation $\lambda(\cdot)$ is a C^∞ function, $\lambda(a) = \lambda + \mathcal{O}(a)$ as $a \to 0+$, where λ is a negative eigenvalue of the operator analogous to (1.44); the latter may contain a twisting term if the cross section is not circular.

28. Adopt the notation of Sect. 1.6 and consider the operator $H_0 = -a^{-2}\partial_u^2 + T_0$. Let P_0 be the eigenprojection of H^0 corresponding to the eigenvalue $E_0 := a^{-2}\kappa_1^2 + \lambda$ and denote by $\hat{R}^0$ the reduced resolvent of H_0 with respect to E_0. The eigenvalue $\epsilon(a)$ from *Theorem* 1.6 can then be expressed as

$$\epsilon(a) = a^{-2}\,\kappa_1^2 + \lambda + \sum_{m=1}^{\infty} \frac{(-1)^m}{m} \sum_{p_1+\cdots+p_m=m-1} \operatorname{tr} \prod_{i=1}^{m} (H - H_0) S^{(p_i)},$$

where $S^0 = P^0$ and $S^{(p_i)} = \hat{R}^0 (E^0)^{p_i}$ for $p_i \in \mathbb{N}$. Work out the lowest orders of the expansion for the perturbation $W \equiv ab(s, u, \partial_s)$ given in the transverse-mode basis by $W_{jk} = \sum_{l=1}^{\infty} (-\partial_s b_{jk}^{(l)} \partial_s + V_{jk}^{(l)}) a^l$ with

$$\begin{aligned}
b_{jk}^{(l)} &:= (l+1)\,(-\gamma)^l \int_{-1}^{1} u^l\, \chi_j(u)\chi_k(u)\, du \\
V_{jk}^{(l)} &:= \frac{1}{4}(l+1)\,(-\gamma)^{l-2} \int_{-1}^{1} u^l \left[-\gamma^4 - l\gamma\ddot{\gamma} - \frac{5}{6}\, l(l-1)\dot{\gamma}^2 \right] \chi_j(u)\chi_k(u)\, du\,.
\end{aligned}$$

Do the same for the three-dimensional tube of *Problem* 27 where we have

$$\begin{aligned}
b_{jk}^{(l)} &:= (l+1)\,(-\gamma)^l \int_{B_1} (r\,\cos(\theta-\alpha))^l\, \chi_j(u)\chi_k(u)\, du \\
V_{jk}^{(l)} &:= \frac{1}{4}(l+1)\,(-\gamma)^{l-2} \int_{B_1} (r\,\cos(\theta-\alpha))^{l-2} r^2 \\
&\quad \times \Bigg\{ \left[-\gamma^4 - l\gamma(\ddot{\gamma} - \gamma\tau^2) - \frac{5}{6}\, l(l-1)\dot{\gamma}^2 \right] \cos^2(\theta-\alpha) \\
&\quad - \left[l\gamma(2\dot{\gamma}\tau + \gamma\dot{\tau}) + \frac{5}{3}\, l(l-1)\gamma\dot{\gamma}\tau \right] \sin(\theta-\alpha)\cos(\theta-\alpha) \\
&\quad - \frac{5}{6}\, l(l-1)\gamma^2\tau^2 \sin^2(\theta-\alpha) \Bigg\} \chi_j(u)\chi_k(u)\, du\,.
\end{aligned}$$

Hint: The perturbative expansion can be found in [Ka66, Sect.2.2], see also [DE95].

29. Prove inequality (1.54).

Hint: Without loss of generality, we may suppose that $J = (-b/2, b/2)$ with some positive b. Define a function g on I by $g(t) = 2|t|/b$ if $|t| \leq b/2$ and $g(t) = 1$ otherwise. Write $\phi = \phi g + (1 - g)\phi)$ and use the fact that $(\phi g)(0) = 0$ to obtain a suitable estimate on the function $(\phi g)(x)$.

30. Assume that the twisting function $\dot{\alpha}$ satisfies the assumptions of *Theorem* 1.7 and prove the following statement: if $V : \mathbb{R} \to \mathbb{R}$ is a bounded function with compact support and such that $\int_{\mathbb{R}} V(x)\,\mathrm{d}x < 0$, then to any $\lambda > 0$ there exists an $n \in \mathbb{N}$ such that the operator $-\Delta_D^{\Omega_\alpha} + \lambda V(x_3 + n)$ in $L^2(\Omega_\alpha)$ has at least one eigenvalue below ν_1.

Hint: Construct a suitable sequence of test functions.

Chapter 2
Transport in Locally Perturbed Tubes

Our next aim is to discuss the systems of the previous chapter from the viewpoint of particle transport. For simplicity we are going to pay most attention to the two-dimensional case where Ω is a Dirichlet strip in the plane. The perturbations of the ideal straight waveguide which we shall consider here are again of a local nature; this allows us to work in the scattering-theory setting where the time evolution is compared to an appropriate free asymptotic dynamics.

2.1 Existence and Completeness

The natural comparison operator is that of a straight tube, $H_0 = -\Delta_D^{\Omega_0}$. In the usual scattering theory for Schrödinger operators we most often compare pairs of operators acting on the same Hilbert space. For waveguides this happens, e.g., if the perturbation is a potential or a measure in the kinetic term which we have discussed in Sect. 1.4. In that case the existence and asymptotic completeness of the ***wave operators*** defined as usual by

$$\Omega_\pm(H, H_0) := \operatorname*{s-lim}_{t\to\pm\infty} \mathrm{e}^{iHt}\mathrm{e}^{-iH_0t}P_{\mathrm{ac}}(H_0)$$

is easily established (Problems 3 and 4). This is not the case for perturbations of a geometric nature, however, such a comparison is still possible if we can replace the Hamiltonian by a unitarily equivalent operator on $L^2(\Omega_0)$.

A prime example of this is given by smoothly bent planar strips where we can use the straightening transformation and pass from $-\Delta_D^\Omega$ to the operator H given by formula (1.7). The scattering problem is then well defined under suitable regularity and asymptotic straightness requirements on the strip Ω.

Theorem 2.1 *Let assumptions (i), (ii)$_2$, and (iii)$_2$ of Sect. 1.1 be valid and $a\|\gamma\|_\infty < 1$. Furthermore, suppose that the functions $\gamma, \dot\gamma^2, \ddot\gamma$ are $\mathcal{O}(|s|^{-1-\delta})$ for*

P. Exner and H. Kovařík, *Quantum Waveguides*,
Theoretical and Mathematical Physics, DOI 10.1007/978-3-319-18576-7_2

some $\delta > 0$ as $|s| \to \infty$. Then the wave operators $\Omega_\pm(H, H_0)$ exist, are complete, and the singularly continuous spectrum of H is empty.

Proof Using the notation of Sect. 1.6 we can write the difference of the two operators as $-\partial_s(b-1)\partial_s + V = B^*A$, where the operator $A : L^2(\Omega_0) \to L^2(\Omega_0) \otimes \mathbb{C}^2$ acts as $\binom{A_0}{-iA_1\partial_s}$ with $A_0 := |V|^{1/2}$, $A_1 := |b-1|^{1/2}$, and B is the analogous operator with the coefficients replaced by $V^{1/2} := |V|^{1/2}\text{sgn}\, V$ and $(b-1)^{1/2}$, respectively. This factorization allows us to employ the smooth-perturbation method similarly as it is done in the case of one-dimensional Schrödinger operators (see the notes).

Putting $\varrho(s) := (1+s^2)^{-(1+\varepsilon)/4}$ for a fixed $\varepsilon \in (0, \delta]$, we infer from the curvature decay assumptions that $\max\{\|A_l\varrho^{-1}\|_\infty, \|B_l\varrho^{-1}\|_\infty\} < \infty$ holds for $l = 0, 1$. The free resolvent $R_0(z) := (H_0-z)^{-1}$ then satisfies the estimate

$$\|A_l(-i\partial_s)^l R_0(z)\| \le \|A_l\varrho^{-1}\|_\infty \sup_j \left\| \varrho(s)(-i\partial_s)^l(-\partial_s^2 + \nu_j - z)^{-1} \right\|$$

for $l = 0, 1$, where $\nu_j = \kappa_1^2 j^2$ are the transverse eigenvalues, and the analogous inequalities hold with B_l. The last factor can be estimated by the product of L^2 norms of the functions g and $p \mapsto p^l(p^2 + \nu_j - z)^{-1}$, cf. Theorem 11.20 of [RS]. Using the first resolvent identity we conclude that the operator-norm limit of $I - A[BR_0(\lambda \pm i\eta)]^*$ as $\eta \to 0$ exists away from the thresholds, i.e. for any $\lambda \ne \nu_j$. In a similar way we derive the inequality

$$\|AR_0(\lambda \pm i\eta)\|^2 + \|BR_0(\lambda \pm i\eta)\|^2 \le c\eta^{-1}$$

with some $c > 0$ for λ in any compact interval I which does not contain any of the points ν_j (Problem 1). Moreover, since $A_l, B_l \in L^2$ by assumption, one can check that the operator $AR_0(z)[BR_0(z')]^*$ is trace class as a product of two Hilbert-Schmidt operators, and thus compact for any non-real z, z'. The rest of the argument proceeds as in the potential scattering case, cf. Theorem 10.5.1 of [Sch], since only a finite number of transverse modes is involved in expressions containing the spectral projection $E_{H_0}(I)$. We arrive thus at the condition

$$\int_{\mathbb{R}} \sup_{-a<u<a} \left\{ |V(s,u)|^2 + |b(s,u) - 1|^2 \right\} \varrho(s)^\alpha < \infty \,,$$

which is satisfied for any $\alpha > 0$ in view of the decay assumptions we made. ∎

Remark 2.1.1 In a similar way one can prove asymptotic completeness for scattering in a three-dimensional smoothly bent tube (Problem 2), as well as for tubes perturbed by a potential or a kinetic-term weight (Problems 3 and 4).

An alternative way to prove, under slightly modified assumptions, that a bent and asymptotically straight tube has no singularly continuous spectrum is to employ ***Mourre's method*** of positive commutator. Let us sketch its main ideas briefly with our purpose in mind; for more information we refer to the literature indicated in the notes. The method is based on a suitable choice of a ***conjugate operator***: one looks for an operator A, self-adjoint on $L^2(\Omega_0)$, such that for a given interval $I \subset \sigma(H)$ there is an operator K, compact in $L^2(\Omega_0)$, and a positive constant c such that

$$E_H(I)\,[H, iA]\,E_H(I) \geq c\,E_H(I) + K\,, \tag{2.1}$$

where $E_H(I)$ denotes the spectral projection of H onto the interval I and the commutator $[iA, H]$ is understood as a bounded operator from $H_0^1(\Omega_0)$ to its dual $(H_0^1(\Omega_0))^*$. Inequality (2.1) is referred to as ***Mourre's estimate***; if it holds with $K = 0$ we say it is strictly valid. This estimate has, under certain conditions, consequences for the structure of the spectrum of H in the interval I. These conditions can be expressed in terms of the regularity of the map

$$\mathbb{R} \ni t \mapsto \mathrm{e}^{itA}(H - i)^{-1}\mathrm{e}^{-itA} \tag{2.2}$$

from $\mathbb{R}$ to $\mathcal{B}(L^2(\Omega_0))$. We say that $H \in C^1(A)$ if the above map is of class C^1 in the strong operator topology; if, moreover, the derivative of (2.2) is Hölder continuous of order $\alpha > 0$, we write $H \in C^{1+\alpha}(A)$. Using these notions one is able to state the following result:

Theorem 2.2 *Suppose that* e^{itA} *leaves the form domain of* H *invariant and that* $H \in C^{1+\alpha}(A)$ *for some* $\alpha > 0$. *If (2.1) holds true on an interval* $I \subset \sigma(H)$, *then the singularly continuous spectrum of* H *on* I *is empty and the interval* I *contains at most finitely many eigenvalues of* H, *each of them being of a finite multiplicity. If, in addition, (2.1) holds with* $K = 0$, *then the spectrum of* H *on* I *is purely absolutely continuous.*

To apply *Theorem* 2.2 to our problem, consider an open interval separated from the transverse thresholds, $I \subset \sigma(H) \setminus T$ with $T = \{\nu_j\}_{j\in\mathbb{N}}$, and choose

$$A = -\frac{i}{2}(s\,\partial_s + \partial_s s)$$

defined initially, say, on $C_0^\infty(\Omega_0)$ and extended to a closed operator on $L^2(\Omega_0)$. It is not difficult to check that

$$(\mathrm{e}^{itA} f)(u, s) = \mathrm{e}^{t/2} f(u, \mathrm{e}^t s) \quad \text{for } t \in \mathbb{R} \text{ and } f \in L^2(\Omega_0)\,,$$

i.e. that A generates the group of dilations in the longitudinal variable. This implies, in particular, that e^{itA} leaves $H_0^1(\Omega_0)$ invariant. Moreover, a direct computation shows that

$$[H, iA] = -2\partial_s(1 + u\gamma(s))^{-2}\,\partial_s - 2\partial_s\, s\,\frac{\dot\gamma(s)}{(1 + u\gamma(s))^3}\,\partial_s - s\,\partial_s V(u, s), \tag{2.3}$$

where $V(u, s)$ is the effective potential (1.8). Under suitable decay assumptions on the curvature one can check that the difference $(H - i)^{-1} - (H_0 - i)^{-1}$ is compact on $L^2(\Omega_0)$ which yields the inequality

$$E_H(I)\,[H, iA]\,E_H(I) = -2\partial_s^2\, E_H(I) + K \tag{2.4}$$

with a compact K. In view of our assumption about I, it is not difficult to see that the operator $-\partial_s^2\, E_H(I)$ is strictly positive, hence if one can show that $H \in C^{1+\alpha}(A)$ holds for some $\alpha > 0$, *Theorem* 2.2 could be applied. It turns out that the needed regularity of the map (2.2) can be demonstrated under appropriate decay assumptions on the curvature γ and its derivatives.

Theorem 2.3 *Let assumptions (i), (ii)$_3$ of Sect. 1.1 hold. Furthermore, suppose that $\gamma(s), \ddot\gamma(s) \to 0$ holds as $|s| \to \infty$ and that $\dot\gamma(s), \dddot\gamma(s)$ are $\mathcal{O}(|s|^{-1-\delta})$ for some $\delta > 0$ as $|s| \to \infty$. Then (a) $\sigma_{\rm ess}(H) = [\nu_1, \infty)$, (b) $\sigma_{\rm sc}(H) = \emptyset$, (c) $\sigma_{\rm p}(H) \cup T$ is countable and closed, and (d) $\sigma_{\rm p}(H) \setminus T$ consists at most of eigenvalues of finite multiplicity which can accumulate only at points of T.*

2.2 The On-Shell S-Matrix: An Example

Full information about scattering requires, of course, more than just checking that the problem is well posed. The central question is to find the ***on-shell scattering operator*** $S(k)$ which describes scattering at a given energy k^2. In general it is not unusual that the space on which $S(k)$ acts depends on energy. In case of waveguide scattering this dependence has a characteristic form: the on-shell space dimension is

$$\sum_{j=1}^{n_a} N_j(k)\,, \tag{2.5}$$

where n_a is the number of tubes leaving the scattering region, for example $n_a = 2$ if Ω is a single locally deformed strip, and $N_j(k)$ is the number of propagating modes in the j-th outgoing tube which obviously coincides with the number of transverse eigenvalues satisfying the inequality $\nu_n^{(j)} \le k^2$, thus $N_j(k) = [k\kappa_{1,j}^{-1}]$ holds if the outgoing channel is an asymptotically straight Dirichlet strip.

Since we consider situations where n_a is finite, the operator $S(k)$ can be regarded as a matrix of the dimension given by (2.5) the elements of which are the reflection and transmission amplitudes understood in the general sense, i.e. taking into account that the particle may leave the scattering region in a state whose transverse component differs from the one with which it entered.

Finding these amplitudes is a difficult task. A class of systems for which it can be accomplished numerically is represented by those Ω which decompose into a union of regions where the corresponding Schrödinger equation can be solved by separation of variables; the global scattering solution is then constructed using mode matching similar to that used in Sects. 1.2 and 1.5. We shall illustrate this method on the example of a pair of window-coupled waveguides having generally different widths d_1, d_2 which we have introduced in Sect. 1.5.1.

For definiteness let us suppose that the incident wave is in the upper channel, being of the form $\chi_j^{(+)}(y)\,\exp(-ik_j^{(+)}x)$, where $k_j^{(\pm)} := \kappa_1\sqrt{k^2 - j^2\varrho^{-(1\mp 1)}}$ are used as symbols for channel momenta. We denote by $r_{jj'}^{(\pm)}$, $t_{jj'}^{(\pm)}$, respectively, the corresponding reflection and transmission amplitudes to the j'-th transverse mode in the upper and lower guide. Due to the mirror symmetry with respect to the line $x = 0$, we can again consider separately the two parities, writing

$$r_{jj'}^{(\pm)} = \frac{1}{2}\left(\rho_{jj'}^{(s,\pm)} + \rho_{jj'}^{(a,\pm)}\right), \quad t_{jj'}^{(\pm)} = \frac{1}{2}\left(\rho_{jj'}^{(s,\pm)} - \rho_{jj'}^{(a,\pm)}\right), \tag{2.6}$$

where $\rho_{jj'}^{(\sigma,\pm)}$, $\sigma = s, a$, are the appropriate reflection amplitudes. In the even case, which corresponds to the Neumann condition at $x = 0$, we seek solutions using for $0 < x \le a$ and $x \ge a$, $y \in \mathcal{C}_+$, respectively, the following Ansatz,

$$\psi(x,y) := \begin{cases} \sum_{\ell=1}^{\infty} a_\ell \frac{\cos(ip_\ell x)}{\cos(ip_\ell a)}\,\eta_\ell(y) \\ \sum_{j'=1}^{\infty}\left(\delta_{jj'}\,\mathrm{e}^{-ik_j^{(+)}(x-a)} + \rho_{jj'}^{(+)}\mathrm{e}^{ik_{j'}^{(+)}(x-a)}\right)\chi_{j'}^{(+)}(y) \\ \sum_{j'=1}^{\infty} \rho_{jj'}^{(-)}\mathrm{e}^{ik_{j'}^{(-)}(x-a)}\,\chi_{j'}^{(-)}(y) \end{cases} \tag{2.7}$$

where p_j is the same as in (1.37). The exterior part can also be written as

$$\psi(x,y) = \sum_{m'=1}^{\infty}\left(\delta_{mm'}\,\mathrm{e}^{-ik_m(x-a)} + \rho_{mm'}\mathrm{e}^{ik_{m'}(x-a)}\right)\xi_{m'}(y)\,,$$

where ξ_m are elements of the ordered basis corresponding to (1.36),

$$\rho_{mm'} := \begin{cases} \rho_{jj'}^{(+)} \;\ldots\; \theta_m = j\,,\; \theta_{m'} = j' \\ \rho_{jj'}^{(-)} \;\ldots\; \theta_m = j\,,\; \theta_{m'} = j'\varrho^{-1} \end{cases}$$

and $k_m := k_j^{(\pm)}$ for $\theta_m = j\,,\, j\varrho^{-1}$, respectively. Matching the functions (2.7) smoothly at $x = a$ we arrive at the equation

$$\sum_{m'=1}^{\infty}\left(ik_\ell + p_{m'}\tan(ip_{m'}a)\right)(\xi_\ell, \eta_{m'})\,a_{m'} = 2ik_\ell\delta_{m\ell}\,, \tag{2.8}$$

where the index m corresponds to the incident wave and the overlap integrals $(\xi_\ell, \eta_{m'})$ are the same as in (1.38); in the odd case corresponding to the Dirichlet condition at $x = 0$ one has to replace tan by $-$cot. The reflection amplitudes are then given by

$$\rho_{m\ell}^{(\pm)} = -\delta_{m\ell} + \sum_{m'=1}^{\infty} a_{m'}^{(\pm)}(\xi_\ell, \eta_{m'});$$

they determine the original quantities via (2.6). In a similar way one finds the reflection and transmission amplitudes in the case when the incident wave is in the lower channel and by that the full on-shell S-matrix; convergence of the truncating approximations is checked as in *Proposition* 1.2.3.

Often it is not the S-matrix itself but a quantity derived from it which is of primary physical interest. When perturbed waveguides are used to model systems of quantum wires coupled to macroscopic reservoirs we are concerned with ***conductance*** (or its inverse quantity, resistance) between a given pair of leads, which is given by the ***Landauer-Büttiker formula***. Suppose, for instance, that we deal with the incoming current in the upper right guide and the outgoing one in the lower left, then the conductance (measured in the standard units e^2/h) is given by

$$G_{l+,r-}(k) = \sum_{j=1}^{N_+(k)} \sum_{j'=1}^{N_-(k)} \frac{k_{j'}^{(-)}}{k_j^{(+)}} |t_{jj'}^{(-)}(k)|^2, \tag{2.9}$$

where k and the current-carrier momenta $k_j^{(\pm)}$ are determined by the Fermi energy and chemical potentials of the reservoirs and $N_\pm(k) = [k\kappa_{1\pm}^{-1}]$ are the number of propagating modes in the considered channels; analogous expressions can be written for conductances between other pairs of leads.

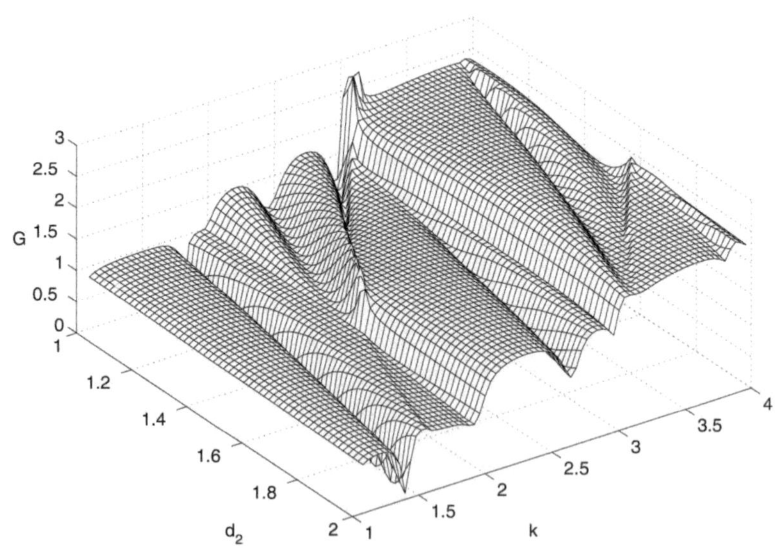

Fig. 2.1 A conductance plot for an asymmetric coupled waveguide system

As an illustration, we show in Fig. 2.1 the conductance plot for transport from the upper right to the upper left channel for $d_1 = \pi$ and $a = 1$ as a function of the momentum k and the lower channel width d_2. If the window was closed, the conductance $G_{l-,r-}(\cdot)$ would simply be a step function with a jump at every threshold. The general steplike pattern is preserved, being modified by the coupling, in particular, we observe pronounced resonances, the positions of which change with the channel width ratio.

2.3 Resonances from Perturbed Symmetry

One of the conspicuous effects in waveguides are scattering resonances, which we are going to discuss in this and the next section, because they typically entail sharp changes in transport properties. There are different mechanisms which can create resonances. The simplest one is based on symmetry violations. If a waveguide supports an eigenvalue embedded in the continuous spectrum which owes its existence to a particular symmetry, it is natural to expect that this eigenvalue turns into a resonance once the symmetry in question is perturbed.

Before discussing this mechanism in more detail, one has to make sure that its basic premise is not empty, i.e. that embedded eigenvalues can exist.

Examples 2.3.1 (a) Let $\Omega := \{\vec{x} \in \mathbb{R}^2 : -g(x) < y < g(x)\}$ be a symmetric strip with a protrusion. Specifically, suppose that g is a piecewise continuous function with $g(x) \geq \frac{1}{2}d$ and that there are sets $U \subset C \subset \mathbb{R}$, respectively open and compact, such that $g(x) > \frac{1}{2}d$ for $x \in U$ and $g(x) = \frac{1}{2}d$ for $x \in \mathbb{R} \setminus C$. By *Theorem* 1.4 we have $\sigma_{\mathrm{ess}}(-\Delta_D^\Omega) = [\epsilon_d, \infty)$. At the same time the operator decomposes into the even and odd part with respect to the strip axis, $y = 0$, the latter being unitarily equivalent to the Dirichlet Laplacian in the halfstrip $\Omega_+ := \{\vec{x} \in \mathbb{R}^2 : 0 < y < g(x)\}$. Consequently, if $-\Delta_D^{\Omega_+}$ has an eigenvalue in $(\epsilon_d, 4\epsilon_d)$, it is an embedded eigenvalue of the original operator (Problem 6).
(b) Let Ω be a pair of strips from Sect. 1.5.3 crossing at a right angle. The operator $-\Delta_D^\Omega$ has an embedded eigenvalue $\approx 3.72\epsilon_d$ (Problem 7).
(c) Similar conclusions can be made about local perturbations of the *Neumann* Laplacian $-\Delta_N^{\Omega_0}$ in the straight strip having $\sigma_{\mathrm{ess}}(-\Delta_N^{\Omega_0}) = [0, \infty)$. The operator H_a obtained by imposing an additional Neumann condition at a segment of the strip axis of length $2a$ has embedded eigenvalues for any $a > 0$ (Problem 3.2b).

Embedded eigenvalues can also be generated by a potential perturbation of a straight waveguide of the type discussed in Sect. 1.4. We shall now use this example to illustrate how the resonances emerge. We start from the unperturbed operator $-\Delta_D^{\Omega_0}$ referring to the straight strip $\Omega_0 = \mathbb{R} \times (-a, a)$ and put

$$H_\lambda := -\Delta_D^{\Omega_0} + V(x) + \lambda U(\vec{x}) , \tag{2.10}$$

where V, U are real-valued functions on $\mathbb{R}$ and Ω_0, respectively, such that

(i) V is attractive, $V(x) \leq 0$, and it does not vanish everywhere. Moreover, it is short-range, $|V(x)| \leq \text{const}\,\langle x\rangle^{-2-\delta}$ for some $\delta > 0$, and it extends to a function analytic in the sector $\mathcal{M}_{\alpha_0} := \{z \in \mathbb{C} : |\arg z| \leq \alpha_0\}$ for some $\alpha_0 > 0$ and obeys the same bound there,

(ii) U is nonzero with similar properties, $|U(\vec{x})| \leq \text{const}\,\langle x\rangle^{-2-\delta}$ for some $\delta > 0$ and all $\vec{x} = (x, y) \in \Omega$, and $U(\cdot, y)$ extends for any fixed $y \in (-a, a)$ to an analytic function in $\mathcal{M}_{\alpha_0}$ and satisfies the same bound there.

Here $\langle x\rangle := \sqrt{1+x^2}$; since the potentials are by assumption continuous and bounded, the right-hand side in (2.10) is well defined. The unperturbed operator H_0 admits a separation of variables and the longitudinal part $h^V := -\partial_x^2 + V(x)$ has in view of *(i)* a nonempty and finite discrete spectrum,

$$\mu_1 < \mu_2 < \cdots \mu_N < 0 ;$$

the normalized eigenfunctions $\phi_n \in L^2(\mathbb{R})$, $n = 1, \dots, N$, associated with these simple eigenvalues are exponentially decaying. On the other hand, the transverse spectrum consists of the eigenvalues $\nu_j = \kappa_j^2 = (\pi j/2a)^2$, $j \in \mathbb{N}$, corresponding to the eigenfunctions (1.10), hence the spectrum of H_0 consists of the continuous part, $\sigma_{\text{ess}}(H_0) = \sigma_{\text{ac}}(H_0) = [\nu_1, \infty)$, and an infinite family of eigenvalues,

$$\sigma_{\text{p}}(H_0) = \left\{ \mu_n + \nu_j : n = 1, \dots, N, \; j = 1, 2, \dots \right\} .$$

Among these a finite subset is isolated, while the rest satisfying the condition

$$\nu_1 < \mu_n + \nu_j \neq \nu_k , \quad k = 2, 3, \dots , \tag{2.11}$$

are embedded in the continuous spectrum away from the thresholds. We rewrite the Hamiltonian (2.10) as an infinite matrix differential operator $\{H_{jk}(\lambda)\}$ on $L^2(\mathbb{R})$ with the elements

$$H_{jk}(\lambda) := \mathcal{J}_j^* H_\lambda \mathcal{J}_k = \left(h^V + \nu_j\right) \delta_{jk} + \lambda U_{jk}(x) , \tag{2.12}$$

where in the last term $U_{jk}(x) := \int_{-a}^{a} U(x, y)\overline{\chi_j}(y)\chi_k(y)\,dy$ and we use the embeddings $\mathcal{J}_k : L^2(\mathbb{R}) \to L^2(\Omega_0)$ and their adjoints $\mathcal{J}_k^* : L^2(\Omega_0) \to L^2(\mathbb{R})$ which act as

$$(\mathcal{J}_k u)(x, y) = u(x)\,\chi_k(y) , \quad (\mathcal{J}_k^* f)(x) = \int_{-a}^{a} f(x, y)\chi_k(y)\,dy .$$

Speaking of resonances we have in mind the most common definition which is based on analytical continuation of the Hamiltonian resolvent across the cut(s) associated with the continuous spectrum into a domain on another sheet of the corresponding energy surface, conventionally to the lower complex halfplane. A ***resonance*** is then identified with a pole in this analytic continuation; it is physically important if the pole is close to the real axis and the respective residue is not negligible. This concept naturally requires a sort of analyticity hypothesis, for instance such as we have made in the above assumptions.

One of the most efficient methods to determine resonances of Schrödinger operators is based on the so-called ***complex scaling***. With a small modification this technique can also be applied to waveguides. In this case one has to scale only the longitudinal variable as we shall now illustrate on the example in question. We begin with the family of unitary operators

$$S_\theta : \; (S_\theta\psi)(x, y) = e^{\theta/2}\psi(e^\theta x, y)\,, \quad \theta \in \mathbb{R}\,, \tag{2.13}$$

on $L^2(\mathbb{R})$ and extend this scaling transformations analytically to $\mathcal{M}_{\alpha_0}$. This is made possible by assumptions *(i), (ii)* according to which the transformed Hamiltonians are of the form

$$\begin{aligned} H_{\theta,\lambda} &:= S_\theta H_\lambda S_\theta^{-1} = H_{\theta,0} + \lambda U_\theta\,, \\ H_{\theta,0} &:= e^{-2\theta}(-\partial_x^2) - \partial_y^2 + V_\theta(x)\,, \end{aligned}$$

where $V_\theta(x) := V(e^{\theta x})$ and $U_\theta(x, y) := U(e^\theta x, y)$. The operators $H_{\theta,0}$ with $\theta \in \mathcal{M}_{\alpha_0}$ clearly constitute a type (A) analytic family of m-sectorial operators. It is straightforward to check that U_θ is relatively bounded with respect to $H_{\theta,0}$, thus the operators $H_{\theta,\lambda}$ with the same θ and $|\lambda|$ small enough constitute again a type (A) analytic family. The free part of the transformed operator still separates variables, hence its spectrum equals

$$\sigma\left(H_{\theta,0}\right) = \bigcup_{j=1}^{\infty} \left\{\nu_j + \sigma\left(h_\theta^V\right)\right\}\,,$$

where $h_\theta^V := -e^{-2\theta}\partial_x^2 + V_\theta(x)$. Since the potential V is dilation analytic by assumption, the discrete spectrum of h_θ^V is independent of θ; we have

$$\sigma\left(h_\theta^V\right) = e^{-2\theta}\mathbb{R}_+ \cup \{\mu_1, \ldots, \mu_N\} \cup \{\rho_1, \rho_2, \ldots\}\,.$$

Here μ_n are eigenvalues of h^V which will turn into resonances as a result of the perturbation. On the other hand, the ρ_r are the "intrinsic" resonances, i.e. complex poles of the resolvent of h_θ^V uncovered by the rotation of the essential spectrum; in view of assumption *(i)* there is at most a finite number of them in any finite part of the lower complex halfplane (see the notes). The two pole types are easily distinguished by their behavior in the limit $\lambda \to 0$ because only the former ones tend to the real axis as the perturbation is removed.

The main insight of the complex scaling method is that moving the essential spectrum we turn the embedded eigenvalues into isolated ones whose perturbation can be treated by usual methods; it is easy when the perturbation is relatively bounded as in our case. Any fixed eigenvalue $\epsilon_0 = \mu_n + \nu_j$ of $H_{\theta,0}$ has a neighborhood containing none of the points $\rho_k + \nu_{j'}$ in which we choose a contour encircling it; for the sake of simplicity we consider only the non-degenerate case, i.e. we suppose that $\mu_n + \nu_j \neq \mu_{n'} + \nu_{j'}$ holds for different pairs of indices.

It is sufficient to consider a purely imaginary scaling parameter, $\theta = i\beta$ with $\beta > 0$. Let P_θ be the projection onto the eigenspace associated with such an ϵ_0 and let $R_\theta(z) := (H_{\theta,0} - z)^{-1}$, then we set

$$S_\theta^{(p)} := \frac{1}{2\pi i} \int_{\mathcal{C}} \frac{R_\theta(z)}{(\epsilon_0 - z)^p}\, \mathrm{d}z$$

for $p \geq 0$, in particular, $P_\theta = -S_\theta^{(0)}$ and $S_\theta^{(1)} = \hat{R}_\theta(\epsilon_0)$ is the reduced resolvent value at the point ϵ_0. The assumption *(ii)* implies the existence of a positive c_θ such that $\max_{z \in \mathcal{C}} \|U_\theta R_\theta(z)\| \leq c_\theta$, and it follows that

$$\|U_\theta S_\theta^{(p)}\| \leq c_\theta \frac{|\mathcal{C}|}{2\pi} \left[\mathrm{dist}\,(\mathcal{C}, \epsilon_0)\right]^{-p}$$

holds for any $p \geq 0$. Thus we can justify the existence of the perturbation expansion,

$$\epsilon(\lambda) = \mu_n + \nu_j + \sum_{m=1}^{\infty} \epsilon_m(\lambda)\,, \tag{2.14}$$

where

$$\epsilon_m(\lambda) = \sum_{p_1 + \cdots + p_m = m-1} \frac{(-\lambda)^m}{m}\, \mathrm{tr} \prod_{i=1}^{m} U_\theta S_\theta^{(p_i)}\,,$$

because $\epsilon_m(\lambda) = \mathcal{O}(\lambda^m)$ and the convergence of the series (2.14) is checked in the same way as in Problem 1.28.

Let us next determine what the leading terms in the expansion look like. The first-order correction, $\epsilon_1(\lambda) = \mathrm{tr}\ (\lambda U_\theta P_\theta)$, is real-valued,

$$\epsilon_1(\lambda) = \left(\overline{\phi_n^\theta \otimes \chi_j}, \lambda U_\theta\, \phi_n^\theta \otimes \chi_j\right) = \left(\phi_n \otimes \chi_j, \lambda U \phi_n \otimes \chi_j\right) = \lambda \left(\phi_n, U_{jj} \phi_n\right)\,,$$

where ϕ_n is the eigenvector of h^V associated with μ_n. Thus, as usual in such situations, it does not contribute to the resonance width. The second-order term is conventionally computed by taking the limit $\beta \to 0$. In this way we get

$$\epsilon_2(\lambda) = -\lambda^2\, \mathrm{tr}\left(P_{j,n} U \hat{R}_1(\epsilon_0 - i0)\, U P_{j,n}\right) = -\lambda^2 \sum_{k=1}^{\infty} \left(U_{jk}\phi_n, \hat{\mathcal{R}}_k U_{jk}\phi_n\right)\,,$$

where $P_{j,n}$ is the projection onto the subspace spanned by $\phi_n \otimes \chi_j$ and $\hat{\mathcal{R}}_k$ is the shorthand for the reduced resolvent obtained by subtracting the pole term from $(h^V - \epsilon_0 + \nu_k - i0)^{-1}$. We are interested primarily in the imaginary part of $\epsilon_2(\lambda)$ which determines the resonance width in the leading order.

Notice first that the imaginary part of the last series is in fact a finite sum. We put $k(\epsilon_0) := \max\{k : \epsilon_0 - \nu_k > 0\}$; if the unperturbed eigenvalue is embedded we have $k(\epsilon_0) \geq 1$, otherwise the set is empty and we put $k(\epsilon_0) = 0$ by definition. It is clear that $\hat{\mathcal{R}}_k^* = \hat{\mathcal{R}}_k$ holds for $k > k(\epsilon_0)$, hence we have

$$\operatorname{Im} \epsilon_2(\lambda) = -\lambda^2 \sum_{k=1}^{k(\epsilon_0)} \left(U_{jk}\phi_n, (\operatorname{Im} \hat{\mathcal{R}}_k)\, U_{jk}\phi_n \right) .$$

To write the right-hand side explicitly we need to express $\operatorname{Im} \hat{\mathcal{R}}_k$. The imaginary part and the relation between the free and full resolvent can be rewritten using the resolvent identities; in this way we get for any $\epsilon > 0$ the formula

$$\operatorname{Im} (h^V - \epsilon - i0)^{-1} = \omega(\epsilon + i0)^* \operatorname{Im} (-\partial_x^2 - \epsilon - i0)^{-1} \omega(\epsilon + i0) \tag{2.15}$$

(Problem 8) in which $\omega(z) := \left(I + V(-\partial_x^2 - z)^{-1}\right)^{-1}$ is the inverse to

$$\omega^{-1}(z) : \left(\omega^{-1}(z)\phi\right)(x) = \phi(x) + \frac{iV(x)}{2\sqrt{z}} \int_{\mathbb{R}} \mathrm{e}^{i\sqrt{z}|x-x'|}\, \phi(x')\, \mathrm{d}x' .$$

By assumption *(i)* which ensures, in particular, that h^V has no positive eigenvalues, the operator $\omega(\epsilon + i0)$ is well defined. Furthermore, we have

$$\operatorname{Im} (-\partial_x^2 - \epsilon - i0)^{-1} = \frac{\pi}{2\sqrt{\epsilon}} \sum_{\sigma=\pm} \tau_\sigma(\epsilon)^* \tau_\sigma(\epsilon) \tag{2.16}$$

for any $\epsilon > 0$ where $\tau_\sigma(\epsilon) : H^1(\mathbb{R}) \to \mathbb{C}$ on the right-hand side is the trace map acting as $\tau_\sigma(\epsilon)\phi = \hat{\phi}(\sigma\sqrt{\epsilon})$ with $\hat{\phi}$ being the Fourier transform of ϕ (Problem 8). The above discussion can be summarized in the following way.

Theorem 2.4 *Assume (i), (ii). Moreover, let $\epsilon_0 = \mu_n + \nu_j$ be a simple eigenvalue of H_0 satisfying conditions (2.11). Then ϵ_0 is also a simple eigenvalue of the operator $H_{\theta,0}$ and a weak potential perturbation $\lambda U(\vec{x})$ in (2.10) moves it to $\epsilon(\lambda)$ with*

$$\operatorname{Im} \epsilon(\lambda) = -\frac{\lambda^2}{2} \sum_{k=1}^{k(\epsilon_0)} \sum_{\sigma=\pm} \frac{\pi}{\sqrt{\epsilon_0 - \nu_k}} \left| \tau_\sigma(\epsilon_0 - \nu_k)\, \omega(\epsilon_0 - \nu_k + i0)\, U_{jk}\phi_n \right|^2 + \mathcal{O}(\lambda^3) ,$$

as $\lambda \to 0$. If the second-order coefficient is nonzero, then $\epsilon(\lambda)$ describes a resonance of the operator H_λ.

Remarks 2.3.1 (a) The imaginary part given above is non-positive for small λ. It may happen, of course, that ϵ_0 persists as an eigenvalue. A trivial example is represented by a potential which preserves the symmetry, $U(\vec{x}) = U_1(x) + U_2(y)$ with suitable functions of which U_1 can be added to the potential V; notice that $k(\epsilon_0) < j$ so the diagonal elements of the matrix potential do not contribute. The leading coefficient may also accidentally vanish for potentials which do not decompose, however, then higher terms of the series may be nonzero.
(b) Notice that the ω introduced above is in fact a wave operator for the pair $(h^V, -\partial_x^2)$. It follows that the squared numbers in the above formula can be formally written as $|(\psi_{\pm\sqrt{\epsilon_0-\nu_k}}, U_{jk}\phi_k)|^2$, where ψ_m is the generalized eigenfunction of h^V with the momentum m. This shows that the leading term of the resonance width expansion is in this case given by ***Fermi's golden rule***.

2.4 Resonances in Thin Bent Strips

The symmetry violation is not the only mechanism which can give rise to resonances. Let us now return to one of our basic examples, a curved planar strip, and discuss it from the present point of view; the role of the perturbation parameter will be played by the strip width d. To explain the idea, we express the Hamiltonian H introduced in Sect. 1.1 in terms of the transverse modes, similarly as we did earlier in *Theorem* 1.6 where, however, we only singled out the lowest transverse mode, or in the previous section using the embedding operators $\mathcal{J}_j$ and their adjoints.

If the strip is asymptotically straight, i.e. the curvature decays fast enough, the spectrum of $-\partial_s^2 - \frac{1}{4}\gamma(s)^2$ consists of a continuous part which is the positive halfline and a nonempty family $\{\lambda_n\}$ of simple negative eigenvalues. Let us define the operator

$$H^0 := A - \partial_u^2, \qquad A := -\partial_s^2 + V^0, \quad V^0 = -\frac{1}{4}\gamma(s)^2 \tag{2.17}$$

acting on the Hilbert space $\mathcal{H} = L^2(\mathbb{R} \times (0, d), \mathrm{d}s\,\mathrm{d}u)$ with Dirichlet conditions at $u = 0, d$. Since the spectrum of the transverse part of H^0 is discrete with the eigenvalues $\nu_j = \kappa_j^2$, $j \in \mathbb{N}$, it is clear that for any $j \geq 2$ and d small enough the numbers $\lambda_n + \nu_j$ are eigenvalues embedded in the continuous spectrum of H^0. If we regard the original operator $-\Delta_D^\Omega$ as a result of perturbing H^0, as we did when discussing the discrete spectrum in Sect. 1.6, one expects that these embedded eigenvalues can turn into resonances provided we impose suitable analyticity assumptions on the curvature, for instance,

(i) γ extends to an analytic function, denoted by the same symbol, in $\mathcal{M}_{\alpha_0,\eta_0} := \{z \in \mathbb{C} : |\arg z| \leq \alpha_0 \text{ or } |\mathrm{Im}\, z| < \eta_0\}$ with $\alpha_0 < \pi/2$ and $\eta_0 > 0$,
(ii) to each $\alpha \in (0, \alpha_0)$ and $\eta \in (0, \eta_0)$ one can find positive $c_{\alpha,\eta}$ and δ such that the inequality $|\gamma(z)| < c_{\alpha,\eta}(1+|z|)^{-1-\delta}$ holds in $\mathcal{M}_{\alpha_0,\eta_0}$.

Proceeding as in the previous section, we can then derive an expansion for the resonance pole position and to estimate the first nonzero contribution to its imaginary part, i.e. the resonance width (cf. Problem 10 and the notes).

Theorem 2.5 *Let H be given by (1.7). Suppose that the strip Ω does not intersect itself and assumptions (i), (ii) are valid. Then for all sufficiently small widths d each eigenvalue $\lambda_n+\nu_j$ of H^0 with $j \geq 2$ gives rise to a resonance $\epsilon_{j,n}(d)$ of H the position of which is given by a convergent series,*

$$\epsilon_{j,n}(d) = \mu_n + \nu_j + \sum_{m=1}^{\infty} \epsilon_m^{(j,n)}(d) ,$$

where $\epsilon_m^{(j,n)}(d) = \mathcal{O}(d^m)$ as $d \to 0$. The first-order term is real-valued and the second-order term satisfies the estimate

$$0 \leq \operatorname{Im} \epsilon_2^{(j,n)}(d) \leq c_{\eta,j}\, \mathrm{e}^{-2\pi\eta\sqrt{2j-1}/d}$$

for any $\eta \in (0, \eta_0)$ and some positive $c_{\eta,j}$ depending on η and j.

The second claim of the theorem shows that $\epsilon_m^{(j,n)}(d)$ may tend to zero much faster than the $\mathcal{O}(d^m)$ rate which such a straightforward argument gives. It is not *a priori* clear whether the lowest order term are dominant as $d \to 0$, however, one can prove similar bounds on the total resonance width:

Theorem 2.6 *Suppose that the strip Ω does not intersect itself and assumptions (i), (ii) are valid. Then for any $\eta \in (0, \eta_0)$, $j \geq 2$, and $n = 1, \dots, N$ there exists a $c_{n,\eta} > 0$ such that*

$$0 \leq -\operatorname{Im} \epsilon_{j,n}(d) \leq c_{\eta,j}\, \mathrm{e}^{-2\pi\eta\sqrt{2j-1}/d} \tag{2.18}$$

holds for all d small enough.

Sketch of the proof: As in the previous case, to demonstrate *Theorem* 2.6 one has to treat the resonances of H as perturbations of a suitable operator with eigenvalues embedded in the continuous spectrum; we write therefore $H = H^0 + W$, where H^0 is given by (2.17) and W is the perturbation. The spectrum of the operator H^0 is of the form

$$\sigma(H^0) = \left\{\lambda + E \,:\, \lambda \in \sigma(A),\ E \in \sigma(-\partial_u^2)\right\},$$

where

$$\sigma(A) = \{\lambda_j\}_{j=1}^{N} \cup [0,\infty), \quad \sigma(-\partial_u^2) = \left\{\nu_j\right\}_{j=1}^{\infty},$$

with $\nu_j := \kappa_j^2$, $\kappa_j := \pi j/d$. Since Ω is not straight, $\gamma \neq 0$, the discrete spectrum of A is nonempty and the eigenvalues λ_j are simple. Moreover, from assumption (ii) it follows that their number N is finite. Then the eigenvalues

$$E_{j,n}^0 = \lambda_n + \nu_j$$

with $j \geq 2$ are embedded in the continuous spectrum of H^0 for d small enough and we expect them to give rise to resonances of the full operator H.

We pass to the unitary equivalent operator by performing the inverse Fourier transformation in the s variable, denoted by F_s^{-1}. We introduce

$$p := F_s^{-1} i\partial_s F_s\,, \qquad D := -i\partial_p = F_s^{-1} s F_s$$

and with a slight abuse of notation we shall employ the usual symbols for all other transformed operators,

$$H = p\, b(D,u) p - \partial_u^2 + V(D,u). \tag{2.19}$$

As in Sect. 2.3 above, we are going to use a complex scaling, this time an *exterior* one defined as

$$p_\theta(t) := \begin{cases} t & \text{if} \quad t \in \Omega_i := (-\omega, \omega) \\ \pm\omega + e^\theta (t \mp \omega) & \text{if} \quad t \in \Omega_e := \mathbb{R} \setminus \bar{\Omega}_i \end{cases} \tag{2.20}$$

where ω is a positive number to be determined later. First we consider $\theta \in \mathbb{R}$ and associate with a given closed operator T the family of operators

$$T_\theta := U_\theta T\, U_\theta^{-1}, \quad U_\theta \varphi := p_\theta'^{1/2} \varphi \circ p_\theta\,.$$

If the function $\theta \mapsto T_\theta$ has an analytic continuation to some strip $\{\theta \in \mathbb{C} : |\operatorname{Im}\theta| < \alpha\}$, we are able to define complex deformations of T. In particular, the family of complex deformations of the Hamiltonian (2.19) is given by

$$H_\theta = H_\theta^0 + W_\theta, \quad H_\theta^0 = A_\theta \otimes I + I \otimes (-\partial_u^2), \tag{2.21}$$

where

$$W_\theta = p_\theta (b-1)_\theta p_\theta + (V - V^0)_\theta.$$

Under the premises of the theorem the operators H_θ and H_θ^0 form self-adjoint analytic families of type (A). From now we set for simplicity $\theta = i\beta$ with $\beta > 0$. It is not hard to check that the spectrum of $H_{i\beta}^0$ equals

$$\sigma(H_{i\beta}^0) = \left\{\lambda + \nu_j \,:\, \lambda \in \left(\{\lambda_n\}_{n=1}^N \cup \varrho \cup \sigma(p_{i\beta}^2)\right),\ j = 1, 2, \dots\right\}, \tag{2.22}$$

where ϱ denotes the (possibly empty) set of resonances of the operators $A_{i\beta}$.

As before the resonances of H are identified with the complex eigenvalues of the non-selfadjoint operator $H_{i\beta}$, and their positions can be estimated with the help of the regular perturbation theory, where the role of the unperturbed operator is played by $H_{i\beta}^0$ and the perturbation is represented by $W_{i\beta}$. We choose a fixed eigenvalue $E_{n,j}^0 = \lambda_n + \nu_j$, $j \geq 2$, of $H_{i\beta}^0$ and define

$$\Gamma := \left\{z \in \mathbb{C} \,:\, |z - E_{n,j}^0| = r\right\}, \quad r = \frac{1}{2}\,\mathrm{dist}(\lambda_n, \sigma(A) \setminus \{\lambda_n\})$$

to be a circular contour around $E_{n,j}^0$ such that no other eigenvalue of $H_{i\beta}^0$ lies within Γ. It is convenient to use the transverse mode decomposition of $H_{i\beta}^0$,

$$H_{i\beta}^0 = \sum_{k\geq 1} \mathcal{J}_k H_{i\beta}^{0,k} \mathcal{J}_k^* \qquad \text{with} \quad H_{i\beta}^{0,k} = \mathcal{J}_k^* H_{i\beta}^0 \mathcal{J}_k \ \text{ in } \ L^2(\mathbb{R}, \mathrm{d}p)$$

where the $\mathcal{J}_k$'s are the natural embedding operators introduced in the previous section. Assume now that $E = E_{n,j}$ is the resonance arising from $E_{n,j}^0$ and that $\phi_{i\beta}$ is the associated eigenfunction,

$$H_{i\beta}\,\phi_{i\beta} = E\,\phi_{i\beta}\,. \tag{2.23}$$

This equation is equivalent to the system

$$\begin{aligned} \left(P_j H_{i\beta} P_j - P_j W_{i\beta} \hat{R}_{i\beta}^j(E) W_{i\beta} P_j\right)\,\phi_{i\beta} &= E\,P_j\phi_{i\beta}, \\ Q_j\,\phi_{i\beta} &= -\hat{R}_{i\beta}^j(E) W_{i\beta} P_j \phi_{i\beta}, \end{aligned}$$

where $P_j := \mathcal{J}_j\mathcal{J}_j^*$, $Q_j := I - P_j$, and $\hat{R}_{i\beta}^j(E) := Q_j\left(Q_j(H_{i\beta} - E)Q_j\right)^{-1} Q_j$. Moreover, it is easy to see that the first equation is further equivalent to

$$\left(H_{i\beta}^j - B_{i\beta}^j(E)\right)\,\phi_{i\beta}^j = E\,\phi_{i\beta}^j\,, \quad B_{i\beta}^j(E) := \mathcal{J}_j^* W_{i\beta} \hat{R}_{i\beta}^j(E) W_{i\beta} \mathcal{J}_j\,, \tag{2.24}$$

in $L^2(\mathbb{R})$, where $H_{i\beta}^j := P_j H_{i\beta} P_j$ and $\phi_{i\beta}^j := P_j\,\phi_{i\beta}$. Taking the imaginary part of equation (2.24) we get

$$\mathrm{Im}\, E\,\|\phi_{i\beta}^j\|^2 = (\mathrm{Im}\,(H_{i\beta}^j - B_{i\beta}^j(E))\phi_{i\beta}^j, \phi_{i\beta}^j). \tag{2.25}$$

Using next the identity $\mathrm{Im}\,(ABA) = 2\,\mathrm{Re}\,(\mathrm{Im}\,(A)BA) + A^*\mathrm{Im}\,(B)A$ in combination with the resolvent equation, we can write $\mathrm{Im}\,B_{i\beta}^j$ as

$$\operatorname{Im} B^j_{i\beta} = Z_{i\beta} + \operatorname{Im} E\, |\hat R^j_{i\beta} W_{i\beta} \mathcal{J}_j|^2$$
$$Z_{i\beta} := \mathcal{J}^*_j \left\{ 2\mathrm{Re}\left(\operatorname{Im} (W_{i\beta} \hat R^j_{i\beta} W_{i\beta} \right) - W^*_{i\beta} (\hat R^j_{i\beta})^* \operatorname{Im} (Q_j H_{i\beta} Q_j) \hat R^j_{i\beta} W_{i\beta} \right\} \mathcal{J}_j .$$

Inserting this into Eq. (2.25) we obtain

$$\operatorname{Im} E \left(\|\phi^j_{i\beta}\|^2 + \|\hat R^j_{i\beta} W_{i\beta} \mathcal{J}_j \phi^j_{i\beta}\|^2 \right) = \|P_j \phi_{i\beta}\|^2_{\mathcal{H}} + \|Q_j \phi_{i\beta}\|^2_{\mathcal{H}} = \|\phi_{i\beta}\|^2_{\mathcal{H}}$$

and since the eigenfunction $\phi_{i\beta}$ is supposed to be normalized, we arrive at

$$\operatorname{Im} E = ((\operatorname{Im} H^j_{i\beta} - Z_{i\beta})\phi^j_{i\beta}, \phi^j_{i\beta}) . \tag{2.26}$$

This equation will yield the desired bound (2.18); to this end we need a couple of definitions. We choose ω in the scaling relation (2.20) to be

$$\omega := \frac{\pi}{d} \sqrt{(2j-1)(1-\xi d)} ,$$

where ξ is a positive parameter. Moreover, we define the function

$$\rho(p) := \eta \int_{\min\{0,p\}}^{\max\{0,p\}} \chi_{\Omega_i \setminus \Omega_*}(t)\, \mathrm{d}t ,$$

where $\Omega_* = (-p_*, p_*)$ and p_* is a suitable positive constant independent of d. Then one can prove (see the notes) that there exists a number C_η such that

$$\left\| \langle p\rangle^{-1} e^{-\rho} (\operatorname{Im} H^j_{i\beta} - Z_{i\beta}) \langle p\rangle^{-1} e^{-\rho} \right\| \le C_\eta\, e^{-\rho(\omega)} , \tag{2.27}$$

where

$$\langle p\rangle = (p^2+\tau)^{1/2}, \quad \tau := \sup \left\{ \|e^\rho\, V^0_{i\beta}\| \,:\, |\beta| \le \alpha_0 \right\} .$$

Since $\operatorname{Im} E$ cannot be positive, insertion of (2.27) into equation (2.26) gives

$$0 \le -\operatorname{Im} E \le C_\eta\, e^{-\rho(\omega)} \left(\|p\, e^\rho\, \phi^j_{i\beta}\|^2 + \tau \|e^\rho\, \phi^j_{i\beta}\|^2 \right) . \tag{2.28}$$

At this point we have to take into account the exponential decay of the complex scaled resonance eigenvectors $\phi^j_{i\beta}$. Indeed, with our definition of the function ρ we have

$$\|p\, e^\rho\, \phi^j_{i\beta}\|^2 \le 2, \quad \|e^\rho\, \phi^j_{i\beta}\|^2 \le 2\, p_* ,$$

see again the notes for more details. Using now the fact that

$$e^{-2\rho(\omega)} = \exp\left\{-\frac{2\pi\eta}{d}\sqrt{2j-1}\,(1+\mathcal{O}(\xi d))\right\} \quad \text{as} \quad d \to 0$$

we get the upper bound (2.18) from (2.28). ∎

2.5 Notes

Section 2.1 The smooth perturbation method used in the proof of *Theorem* 2.1 is due to Kato [Ka66]—see, e.g., Sect. 8.7 of [RS], or [Sch], Chap. 10. The abstract result we refer to here is contained in Theorem 10.2.2 of [Sch].

Mourre's method naturally does not require the Hilbert space to be $L^2(\Omega_0)$. The idea to use positive commutators is based on an analogy with the classical Poison bracket of some coordinate q and the Hamiltonian H_{cl}. If one can show that for some trajectory $\{q, H_{\text{cl}}\} = \partial_t q \geq \delta > 0$, then the motion along this trajectory is extended in the coordinate q. The simplest application in quantum mechanics yields a criterium for the absence of eigenvalues. Indeed, if ψ is an eigenfunction of the Hamiltonian H, then by the virial theorem $(\psi, [i\Pi, H]\psi) = 0$ holds for any self-adjoint operator Π satisfying certain regularity properties. E. Mourre proved in [Mou81] that under yet stronger regularity assumptions the positivity of the commutator implies not only $\sigma_{\text{p}}(H) \cap I = \emptyset$ but also the absence of the entire singular spectrum of H in the interval I, that is, the version of *Theorem* 2.2 with $\alpha = 1$. For a proof and further generalizations see the monographs [ABG,CFKS]. *Theorem* 2.3 is taken from [KrT04], where the result is also extended to bent tubes in any dimensions provided δ is large enough. Mourre's method has also recently been applied to the analysis of scattering in twisted three-dimensional waveguides, see [BKR14] for details.

Section 2.2 It is a matter of convention whether we regard threshold states, i.e. those with $\nu_n^{(j)} = k^2$, as propagating modes in the definition of $N_j(k)$. The example discussed here comes from [EŠTV96], in a similar way one can treat scattering in a double waveguide separated by a leaky barrier of Sect. 1.5.2 (Problem 5). Many other examples of waveguide scattering treated by mode matching can be found in [LCM].

The relation between the conductance of a perturbed channel and the corresponding quantum mechanical scattering problem was first formulated by R. Landauer [La70], later extended by M. Büttiker [Bü88] to systems with an arbitrary finite number of outgoing channels. In practical applications one usually adds a factor of two which accounts for the spin states of the electron, in other words the right-hand side of (2.9) is multiplied in the standard units by $2e^2/h$. A rigorous derivation of the Landauer-Büttiker formula together with a bibliography can be found in [CJM05]. Let us add that such a description of transport contains two simplifying assumptions. First, it supposes that the potential difference between the heat baths connected by the waveguide is infinitesimally small—one usually speaks in this connection about *linear response theory* – and secondly, the transport occurs at temperature zero. More generally, the current flowing through the guide is expressed by the formula

$$I = \frac{2e^2}{h} \int_{\mathbb{R}} \left[f_\beta(k^2 - \mu_2) - f_\beta(k^2 - \mu_1) \right] |t(k)|^2 \mathrm{d}k^2 ,$$

where $f_\beta(\epsilon) = (\mathrm{e}^{\beta\epsilon} + 1)^{-1}$ is the Fermi-Dirac distribution function at temperature β^{-1}, μ_j are the chemical potentials in the reservoirs, and for simplicity we left out the factor describing the possibly different incoming and outgoing velocities; differentiating this expression and putting $\beta = \infty$ we get the conductance mentioned above.

Mode matching also offers other insights into the scattering process. Using the Ansatz (2.7) with the coefficients obtained by solving the matching conditions (2.8) we find what the generalized eigenvectors at energy k^2 look like. Then one can compute, in particular, the probability flow distribution $\vec{j}(\vec{x}) := -i\bar{\psi}(\vec{x})\vec{\nabla}\psi(\vec{x})$, for examples see again [EŠTV96], [EKr99], [LCM]. The flow patterns can reveal some features of the scattering, for example, a pronounced vortex suggests the existence of a resonance. On the other hand, vortices in transport of charged particles give rise to a nonzero magnetic moment which is in principle measurable [ESŠF98].

Section 2.3 The embedded eigenvalue in the crossed strips of Example 2.3.1b was noticed first in [SRW89]. On the other hand, the conclusion of Example 2.3.1c extends to a class of more general symmetric obstacles in *Neumann* waveguides —see [ELV94] and [DP98] where some conditions for the nonexistence of such eigenvalues were also derived. Resonances coming from mirror symmetry violations in strips with rectangular protrusions were investigated by mode matching in [AVD95], the analogous question for obstacle-induced eigenvalues in a Neumann waveguide was addressed in [APV00]. For an analysis of resonances coming from symmetry breaking associated with twisting of a three-dimensional waveguide we refer to [KS07].

The resonance system with the Hamiltonian (2.10) is a modification of Nöckel's model [Nö92] which will be discussed in Sect. 7.1.3; the material is taken from [DEM01]. Similar conclusions can be made if a hard-wall strip is replaced by a "soft" waveguide in which the confinement is due to a transverse potential (Problem 9). The most common definition of a resonance used here, in terms of poles of an analytically continued resolvent, is discussed in many places—see, e.g., Chap. 3 of [Ex] and the bibliography given there. Alternatively one can associate resonances, for instance, with poles of the analytically continued scattering matrix. Since the former definition expresses a property of the Hamiltonian alone while the latter concerns a pair of operators which we compare, it is clear that the objects they describe are in general different. On the other hand, it is true that for a "natural" choice of the free and full dynamics both types of resonances usually coincide, but this is a fact which one has to check it in each particular case.

The complex scaling method was formulated in the paper [AC71]. With several modifications and generalizations, cf. Chap. 8 of [CFKS], it developed into a powerful method for treating resonances in atomic and molecular systems—for a review with a bibliography see [Mo98]. The application of longitudinal complex scaling to resonances in waveguides was proposed in [DEŠ95]. For the definition and properties of analytic operator families see Chap. 7 of [Ka]. The "intrinsic" singularities coming from resonances of h^V do not accumulate in $\mathcal{M}_{\alpha_0}$; under the assumption *(i)* this follows from [AC71] or [Je78]. The method used here to evaluate the second-order coefficient in *Theorem* 2.4 is standard—see, e.g., Sect. 8.6 in [RS].

Section 2.4 *Theorem* 2.5 comes from [DEŠ95]. The bound on the imaginary part of the pole positions corresponds to the heuristic semiclassical picture – see [LL], Sect. 7.51—according to which the rate of exponential decay is proportional to

$$2\,\mathrm{Im}\int_0^{i\eta_0}\left(\sqrt{\epsilon-V_{0,j}(\zeta)}-\sqrt{\epsilon-V_{0,j-1}(\zeta)}\right)\mathrm{d}\zeta=\frac{2\pi\eta_0}{d}\sqrt{2j-1}+\mathcal{O}(d^0),$$

where $V_{0,j}=\frac{1}{4}\gamma^2+\nu_j$ and $\epsilon=\lambda_n+\nu_j+\mathcal{O}(d)$. *Theorem* 2.6 showing that the total resonance width has the same exponential bound as the lowest nontrivial term in the expansion of *Theorem* 2.5 comes from [DEM98], an analogous result was proved in [Ne97]. We refer to these papers for some technical statements made in the proof.

2.6 Problems

1. Fill in the details of the proof of *Theorem* 2.1.
Hint: Compute the integral $\int_0^\infty p^{2l}[(p^2+\mu)^2+\eta^2]^{-1}\mathrm{d}p$.

2. Modify *Theorem* 2.1 for the case when H refers to a bent tube in $\mathbb{R}^3$ satisfying Tang's condition (1.18) together with the other assumptions of Sect. 1.3.

3. Let H be the self-adjoint operator associated with quadratic form (1.24). Suppose that the potential V satisfies the assumptions of *Proposition* 1.4.1 and in addition, that $|V(\vec{x})|<c|x|^{-1-\varepsilon}$ holds if $|x|>x_0$ for some positive c, x_0, and ε. Then the wave operators $\Omega_\pm(H,H_0)$ exist, are complete, and $\sigma_{sc}(H)=\emptyset$.
Hint: Proceed as in the proof of *Theorem* 2.1.

4. Check the asymptotic completeness for the pair H, H_0 where H is associated with the form (1.26) and the function $\rho(\cdot)-1$ has a compact support.

5. Find by mode matching the on-shell S-matrix for double waveguides of Sect. 1.5.2.
Hint: Modify the argument of Sect. 2.2—cf. [EKr99].

6. Suppose that the protrusion in Example 2.3.1a is of rectangular shape, $g(x)=\frac{1}{2}d_1\in(\frac{1}{2}d,d)$ for $|x|\le\frac{1}{2}L$ and $g(x)=\frac{1}{2}d$ otherwise. Check that $-\Delta_D^\Omega$ has an embedded eigenvalue whenever $L>dd_1/\sqrt{d_1^2-d^2}$. Show that to a given $n\in\mathbb{N}$ one can find a protruded strip Ω such that $-\Delta_D^\Omega$ has at least n embedded eigenvalues.
Hint: Use bracketing estimates.

7. Prove that the crossed strips of Example 2.3.1b support an embedded eigenvalue.
Hint: Use the symmetry of the problem and *Proposition* 1.2.3.

8. Prove relations (2.15) and (2.16).
Hint: For the latter use the momentum representation.

9. The conclusions of Sect. 2.3 can be modified to the case of a potential confinement, i.e. for the operator $H_\lambda := -\Delta + V(x) + W(y) + \lambda U(\vec{x})$ on $L^2(\mathbb{R}^2)$, where U, V are similar as before and W satisfies, e.g., the inequality $W(y) \geq cy^2$ for some $c > 0$.

10. Prove *Theorem* 2.5.
Hint: To estimate the imaginary part of $\epsilon_2^{(j,n)}(d)$ use the analytic continuation of the group of shifts in the longitudinal variable—cf. [DEŠ95].

Chapter 3
More About the Waveguide Spectra

3.1 Spectral Estimates

We have seen in Sect. 1.6 that for thin bent tubes perturbation theory is rather efficient in determining the bound-state energies. If such a tube is thick or the mode coupling does not come from bending, we have to use other tools, some of them crude, others more sophisticated.

3.1.1 Simple Bounds

The first question concerns the number of bound states. If we are able to map Ω onto a straight tube as, for instance, in the case of bent ducts, it is possible to estimate the transformed operator by the one with decoupled variables and to employ known results about one-dimensional Schrödinger operators (see the notes). To describe how this can be done we define $g_\pm := (1 \pm a\|\gamma\|_\infty)^2$, where $a := \frac{1}{2}\,\mathrm{diam}\, M$ for $d = 3$, and introduce functions by which one can estimate the effective potentials (1.8) and (1.21), for convenience with the switched sign,

$$\tilde{W}_1(s) := \frac{\gamma(s)^2}{4g_-} + \frac{a|\ddot{\gamma}(s)|}{2g_-^{3/2}} + \frac{5a^2\dot{\gamma}(s)^2}{4g_-^2}$$

for $d = 2$, and an analogous expression for $d = 3$ in which the last two numerators are replaced by $a(|\ddot{\gamma}-\gamma\tau^2| + |2\dot{\gamma}\tau+\gamma\dot{\tau}|)$ and $5a^2(|\gamma\tau| + |\dot{\gamma}|)^2$, respectively. To deal with contributions from the higher modes, $j = 2, 3, \ldots$, we introduce

$$\tilde{W}_j(s) := \max\left\{0,\ \tilde{W}_1(s) - \nu_j + \nu_1\right\}.$$

P. Exner and H. Kovařík, *Quantum Waveguides*,
Theoretical and Mathematical Physics, DOI 10.1007/978-3-319-18576-7_3

It is clear that only a finite number of them is nonzero; the index j runs up to the largest integer for which $\nu_j < \|\tilde{W}\|_\infty + \nu_1$ so in thin tubes this part is missing at all. With these prerequisites we are able to formulate the result.

Proposition 3.1.1 *Adopt the hypotheses of Proposition 1.1.2 for $d = 2$ and of Problem 1.15 for $d = 3$, then $N(-\Delta_D^\Omega) \equiv N(-\Delta_D^\Omega, \nu_1) := \sharp\sigma_{\text{disc}}(-\Delta_D^\Omega)$ satisfies the inequality*

$$N(-\Delta_D^\Omega) \le 1 + g_+ \mathcal{I}(\tilde{W}_1) + \sum_{j=2}^\infty \frac{\sqrt{g_+}}{\sqrt{\nu_j - \nu_1}} \int_{\mathbb{R}} \tilde{W}_j(s)\, \mathrm{d}s\,,$$

$$\mathcal{I}(\tilde{W}_1) := \min\left\{ \int_{\mathbb{R}} |s|\, \tilde{W}_1(s)\, \mathrm{d}s\,,\; \frac{\int_{\mathbb{R}^2} \tilde{W}_1(s)|s-t|\, \tilde{W}_1(t)\, \mathrm{d}s\, \mathrm{d}t}{\int_{\mathbb{R}} \tilde{W}_1(s)\, \mathrm{d}s} \right\}.$$

Proof Replacing the curvature-induced effective potential V by $-\tilde{W}_1$, and the factors $(1+u\gamma)^{-2}$, $(1+r\gamma\cos(\theta-\alpha))^{-2}$ by g_+^{-1} for $d = 2, 3$, respectively, we get an estimate on $-\Delta_D^\Omega$ from below by operators with separated variables, which implies

$$-\Delta_D^\Omega - \nu_1 \ge \bigoplus_{j=1}^\infty \left(-\frac{1}{g_+} \frac{\mathrm{d}^2}{\mathrm{d}s^2} - \tilde{W}_j(s) \right).$$

Each operator in the orthogonal sum acts on $L^2(\mathbb{R})$. The quantity $N(-\Delta_D^\Omega)$ is equal to the number of negative eigenvalues of $-\Delta_D^\Omega - \kappa_1^2$, hence by the minimax principle we have

$$N(-\Delta_D^\Omega) \le N\left(-\frac{\mathrm{d}^2}{\mathrm{d}s^2} - g_+ \tilde{W}_1(s),\, 0 \right) + \sum_{j=2}^\infty N\left(-\frac{\mathrm{d}^2}{\mathrm{d}s^2} - g_+ \tilde{W}_j(s),\, 0 \right), \tag{3.1}$$

where $N(T, \tau)$ stands as usual for the number of eigenvalues of the operator T less than τ; we have used the fact that multiplying an operator by a positive constant does not change the number of its negative eigenvalues. The first term on the right-hand side of (3.1) can be estimated by a Bargmann-type inequality (see the notes),

$$N\left(-\frac{\mathrm{d}^2}{\mathrm{d}s^2} - g_+ \tilde{W}_1(s),\, 0 \right) \le 1 + g_+ \int_{\mathbb{R}} |s|\, \tilde{W}_1(s)\, \mathrm{d}s. \tag{3.2}$$

The remaining terms on the right-hand side of (3.1) can be controlled with the help of the Birman-Schwinger principle; the j-th term is bounded from above by

$$g_+ \operatorname{Tr}\left[\sqrt{\tilde{W}_1} \left(-\frac{\mathrm{d}^2}{\mathrm{d}s^2} + g_+(\nu_j - \nu_1) \right)^{-1} \sqrt{\tilde{W}_1} \right] = \frac{\sqrt{g_+}}{\sqrt{\nu_j - \nu_1}} \int_{\mathbb{R}} \tilde{W}_1(s)\, \mathrm{d}s\,.$$

Alternatively, the bound (3.2) can be replaced by the Birman-Schwinger estimates, as modified to the one-dimensional case by Seto, Klaus, and Newton (see the notes); this yields the second expression appearing in $\mathcal{I}(\tilde{W}_1)$. ∎

The Bargmann and Birman-Schwinger bounds are known for being inaccurate in the case of a strong coupling. In Schrödinger operator theory this is manifested by the power of the coupling constant which does not match the Weyl asymptotics. Such a simple approach cannot be used here, because the effective potentials (1.8) and (1.21) have a particular structure given by the geometry of Ω. The role of the coupling constant is played rather by the norm $\|\gamma\|_{L^1(\mathbb{R})}$ which in view of (1.4) is for $d = 2$ the total amount of bending taken in the absolute value, and has a similar meaning for $d = 3$; we will say that the tube Ω is ***strongly bent*** if $\int_{\mathbb{R}} |\gamma(s)|\, ds \gg 1$. The strong-coupling asymptotic regime means to compare tubes with a fixed cross section and a family of generating curves Γ_λ which are changing with respect to a parameter in such a way that $\|\gamma_\lambda\|_{L^1(\mathbb{R})} \to \infty$ as $\lambda \to \infty$. An example is provided by planar curves with a scaled curvature, $\gamma_\lambda(s) = \gamma(s/\lambda)$. Needless to say, one has to check in particular cases whether the corresponding family of tubes $\{\Omega_\lambda\}$ is free of self-intersections (unless we bypass the injectivity requirement as in Remark 1.1.1).

Proposition 3.1.2 *Let $\Omega_{\Gamma,M}$ be built over a strongly curved Γ. Then under the assumptions of the previous proposition, the asymptotic relation*

$$N(-\Delta_D^\Omega) \lesssim \frac{g_+}{\pi} \sum_{j=1}^{\infty} \int_{\mathbb{R}} \tilde{W}_j(s)^{1/2}\, ds$$

holds in the sense that for a family of Ω for which some of the integrals diverge the ratio of $N(-\Delta_D^\Omega)$ to the right-hand side is asymptotically bounded by one.

Proof It is sufficient to use the following simple estimate,

$$H - \nu_1 \ge -g_+^{-1}\partial_s^2 - \Delta_D^M - \tilde{W}_1 - \nu_1 \ge \bigoplus_{j=1}^{\infty} \left(-g_+^{-1}\partial_s^2 - \tilde{W}_j\right) \otimes I_M\,,$$

and to apply the standard Weyl asymptotics for one-dimensional Schrödinger operators [RS, Theorem. XIII.80] to its right-hand side in combination with the appropriate scaling in the longitudinal variable. ∎

The above claim is somewhat vague since we do not attempt to specify in general tube families with the said property. Nevertheless, it is clear that semiclassical bounds can be expected to hold in strongly bent tubes, in particular, in view of the inequality $\tilde{W}_1(s)^{1/2} \ge \frac{1}{2}\, g_-^{-1/2}|\gamma(s)|$. Moreover, in a similar way one can derive a rough *lower* semiclassical bound to the eigenvalue number.

Other bounds can be obtained without a coordinate transformation by a direct use of the bracketing technique. Its efficiency depends substantially on the geometry of the problem and our ability to analyze the spectrum of the Dirichlet Laplacians used in the estimates. Sometimes we can get in this way a lot of information, not only about the number of bound states, but also about the location of the eigenvalues in the interval below the first transverse eigenvalue determining the essential spectrum threshold. We have already seen several examples concerning, in particular, coupled waveguides in Sect. 1.5. However, the bracketing technique can also provide more general results. As an example, let us mention a rough lower bound for the ground state.

Theorem 3.1 *Let $\Omega \subset \mathbb{R}^d$, $d = 2, 3$, be a bent tube satisfying the hypotheses of Propositions 1.1.1 and 1.3.1, respectively. Suppose that $-\Delta_D^\Omega$ has exactly N eigenvalues, then ϵ_1, the lowest of them, is bounded from below as follows,*

$$\epsilon_1 \geq 3^{1-N} c_d \nu_1 \,,$$

where the constants in this estimate are expressed in terms of Bessel function zeros as $c_2 = \left(\frac{j_{0,1}}{j_{1,1}}\right)^2 \approx 0.394$ and $c_3 = \left(\frac{\pi}{j_{3/2,1}}\right)^2 \approx 0.489$.

Proof Adding Dirichlet boundaries at $s = \pm\ell$ we estimate $-\Delta_D^\Omega$ from above by the decoupled operator $H_\ell^{(-)} \oplus (-\Delta_D^{\Omega_\ell}) \oplus H_\ell^{(-)}$, where Ω_ℓ is the cut tube of the axis of which has length 2ℓ. The tail operators can by further estimated from below by switching from the Dirichlet to the Neumann condition at the cuts, so we know from the indicated propositions that for any $\varepsilon > 0$ one can achieve that $\inf \sigma(H_\ell^{(\pm)}) \geq \nu_1 - \varepsilon$ by choosing ℓ large enough. Consequently, the part of the spectrum in $(0, \nu_1)$ is in the limit $\ell \to \infty$ controlled by the middle part only. The latter has a purely discrete spectrum with the known domain dependence: the eigenvalues $\epsilon_n(\ell)$ are continuously decreasing as ℓ increases.

On the other hand, for any fixed ℓ the eigenvalues satisfy bounds on their ratios. A result of Ashbaugh and Benguria which concluded a long array of papers starting from the PPW-conjecture (see the notes) tells us that $\epsilon_2(\ell)/\epsilon_1(\ell)$ in dimension d is bounded by the same ratio for the ball in $\mathbb{R}^d$, which is $(j_{1,1}/j_{0,1})^2$ for $d = 2$ and $(j_{3/2,1}/\pi)^2$ for $d = 3$. Similar bounds hold for $\epsilon_{n+1}(\ell)/\epsilon_1(\ell)$ in which case the factor 3^{n-1} has to be added. Being valid for any ℓ the bounds are naturally preserved in the limit $\ell \to \infty$. Thus if $-\Delta_D^\Omega$ has N isolated eigenvalues, then in view of the monotonicity the numbers $\epsilon_n(\ell)$, $n \leq N$, tend to them, while $\lim_{\ell\to\infty} \epsilon_{N+1}(\ell) = \nu_1$ which yields the result. ∎

Remarks 3.1.1 (a) For simplicity we have formulated the result for bent tubes. It is clear that it applies to polygonal ducts, and more generally, to any Ω having the form of a compact set from which a finite number of straight or asymptotically straight channels are emanating. It is common for universal bounds like this one that they are not very precise. Recall the L-shaped strip where the eigenvalue $\approx 0.93\kappa_1^2$ is much closer to the continuum than to $c_2\kappa_1^2$. However, in the cross-shaped guide of Problem 1.26 there is a single eigenvalue $\approx 0.66\kappa_1^2$—and imagine how far the cross is from the circle of radius $a \approx 0.77d$!

(b) If there are more than a few eigenvalues the bound becomes useless. On the other hand, if we have a waveguide system depending on a parameter which controls the discrete spectrum, the result tells us how low the ground state can go maximally before the next eigenvalue emerges from the continuum. For instance in the symmetric case of laterally coupled waveguides of Sect. 1.5.1 we have $\epsilon_1(a_2) \approx 0.51\kappa_1^2$ which is not that far from $c_2\kappa_1^2$.

Let us finally mention a simple lower bound to the spectrum of a bent strip, which is of a local nature. Consider a circular annulus $\mathcal{A} \equiv \mathcal{A}(r,a)$ with outer and inner radius $r_\pm := r \pm a$, respectively, and denote by $k(r,a)^2$ the ground state eigenvalue of the Dirichlet Laplacian on $\mathcal{A}$,

$$k(r,a)^2 = \inf\left\{ \|\nabla\psi\|^2 : \psi \in C_0^\infty(\mathcal{A}),\ \|\psi\| = 1 \right\}.$$

Given a bounded function $\gamma : \mathbb{R} \to \mathbb{R}$ we define

$$\kappa_1[\gamma,a] := \min\left\{\kappa_1,\ \inf\{k(|\gamma(s)|^{-1},a) :\ \gamma(s) \neq 0\}\right\},$$

where $\kappa_1 := \pi/2a$ as usual. Then we have the following result.

Proposition 3.1.3 *Suppose that assumption (i) of Sect. 1.1 is satisfied, and that the curvature γ is bounded and such that $a\|\gamma\|_\infty < 1$. Then*

$$\inf\sigma(-\Delta_D^\Omega) \geq \kappa_1[\gamma,a]^2 > \left(\frac{j_{0,1}}{2a}\right)^2.$$

Proof $k(r,a)$ can be found using polar coordinates. The radial part of the operator is unitarily equivalent to $-\partial_u^2 - \frac{1}{4}(r+u)^{-2}$ with the Dirichlet boundary conditions on $L^2(r-a, r+a)$, which shows that $k(\cdot,a)$ is monotonously increasing between the values $j_{0,1}/2a$ for $r \to a-$ and κ_1 for $r \to \infty$. Neglecting the non-negative longitudinal term, we can estimate the quadratic form of $-\Delta_D^\Omega$ by

$$\begin{aligned} t[\psi] &= \|\tilde{H}^{1/2}\psi\|_g^2 \geq \int_{\mathbb{R}} \mathrm{d}s \int_{-a}^{a} du\, (1+u\gamma(s))|\partial_u\psi(s,u)|^2 \\ &\geq \int_{\mathbb{R}} \mathrm{d}s\, k(|\gamma(s)|^{-1},a)^2 \int_{-a}^{a} \mathrm{d}u\, (1+u\gamma(s))|\psi(s,u)|^2 \\ &\geq \kappa_1[\gamma,a]^2 \int_{\mathbb{R}} \mathrm{d}s \int_{-a}^{a} du\, (1+u\gamma(s))|\psi(s,u)|^2 = \kappa_1[\gamma,a]^2\|\psi\|_g^2 \end{aligned}$$

for any ψ from the form domain of $\tilde{H}$, where the second inequality follows from the definition of $k(r,a)$ written explicitly in polar coordinates and the third one from its monotonicity with respect to the radius. ∎

Remark 3.1.2 The lower bound thus corresponds to the most curved part of the strip. Being local it does not require the curvature γ to decay, so it is valid irrespective of the character which the spectrum has at its bottom. The worst bound given by *Proposition* 3.1.3 is $\approx 0.586\kappa_1^2$, hence for bent strips the present method yields a result better than *Theorem* 3.1. Using the quadratic form (1.19) one can find in the same way, e.g., a lower bound for a bent circular tube in $\mathbb{R}^3$ through the ground state eigenvalue of the Dirichlet Laplacian in a toroidal region (see the notes).

3.1.2 Lieb-Thirring Inequalities

Consider now Schrödinger operators in a straight tube Ω_0 of the form (1.24) with a potential the regularity properties of which will be specified below. From the start, however, we fix its sign (we use here the symbols $V_\pm := \frac{1}{2}(|V| \pm V)$ for the positive and negative parts; elsewhere the subscripts may have a different meaning). Our aim is to find upper bounds on moments of eigenvalues relative to the threshold of the essential spectrum, and since the inequality $t_V \geq t_{-V_-}$ implies the same relation between the respective eigenvalues by the minimax principle, we shall restrict ourselves to potentials which are non-positive. Furthermore, since we want to compare the behavior at different coupling strengths it is useful to introduce a coupling constant, i.e. to study the operators

$$H_{\lambda V}^{\Omega_0} := -\Delta_D^{\Omega_0} - \lambda V \quad \text{with} \quad V \geq 0\,,\ \lambda \geq 0\,. \tag{3.3}$$

Propositions 1.4.1 and 1.4.2 specify conditions under which such an operator has a nontrivial discrete spectrum; due to the sign definiteness the second one requires just that V is nonzero. Thus let $\{\epsilon_n(\lambda)\}$ be the sequence of these eigenvalues (counting multiplicities) arranged in the ascending order. We seek bounds on the eigenvalue moments $S_\delta(\lambda) \equiv S_\delta^{\Omega_0}(\lambda V)$ defined as

$$S_\delta(\lambda) := \mathrm{tr}\left(H_{\lambda V}^{\Omega_0} - \nu_1\right)_-^\delta = \sum_n (\nu_1 - \epsilon_n(\lambda))^\delta\,, \quad \delta \geq 0\,. \tag{3.4}$$

Before we address this question, let us recall briefly the usual Lieb-Thirring inequalities for Schrödinger operators $H_\lambda = -\Delta - \lambda V$ on $L^2(\mathbb{R}^d)$ with λ, V of the same signs as above—for more information see the notes. The moments $S_\delta(\lambda) := \mathrm{tr}(H_\lambda)_-^\delta$ are estimated by means of the phase-space quantity

$$S_{\delta,d}^{\mathrm{cl}}(\lambda) := \int_{\mathbb{R}^{2d}} (|\xi|^2 - \lambda V(x))_-^\delta \, \frac{\mathrm{d}\xi \mathrm{d}x}{(2\pi)^d} = \lambda^{\delta+\frac{d}{2}} L_{\delta,d}^{\mathrm{cl}} \int_{\mathbb{R}^d} V(x)^{\delta+\frac{d}{2}}\, \mathrm{d}x\,,$$

where $L^{\rm cl}_{\delta,d} := \Gamma(\delta+1)\left[2^d\pi^{d/2}\Gamma\left(\delta+\frac{d}{2}+1\right)\right]^{-1}$. The power δ for which the estimate is possible is dimension dependent: assuming that $\delta\geq 1/2$ if $d=1$, $\delta>0$ if $d=2$, and $\delta\geq 0$ if $d\geq 3$, the ***Lieb-Thirring inequality***

$$S_\delta(\lambda)\leq R(\delta,d)S^{\rm cl}_{\delta,d}(\lambda) \tag{3.5}$$

holds with an $R(\delta,d)\geq 1$, while for other pairs (δ,d) the estimate (3.5) fails.

If we try to find an analogous inequality for the operators (3.3) we have to realize that the problem has mixed dimensionality and in the weak-coupling case its one-dimensional character dominates; we have already remarked upon this in Sect. 1.4 and we will discuss this problem again in detail in Sect. 6.1.

Theorem 3.2 *Let $V\in L^{\delta+\frac{d}{2}}(\Omega_0)\cap L^{\delta+\frac{1}{2}}\left(\mathbb{R},\mathrm{d}x;L^2(M,\chi_1(y)^2\mathrm{d}y)\right)$, then for any $\delta\geq 1/2$, $\lambda\geq 0$ and all $\varepsilon>0$ the quantity (3.4) satisfies the bound*

$$S_\delta(\lambda)\leq c_1\lambda^{\delta+\frac{1}{2}}\int_{\mathbb{R}}\left(\int_M V(x,y)\,\chi_1(y)^2\,\mathrm{d}y\right)^{\delta+\frac{1}{2}}\mathrm{d}x+c_2S^{\rm cl}_{\delta,d}(\lambda) \tag{3.6}$$

with the constants

$$c_1\leq(1+\varepsilon)^{\delta+\frac{1}{2}}\,r(\delta,1)\,L^{\rm cl}_{\delta,1}\,,$$
$$c_2\leq(1+\varepsilon^{-1})^{\delta+\frac{d}{2}}\left(\frac{\nu_2}{\nu_2-\nu_1}\right)^{\frac{d-1}{d}}r(\delta,1)\,R(\delta+\frac{1}{2},d-1)\,,$$

where $R(\delta+\frac{1}{2},d-1)$ and $r(\delta,1)$ are the constant involved in (3.5) and its operator generalizations described in the notes.

Proof The idea is to reduce the problem to integrals of spectral estimates of the operator

$$W_M(\lambda;x):=-\Delta^M_d-\lambda V(x,\cdot)$$

on $L^2(M)$. To this end, we evaluate the quadratic form of $H^{\Omega_0}_{\lambda V}$ on its form core $C_0^\infty(\Omega_0)$, further we employ Fubini's theorem and an easy estimate concerning the negative part. This yields

$$S_\delta(\lambda)\leq\mathrm{tr}_{L^2(\mathbb{R})\otimes L^2(M)}\left(-\partial_x^2\otimes I_M-(W_M(\lambda;x)-\nu_1)_-\right)_-^\delta\,,$$

where I_M is the identity operator on $L^2(M)$. Now the operator version of the Lieb-Thirring inequality mentioned in the notes in which we set $d=1$, $\mathcal{G}=L^2(M)$, and $W(x)=(W_M(\lambda;x)-\nu_1)_-$ gives

$$S_\delta(\lambda)\leq r(\delta,1)\,L^{\rm cl}_{\delta,1}\int_{\mathbb{R}}\mathrm{tr}(W_M(\lambda;x)-\nu_1)_-^{\delta+\frac{1}{2}}\,\mathrm{d}x\,, \tag{3.7}$$

hence it remains to estimate the integrated function. Let $P = (\chi_1, \cdot)\chi_1$ be the projection onto the subspace spanned by χ_1 and let $Q := I_M - P$. Since $V \geq 0$ by assumption, one has $(Q-\varepsilon P)V(Q-\varepsilon P) \geq 0$, and therefore

$$W_M(\lambda; x) \geq PW_M((1+\varepsilon)\lambda; x)P + QW_M((1-\varepsilon^{-1})\lambda; x)Q\,.$$

By definition $P(-\Delta_D^M+\nu_1)P = 0$, so we get for any $\varepsilon > 0$ the estimate

$$W_M(\lambda; x) \geq \hat{W}_M((1+\varepsilon)\lambda; x) + \tilde{W}_M((1+\varepsilon^{-1})\lambda; x)\,,$$

where $\hat{W}_M(\lambda; x) := -\lambda PV(x, \cdot)P$ and $\tilde{W}_M(\lambda; x) := Q(W_M(\lambda; x)-\nu_1)Q$. Since these operators act in the ranges of the projections P and Q, respectively, we can estimate the expression $\mathrm{tr}(W_M(\lambda; x)-\nu_1)_-^{\delta+\frac{1}{2}}$ from above by

$$\mathrm{tr}(\hat{W}_M((1+\varepsilon)\lambda; x)-\nu_1)_-^{\delta+\frac{1}{2}} + \mathrm{tr}(\tilde{W}_M((1+\varepsilon^{-1})\lambda; x)-\nu_1)_-^{\delta+\frac{1}{2}}\,. \tag{3.8}$$

Since P is a one-dimensional projection and $V \geq 0$, the part containing $\hat{W}_M$ gives rise to the first term on the right-hand side of (3.6). In the second part we use the inequalities $\nu_n - \nu_1 \geq \nu_2^{-1}(\nu_2-\nu_1)\nu_n$ valid for $n \geq 2$, which give

$$Q(-\Delta_d^M - \nu_1)Q = \sum_{n=2}^{\infty}(\nu_2-\nu_1)\chi_n\,(\chi_n, \cdot)_{L^2(M)} \geq \frac{\nu_2-\nu_1}{\nu_2}\,Q(-\Delta_d^M)Q\,.$$

Thus denoting $\tau := \nu_2^{-1}(\nu_2-\nu_1)$ and $\rho := \tau^{-1}\lambda(1+\varepsilon^{-1})$, we are able to estimate the second term in the expression (3.8) from above by

$$\mathrm{tr}(Q(-\tau\Delta_D^M - \tau\rho V(x, \cdot))Q)_-^{\delta+\frac{1}{2}} \leq \tau^{\delta+\frac{1}{2}}\mathrm{tr}(QW(\rho; x)Q)_-^{\delta+\frac{1}{2}}\,,$$

which is further bounded by $\tau^{\delta+\frac{1}{2}}\mathrm{tr}(W(\rho; x))_-^{\delta+\frac{1}{2}}$. By a bracketing argument the negative eigenvalues of the operator $W(\rho; x)$ on $L^2(M)$ are estimated from below by the respective eigenvalues of the operator $H_\rho(x) := -\Delta - \rho V(x, \cdot)$ on $L^2(\mathbb{R}^{d-1})$, to which we may apply the inequality (3.5) with $\delta' = \delta + \frac{1}{2}$ and $d' = d-1$. In view of this the second term in (3.8) is bounded by

$$\tau^{\delta'} S_{\delta'}(\lambda) = R(\delta', d')\, L_{\delta',d'}^{\mathrm{cl}}\, \tau^{\delta'}\, \rho^{\delta+\frac{1}{2}} \int_M V(x, y)^{\delta+\frac{d}{2}}\, \mathrm{d}y\,;$$

we made use here of the identity $\delta + \frac{d}{2} = \delta' + \frac{d'}{2}$. It remains to insert the last expression into (3.7) and to employ another identity, $L_{\delta',d'}^{\mathrm{cl}} L_{\delta,1}^{\mathrm{cl}} = L_{\delta,d}^{\mathrm{cl}}$, to get the sought inequality (3.6). ∎

The obtained result can be used, in particular, to derive Lieb-Thirring-type bounds for bent tubes. Again let $g_+ := (1+a\|\gamma\|_\infty)^2$ and define the eigenvalue moments $S_\delta(\Omega) \equiv S_\delta(\Omega_{\Gamma,M}) := \mathrm{tr}\left(-\Delta_D^\Omega - \nu_1\right)_-^\delta$ in analogy with (3.4). To unify the notation for $d=2,3$ we write both the effective potentials (1.8) and (1.21) as $V(s,\omega)$, where $\omega = u$ and $\omega = (r,\theta)$, respectively, and the transverse volume element is denoted by $\mathrm{d}\omega$.

Corollary 3.1.1 *Under the assumptions of Proposition 3.1.1, the following bound,*

$$S_\delta(\Omega) \le c_1 g_+ \int_{\mathbb{R}} \left(\int_M \chi_1(\omega)^2\, V(s,\omega)_-\, \mathrm{d}\omega \right)^{\delta+\frac{1}{2}} \mathrm{d}s + c_2 g_+ S_{\delta,d}^{\mathrm{cl}}(\Omega)\,, \tag{3.9}$$

holds, where c_1, c_2 are constants of Theorem 3.2 and

$$S_{\delta,d}^{\mathrm{cl}}(\Omega) := L_{\delta,d}^{\mathrm{cl}} \int_\Omega V(s,\omega)_-^{\delta+\frac{d}{2}}\, \mathrm{d}s\mathrm{d}\omega\,.$$

Proof The bound $H \ge -g_+^{-1}\partial_s^2 - \Delta_D^M - V(s,\omega)_-$ together with the scaling in the longitudinal variable gives the result. ∎

Notice that if we use a rougher bound by the potential $-\tilde{W}_1$ independent of the transverse variables which we have introduced above, then the first term in the inequality (3.9) will simplify to $c_1 g_+ \int_{\mathbb{R}} \tilde{W}_1(s)^{\delta+\frac{1}{2}}\, \mathrm{d}s$.

3.1.3 The Number of Eigenvalues in Twisted Waveguides

Let us look next at a related problem, this time concerning the interplay between a twisting and a potential in a straight tube. A particular case of the inequalities (3.5) gives a bound on the number of negative eigenvalues of a Schrödinger operator $-\Delta - V$ on $L^2(\mathbb{R}^d)$ with $d \ge 3$, known as the ***Cwickel-Lieb-Rosenbljum inequality***,

$$N(-\Delta - V) \le C_d \int_{\mathbb{R}^d} V_+^{d/2}(x)\, \mathrm{d}x\,, \quad d \ge 3\,, \tag{3.10}$$

for some $C_d = R(0,d)$. The bound does not hold for $d = 1,2$ and the same is true for the Laplacian in a straight tube: since the operator $H_{\lambda V}^{\Omega_0}$ has weakly coupled bound states—cf. *Theorem* 6.1 below—a restriction to $N(H_V^{\Omega_0}, \nu_1)$ similar to (3.10) must obviously fail.

On the other hand, we know from Sect. 1.7.1 that a local twisting of a straight tube $\Omega_0 = \mathbb{R} \times M \subset \mathbb{R}^3$ prevents to a certain extent the existence of weakly coupled bound states, and we are going to show that as a consequence, a bound analogous to (3.10) does hold in a locally twisted waveguide. We use the notation of Sect. 1.7.1 denoting by Ω_α the tube which results from the twisting of Ω_0. Correspondingly,

$H_V^{\Omega_\alpha}$ is the Schrödinger operator on $L^2(\Omega_\alpha)$ with a potential $-V$. To avoid confusion, we will denote in this section by $\vec{x}$ the position variable in the straight tube Ω_0 and by boldface $\vec{\boldsymbol{x}}$ the variable in the twisted tube Ω_α. With this notation, we can make the following claim.

Theorem 3.3 *Let α and M satisfy the assumptions of Theorem 1.7. Then there is a constant C_α such that for any $0 \le V \in L^{3/2}(\Omega_\alpha, (1+x_3^2)\,\mathrm{d}\vec{\boldsymbol{x}})$ we have*

$$N(H_V^{\Omega_\alpha}, \nu_1) \le C_\alpha \int_{\Omega_\alpha} V(\vec{\boldsymbol{x}})^{3/2} (1+x_3^2)\,\mathrm{d}\vec{\boldsymbol{x}}\,. \tag{3.11}$$

The proof requires an auxiliary result using the notation introduced in Sect. 1.7.1.

Lemma 3.1.3 *Let $U : L^2(\Omega_\alpha) \to L^2(\Omega_0)$ be the unitary map (1.47). Then there is a constant $c > 0$ such that*

$$U(-\Delta_D^{\Omega_\alpha})U^{-1} - \nu_1 \ge c\,(-\Delta_D^{\Omega_0} - \nu_1)$$

holds true in the sense of quadratic forms in $L^2(\Omega_0)$.

Proof Let u be a test function and write $u = f\psi_1$ with ψ_1 being the ground state eigenfunction of $-\Delta_D^M$. By *Lemma* 1.7.2 it follows that for any $a < 1$ there exists a constant c_a such that

$$|T_{2,3}[f]| \le c_a\, T_1[f] + a\, T_2[f] + T_3[f]\,. \tag{3.12}$$

Combining this estimate with the fact that $T_2[f] + T_3[f] - |T_{2,3}[f]| \ge 0$ we obtain

$$\begin{aligned} T_1[f] + T_2[f] + T_3[f] + T_{2,3}[f] &\ge \frac{1}{2}\, T_1[f] + \left(1 - \frac{a}{2\,c_a}\right) T_2[f] \\ + \left(1 - \frac{1}{2\,c_a}\right) T_3[f] - \left(1 - \frac{1}{2\,c_a}\right) |T_{2,3}[f]| &\ge \frac{1}{2}\, T_1[f] + \frac{1-a}{2\,c_a}\, T_2[f] \\ \ge \min\left\{\frac{1}{2}, \frac{1-a}{2\,c_a}\right\} \int_{I\times M} \left(|\nabla f|^2 \psi_1^2\right)(\vec{x})\,\mathrm{d}\vec{x}\,. \end{aligned}$$

Consequently, $\left(u, \left(U(-\Delta_D^{\Omega_\alpha})U^{-1} - \nu_1\right)u\right)$ is bounded from below by

$$c \int_{\Omega_0} \left(|\nabla f|^2 \psi_1^2\right)(\vec{x})\,\mathrm{d}\vec{x} = c \int_{\Omega_0} \left(|\nabla u|^2 - \nu_1 u^2\right)(\vec{x})\,\mathrm{d}\vec{x}\,;$$

this concludes the proof. ∎

Proof of Theorem 3.3 We again employ the notation of Sect. 1.7.1 and define $\hat{V}(\vec{x}) = V(r_\alpha(x_3)\vec{x})$ for any $\vec{x} \in \Omega_0$. In view of *Proposition* 1.7.1 it follows that $U(-\Delta_D^{\Omega_\alpha})U^{-1} - \nu_1 \ge c\,\dot{\alpha}^2$ holds in the sense of quadratic forms in $L^2(\Omega_0)$, thus

$$\begin{aligned} N(H_V^{\Omega_\alpha}, \nu_1) &= N(U(-\Delta_D^{\Omega_\alpha})U^{-1} - \hat{V} - \nu_1,\, 0) \\ &\le N\Big(\frac{1}{2}(U(-\Delta_D^{\Omega_\alpha})U^{-1} - \nu_1) + c_1\dot{\alpha}^2 - \hat{V}\Big) \end{aligned} \tag{3.13}$$

for some constant $c_1 > 0$; as usual we write $N(T, 0) = N(T)$. Since multiplication of an operator by a positive constant does not change the number of its negative eigenvalues, we get from (3.13) and *Lemma* 1.7.3 the estimate

$$N(H_V^{\Omega_\alpha}, \nu_1) \le N(A - c\hat{V})\,, \tag{3.14}$$

where $A = -\Delta_D^{\Omega_0} - \nu_1 + c_1\dot{\alpha}^2(x_3)$. Its heat kernel $e^{-\tau A}(\vec{x}, \vec{x}')$ can be by separation of variables expressed as

$$e^{-\tau A}(\vec{x}, \vec{x}') = \sum_{j=1}^{\infty} e^{\tau(\nu_1 - \nu_j)}\, \psi_j(\vec{x}_t)\, \psi_j(\vec{x}_t')\, q(t, x_3, x_3')\,,$$

where ψ_j are normalized real-valued eigenfunctions of $-\Delta_D^M$ with eigenvalues ν_j and $q(\tau, \cdot, \cdot)$ is the heat kernel of the one-dimensional Schrödinger operator $-\frac{\mathrm{d}^2}{\mathrm{d}r^2} + c_1\dot{\alpha}^2(r)$ on $L^2(\mathbb{R})$. We use its decay properties: by [GS09, Sect. 8] there is a constant c_2 such that

$$q(t, r, r) \le \frac{c_2\,(1 + r^2)}{t^{3/2}} \quad \text{if } t \ge 1\,, \quad q(t, r, r) \le \frac{c_2}{\sqrt{t}} \quad \text{if } 0 < t < 1\,.$$

On the other hand, to estimate the transverse part we note that by [Da, Corollary 4.6.3] there is a $c_3 > 0$ such that $|\psi_j(x_t)| \le c\,\nu_j\,\psi_1(x_t)$ holds for all $x_t \in M$. Since the cross section M is two-dimensional, we have $\nu_j = \mathcal{O}(j)$ as $j \to \infty$, and

$$\sum_{j=1}^{\infty} e^{t(\nu_1 - \nu_j)}\, \psi_j^2(x_t) \le c_4\, \psi_1^2(x_t) \quad \text{for} \quad t \ge 1\,.$$

Finally, by [Da, Theorem 2.4.4] the same expression can be estimated by $c_4 t^{-1}$ in the case $0 < t < 1$; combining all the above estimates we get

$$e^{-tA}(\vec{x}, \vec{x}) \le c_5\,(1 + x_3^2)\, t^{-3/2}. \tag{3.15}$$

Next we employ Lieb's inequality (see the notes) which says that

$$N(A - c\,\hat{V}) \le C \int_{\Omega_0} \int_0^\infty e^{-\tau A}(\vec{x}, \vec{x})\, \tau^{-1}\, (\tau\, \hat{V}(\vec{x}) - 1)_+\, \mathrm{d}\tau\, \mathrm{d}\vec{x}\,;$$

inserting the estimate (3.15) into it we get after a simple calculation

$$N(A - c\,\hat{V}) \le C \int_{\Omega_0} \hat{V}(\vec{x})^{3/2}\,(1 + x_3^2)\,\mathrm{d}\vec{x} = C \int_{\Omega_\alpha} V(\vec{x})^{3/2}\,(1 + x_3^2)\,\mathrm{d}\vec{x}\,,$$

which in view of (3.14) completes the proof. ∎

3.2 Related Results

3.2.1 Combined Boundary Conditions

We have already discussed an example in which the Dirichlet boundary condition in a finite part of the boundary is changed to the Neumann condition. Now we want to consider situations where such a Neumann boundary segment is of infinite length, thus changing the essential spectrum.

Let $\Omega \equiv \Omega_{\Gamma,a}$ be the curved strip of Sect. 1.1 generated by a curve Γ. Consider the operator $-\Delta_{DN}^{\Omega}$ which acts as Laplacian with Dirichlet and Neumann condition at the $\Gamma_{\mp a}$ parts of the boundary referring to $u = \mp a$, respectively. Using the curvilinear coordinates we can replace $-\Delta_{DN}^{\Omega}$ in analogy with (1.6) by the unitarily equivalent operator H on $L^2(\Omega_0, g^{1/2}\mathrm{d}s\,\mathrm{d}u)$ which is associated with the following quadratic form,

$$t[\psi] := \|g^{-1/4}\partial_s\psi\|^2 + \|g^{1/4}\partial_u\psi\|^2\,,$$

the domain of which is $\{\psi \in H^1(\Omega_0) : \psi(s,-a) = 0 \text{ for a.a. } s \in \mathbb{R}\}$. For a straight strip the spectrum is $[\kappa_1^2, \infty)$, with $\kappa_1 := \pi/4a$ in this case.

It appears, however, that the spectral properties of such a non-symmetric waveguide depend crucially on the sign of γ. Specifically, a discrete spectrum due to a bend of Ω exists only if the Neumann boundary is at its *outer* edge.

Theorem 3.4 *Adopt assumptions (i), (ii)$_1$, and (iii)$_1$ of Sect. 1.1. In addition, assume that Γ is not a straight line and $a\|\gamma\|_\infty < 1$.*
(a) $\inf\sigma(H) < \kappa_1^2$ holds if the total bending $\beta_\Gamma \ge 0$.
(b) On the other hand, $\gamma(s) \le 0$ for all $s \in \mathbb{R}$ implies $\inf\sigma(H) \ge \kappa_1^2$.

Corollary 3.2.1 *Assume (i) and (ii)$_1$ together with $a\|\gamma\|_\infty < 1$. The discrete spectrum of H is non-empty if γ has a compact support and $\beta_\Gamma > 0$.*

Proof Since Ω is straight outside a compact region, $\inf\sigma_{\mathrm{ess}}(H) = [\kappa_1^2, \infty)$, thus the result follows from part (a) of the theorem. ∎

Proof of Theorem 3.4 As is the case for pure Dirichlet boundary we are looking for ψ from the form domain of H such that

$$q[\psi] := t[\psi] - \kappa_1^2 \|g^{1/4}\psi\|^2 < 0\,.$$

We choose $\psi = \phi_\lambda \psi_1 + \varepsilon f$, where ϕ_λ is the function defined by (1.11) and $f(s,u) = \jmath(s)\psi_1(u)$ with $\jmath$ being a smooth function supported in $(-s_0, s_0)$. Using the explicit form of the lowest transverse mode, $\psi_1(u) = a^{-1/2} \sin \kappa_1(u+a)$, we get by a straightforward computation

$$\begin{aligned} q[\phi_\lambda \psi_1 + \varepsilon f] \le \frac{\lambda \|\dot{\phi}\|^2}{1 - a\|\gamma\|_\infty} - \frac{1}{2a}\int_{\mathbb{R}} \gamma(s)\,\mathrm{d}s - \frac{\varepsilon}{2a}\int_{\mathbb{R}} \jmath(s)\gamma(s)\,\mathrm{d}s \\ + \varepsilon^2 \int_{\mathbb{R}} \left(\dot{\jmath}(s)^2 - \frac{\gamma(s)}{2a}\jmath(s)^2 \right) \mathrm{d}s\,. \end{aligned}$$

If $\beta_\Gamma > 0$ it suffices to choose $\varepsilon = 0$ and λ small enough. If $\beta_\Gamma = 0$ we can always pick $\jmath$ for which the term linear in ε is nonzero. Hence the form q takes a negative value for suitable ε and λ, and part (a) is proved. On the other hand, to check (b) one can employ a local lower bound similar to that of *Proposition* 3.1.3 (Problem 1). ∎

Combined boundary conditions can also give rise to nontrivial spectral properties in straight waveguides. If one boundary of such a strip is Neumann and the other Dirichlet with a finite Neumann segment, one can study the discrete spectrum by simple modification of the methods used in Sect. 1.5.1 (Problem 2). Other combinations may lead to different spectral properties. As an example we mention in Problem 4 the situation where at each boundary the boundary condition switches from Dirichlet to Neumann at a fixed point, with an opposite orientation at the two sides of the strip.

3.2.2 Robin Boundary Conditions

Robin, or mixed, boundary conditions interpolate between the Dirichlet and Neumann conditions. They are characterized by a real parameter α. We will consider them on a curved strip Ω defined as above; the corresponding Robin Laplacian $-\Delta_\alpha^\Omega$ will be then associated with the quadratic form

$$\int_\Omega |\nabla\phi|^2(\vec{x})\,\mathrm{d}\vec{x} + \alpha \int_{\partial\Omega} |\phi|^2(s)\,\mathrm{d}s\,, \qquad \phi \in H^1(\Omega)\,.$$

Consequently, the functions from the domain of $-\Delta_\alpha^\Omega$ satisfy (in the weak sense) the Robin boundary conditions

$$\partial_n \varphi + \alpha\,\varphi = 0 \qquad \text{on } \partial\Omega,$$

where ∂_n denotes the outer normal derivative. In particular $\alpha = 0$ yields the Neumann condition; the Dirichlet condition formally corresponds to $\alpha = \infty$.

For simplicity we shall suppose that the boundary condition is repulsive, $\alpha > 0$. We also note that α may in general vary along the boundary of Ω, in fact the Robin condition can be regarded as a one-sided analogue of the leaky barrier discussed in Sect. 1.5.2. Here, however, we will keep α constant; we will be interested in the effect of the curvature on the spectrum of $-\Delta_\alpha^\Omega$.

If we assume that the curvature γ of the reference curve Γ is compactly supported, it is easy to see that $\sigma_{\mathrm{ess}}(-\Delta_\alpha^\Omega)$ is the interval $[\lambda_\alpha, \infty)$, where λ_α is the lowest eigenvalue of the Robin Laplacian on the strip cross section,

$$\lambda_\alpha = \inf_{f\in H^1(-a,a)} \frac{\int_{-a}^a |f'|^2(u)\,\mathrm{d}u + \alpha\,|f(a)|^2 + \alpha\,|f(-a)|^2}{\int_{-a}^a |f|^2(u)\,\mathrm{d}u}\,.$$

It turns out that the curvature again gives rise to a nonempty discrete spectrum of the operator $-\Delta_\alpha^\Omega$ as long as α is nonzero.

Theorem 3.5 *Assume again (i) and (ii)$_1$. Let the curvature γ be nonzero, compactly supported, and $a\|\gamma\|_\infty < 1$. If $\alpha \neq 0$, then $-\Delta_\alpha^\Omega$ has at least one eigenvalue of finite multiplicity below λ_α.*

Proof By using the curvilinear coordinates as in Sect. 1.1 we find that $-\Delta_\alpha^\Omega$ is unitarily equivalent to the operator H_α on $L^2(\Omega_0, g^{1/2}\mathrm{d}s\,\mathrm{d}u)$ associated with the quadratic form

$$Q_\alpha[\psi] = \|g^{-1/4}\partial_s\psi\|^2 + \|g^{1/4}\partial_u\psi\|^2 + \alpha\sum_{\iota=\pm}\int_{\mathbb{R}} (1+\iota a\gamma(s))\,|\psi(\iota a, s)|^2\,\mathrm{d}s\,.$$

Let χ_α be the positive normalized eigenfunction of the Robin Laplacian in $L^2(-a, a)$ relative to the lowest eigenvalue λ_α. We have

$$-\chi_\alpha'' = \lambda_\alpha\,\chi_\alpha\,, \qquad \chi_\alpha'(-a) = \alpha\,\chi_\alpha(-a)\,, \quad \chi_\alpha'(a) = -\alpha\,\chi_\alpha(a)\,. \tag{3.16}$$

Pick s_0 large enough so that $\gamma(s) = 0$ holds for $|s| \geq s_0$, and define

$$\psi_\delta(s,u) = \chi_\alpha(u)\,\varphi_\delta(s)\,, \quad \varphi_\delta(s) := \min\left\{1,\ \mathrm{e}^{-\delta(s-s_0)},\ \mathrm{e}^{\delta(s+s_0)}\right\}$$

with $\delta > 0$. We use a trial function of the form $\psi_\delta(s,u) + \varepsilon\,\phi(s)\,u\,\chi_\alpha(u)$, where $\phi \in C_0^\infty(\mathbb{R})$ with the support contained in the interval $(-s_0, s_0)$ and $\varepsilon \in \mathbb{R}$. A direct calculation using (3.16) and the fact that χ_α is even (Problem 3) gives

$$\begin{aligned} &Q_\alpha[\psi_\delta + \varepsilon\,\phi\,u\,\chi_\alpha] - \lambda_\alpha\|\psi_\delta + \varepsilon\,\phi\,u\,\chi_\alpha\|^2 \\ &\quad = \delta + C_\alpha\,\varepsilon^2 - 2\varepsilon\int_{-a}^a u\,\chi_\alpha'(u)\,\chi_\alpha(u)\,\mathrm{d}u \int_{\mathbb{R}} \phi(s)\,\gamma(s)\,\mathrm{d}s\,, \end{aligned}$$

where $C_\alpha > 0$ is independent of δ and ε. Since $u\,\chi_\alpha'(u)\,\chi_\alpha(u) < 0$ on $(-a, a)$ (Problem 3) we can choose the function ϕ in such a way that the last term in the above equation is nonzero. Hence by taking δ and $|\varepsilon|$ small enough and choosing the sign of ε in an appropriate way we can achieve that

$$Q_\alpha[\psi_\delta + \varepsilon\,\phi\, u\, \chi_\alpha] - \lambda_\alpha \|\psi_\delta + \varepsilon\,\phi\, u\, \chi_\alpha\|^2 < 0,$$

which completes the proof. ∎

Remark 3.2.1 Notice that the requirement $\alpha \neq 0$ in *Theorem* 3.5 cannot be abandoned. Indeed, for $\alpha = 0$ we have $\lambda_\alpha = 0$, and therefore $\sigma_{\mathrm{ess}}(-\Delta_0^\Omega) = [0, \infty)$. On the other hand, $-\Delta_0^\Omega$ is positive, thus it cannot have isolated eigenvalues. This shows that the Neumann boundary condition is special in this context.

3.2.3 An Isoperimetric Problem

While our main concern in this chapter is bound states in infinitely long tubes, the bent-tube techniques developed here can also be used to investigate spectral properties in compact regions. As an illustration we shall mention an isoperimetric problem for a strip $\Omega \equiv \Omega_{\Gamma,a}$ which is a tubular neighborhood of a closed planar curve Γ. We take a class of such strips with a fixed halfwidth a and the perimeter L of Γ, and ask about the form for which the ground-state eigenvalue ϵ_1 of $-\Delta_D^\Omega$ reaches an extremal value.

Proposition 3.2.1 *Given positive a, L consider all strips $\Omega_{\Gamma,a}$ such that Γ is a closed C^2-smooth curve without self-intersections and the condition $a\|\gamma\|_\infty < 1$ is satisfied. Within this class ϵ_1 is uniquely maximized when Γ is a circle.*

Proof In view of the assumptions one can introduce in the strip Ω locally orthogonal coordinates s, u in the same way as in Sect. 1.1; then the principal eigenvalue of $-\Delta_D^\Omega$ is given by

$$\epsilon_1 = \inf_{\|\psi\|=1} \int_0^L \mathrm{d}s \int_{-a}^a \mathrm{d}u \Big[(1+u\gamma(s))^{-1} |\partial_s\psi(s,u)|^2 + (1+u\gamma(s))|\partial_u\psi(s,u)|^2 \Big],$$

where ψ runs through a core of the operator in question. In particular, taking a smooth trial function ψ independent of s we get

$$\epsilon_1 \le \int_0^L \mathrm{d}s \int_{-a}^a \mathrm{d}u (1+u\gamma(s))|\partial_u\psi(u)|^2 = \int_{-a}^a (L+2\pi u)\, |\partial_u\psi(u)|^2\, \mathrm{d}u\,,$$

where in the second step we have used the fact that Γ is a smooth loop and thus it satisfies $\beta_\Gamma = 2\pi$; the inequality would be sharp if the true ground state were to depend on s. Taking the infimum of the right-hand side over the trial functions independent of s we find $\epsilon_1 \le \tilde{\epsilon}_1$, where the last symbol means the ground state eigenvalue for the corresponding circular annulus, $\tilde{\epsilon}_1 := k(L/2\pi, a)^2$ in the notation of *Proposition* 3.1.3. It remains to check the uniqueness. Since $\epsilon_1 = \tilde{\epsilon}_1$ requires the ground-state eigenfunction ψ_1 to be independent of s, the longitudinal term on the left-hand side of $-\Delta_D^\Omega \psi_1 = \tilde{\epsilon}_1 \psi_1$ vanishes. On the other hand, the derivative $\partial_u \psi_1$ can vanish at an isolated value of u only, so $\gamma(s)(1+u\gamma(s))^{-1}$ must be s-independent for almost all $u \in (-a, a)$ once the curve Γ maximizes ϵ_1, which is possible only if Γ is a circle. ∎

3.2.4 Higher Dimensions

Our basic physical motivation is to study quantum dynamics constrained to particular regions of the configuration space. This means that we are interested mostly in tubes in $\mathbb{R}^d$, $d = 2, 3$, and also in layers in the three-dimensional space which we will discuss in Chap. 4 and further. Nevertheless, it is natural to ask whether the results of this chapter have higher-dimensional analogs.

Consider thus a curve Γ in $\mathbb{R}^d$. As usual we parametrize it by its arc length and suppose that as a function $s \mapsto \Gamma(s)$ it is C^d-smooth. We also assume that Γ has a global positively oriented Frenet frame $\{e_1, \dots, e_d\}$, where $e_1 = \dot{\Gamma}$ is the tangent vector, and moreover, that the functions $e_i \in C^1(\mathbb{R}, \mathbb{R}^d)$ and each $\dot{e}_i(s)$ lies in the span of $e_1(s), \dots, e_{i+1}(s)$. For a fixed cross section $M \subset \mathbb{R}^{d-1}$ assumed to be an open connected set, containing zero, and such that $\sup_{x \in M} |x| < \infty$, we define a tube Ω as $f(\mathbb{R} \times M)$, where

$$f(s, u_2, \dots, u_d) := \Gamma(s) + \sum_{\mu,\nu=2}^{d} \mathcal{R}_{\mu\nu}(s) u_\mu e_\nu(s) \tag{3.17}$$

and $\{\mathcal{R}_{\mu\nu}(s)\}$ is a family of $(d-1) \times (d-1)$ orthogonal matrices, continuous in the variable s; as in Sect. 1.3 we assume that Ω does not intersect itself.

The derivatives of the moving frame vectors are given by Serret-Frenet formulæ, $\dot{e}_i = \sum_j \gamma_{ij} e_j$, where the skew-symmetric matrix γ has the elements

$$\gamma_{ij} = \delta_{i,j-1}\gamma_i - \delta_{i-1,j}\gamma_j$$

with $\gamma_i : \mathbb{R} \to \mathbb{R}$ being the i-th curvature of Γ. To define Tang coordinate systems in the d-dimensional situation we use the system of first-order equations

$$\dot{\mathcal{R}}_{\mu\nu} + \sum_{\rho=2}^{d} \mathcal{R}_{\mu\rho}\gamma_{\rho\nu} = 0\,. \tag{3.18}$$

It is not difficult to check that under the stated assumptions a continuous matrix solution exists and is orthogonal for each $s \in \mathbb{R}$ (Problem 9a). Then we define a new frame rotating with the respect to the Frenet frame by $e_i^{\mathrm{T}} := \sum_j \mathcal{R}_{ij} e_j$

using the $d \times d$ matrix $\mathcal{R} := \operatorname{diag}\big(1, (\mathcal{R}_{\mu\nu})\big)$. We will again suppose that the cross section of the tube Ω remains fixed with respect to this frame—cf. Remark 1.3.1a—which means that (3.17) maps $(s, u_2, \dots, u_d)$ into $\Gamma(s) + \sum_\mu u_\mu e_\mu^{\mathrm{T}}(s)$. Then these curvilinear coordinates allow us to rewrite the corresponding Dirichlet Laplacian $-\Delta_D^\Omega$ decoupling the longitudinal and transverse coordinates (Problem 9b), and consequently, to prove the following claim (Problem 11).

Theorem 3.6 *In addition to the assumptions made above, suppose that the first curvature γ_1 is bounded and $a\|\gamma_1\|_\infty < 1$. Then*
(a) $\sigma_{\mathrm{ess}}(-\Delta_D^\Omega) = [\nu_1, \infty)$ holds if $\lim_{|s|\to\infty} \gamma_1(s) = 0$,
(b) $\inf \sigma_{\mathrm{ess}}(-\Delta_D^\Omega) < \nu_1$ holds whenever γ_1 is nonzero, in particular, $-\Delta_D^\Omega$ has at least one eigenvalue below ν_1 provided the tube is asymptotically straight.

3.3 Interacting Particles

So far we have considered properties of a single confined particle. Some of the results can be used to describe many-particle states in quantum waveguides provided the particles are not interacting or the interaction can be neglected as in the case of a dilute electron gas in a quantum wire. For instance, *Proposition* 3.1.1 then gives a bound on the number of particles which a bent duct can bind if the latter are fermions which occupy the one-particle bound states in accordance with the Pauli principle; the right-hand side must be multiplied by two, or more generally by the number $2s + 1$ of spin states.

The situation is more complicated if the mutual interactions must be taken into account. Having in mind electrons, we shall consider N particles with spin $\frac{1}{2}$ and the charge $-e$, which interact therefore by electrostatic repulsion. We suppose that they are confined within a hard-wall bent strip $\Omega_{\Gamma,a}$ with the axis determined by its signed curvature γ which, for simplicity, we shall assume here to be compactly supported. The Hamiltonian $H_N \equiv H_N(\gamma, a, e)$ can be rewritten in a unitarily equivalent form using the curvilinear coordinates introduced in Sect. 1.1; then it acts as

$$H_N = \sum_{j=1}^{N} \Big\{ -\partial_{s_j}(1 + u_j\gamma(s_j))^{-2}\partial_{s_j} - \partial_{u_j}^2 + V(s_j, u_j) \Big\} + e^2 \sum_{1\le j<l\le N} |\vec{r}_j - \vec{r}_l|^{-1}$$

on the Hilbert space $\mathcal{H}_N := \bigotimes_{j=1}^N L^2(\Omega_0)$, where V is the curvature-induced potential (1.8) and $\vec{r}_j = \vec{r}_j(s_j, u_j)$ are the Cartesian coordinates of the N-th electron. The true state space of the system is, of course, the subspace of $\mathcal{H}_N$ consisting of functions antisymmetric with respect to particle exchanges, but it is convenient to take the Pauli principle into account only later. The operator H_N is essentially self-adjoint on the N-fold algebraic tensor product of $U\,\mathrm{Dom}\,(-\Delta_D^\Omega)$, where U is the unitary operator appearing in (1.7); alternatively one can use any core of $-\Delta_D^{\Omega_0}$ (cf. Remark 1.1.2a). Moreover, we have

$$\inf \sigma_{\text{ess}}(H_N) \le \kappa_1^2 N \tag{3.19}$$

and the equality holds if and only if $\gamma = 0$ (Problem 12). The threshold depends, in general, on the electrostatic interaction because the latter influences the bound state energies of clusters to which the N-particle family can be decomposed; thus one-particle methods are no longer useful.

At the same time, an implicit bound can be derived by a method used in atomic physics to find ionization properties of Coulomb Hamiltonians. To formulate it we first introduce some notation. Let $\{\lambda_m\}_{m=1}^{\infty}$ be the spectrum of $-\Delta_D^{R_\beta}$ for the rectangle $R_\beta := \left(-\frac{3}{2}\beta g_+^{1/2}, \frac{3}{2}\beta g_+^{1/2}\right) \times (-a, a)$ with a fixed $\beta > 0$, where $g_\pm$ were introduced in Sect. 3.1.1 together with the function $\tilde{W}_1$ estimating the potential (1.8). Then we define

$$T_\beta(N) := \sum_{m=1}^{N} \lambda_{\left[\frac{m+1}{2}\right]}$$

which allows us to state the following inequality.

Theorem 3.7 *Adopt assumptions (i), (ii)$_2$ of Sect. 1.1. In addition, suppose that the curvature is compactly supported,* $\gamma(s) = 0$ *for* $|s| > b > \frac{1}{2}a$, *and* $a\|\gamma\|_\infty < 1$. *Then* $\sigma_{\text{disc}}(H_N) = \emptyset$ *holds for* $N \ge 2$ *if the condition*

$$T_\beta(N) + \frac{e^2}{2\beta\sqrt{13}}\, N(N-1) \ge \|\tilde{W}_1\|_\infty N + \kappa_1^2 N + \frac{e^2}{19\beta\sqrt{2}} \tag{3.20}$$

holds for some $\beta \ge \max\{2b, 629\, e^{-2}\}$.

Proof We use a variational argument based on a suitable decomposition of the configuration space, the points of which are (s, u) with $s = \{s_1, \dots, s_N\}$ and $u = \{u_1, \dots, u_N\}$. Consider a pair of smooth functions $v,\, h : \mathbb{R}_+ \to [0, 1]$ such that v interpolates between $v(t) = 0$ for $t \le 1$ and $v(t) = 1$ for $t \ge \frac{3}{2}$, and $v(t)^2 + h(t)^2 = 1$; then we construct such a decomposition using the functions

$$s \mapsto v(\|s\|_\infty \beta^{-1}),\ h(\|s\|_\infty \beta^{-1}),$$

where $\|s\|_\infty := \max\{s_1, \dots, s_N\}$ and $\beta > 2b > a$ will be specified later. Abusing the notation, we also employ the symbols v, h for the corresponding operators of multiplication. For a fixed $\psi \in \text{Dom}\,(H_N)$ we evaluate $(v\psi, [H_N, v]\psi)$ and the analogous expression for the function h; in both cases only the longitudinal term in the kinetic part of H_N contributes. Thus $(\psi, H_N\psi)$ equals

$$(v\psi, H_N v\psi) + (h\psi, H_N h\psi) + \sum_{j=1}^{N} \left\{ \left\|(1+u_j\gamma_j)^{-1} v_j \psi\right\|^2 + \left\|(1+u_j\gamma_j)^{-1} h_j \psi\right\|^2 \right\},$$

where we use the shorthands $v_j := \frac{\partial v}{\partial s_j}$, $h_j := \frac{\partial h}{\partial s_j}$, and $\gamma_j := \gamma(s_j)$. The factors $(1+u_j\gamma_j)^{-1}$ may be neglected, because v_j, h_j are nonzero only if $s_j \geq \beta > 2b$ in which case $\gamma_j = 0$ holds by assumption. Furthermore, with the exception of the hyperplanes where two or more coordinates coincide (which is a zero measure set) the norm $\|s\|_\infty$ is equal to one of the coordinates $s_1, \dots, s_n$, and therefore

$$\sum_{j=1}^{N} \left\{ \left\| v_j\psi \right\|^2 + \left\| h_j\psi \right\|^2 \right\} \leq \|\psi\|^2 \max_{1\leq j\leq N} \left\{ \left\| v_j \right\|_\infty^2 + \left\| h_j \right\|_\infty^2 \right\} \leq \beta^{-2} C_0 \|\psi\|^2 ,$$

where $C_0 := \|v'\|_\infty^2 + \|h'\|_\infty^2$. This yields the estimate

$$(\psi, H_N\psi) \geq L_1[v\psi] + L_1[h\psi], \quad L_1[\phi] := (\phi, H_N\phi) - \frac{C_0}{\beta^2} \|\phi\|_{N_\beta}^2, \tag{3.21}$$

where the last index means the norm of the vector ϕ restricted to the subset $N_\beta := \left\{ s :\, \beta \leq \|s\|_\infty \leq \frac{3}{2}\beta \right\}$ of the configuration space.

Next one has to estimate separately contributions from the inner and outer parts. Consider first the exterior. We introduce the functions

$$f_j(s) := v\left(2s_j \|s\|_\infty^{-1}\right) \prod_{n=1}^{j-1} h\left(2s_n \|s\|_\infty^{-1}\right), \quad j = 1, \dots, N-1$$

(an empty product equals one), and $f_N(s) := \prod_{n=1}^{N-1} h\left(2s_n \|s\|_\infty^{-1}\right)$, which satisfy the relation $\sum_{j=1}^{N} f_j(s)^2 = 1$. Moreover, the functions $s_j \mapsto v(2s_j\|s\|_\infty), h(2s_j\|s\|_\infty)$ have nonzero derivative only if $|s_j| \geq \frac{1}{2}\|s\|_\infty^{-1}$. In particular, on the support of $s \mapsto v(\|s\|_\infty \beta^{-1})$ the derivative is nonzero only if $|s_j| \geq \frac{1}{2}\beta > b$; in other words, the function $s \mapsto f_j(s)^2 v(\|s\|_\infty \beta^{-1})$ has zero derivative in all parts of the configuration space such that at least one of the electrons dwells in the curved part of the waveguide. Commuting now H_N with f_j, we get in the same way as above the identity

$$L_1[v\psi] = \sum_{j=1}^{N} \left\{ L_1[f_j v\psi] - \|(\nabla_s f_j) v\psi\|^2 \right\} ,$$

where $\nabla_s := \left(\partial_{s_1}, \dots, \partial_{s_N}\right)$. To deal with the last part, we need a pointwise upper bound to the gradient term. Letting $\sigma_j := 2s_j\|s\|_\infty^{-1}$ and using the definition of f_j we can express this quantity; after a partial resummation using the relation $v^2 + h^2 = 1$ we find that $\sum_{j=1}^{N} |(\nabla_s f_j)(s)|^2$ equals

$$\frac{4}{\|s\|_\infty^2}\left\{v'(\sigma_1)^2+h'(\sigma_1)^2+h(\sigma_1)^2h'(\sigma_2)^2+\cdots+h(\sigma_1)^2\ldots h(\sigma_{N-1})^2h'(\sigma_N)^2\right\}$$
$$\leq\frac{4}{\|s\|_\infty^2}\left\{v'(\sigma_1)^2+\sum_{j=1}^N h'(\sigma_j)^2\right\}\leq\frac{4NC_0}{\|s\|_\infty^2},$$

and therefore

$$L_1[v\psi]\geq\sum_{j=1}^N L_1[f_jv\psi]-4NC_0\left\|v\psi\|s\|_\infty^{-1}\right\|^2=\sum_{j=1}^N L_2[f_jv\psi],$$

where $L_2[\phi]:=L_1[\phi]-4NC_0\left\|\phi\|s\|_\infty^{-1}\right\|^2$. Hence the task is reduced to finding a lower bound to $L_2(\psi_j)$ with $\psi_j:=f_jv\psi$. Since $s_j\geq\frac12\|s\|_\infty\geq\frac12\beta>b$ holds on the support of ψ_j, we have $V(s_j,u_j)=0$ there. This means, in particular, that we can rewrite $(\psi_j,H_N\psi_j)$ in the following way,

$$(\psi_j,H_{N-1}\psi_j)+\left\|\partial_{s_j}\psi_j\right\|^2+\left\|\partial_{u_j}\psi_j\right\|^2+e^2\sum_{j\neq l=1}^N\left(\psi_j,|\vec r_j-\vec r_l|^{-1}\psi_j\right),$$

where H_{N-1} refers to the system with the j-th electron excluded. The sum of the first three terms is bounded from below by $(\mu_{N-1}+\kappa_1^2)\|\psi_j\|^2$, and since $|\vec r_j-\vec r_l|\leq\sqrt{(s_j-s_l)^2+4a^2}\leq2\sqrt{\|s\|_\infty^2+a^2}$, we have

$$(\psi_j,H_N\psi_j)\geq\left(\mu_{N-1}+\kappa_1^2\right)\|\psi_j\|^2+\frac{e^2(N-1)}{2}\left(\psi_j,(\|s\|^2+a^2)^{-1/2}\psi_j\right). \tag{3.22}$$

The sought lower bound then follows from the definition of the functional L_2 in combination with (3.21); we get it by subtracting the expression

$$4NC_0\left\|\psi_j\|s\|_\infty^{-1}\right\|^2+C_0\beta^{-2}\|\psi_j\|_{N_\beta}^2\leq\frac{C_0(8N+3)}{2\beta}\left\|\psi_j\|s\|_\infty^{-1/2}\right\|^2$$

from the right-hand side of (3.22); recall that $N_\beta:=\left\{s\,:\,\beta\leq\|s\|_\infty\leq\frac32\beta\right\}$. Moreover, $\|s\|_\infty\geq\beta>2b>a$ yields $(\|s\|^2+a^2)^{1/2}\leq\sqrt2\,\|s\|_\infty$, and therefore

$$L_2[\psi_j]\geq\left(\mu_{N-1}+\kappa_1^2\right)\|\psi_j\|^2+\left(\frac{e^2(N-1)}{2\sqrt2}-\frac{C_0(8N+3)}{2\beta}\right)\left\|\psi_j\|s\|_\infty^{-1/2}\right\|^2.$$

Since $N\geq2$, the last term is positive for $\beta>19\sqrt2C_0e^{-2}$. In such a case we have $L_1[v\psi]\geq(\mu_{N-1}+\kappa_1^2)\|\psi_j\|^2$, so the exterior part of the wave function does not contribute to the discrete spectrum.

Let us turn to the inner part. The functional $L_1[h\psi]$ in the decomposition (3.21) can be estimated using the explicit form of the operator H_N as follows

$$L_1[h\psi] \geq g_+^{-1}\|\nabla_s h\psi\|^2 + \|\nabla_u h\psi\|^2 + \sum_{j=1}^{N} \left(h\psi, V(s_j,u_j)h\psi\right)$$
$$+e^2 \sum_{1\leq j<k\leq N} \left(h\psi, |\vec{r}_j-\vec{r}_k|^{-1}h\psi\right) - \frac{C_0}{\beta^2}\|h\psi\|^2.$$

Since $|V(s_j,u_j)| \leq \|\tilde{W}_1\|_\infty$ holds for $j = 1,\dots,N$, the potential term can be estimated by $N\|\tilde{W}_1\|_\infty\|h\psi\|^2$. Furthermore, on the support of h we have

$$|\vec{r}_j-\vec{r}_k| \leq 2\sqrt{\|s\|_\infty^2+a^2} \leq \sqrt{9\beta^2+4a^2},$$

because $\|s\|_\infty \leq \frac{3}{2}\beta$ holds there. At the same time, $\beta > 2b > a$, so we arrive at the estimate $|\vec{r}_j-\vec{r}_k| \leq \beta\sqrt{13}$ which yields

$$\sum_{1\leq j<k\leq N} \left(h\psi, |\vec{r}_j-\vec{r}_k|^{-1}h\psi\right) \geq \frac{N(N-1)}{2\beta\sqrt{13}}\|h\psi\|^2.$$

Combining these two bounds with the inequality $C_0/\beta < e^2/19\sqrt{2}$ we find that $L_1[h\psi]$ is bounded from below by the expression

$$g_+^{-1}\|\nabla_s h\psi\|^2 + \|\nabla_u h\psi\|^2 + \left[-N\|\tilde{W}\|_\infty + \frac{e^2N(N-1)}{2\beta\sqrt{13}} - \frac{e^2}{19\beta\sqrt{2}}\right]\|h\psi\|^2.$$

Together with the exterior estimate this tells us that the discrete spectrum of H_N is empty for $N \geq 2$ if the above expression is not smaller that $\kappa_1^2 N\|h\psi\|^2$ for some β which satisfies the condition $\beta \geq \max\left\{2b, \frac{19\sqrt{2}C_0}{e^2}\right\}$. The first two terms represent the quadratic form of the $2N$-dimensional Dirichlet Laplacian in R_β^N. Now finally the Pauli principle enters the game; it implies that each eigenvalue may appear only twice for spin $\frac{1}{2}$. Thus one has to take the orthogonal sum of two copies of the Laplacian on R_β and to sum the first N eigenvalues of such an operator; this is exactly the quantity which we have denoted by $T_\beta(N)$.

To finish the proof, it remains to find C_0 which appears in the above bounds. Without trying to optimize it we put $v(\xi) := \sin\left(4\pi\xi^2(1-2\xi^2)\right)$ for $t-1 =: \xi \in \left(0, \frac{1}{2}\right)$. This gives $v'(\xi)^2 + g'(\xi)^2 = (8\pi)^2\xi^2(1-4\xi^2)^2$, so we may choose $C_0 = (8\pi)^2/27 \approx 23.39$ obtaining thus the stated bound on β. ∎

Corollary 3.3.1 $\sigma_{\rm disc}(H_N) = \emptyset$ *holds for a fixed* $N \geq 2$ *if* e *is large enough.*

Proof We have $\frac{1}{2\sqrt{13}}N(N-1) - \frac{1}{19\sqrt{2}} > 0$ and the remaining terms in condition (3.20) are independent of e. ∎

In this way Coulomb repulsion may prevent a curved guide of a fixed geometry to trap more than one charged particle. On the other hand, one can always find a strip $\Omega_{\Gamma,a}$ which binds a given number of electrons (Problem 13).

3.4 Acoustic Waveguides

So far we have paid attention mostly to waveguides with Dirichlet boundaries even if other conditions have appeared from time to time, for instance, in Sects. 1.5 or 3.2. There are situations where the choice of such conditions have a deep physical motivation. A prime example is given by ***acoustic waveguides***. The sound dynamics in a tube or a system of tubes is described by the wave equation rather than Schrödinger's equation but in the stationary case the problem is reduced again to the study of the Laplacian in infinite cylindrical regions, however, this time with Neumann boundary conditions.

It is not surprising that the spectral properties of such operators differ from those of Dirichlet waveguide Hamiltonians. Their spectrum typically covers the positive real halfline, and since the Neumann Laplacian is positive by definition, there is no room for a discrete spectrum. Nevertheless, acoustic waveguides can exhibit localized modes corresponding eigenvalues of the operator embedded in the continuous spectrum which typically come from a symmetry. The situation calls to mind the Nöckel-type model discussed in Sect. 2.3; we will show that such eigenvalues again turn into resonances when the symmetry is violated.

3.4.1 Eigenvalues of the Neumann Laplacian in Tubes

The spectrum of the Neumann Laplacian in a straight tube is naturally purely absolutely continuous. The situation may change when obstacles are placed into the guide. Let $\Omega = \mathbb{R} \times M$, where as before M is the cross section of the waveguide, and let $\Sigma^c \subset \Omega$ be a bounded obstacle. For simplicity we shall consider the two-dimensional case when M is a line segment. We are interested in eigenvalues of the operator $-\Delta_N^\Sigma$ acting in $L^2(\Sigma)$ with $\Sigma := \Omega \setminus \Sigma^c$; as indicated above all such eigenvalues have to be embedded in $\sigma_{\mathrm{ess}}(-\Delta_N^\Sigma) = \mathbb{R}^+$.

As mentioned in Sect. 2.3 such eigenvalues often arise from some symmetry of the system—see, in particular, Examples 2.3.1. It turns out that the same effect occurs in the waveguide described above provided the obstacle Σ^c is mirror-symmetric with respect to the axis of the strip.

Theorem 3.8 *Let $M = (-1, 1)$ and suppose that the obstacle is of the form $\Sigma^c = \{(x, y) \in \Omega : |y| < f(x)\}$, where $f \in C_0^\infty(\mathbb{R})$ is such that $0 \leq f < 1$. If f is nonzero, then $-\Delta_N^\Sigma$ has at least one eigenvalue in the interval $\left(0, \frac{1}{4}\pi^2\right)$.*

Proof We again employ the symmetry decomposition mentioned in Sect. 1.5 introducing the subspaces of functions of a given transverse parity,

$$\mathcal{H}_j := \{u \in L^2(\Sigma) :\ u(x,-y) = (-1)^j\, u(x,y) \ \text{ for } \ (x,y) \in \Sigma\}, \quad j = 1,2,$$

such that $L^2(\Sigma) = \mathcal{H}_1 \oplus \mathcal{H}_2$. A simple calculation shows that $\mathcal{H}_j,\ j = 1,2$, are invariant subspaces of $-\Delta_N^\Sigma$. Since $-\Delta_N^\Sigma\restriction_{\mathcal{H}_1}$ is subject to Dirichlet boundary conditions on $\{(x,y) \in \Sigma : y = 0\}$, and since $\mathrm{supp}\, f$ is compact by assumption, it is easy to see that $\inf \sigma_{\mathrm{ess}}\big(-\Delta_N^\Sigma\restriction_{\mathcal{H}_1}\big) = \frac14\pi^2$. Therefore it suffices to prove that $-\Delta_N^\Sigma\restriction_{\mathcal{H}_1}$ has at least one isolated eigenvalue. We again employ variational principle with a suitable choice of test functions; we consider the family

$$u_\varepsilon(x,y) := \varphi(\varepsilon x)\, \sin\left(\frac{\pi y}{2}\right),$$

where $\varphi \in C_0^\infty(-2,2)$ is such that $\varphi(x) = 1$ holds for $x \in [-1,1]$ and $\varepsilon > 0$ is a small parameter. Then $u_\varepsilon \in \mathcal{H}_1 \cap H^1(\Sigma)$ for any $\varepsilon > 0$ and we find that

$$\int_\Sigma |\nabla u_\varepsilon|^2(x,y)\,\mathrm{d}x\,\mathrm{d}y - \frac{\pi^2}{4}\int_\Sigma |u_\varepsilon|^2(x,u)\,\mathrm{d}x\,\mathrm{d}y$$
$$= 2\varepsilon \int_{\mathbb{R}} |\varphi'|^2(x)\,\mathrm{d}x - \frac{\pi^2}{2}\int_{\mathrm{supp}\, f} \sin(\pi f(x))\,\mathrm{d}x\,.$$

As $\varepsilon \to 0$ the first term on the right-hand side vanishes while the second term is a negative constant. Hence taking ε small enough we achieve that the right-hand side in the last relation is negative; this proves the existence of an embedded eigenvalue of $-\Delta_N^\Sigma$ in the interval $(0, \frac14\pi^2)$. ∎

A similar result also holds when the obstacle has the form of a Neumann cut along a segment of the waveguide axis (Problem 2b).

3.4.2 Resonances in Acoustic Waveguides

We have already pointed out in Sect. 2.3 that if a system has embedded eigenvalues due to a symmetry, a perturbation which violates the latter will in general turn those eigenvalues into resonances. We are now going to discuss from this point of view acoustic waveguides with obstacles considered in the previous section. We will not tackle the resonances, however, through a meromorphic continuation of the resolvent, which is not trivial to construct in this model, but rather as complex poles of the scattering matrix (see the notes to Sect. 2.3).

Let us first describe the problem setting. Similarly as in *Theorem* 3.8 we consider a Neumann waveguide with a symmetric obstacle Σ_0^c, now as an unperturbed system. Let L be the length of the obstacle boundary and assume that $\partial\Sigma_0^c$ is parametrized by its arc length s as follows,

$$x = X(s)\,, \quad y = Y(s)\,, \qquad 0 \le s \le L\,,$$

which allows us to express the boundary curvature as $\gamma(s) = (\dot{X}\ddot{Y} - \dot{Y}\ddot{X})(s)$. Now we move the obstacle in the direction perpendicular to the guide axis by a distance ε and denote the shifted obstacle by Σ_ε^c; we keep ε small enough to ensure that $\Sigma_\varepsilon^c \subset \Omega$. We put $\Sigma_\varepsilon := \Omega \setminus \Sigma_\varepsilon^c$ and consider the following boundary value problem,

$$\Delta u + \omega^2 u = 0 \quad \text{in } \Sigma_\varepsilon, \qquad \partial_n u = 0 \quad \text{on } \partial\Sigma_\varepsilon, \tag{3.23}$$

where ∂_n denotes the outer normal derivative and ω is the spectral parameter; the square is due to the physical nature of the problem which is associated with the wave rather than Schrödinger equation. First we have to specify how to define resonances in this setting. Given an ω with $\operatorname{Im}\omega < 0$ we set $\xi_j = -\sqrt{(\pi j/2)^2 - \omega^2}$, $j = 0, 1, 2 \dots$. Moreover, let $\eta_0(y) := 1/\sqrt{2}$ and $\eta_j(y) := \cos\left(\frac{\pi j(y+1)}{2}\right)$, $j \ge 1$, be elements of the transverse-mode basis, and define

$$e_j^{\pm}(x) = \begin{cases} \mathrm{e}^{-\xi_j |x|} & \dots \ \pm x > 0 \\ 0 & \dots \ \text{otherwise} \end{cases}, \quad \tilde{e}_j^{\pm}(x) = \begin{cases} \mathrm{e}^{\xi_j |x|} & \dots \ \pm x > 0 \\ 0 & \dots \ \text{otherwise} \end{cases}$$

Note that in contrast to (1.10) the functions η_j satisfy Neumann conditions at $y = \pm 1$. Next we take R large enough so that $\operatorname{supp} f \subset (-R, R)$ and consider the solutions $u_j^{\pm}(x, y; \omega)$ of Eq. (3.23) which for $|x| > R$ have the form

$$u_j^{\mu}(x, y; \omega) = e_j^{\mu}(x)\, \eta_j(y) + \sum_{\nu=\pm} \sum_{k=0}^{\infty} T_{jk}^{\mu\nu}(\omega)\, \tilde{e}_j^{\nu}(x)\, \eta_k(y),$$

where μ and ν are indices taking values $\pm$ and $T_{jk}^{\mu\nu}(\omega)$ are complex-valued functions of ω which are uniquely determined by the above Ansatz and relations (3.23). The infinite matrix

$$T(\omega) = \left\{T_{jk}^{\mu\nu}(\omega)\right\}_{\mu,\nu=\pm,\; j,k=0,1,\dots}$$

is for real ω called the ***scattering matrix*** of the problem (3.23). It extends to the complex plane being holomorphic for $\operatorname{Im}\omega < 0$, and furthermore, it admits a meromorphic continuation onto the upper half-plane $\operatorname{Im}\omega > 0$ accessed from the lower half-plane through the interval $(-\pi/2, \pi/2)$; we identify ***resonances*** of the problem (3.23) with the poles of such a meromorphic continuation of $T(\omega)$ onto the upper half-plane.

To describe the behavior of those resonances, we start from the unperturbed system corresponding to $\varepsilon = 0$. We know that in that case $-\Delta_N^{\Sigma_0}$ has at least one eigenvalue ω_0^2 with $\omega_0 \in \mathbb{R}$ and a normalized eigenfunction u_0. Let us denote by $u_0^{\pm}(\cdot,\cdot)$ two generalized eigenfunctions relative to ω_0, i.e. two solutions of (3.23) with $\varepsilon = 0$, singled out by the asymptotic expansions

$$u_0^{\pm}(x,y) = e^{\pm i\omega_0 x} + T_{00}^{\pm\pm}(\omega_0)\, e^{\mp i\omega_0 x} + \mathcal{O}(e^{-\delta|x|}) \quad \dots \quad x \to +\infty$$

$$u_0^{\pm}(x,y) = T_{00}^{\pm\mp}(\omega_0)\, e^{\pm i\omega_0 x} + \mathcal{O}(e^{-\delta|x|}) \quad \dots \quad x \to -\infty$$

where δ is a positive constant, together with the orthogonality condition

$$\int_{\Sigma_0} \overline{u_0(x,y)}\, u_0^{\pm}(x,y)\, \mathrm{d}x\, \mathrm{d}y = 0\,.$$

Now we shift the obstacle in the vertical direction by ε and ask about the asymptotical behavior of the corresponding resonances as $\varepsilon \to 0$; it is expressed by the following result (see the notes) which can be regarded as a version of Fermi's golden rule in the present case.

Theorem 3.9 *Let ω_0^2 with $\omega_0 > 0$ be a simple eigenvalue of $-\Delta_N^{\Sigma_0}$ and let u_0 be the normalized eigenfunction associated with ω_0^2. Then for ε small enough there is a unique resonance ω_ε of the problem (3.23) satisfying* $\mathrm{Re}\,\omega_\varepsilon \to \omega_0$ *and*

$$\mathrm{Im}\,\omega_\varepsilon = \left(|\alpha_+|^2 + |\alpha_-|^2\right) \varepsilon^2 + \mathcal{O}(\varepsilon^3) \quad \text{as} \quad \varepsilon \to 0\,,$$

with

$$\alpha_\pm := \frac{1}{4\omega_0} \int_0^L \overline{u_0^{\pm}(s)} \left(\dot{X}(s)\big(\partial_s^2 u_0(s) + \omega_0^2 u_0(s)\big) - \gamma(s)\dot{Y}(s)\partial_s u_0(s)\right) \mathrm{d}s\,,$$

where $u_0(s) \equiv u_0(X(s), Y(s))$ and $u_0^{\pm}(s) \equiv u_0^{\pm}(X(s), Y(s))$.

3.5 Notes

Section 3.1 *Proposition* 3.1.1 is taken essentially from [EV99] where it was stated for $d = 2$. For the Bargmann inequality see [Barg52, BSh91, Theorem 2.5.3].

The Birman-Schwinger estimate on the number of bound states in dimension one and two requires a trick which was independently discovered in [Se74, Kl77, Ne83] and consists of splitting off the resolvent singularity responsible for the one bound state, which is always present for the attractive interaction, and estimating the trace norm of the rest. On the other hand, one does not use in this way the full strength of the BS theory manifested in the weak-coupling analysis of Chap. 6, and as usual with the BS technique, the result is not optimal for a strong coupling. *Theorem* 3.1 is taken from [AE90]. It relies on inequalities the most simple of which was known as the Payne-Pólya-Weinberger conjecture after the authors who formulated it first in [PPW55]. It was a subject of intense mathematical study for more than three decades until it was finally proven in [AB92a, AB92b]; in the first of these papers the reader can also find the history of the problem. *Proposition* 3.1.3 comes from the paper [EFK04] where such a result is proved generally for a class of curved tubes in $\mathbb{R}^d$. Notice that in the non-circular case this class is not related to Tang's system and its higher dimensional analogues—cf. Sect. 3.2.4—rather it requires the cross section to be fixed with respect to the Frenet frame of the generating curve.

Lieb-Thirring inequalities were introduced in [LT76] as a generalization of the CLR-bound for the number of bound states of a Schrödinger operator [RS, Theorem XIII.12]. The present-day knowledge about the constants involved is as follows,
(a) $R(\delta, d) = 1$ if $\delta \geq 3/2,\ d \in \mathbb{N}$,
(b) $R(\delta, d) \leq 2$ if $1 \leq \delta < 3/2,\ d \in \mathbb{N}$ or $1/2 \leq \delta < 1,\ d = 1$,
(c) $R(\delta, d) \leq 4$ if $1/2 \leq \delta < 1,\ d \geq 2$,
see [HLT98, HLW00, LW00] where one can also find other related results together with a rich bibliography. Recall that the constants are determined by the weak-coupling behavior, because for all $\delta \geq 0$ and reasonable regular V the Weyl asymptotics, $S_\delta(\lambda)[S^{\mathrm{cl}}_{\delta,d}(\lambda)]^{-1} \to 1$ as $\lambda \to \infty$, is valid which gives $R(\delta, d) \geq 1$. What is important here is that (3.5) allows a generalization to potentials with values in non-negative compact operators on a separable Hilbert space $\mathcal{G}$. Let $A := -\Delta \otimes I_{\mathcal{G}} - W(x)$ be such a generalized Schrödinger operator on $L^2(\mathbb{R}^d) \otimes \mathcal{G}$, then for $\delta \geq 1/2$ if $d = 1$ and $\delta > 0$ if $d \geq 2$ the following inequality holds,

$$\mathrm{tr}_{L^2(\mathbb{R}^d)\otimes\mathcal{G}} A_-^\delta \leq r(\delta, d) \int_{\mathbb{R}^d} \mathrm{tr}_{\mathcal{G}} W(x)^{\delta+\frac{d}{2}}\, \mathrm{d}x\,,$$

provided the right-hand side makes sense. The proof for $d = 1$ relies on the analysis of the corresponding ODEs and a matrix approximation of A, the result is then extended by induction to higher dimensions—see [HLW00, LW00]. The constants have the properties analogous to (a)–(c) and satisfy the bound $r(\delta, d) \geq R(\delta, d)$. The main result of this section, *Theorem* 3.2, was proved in [EW02]. A similar decomposition can be used to prove LT-type inequalities also in situations when the perturbation is not given by a potential but rather it is due to deformation and/or boundary conditions—see Problems 5–7 and [ELW04] for a more detailed discussion.

The bound on the number of potential-induced bound states in locally twisted tubes, *Theorem* 3.3, comes from [GKP14], where it was proved in a different way. In the proof of *Theorem* 3.3 we have employed bounds derived in [Da] and [GS09] where other useful properties of heat kernels can be found; for Lieb's inequality we refer to [Li76, RS, Sect. VIII.3]. Note that the bound (3.11) has the correct semiclassical behavior, $N(H^{\Omega_\alpha}_{\lambda V}, \nu_1) \simeq \lambda^{3/2}$ as $\lambda \to \infty$; as in the case of *Theorem* 3.2 the three-dimensional nature of the tube dominates at high energies. Finally, it is also worth mentioning that the spectrum of a quantum waveguide of non-constant cross section might be purely discrete. Such a situation occurs, for example, if the waveguide has cusps, i.e. its width tends to zero sufficiently fast at infinity; eigenvalues of such waveguides have been studied, e.g., in [JMS92, GW11, Ko11, EB13].

Section 3.2 The usual definition of the Neumann Laplacian through the appropriate quadratic form can be found, e.g., in [RS, Sect. XIII.15] or [Da, Chap. 7]. For regions with a sufficiently regular boundary the operator domain is specified by the *Neumann boundary condition*, i.e. vanishing of the normal derivative—cf. [DKř02b]. The idea of *Theorem* 3.4 and of its corollary comes from [DKř02a], where this result was proved under stronger assumptions. Notice that in the case $\beta_\Sigma^c > 0$ the product $\phi_\lambda \chi_1$ itself yields a suitable trial function. A similar effect can be seen in other situations, e.g., for curved Dirichlet layers which we will discuss in Sect. 4.1 or for strips embedded into a non-Euclidean space [Kr03]. A numerical analysis of a bent waveguide of the bookcover form with mixed Dirichlet and Neumann conditions was performed in [OM03]. Analogous effects can also be demonstrated for other equations, for instance, those modelling shelf waves along an "outward" curved coast [JLP05].

The straight Dirichlet-Neumann waveguides described in Problems 2 and 4 have been discussed in [DKř02b]. The case of a "window" in the Dirichlet boundary appeared earlier in [ELV94] within the context of a Neumann guide with an obstacle. The main physical motivation comes in this case from acoustics; more general results on embedded eigenvalues in straight channels due to symmetric Neumann obstacles, not necessarily of zero measure, can be found in [DP98] or [KPV00]. The spectrum in the mixed-condition situation of Problem 4, which can be regarded as a two-dimensional version of "twisted" boundary conditions, was found numerically in [DKř02b] using mode matching; the result suggests that the first critical value is $a_1 \approx 0.26\,d$, more recent results on such systems can be found in [BC11, BC12].

Waveguides with combined Dirichlet-Neumann boundary conditions have also been studied from the homogenization point of view: results on the bottom of the spectrum in the situation when the boundary conditions are frequently alternating were established in [BBC10, BBC11a, BBC11b]. For waveguides with Robin boundary conditions one also has various results in addition to that of *Theorem* 3.5, one can mention, for instance, [FKr06, Jí07, Kr09, CDN10, OM10, BMT12].

Proposition 3.2.1 comes from [EHL99]. Notice that this result contrasts with the Faber-Krahn inequality (1.22), because the circular shape *maximizes* the ground-state eigenvalue. A similar behavior can be observed in other problems involving regions which are not simply connected—see, e.g., [HKK01]. In the paper [EHL99] the reader can find a discussion of various related isoperimetric problems. *Theorem* 3.6 comes from [CDFK05], for notions related to curves in $\mathbb{R}^d$ the reader can consult, e.g., [Kli].

Section 3.3 *Theorem* 3.7 comes from [EV99]. The argument also yields a bound on the number of charged bosons which a bent strip can bind, just the first term in (3.20) has to be replaced by the lowest Dirichlet eigenvalue in R_β^N. For fermions we naturally get a stronger restriction because $T_\beta(N) = \mathcal{O}(N^2)$ as $N \to \infty$ and locally it may grow even faster if a is small. It is an open problem whether for large N one can formulate an effective theory analogous to the Thomas-Fermi model of quantum dots [Y99]. Let us mention also that the scattering counterpart of *Corollary* 3.3.1 is usually called a *Coulomb blockade:* a bound state of a charged particle (or a certain number of particles) creates an electrostatic barrier which can make transport of another charged particle through the guide impossible.

Section 3.4 *Theorem* 3.8 is taken from [ELV94], its generalization to n-dimensional waveguides and more general obstacles was established in [DP98]. The proof of the asymptotic behavior of obstacle resonances given in *Theorem* 3.9 can be found in [APV00], where the result was established. An analogous behavior should be observed in the situation when the obstacle is a Neumann cut along a segment parallel to the waveguides axis mentioned in Problem 2b. One can conjecture that resonances would also arise if such a cut placed originally at the strip axis is rotated but the proof of this claim remains an open problem. Various properties of eigenvalues embedded in the continuous spectrum of acoustic waveguides were studied in [CNR12, GB13, Na11a]. In particular, it was shown in [Na11a] that under an appropriate choice of local smooth perturbations of the waveguide it is possible to create an eigenvalue located near any given inner threshold.

3.6 Problems

1. Prove the claim (b) of *Theorem* 3.4.
Hint: Estimate the quadratic form $t[\cdot]$ from below using the ground state eigenvalue of annular regions with Neumann and Dirichlet condition at the inner and outer boundary, respectively—cf. [DKř02a].

2. Let $-\Delta_{DN}^{\Omega_a}$ be the Laplace operator in the strip $\{\vec{x} \,:\, 0 < y < d,\; x \in \mathbb{R}\}$ with the domain consisting of all $\psi \in H^1(\Omega_a)$ with $-\Delta\psi \in L^2$ satisfying the Dirichlet boundary condition if $|x| > a,\; y = d$, and the Neumann condition, $\partial_y\psi = 0$, at the remaining part of the boundary $\partial\Omega_a$.
(a) Show that $\sigma_{\mathrm{ess}}(-\Delta_{DN}^{\Omega_a}) = [\frac{1}{4}\epsilon_d, \infty)$ and $\sigma_{\mathrm{disc}}(-\Delta_{DN}^{\Omega_a}) \neq \emptyset$ for any $a > 0$. Furthermore, the eigenvalues $\epsilon_m(a) \in (0, \frac{1}{4}\epsilon_d)$, $m = 1, \dots, N$, decrease continuously in a and satisfy the bounds $\left(\frac{\pi}{4a}\right)^2 (m-1)^2 \le \epsilon_m(a) < \left(\frac{\pi}{4a}\right)^2 m^2$. Their number is estimated by $N_a \le N \le N_a + 1$, where $N_a := \min\{1, [a/d]\}$, and the critical "window" widths a_m, $m = 2, \dots,$ by $m - 1 \le a_m < m$.
(b) Use this result to demonstrate the existence of embedded eigenvalues for the Neumann Laplacian in a strip of width $2d$ due to an obstacle formed by an additional Neumann condition at a segment of the strip axis.
Hint: Cf. [ELV94, ELU93] and [DKř02b].

3. Let λ_α and χ_α be the lowest eigenvalue λ_α and the associated eigenfunction χ_α, respectively, of the Robin Laplacian on $(-a, a)$ defined in Sect. 3.2.2.
(a) Suppose that $\alpha > 0$. Check that $\lambda_\alpha = k^2$, where k is the first positive root of the implicit equation $k\,\tan(ka) = \alpha$ and $\chi_\alpha(u) = \cos(\sqrt{\lambda_\alpha}u)$ up to a normalization constant.
(b) On the other hand, let $\alpha < 0$. Show that $\lambda_\alpha = -\kappa^2$ where κ is the first positive root of the implicit equation $\kappa\,\tanh(\kappa a) = -\alpha$, and that $\chi_\alpha(u) = \cosh(\kappa u)$, again up to a normalization.

4. Let $-\Delta_{DN}^{\Omega^a}$ be the Laplace operator in the same strip as in the previous problem, however, with Dirichlet boundary condition for $x \operatorname{sgn}\left(y - \frac{1}{2}d\right) > a$ for $y = 0, d$ and Neumann otherwise. We have again $\sigma_{\text{ess}}(-\Delta_{DN}^{\Omega_a}) = [\frac{1}{4}\epsilon_d, \infty)$. Moreover, there is an $a_1 \in (0, d)$ such that $\sigma_{\text{disc}}(-\Delta_{DN}^{\Omega_a}) = \emptyset$ holds for $a \le a_1$, while for $a > a_1$ there is at least one eigenvalue in $(0, \frac{1}{4}\epsilon_d)$. The remaining properties of the discrete spectrum are the same as in Problem 1a, the only modification to do is to put $N_a := [a/d]$.
Hint: The nonexistence of bound states for a sufficiently small positive a can be demonstrated by the variational method which we will describe later in the proof of *Theorem* 6.10—cf. [DKř02b] for details.
5. Let Ω satisfy the assumptions of *Theorem* 1.4. Then the following inequality,

$$\operatorname{tr}(-\Delta_D^\Omega - \nu_1)_-^\delta \le r(\delta, 1)\, L_{\delta,1}^{\text{cl}} I_{\Omega,\delta},$$

holds, where

$$I_{\Omega,\delta} := \int_{\mathbb{R}} \operatorname{tr}\left(-\Delta_D^{M_x} - \nu_1\right)_-^{\delta+1/2} \mathrm{d}\xi = \int_{\mathbb{R}} \sum_n \left(\nu_{n,x} - \nu_1\right)_-^{\delta+1/2} \mathrm{d}\xi .$$

Hint: Use the decomposition $-\Delta_D^\Omega = \int^\oplus (-\Delta_D^{M_x})\,\mathrm{d}x$ similarly as in *Theorem* 3.2.
6. Assume in addition that the waveguide may also have Neumann obstacles or windows, namely there is a precompact set $\mathcal{N} \subset \overline{\Omega}$ with $\mathcal{N} \setminus \Gamma$ open and the longitudinal projection $M_{\mathcal{N}} := \{y \in \mathbb{R}^{d-1} : \exists x \in \mathbb{R} \text{ such that } (x, y) \in \mathcal{N}\}$ of zero measure in $\mathbb{R}^{d-1}$. By $-\Delta_{D\mathcal{N}}^\Omega$ we denote the Laplace operator with Dirichlet boundary conditions on $\partial\Omega \setminus \mathcal{N}$ and Neumann conditions on $\mathcal{N}$ which is associated with the quadratic form $t_{\Omega,\mathcal{N}} := \int_{\Omega\setminus\mathcal{N}} |\nabla\psi|^2(\vec{x})\,\mathrm{d}\vec{x}$ defined on the H^1-closure of the set of all smooth functions in $\Omega\setminus\mathcal{N}$ which vanish for large $|x|$ and in the vicinity of $\partial\Omega\setminus\mathcal{N}$ and which are square integrable together with their first partial derivatives. Furthermore, let n_x denote the transverse cut of $\mathcal{N}$ at $x \in \mathbb{R}$ and $-\Delta_{D,n_x}^{M_x}$ be the corresponding transverse component of the Hamiltonian. We suppose that the spectrum of $-\Delta_{D,n_x}^{M_x}$ below ν_1 is discrete (which is true unless n_x is too "wild"—cf. [HSS91, Si92]) and the functions $x \mapsto \min\{\nu_{n,x}, \nu_1\}$ are measurable.
Assume now that the spectrum of $-\Delta_{D\mathcal{N}}^\Omega$ below ν_1 is discrete and nonempty. The result of the previous problem then generalizes to the present situation as

$$\operatorname{tr}(-\Delta_{D\mathcal{N}}^\Omega - \nu_1)_-^\delta \le r(\delta, 1)\, L_{\delta,1}^{\text{cl}} I_{\Omega,\mathcal{N},\delta},$$

where

$$I_{\Omega,\mathcal{N},\delta} := \int_{\mathbb{R}} \operatorname{tr}\left(-\Delta_{D,n_x}^{M_x} - \nu_1\right)_-^{\delta+1/2} \mathrm{d}\xi = \int_{\mathbb{R}} \sum_n \left(\nu_{n,x} - \nu_1\right)_-^{\delta+1/2} \mathrm{d}\xi .$$

Hint: Follow the same idea—cf. [ELW04].

7. Apply the results of Problem 6 to a strip with a Neumann window, or the symmetric part of the problem considered in Sect. 1.5.1. Show that one has

$$\operatorname{tr}(-\Delta_D^\Omega - \epsilon_d)_-^\delta \le r(\delta, 1)\, L_{\delta,1}^{\mathrm{cl}} \ell \left(\frac{3}{4}\epsilon_d\right)^{\delta+1/2},$$

which is by *Theorem* 1.5 asymptotically correct as $\ell \to \infty$ apart from the factor $r(\delta, 1)$.

8. Show that the weight $(1 + x_3^2)$ in (3.11) cannot be improved in the power-like scale. More precisely, prove that if the function $V \ge 0$ is such that $V(x) \simeq |x_3|^{-2+\varepsilon}$ holds for some $\varepsilon > 0$ as $|x_3| \to \infty$, then $N(H_V^{\Omega_\alpha}, \nu_1) = \infty$.
Hint: One can modify the proof used in [RS, Theorem XIII.6].

9. (a) Prove that the system (3.18) of differential equations determines a continuous family of $(d-1)$-dimensional rotations, unique up to a fixed rotation. Check that (3.18) reduces to the condition (1.18) when $d = 3$ and we put $\mathcal{R} = \begin{pmatrix} \cos\alpha & -\sin\alpha \\ \sin\alpha & \cos\alpha \end{pmatrix}$.
(b) Show that under (3.18) the metric tensor associated with the map (3.17) becomes diagonal, $(g_{ij}) = \operatorname{diag}(h^2, 1, \dots, 1)$, where $h(s,u) := 1 - \gamma_1(s) \sum_\mu \mathcal{R}_{\mu 2}(s) u_\mu$. Using this result check that $-\Delta_D^\Omega$ is unitarily equivalent to the operator $-\partial_s h^{-2} \partial_s - \Delta_D^M + V$ on $L^2(\mathbb{R} \times M)$, where the effective curvature-induced potential is given by

$$V := -\frac{\gamma_1^2}{4h^2} + \frac{1}{2}\frac{h_{ss}}{h^3} - \frac{5}{4}\frac{h_s^2}{h^4}.$$

10. Let Ω be a tube in $\mathbb{R}^d$ described in Sect. 3.2.4, then λ belongs to $\sigma_{\mathrm{ess}}(-\Delta_D^\Omega)$ *iff* there is a sequence $\{\psi_n\} \subset Q(-\Delta_D^\Omega)$ of unit vectors (in the $\|\cdot\|_g$ norm) such that $\operatorname{supp}\psi_n$ is disjoint with $[-n, n] \times M$ and $(-\Delta_D^\Omega - \lambda)\psi_n \to 0$ in $Q(-\Delta_D^\Omega)^*$ as $n \to \infty$.
Hint: Use the form version of Weyl's criterion—cf. Lemma 4.2 in [DDI98].

11. Prove *Theorem* 3.6.
Hint: Use the result of the previous problem and the GJ-argument—cf. [CDFK05].

12. Check that cores of $-\Delta_D^{\Omega_0}$ define through the algebraic tensor product a core for the operator H_N of Sect. 3.3. Prove inequality (3.19) and show that it is sharp if and only if the strip Ω is nontrivially curved.
Hint: Use positivity of the interaction term as in [RS, Theorem XIII.28]. Modify the proof of the standard HVZ theorem to check the inequality $\inf\sigma_{\mathrm{ess}}(H_N) \le \mu_{N-k} + k\kappa_1^2$ with $\mu_j := \inf\sigma(H_N j)$ for $k = 1$, and extend it by induction.

13. Given $N \ge 2$, $a > 0$, and the absolute bending $\int_{\mathbb{R}} |\gamma(s)|\,\mathrm{d}s > 0$, construct a strip $\Omega_{\Gamma,a}$ such that the operator H_N has a nonempty discrete spectrum.
Hint: Take a strip with N bends separated by straight segments long enough to make the repulsion much smaller than the one-particle binding energy in a single bend. Construct a trial function using the result of Problem 1.22.

Chapter 4
Dirichlet Layers

Since our world is three-dimensional, tubular regions are not the only way to confine a quantum particle in a region of the configuration space which is infinitely extended but in some directions only. Here we shall discuss another situation in which the region is a layer Ω obtained by local perturbations of a straight one, $\Omega_0 = \mathbb{R}^2 \times (-a, a)$. We will again suppose that the layer boundary is a hard wall and neglecting unimportant constants we will identify the corresponding Dirichlet Laplacian $-\Delta_D^\Omega$ with the Hamiltonian of the problem.

4.1 Layers of Non-positive Curvature

We are going to discuss first the situation when the mode-coupling perturbation comes from curvature. Physically it may sometimes be easier to prepare such a curved layer than to make a bent tube, but from the mathematical point of view curved layers are more difficult than the strips and tubes discussed in the previous chapters for the simple reason that the geometry of surfaces is more complicated than that of curves. In this section we shall discuss a class of curved layers to which the methods of Sects. 1.1 and 1.3 can be directly adapted.

4.1.1 Geometric Preliminaries

In analogy with the tube case we determine the layer as an a-neighborhood of a surface Σ, so it is natural to start from the parametrization of the latter. We want to define Σ via a map from a suitable set Σ_0, which here will be the plane $\mathbb{R}^2$, to $\mathbb{R}^3$. A new feature is that there is no natural system of coordinates, and in fact, the very assumption that Σ can be covered by an atlas consisting of a single chart, being diffeomorphic to $\mathbb{R}^2$, means a non-negligible restriction.

P. Exner and H. Kovařík, *Quantum Waveguides*,
Theoretical and Mathematical Physics, DOI 10.1007/978-3-319-18576-7_4

We shall go a step further and suppose in the present section that Σ is a C^2-smooth surface which can be equipped with ***geodesic polar coordinates***. This requires the existence of at least one ***pole***, i.e. a point $o \in \Sigma$ such that the exponential mapping $\exp_o : T_o\Sigma \to \Sigma$ is a diffeomorphism (see the notes); in that case Σ is diffeomorphic to $\mathbb{R}^2$ and as such simply connected and non-compact. The coordinate lines are the geodesics emanating from the pole and the geodesic circles which connect points with the same geodesic distance from the pole; let us stress that even at surfaces diffeomorphic to $\mathbb{R}^2$ such a coordinate system may not exist (Problem 1).

Under this assumption we can parametrize the surface Σ, except for the pole, by a map $p : \Sigma_0 \to \mathbb{R}^3$, where $\Sigma_0 := (0,\infty) \times S^1$, with S^1 being the unit circle, is the plane equipped with polar coordinates, $q = (s,\theta)$. The tangent vectors $p_{,\mu} := \partial p/\partial q^\mu$ are linearly independent and their cross-product defines a unit normal field n on Σ. In analogy with (1.5) and (1.16) we define a ***layer*** $\Omega := \mathcal{L}(\Omega_0)$ of width $d = 2a > 0$ over the surface Σ as the image of the straight layer $\Omega_0 := \Sigma_0 \times (-a,a)$ by the function $\mathcal{L} : \Omega_0 \to \mathbb{R}^3$ defined through

$$\mathcal{L}(q,u) := p(q) + un(q)\,. \tag{4.1}$$

To be able to classify layers determined in this way by their metric properties, let us first inspect the geometry of Σ. The surface metric tensor, $g_{\mu\nu} := p_{,\mu} \cdot p_{,\nu}$, has in the geodesic polar coordinates a diagonal form, $(g_{\mu\nu}) = \mathrm{diag}(1, r^2)$, where $r^2 \equiv g := \det(g_{\mu\nu})$ is the square of the Jacobian of the exponential mapping which satisfies the classical Jacobi equation

$$\ddot{r}(s,\theta) + K(s,\theta)\, r(s,\theta) = 0 \quad \text{with} \quad r(0,\theta) = 1 - \dot{r}(0,\vartheta) = 0\,, \tag{4.2}$$

where $\dot{r}$ denotes the partial derivative of r with respect to s. The ***Gauss curvature*** K which appears in (4.2) together with the ***mean curvature*** M are determined in the usual way: the second fundamental form $h_{\mu\nu} := -n_{,\mu} \cdot p_{,\nu}$ gives rise to the Weingarten tensor $h_\mu^{\ \nu} := h_{\mu\rho} g^{\rho\nu}$, which in turn defines the said two curvatures by $K := \det(h_\mu^{\ \nu})$ and $M := \frac{1}{2}\mathrm{tr}(h_\mu^{\ \nu})$. We will also need the corresponding global quantities given by integration with respect to the invariant surface element, $d\sigma := g^{1/2} dq$, the total Gauss curvature $\mathcal{K}$ and the quantity $\mathcal{M}$, defined respectively by

$$\mathcal{K} := \int_{\Sigma_0} K(q)\, d\sigma\,, \quad \mathcal{M} := \left(\int_{\Sigma_0} M(q)^2\, d\sigma \right)^{1/2}.$$

The latter always exists, being possibly infinite, while $\mathcal{K}$ requires the former integral to make sense which will be a matter of an assumption made below; in that case one can derive a useful estimate of the Jacobian (Problem 2).

Next we have to specify the layer geometry. It is convenient to distinguish the tensor component indices using Greek letters for the surface variables and Latin for those of Ω including $q^3 := u$. It follows from (4.1) that the metric tensor of the layer, as of a submanifold in $\mathbb{R}^3$, has the block form

$$(G_{ij}) = \begin{pmatrix} (G_{\mu\nu}) & 0 \\ 0 & 1 \end{pmatrix}, \quad G_{\nu\mu} = (\delta_\nu^\sigma - uh_\nu{}^\sigma)(\delta_\sigma^\rho - uh_\sigma{}^\rho) g_{\rho\mu} .$$

From here the determinant $G := \det(G_{ij})$ is easily computed. Recall that the eigenvalues of the Weingarten map matrix are the ***principal curvatures*** k_1, k_2 through which the Gauss and mean curvatures are expressed as $K = k_1 k_2$ and $M = \frac{1}{2}(k_1 + k_2)$, respectively. It follows that

$$G = g[(1-uk_1)(1-uk_2)]^2 = g(1-2Mu+Ku^2)^2.$$

In particular, this expression defines the volume element by $d\omega := G^{1/2} dq\, du$. As with the tubes, we start the list of assumptions about layer geometry with the requirement that the defining map $\mathcal{L}$ is injective, i.e.

(*i*) Ω is not self-intersecting.

In particular, this imposes a local restriction on the layer thickness which we make here a separate assumption,

(*ii*) $a < \rho_m := (\max\{\|k_1\|_\infty, \|k_2\|_\infty\})^{-1}$, with the principal curvatures assumed to be uniformly bounded, $\|k_j\|_\infty < \infty$ for $j = 1, 2$,

where the number ρ_m has the natural interpretation of the minimal normal curvature radius of Σ; it ensures that the factor $1–2Mu+Ku^2$ is bounded from below on Ω_0 by a positive number. Moreover, $C_- \le 1–2Mu+Ku^2 \le C_+$, where the constants $C_\pm := \left(1 \pm a\rho_m^{-1}\right)^2$ satisfy $0 < C_- < 1 < C_+ < 4$. Together with the C^2-smoothness of Σ this means that $\mathcal{L}$ is a global diffeomeorphism. Another consequence is that $G_{\mu\nu}$ can be estimated by the surface metric,

$$C_- g_{\mu\nu} \le G_{\mu\nu} \le C_+ g_{\mu\nu} . \tag{4.3}$$

This bound is important, because in contrast to the tube case the "straightening" transformation does not allow us here to get rid of the geometry of the generating manifold fully—one cannot unfold Σ into a plane.

Let us further examine how the described parametrization of Ω will be manifested in our model, in which the Hamiltonian is $H = -\Delta_D^\Omega$ with $Q(H) = H^1(\Omega)$. In the coordinates (q, u) it acquires Laplace-Beltrami form

$$\tilde{H} := -G^{-1/2} \partial_i G^{1/2} G^{ij} \partial_j \tag{4.4}$$

acting on $L^2(\Omega_0, G^{1/2} dq\, du)$, where the usual relation between covariant and contravariant tensor components, $G_{ij} G^{jk} := \delta_i^k$, is employed. In other words, we define $\tilde{H} = \tilde{U}(-\Delta_D^\Omega)\tilde{U}^{-1}$, where $\tilde{U} : L^2(\Omega) \to L^2(\Omega_0, d\omega)$ is the unitary operator acting as $\psi \mapsto \psi \circ \mathcal{L}$. Under the C^2-smoothness assumption, $\tilde{H}$ makes sense as the self-adjoint operator associated with the quadratic form

$$t[\psi] \equiv \|\tilde{H}^{1/2}\psi\|_G^2 := (\psi_{,i}, G^{ij}\psi_{,j})_G , \quad Q(\tilde{H}) = H^1(\Omega_0, d\omega) , \tag{4.5}$$

where the subscript indicates the norm and inner product in $L^2(\Omega_0, d\omega)$. Treating $\tilde{H}$ as a partial differential operator defined by means of (4.4) requires in addition the surface Σ to be C^3. Using the block form of the metric tensor, we can write $\tilde{H}$ as a sum of two parts, $\tilde{H} = \tilde{H}_1 + \tilde{H}_2$, with

$$\begin{aligned} \tilde{H}_1 &:= -G^{-1/2}\partial_\mu G^{1/2} G^{\mu\nu}\partial_\nu = -\partial_\mu G^{\mu\nu}\partial_\nu - 2F_{,\mu}G^{\mu\nu}\partial_\nu\,, \\ \tilde{H}_2 &:= -G^{-1/2}\partial_3 G^{1/2}\partial_3 = -\partial_3^2 + 2\,\frac{M-Ku}{1-2Mu+Ku^2}\,\partial_3\,, \end{aligned} \tag{4.6}$$

where we have introduced $F := \ln G^{1/4}$ and expressed $F_{,3}$ explicitly in $\tilde{H}_2$.

Investigating tube Hamiltonians, we have proceeded next to an operator on a straight tube removing the Jacobian by a subsequent unitary transformation. This cannot be done for layers in general, but at least we are able to get rid of the "transverse" factor $1-2Mu+Ku^2$ in the inner-product weight by means of the unitary operator $\hat{U} \,:\, L^2(\Omega_0, d\omega) \to L^2(\Omega_0, d\sigma\, du)$ which acts as

$$\hat{U}\psi := (1-2Mu+Ku^2)^{1/2}\psi\,.$$

Abusing the notation we again employ the symbol H for the transformed operator $U(-\Delta_D^\Omega)U^{-1}$ on $L^2(\Omega_0, d\sigma\, du)$, where $U := \hat{U}\tilde{U}$. For the sake of definiteness the norm and inner product in the corresponding Hilbert space will be indicated by the subscript "g". Computing the operator H explicitly we find that it contains an effective potential which can be conveniently written using the quantity $J := \frac{1}{2}\ln(1-2Mu+Ku^2)$. We have

$$H = -g^{-1/2}\partial_i g^{1/2} G^{ij}\partial_j + V\,, \quad V := g^{-1/2}(g^{1/2}G^{ij}J_{,j})_{,i} + J_{,i}G^{ij}J_{,j}\,, \tag{4.7}$$

which is well defined as a self-adjoint operator under a still stronger smoothness requirement, namely that Σ is C^4. Using again the block form of the metric tensor, we can write the operator as a sum of two parts, $H = H_1 + H_2$, where H_1 contains the part of (4.7) with the summation over Greek indices, and

$$H_2 := -\partial_3^2 + V_2\,, \quad V_2 = \frac{K-M^2}{(1-2Mu+Ku^2)^2}\,. \tag{4.8}$$

An advantage of this form of the Hamiltonian is that it allows us to see how the transverse and longitudinal variable become asymptotically decoupled in the thin layer situation when $a \ll \rho_m$. Using (4.3) one finds

$$H = -g^{-1/2}\partial_\mu g^{1/2} g^{\mu\nu}\partial_\nu - \partial_3^2 + K - M^2 + \mathcal{O}(a) \tag{4.9}$$

(Problem 3). In contrast to (1.9) the first term remembers the geometry of the surface being the Laplace-Beltrami operator Δ_g on Σ. However, as in the tube case the mode-coupling terms vanish in the limit $a \to 0$. To assess the behavior of the leading term of the effective potential, $K - M^2$, it is useful to rewrite it in terms of the principal curvatures as $-\frac{1}{4}(k_1 - k_2)^2$. This expression can vanish not only when Σ is planar, but also if it is locally spherical, i.e. $k_1 = k_2 \neq 0$. Nevertheless, a non-compact and non-planar surface Σ cannot be spherical everywhere. Thus at some parts of such a Σ the curvature-induced interaction is attractive and the discrete spectrum may be nonempty.

4.1.2 Curvature-Induced Bound States

The first question is again to localize the essential spectrum of $-\Delta_D^\Omega$. For a planar layer, i.e. such that $K, M = 0$ holds identically, it coincides clearly with the interval $[\kappa_1^2, \infty)$. We shall call Ω ***asymptotically planar*** if the curvatures vanish at a large geodetic distance from the pole,

(iii) $K(s,\theta),\ M(s,\theta) \to 0$ as $s \to \infty$.

This ensures that the essential spectrum threshold is not pushed down.

Proposition 4.1.1 $\inf \sigma_{\mathrm{ess}}(-\Delta_D^\Omega) \geq \kappa_1^2$ *holds under assumptions (i)–(iii).*

Proof We employ Neumann bracketing dividing Ω into an exterior and interior part, Ω_{ext} and $\Omega_{\mathrm{int}} := \Omega \setminus \bar{\Omega}_{\mathrm{ext}}$, respectively, which are the $\mathcal{L}$ images of $\Omega_{0,s_0} := \Sigma_{0,s_0} \times (-a, a)$ with $\Sigma_{0,s_0} := (s_0, \infty) \times S^1$ for a fixed $s_0 > 0$, and its complement. This yields a lower estimate by a decoupled operator, $H \geq H^N = H^N_{\mathrm{int}} \oplus H^N_{\mathrm{ext}}$, in the sense of quadratic forms. Since the spectrum of H^N_{int} is purely discrete by [Da, Chap. 7], in view of the minimax principle it is sufficient to estimate $\inf \sigma(H^N_{\mathrm{ext}})$. However, for all $\psi \in Q(H^N_{\mathrm{ext}})$ we have

$$t^N_{\mathrm{ext}}[\psi] \geq \|\psi_{,3}\|^2_{G,\mathrm{ext}} \geq \inf_{\Omega_{0,s_0}} \{1 - 2Mu + Ku^2\}\, \|\psi_{,3}\|^2_{L^2(\Omega_0, \mathrm{d}\sigma \mathrm{d}u),\mathrm{ext}}$$

$$\geq \frac{1 - \sup_{\Sigma_{0,s_0}} \{2a|M| + a^2|K|\}}{1 + \sup_{\Sigma_{0,s_0}} \{2a|M| + a^2|K|\}}\, \kappa_1^2\, \|\psi\|^2_{G,\mathrm{ext}} =: (1 + \eta(s_0))\, \kappa_1^2\, \|\psi\|^2_{G,\mathrm{ext}},$$

and $\eta(s_0)$ goes to zero as $s_0 \to \infty$ by assumption *(iii)*. ∎

Remarks 4.1.1 (a) While this lower bound is sufficient for the discussion of the existence of curvature-induced bound states which will follow, one is naturally interested in other properties of the essential spectrum. In *Proposition* 4.2.1 below we will show that under weaker assumptions the opposite inequality is also valid. Moreover, to demonstrate invariance with respect to the curvature-induced perturbation, $\sigma_{\mathrm{ess}}(-\Delta_D^\Omega) = [\kappa_1^2, \infty)$, one can construct a Weyl sequence in analogy with Problem 1.3. To prove its convergence, however, the above assumptions are not sufficient and one has to add requirements which involve derivatives of the Weingarten tensor as well. Such supplementary conditions may be, for instance, the following: Σ is

C^3-smooth, $(h_\mu{}^\nu)_{,\rho} \to 0$ and $k_g := \dot{r}r^{-1} \to 0$ holds as $s \to \infty$ (see the notes). In particular, the last condition is non-covariant with respect to a coordinate change.
(b) On the other hand, there are situations when the essential spectrum invariance is easily verified, for example, layers which are curved in a compact part only. In such a case, of course, the external planar part is important and neither the existence of a pole nor the simple connectedness are needed.

In the tube case *any* nontrivial bending, under mild regularity assumptions, pushes the spectrum threshold down as we demonstrated by variational technique in *Theorems* 1.1 and 1.3. No such universal result is available for curved layers, however, various sufficient conditions can be derived in a similar way: we will seek a function ψ from the form domain of the transformed Hamiltonian $\tilde{H}$ such that the form (4.5) shifted by the threshold value is negative,

$$q[\psi] := t[\psi] - \kappa_1^2 \, \|\psi\|_G^2 < 0\,.$$

The original GJ-idea to construct a trial function starting from a product of the lowest transverse mode eigenfunction and a mollifier applies, however, only to a particular class of curved layers. We shall need the above mentioned assumption concerning the existence of the total Gauss curvature, namely

(*iv*) $K \in L^1(\Sigma_0, \mathrm{d}\sigma)$.

It appears that the sign of $\mathcal{K}$ plays a role here.

Theorem 4.1 *Suppose that Σ is not planar and assumptions (i), (ii), and (iv) are satisfied. If the total Gauss curvature is non-positive, $\mathcal{K} \le 0$, then*

$$\inf \sigma(-\Delta_D^\Omega) < \kappa_1^2\,.$$

In particular, we have $\sigma_{\mathrm{disc}}(-\Delta_D^\Omega) \neq \emptyset$ provided the layer is asymptotically planar, i.e. the assumption (iii) holds in addition.

Proof Following the indicated idea, we begin the construction from the function $\psi : \psi(s,\theta,u) = \phi(s)\chi_1(u)$, where ϕ is a function radially symmetric in the geodesic polar coordinates and otherwise arbitrary for the moment; we use the functions (1.10) as the transverse basis. It is convenient to split the quadratic form t into two parts, $t = t_1 + t_2$, associated with the above mentioned decomposition of $\tilde{H}$. Using the explicit form of $\tilde{H}_2$ we get by a straightforward computation

$$t_2[\psi] - \kappa_1^2 \|\psi\|_G^2 = (\phi, K\phi)_g\,.$$

On the other hand, the longitudinal part $t_1(\psi)$ can be estimated by means of (4.3) and Problem 2 as

$$t_1[\psi] \le C_1 \int_0^\infty |\dot{\phi}(s)|^2 \, s \, \mathrm{d}s\,,$$

where the right-hand side depends on the surface geometry through the constant $C_1 := (C_+/C_-)^2(2\pi+\|K\|_{g,1})$ only, which is finite by assumption *(iv)*. This represents a positive contribution from the trial function tails which we can make small by a suitable choice of the mollifier. In two dimensions we cannot scale an arbitrary function as in (1.11); instead we take a family $\{\phi_\lambda\}$ with

$$\phi_\lambda(s) := \min\left\{1, \frac{K_0(\lambda s)}{K_0(\lambda s_0)}\right\} \tag{4.10}$$

for some $s_0 > 0$ and $\lambda > 0$, where K_0 is the Macdonald function. These functions are not smooth, of course, but the corresponding $\psi_\lambda := \phi_\lambda\chi_1$ is an admissible trial function belonging to $Q(\tilde{H})$. By a direct integration we get

$$\int_0^\infty |\dot{\phi}_\lambda(s)|^2\, s\, \mathrm{d}s < \frac{C_2}{|\ln \lambda s_0|} \tag{4.11}$$

for all λs_0 small enough and some $C_2 > 0$ (Problem 4); hence $t_1[\psi_\lambda]$ can be made arbitrarily small if λ tends to zero. On the other hand, by dominated convergence we get $t_2[\psi_\lambda] - \kappa_1^2\|\psi_\lambda\|_G^2 \to \mathcal{K}$ in the same limit; thus it is enough to choose λ sufficiently small to get the result if $\mathcal{K} < 0$.

To deal with the case $\mathcal{K} = 0$ we add again a small deformation in the central part of the layer setting $\psi_{\lambda,\varepsilon} := \psi_\lambda + \varepsilon f$, where $f(q,u) := j(q)u\chi_1(u)$ with $j \in C_0^\infty((0,s_0)\times S^1)$. Since f obviously belongs to $Q(\tilde{H})$, we can write

$$q[\psi_{\lambda,\varepsilon}] = q[\psi_\lambda] + 2\varepsilon q(f,\psi_\lambda) + \varepsilon^2 q[f]$$

using the fact that the scaling acts outside the support of j. The coefficient of the linear term, $2q(f,\psi_\lambda) = -2(j,M)_g$, can be made nonzero by choosing j supported on a compact set where M does not change sign; such a set certainly exists because Σ is not a plane and the parameter s_0 can be chosen arbitrarily large. This concludes the proof in the same way as in *Theorem* 1.1. ∎

The fact that the negative Gauss curvature leads to a non-void discrete spectrum is intuitively acceptable when we observe relation (4.9): the leading term of the effective potential is more attractive in the parts of Σ where the two principal curvatures have different signs, which means $K < 0$. While being non-universal, the proved theorem still covers a wide class of layers.

Example 4.1.1 (*locally curved layers*) Consider the situation as in Remark 4.1.1b, i.e. a layer built over a surface Σ which is planar outside a compact set. By the Gauss-Bonnet theorem we have $\mathcal{K} = 0$ so the theorem applies. Since the polar coordinate system in the planar region trivially exists, the proof can be extended to surfaces without a pole (Problem 1), or even those which are not diffeomorphic to $\mathbb{R}^2$. We will say more about such extensions in the next section.

Another wide family covered by the hypotheses involves layers built over the ***Cartan-Hadamard surfaces***, which are by definition complete simply connected non-compact surfaces with non-positive Gauss curvature. In view of Cartan-Hadamard theorem each point of such a Σ is a pole so there are infinitely many geodesic polar coordinate systems. Excluding the trivial planar case, the total Gauss curvature is always strictly negative, so such layers possess at least one bound state provided they are asymptotically planar, $\mathcal{K}$ is finite, and assumptions *(i), (ii)* are satisfied.

Examples 4.1.2 (a) *Hyperbolic paraboloid:* the simple quadric in $\mathbb{R}^3$ given by the equation $z = x^2 - y^2$ is an asymptotically planar surface with $\mathcal{K} = -2\pi$, thus *Theorem* 4.1 applies for any layer halfwidth $a < 1/2$.
(b) *Monkey saddle* defined by $z = x^3 - 3xy^2$. This surface is also asymptotically planar, its total Gauss curvature equals -4π, and there is an $a_0 > 0$ such that the corresponding layers are compatible with *(ii)* for $a < a_0$ (Problem 5).

4.2 More General Curved Layers

The reasoning of the previous section was illustrative and directly extended the methods used in Chap. 1. Now we want to discard unnecessary assumptions and include cases which have no analogue in lower dimensions; we shall suppose that Σ is a C^2-smooth *connected orientable* surface embedded in $\mathbb{R}^3$, which is *noncompact* and ***complete***, i.e. no geodetic on Σ is terminated.

4.2.1 Other Sufficient Conditions

We keep the assumptions made in Sect. 4.1.1 and note that the hypotheses about the existence of a pole and simple connectedness were used only in proofs of *Proposition* 4.1.1 and *Theorem* 4.1. If we want to include into considerations generating surfaces for which the atlas need no longer consist of a single chart, it is useful to employ the surface itself as a basis for the parametrization, i.e. to identify the set Σ_0 of the previous section with Σ. We shall again be interested in *asymptotically planar* layers which are in this more general setting specified by the condition

(iii) $K(x),\, M(x) \to 0$ at infinity,

replacing assumption (*iii*) of Sect. 4.1.2; recall that a function f on a noncompact manifold Σ is said to vanish at infinity if to any $\varepsilon > 0$ there are $R_\varepsilon > 0$ and a point $x_\varepsilon \in \Sigma$ such that $|f(x)| < \varepsilon$ holds for any $x \in \Sigma$ with the geodetical distance from x_ε not smaller that R_ε.

Let us first show that the essential spectrum threshold remains preserved.

Proposition 4.2.1 $\inf \sigma_{\text{ess}}(-\Delta_D^\Omega) = \kappa_1^2$ *holds under assumptions (i)–(iii).*

Proof The lower bound is obtained as in *Proposition* 4.1.1 (Problem 6). To prove $\inf \sigma_{\text{ess}}(-\Delta_D^\Omega) \le \kappa_1^2$ notice first that the spectral threshold of the Laplace-Beltrami operator on Σ is zero provided $K \to 0$ at infinity [Do81], hence to any $\varepsilon > 0$ there is an infinite-dimensional subspace $\mathcal{N}_g \subset C_0^\infty(\Sigma)$ such that the inequality $\|\nabla_g \phi\|_g \le \varepsilon \|\phi\|_g$ holds for any $\phi \in \mathcal{N}_g \subset C_0^\infty(\Sigma)$, where ∇_g denotes the vector of covariant derivatives. We will use the simple identity

$$\|\nabla \phi \chi_1\|^2 = \||\nabla \phi|\, \chi_1\|^2 - (\phi \Delta \chi_1, \phi \chi_1) \tag{4.12}$$

valid for all $\phi \in \mathcal{N}_g \subset C_0^\infty(\Sigma)$. By (4.3) the first term on the right-hand side can be estimated as follows,

$$\||\nabla \phi|\, \chi_1\|^2 \le (C_+/C_-^2)\, \|\nabla_g \phi\|_g \le (C_+/C_-^2)\, \varepsilon^2\, \|\phi\, \chi_1\|^2,$$

while for the second one we get

$$-(\phi \Delta \chi_1, \phi \chi_1) = \kappa_1^2\, \|\phi\, \chi_1\|^2 + (\phi \chi_1', 2M_u \phi \chi_1)$$

with M_u defined in Problem 7. Integrating the last term by parts in the variable u and neglecting the negative term coming from the integration we infer from (4.12) that to any $\varepsilon > 0$ there is an $\mathcal{N} := \mathcal{N}_g \otimes \{\chi_1\} \subset C_0^\infty(\Omega)$ such that

$$\|\nabla \psi\|^2 - (\psi, K_u \psi) \le \left(\kappa_1^2 + (C_+/C_-^2)\, \varepsilon^2\right) \|\psi\|^2$$

holds for any $\psi \in \mathcal{N}$, where

$$K_u := \frac{K}{1 - 2Mu + Ku^2}$$

is the Gauss curvature of the parallel surface $\mathcal{L}(\Sigma \times \{u\})$. This proves that $\inf \sigma_{\text{ess}}(-\Delta_D^\Omega - K_u) \le \kappa_1^2$. Since K_u vanishes at infinity by assumption *(iii)*, the operator $K_u(-\Delta_D^\Omega + 1)^{-1}$ is compact in $L^2(\Omega)$, and therefore by Weyl's theorem the result holds for the operator $-\Delta_D^\Omega$ as well. ∎

The place in the proof of *Theorem* 4.1 where geodetic polar coordinates were needed was the choice of the mollifier (4.10). One can use instead the following abstract result which is valid under the assumption *(iv)*.

Lemma 4.2.1 *Let* $K \in L^1(\Sigma)$, *then there is a sequence* $\{\phi_n\}_{n=1}^\infty$ *of smooth functions with compact supports in* Σ *such that*

(a) $0 \le \phi_n \le 1$ *holds for all* $n \in \mathbb{N}$ *and* $x \in \Sigma$,
(b) $\|\nabla_g \phi_n\|_g \to 0$ *as* $n \to \infty$,
(c) $\phi_n(x) \to 1$ *as* $n \to \infty$ *uniformly on compact subsets of* Σ.

Proof Under the stated integrability assumption it follows from Huber's lemma that (Σ, g) is conformally equivalent to a closed surface with a finite number of points removed. However, the integral expressing the norm $\|\nabla_g \varphi_n\|_g$ is a conformal invariant, and it is easy to find a sequence with the required properties on such a pierced closed surface. ∎

Armed with this tool we are ready to prove the claim of *Theorem* 4.1 under weaker assumptions. At the same time we will derive two other sufficient conditions which also apply to surfaces of positive total Gauss curvature. To this end we can no longer construct a trial function from the threshold-resonance function $(x, u) \mapsto \chi_1(u)$, and have to consider from the beginning functions in which the longitudinal and transverse variables are coupled.

Theorem 4.2 *Let Σ be a C^2-smooth and non-planar surface with the properties described at the beginning of this section, such that $K \in L^1(\Sigma)$. Let further the layer Ω built over Σ satisfy assumptions (i) and (ii), then the inequality* $\inf \sigma(-\Delta_D^\Omega) < \kappa_1^2$ *is valid if any of the following conditions holds true:*
(a) $\mathcal{K} \le 0$,
(b) the layer is sufficiently thin, i.e. a is small enough, and $\nabla_g M \in L^2_{\mathrm{loc}}(\Sigma)$,
(c) $\mathcal{M} = \infty$ and $\nabla_g M \in L^2(\Sigma)$.
Consequently, $\sigma_{\mathrm{disc}}(-\Delta_D^\Omega) \ne \emptyset$ holds under the assumption (iii).

Before proving this theorem let us demonstrate its consequences for layers having a nontrivial topology, which means that the generating surface Σ has handles or ends. Recall that an open set $\mathcal{E} \subset \Sigma$ is called an ***end*** of Σ if it is connected, unbounded and its boundary $\partial\mathcal{E}$ is compact (in particular, it may be empty).

Corollary 4.2.1 *Under the hypotheses of Theorem 4.2, we have* $\inf \sigma(-\Delta_D^\Omega) < \kappa_1^2$ *whenever Σ is not conformally equivalent to the plane, in particular, for any surface Σ which is not simply connected.*

Proof Indeed, the Cohn-Vossen inequality says that $\mathcal{K} \le 2\pi\,(2 - 2h - e)$, where h is the genus of Σ, i.e. the number of handles, and e is the number of ends. ∎

Proof of Theorem 4.2 Part (a) is proved as above with the family (4.10) replaced by that of *Lemma* 4.2.1, i.e. we take $\psi_n := \phi_n \chi_1$ which gives

$$q[\psi_n] = \||\nabla \phi_n|\, \chi_1\|^2 + (\phi_n, K\phi_n)_g\ .$$

Since $|\nabla \varphi_n|$ can be estimated by $|\nabla_g \varphi_n|_g$ by means of inequalities (4.3), the first term on the right-hand side tends to zero as $n \to \infty$, while the second one gives $\mathcal{K}$ in the limit by *Lemma* 4.2.1 and dominated convergence. This concludes the proof in the situation when $\mathcal{K} < 0$.

In the critical case, $\mathcal{K} = 0$, we again add a small deformation, $\psi_{n,\varepsilon} := \psi_n + \varepsilon\theta$, where now $\theta(x, u) := j(x) u \chi_1(u)$ with j being a C_0^∞ function supported in a region where M is nonzero and does not change sign. This yields

$$q[\psi_{n,\varepsilon}] = q[\psi_n] + 2\varepsilon\, q(\theta, \psi_n) + \varepsilon^2 q[\theta]\,.$$

Since $\mathcal{K} = 0$, the first term on the right-hand side tends to zero as $n \to \infty$. The form in the second term now requires more attention, because the supports of θ and $\nabla\psi_n$ are not disjoint in general. We express it as

$$q(\theta, \psi_n) = (\theta, 2M_u\, \phi_n\chi_1') + (\nabla\psi_n, \nabla\phi_n) - 2\,(\theta\nabla\chi_1, \nabla\phi_n)\ ,$$

where the last two terms tend to zero as $n \to \infty$ by the the Schwarz inequality, the estimates (4.3) and *Lemma* 4.2.1, while the first integral is calculated to be $-(j, M\phi_n)_g$ as before, with a nonzero limit $-(j, M)_g$. This means that one can conclude the argument in the same way as in *Theorem* 4.1.

To prove the claim under conditions (b) or (c) we modify the trial function ψ_n from the first part of the proof by a multiplicative variable-mixing factor taking $\tilde{\psi}_\lambda(x, u) := (1+M(s, \theta)u)\, \phi_n(x)\chi_1(u)$. Since $\nabla\psi_n(\cdot, u)$ equals

$$(1+Mu)(\nabla\varphi_n)\chi_1(u)+(\nabla M)u\, \varphi_n\chi_1(u)+ \big((1+Mu)\kappa_1\, \varphi_n\chi_1'(u)+M\varphi_n\chi_1(u)\big)\nabla u\,,$$

it is easy to see that $\tilde{\psi}_n \in \mathrm{Dom}\,(Q)$ provided $\nabla_g M \in L^2_{\mathrm{loc}}(\Sigma)$; recall that the curvatures K and M are uniformly bounded by assumption *(ii)*. We have

$$\begin{aligned} q[\tilde{\psi}_n] \le 2\Big((1 + a\|M\|_\infty)^2\, \||\nabla\phi_n|\,\chi_1\|^2 + a^2\||\nabla M|\,\phi_n\chi_1\|^2\Big) \\ + \Big(\phi_n, (K - M^2)\phi_n\Big)_g + \frac{\pi^2-6}{12\kappa_1^2}\Big(\phi_n, KM^2\phi_n\Big)_g\,, \end{aligned}$$

where the inequality giving the factor two comes from the first two terms in $\nabla\psi_n$ as a consequence of (4.3) and Minkovski's inequality; the second line follows from a direct computation using the other two terms of the gradient.

Consider first the condition (c). If $\nabla_g M \in L^2(\Sigma)$ and $K \in L^1(\Sigma)$, then all terms on the right-hand side of the above estimate have finite limits as $n \to \infty$, except for the term containing $K - M^2$ which tends to $-\infty$, hence there is an n_0 such that $q[\tilde{\psi}_{n_0}] < 0$. The sufficiency of the condition (b) follows from two observations. First of all, the integral containing $K - M^2$ is always negative for any non-planar and non-compact surface in view of the remark made at the end of Sect. 4.1.1. Furthermore, the first term in the estimate tends to zero as $n \to \infty$ because of (4.3) and *Lemma* 4.2.1, and the remaining ones vanish for any fixed n as $a \to 0$; recall that $\kappa_1^{-2} = 4a^2/\pi^2$. Hence we can find n_0 large enough so that the sum of the first and the third term is negative, and then choose the layer halfwidth a so small that $q[\tilde{\psi}_{n_0}] < 0$. ∎

Notice that the condition (b) confirms the heuristic expectation based on the formula (4.9), namely that curved layers do have bound states as long as they are asymptotically planar and thin enough. On the other hand, this tells us nothing if we have a fixed layer built over a surface of positive curvature, because we do not know how thin the layer must be for the condition (b) to apply. Sometimes condition (c) can be used.

Example 4.2.1 An elliptic paraboloid is another simple quadric in $\mathbb{R}^3$, given now by the equation $z = \frac{x^2}{b^2} + \frac{y^2}{c^2}$ with $b, c > 0$. It is asymptotically planar with $\mathcal{K} = 2\pi$ and $\mathcal{M} = \infty$, so the theorem applies for any $a < \min\left(\frac{1}{2}b^2, \frac{1}{2}c^2\right)$ (Problem 8). For the particular case $b = c$ we are going to derive another sufficient condition in the following section.

4.2.2 Layers with a Cylindrical Symmetry

There is another class of layers with non-negative total Gauss curvature for which one can establish the existence of the discrete spectrum with no restriction on the width other than the local injectivity assumption *(ii)*. They are characterized by a particular symmetry, namely invariance with respect to rotations around a fixed axis. Suppose thus that Σ is a surface of revolution parametrized by a function $p : \Sigma_0 \to \mathbb{R}^3$ of the form

$$p(s, \theta) := (r(s)\cos\theta, r(s)\sin\theta, z(s)) ,$$

where $r, z \in C^2((0, \infty))$ and $r > 0$. This will be a geodesic polar coordinate chart if we impose the condition $\dot{r}^2 + \dot{z}^2 = 1$, which means that s is the arc length of the curves which are radial cuts of Σ. In such a case we also have $\dot{r}\ddot{r} + \dot{z}\ddot{z} = 0$. An explicit calculation shows that the Weingarten tensor has a diagonal form, $(h_\mu{}^\nu) = \mathrm{diag}(k_s, k_\theta)$, with the principal curvatures $k_s = \dot{r}\ddot{z} - \ddot{r}\dot{z}$ and $k_\theta = \dot{z}r^{-1}$. In fact, it is sufficient to know the function $s \mapsto k_s(s)$ only, because r, z can then be reconstructed using relations analogous to (1.5) with k_s in place of the curvature γ. Recall that the total Gauss curvature of a cylindrically symmetric Σ cannot be negative in view of the Gauss-Bonnet theorem,

$$\mathcal{K} + 2\pi\dot{r}(\infty) = 2\pi\,, \quad \dot{r}(\infty) := \lim_{s\to\infty} \dot{r}(s)\,, \tag{4.13}$$

since $\dot{r}(\infty)$ cannot exceed one in the chosen parametrization. On the other hand, we assume from the beginning that the total Gauss curvature exists which means that the limit value $\dot{r}(\infty)$ makes sense. In addition, r is bounded from below (being positive) which requires $\mathcal{K} \le 2\pi$. Since the case $\mathcal{K} = 0$ is covered by *Theorem* 4.1 we restrict our attention in the following to situations in which $\mathcal{K} \in (0, 2\pi]$, or equivalently, $0 \le \dot{r}(\infty) < 1$.

The last assumption yields a useful estimate of the angular curvature in terms of r^{-1} (Problem 9). In combination with Problem 2 this implies that k_θ does not belong to $L^1(\mathbb{R}_+)$. On the other hand, the meridian curvature k_s is integrable under assumption (iv) as the following estimate shows

$$\int_{s_0}^{\infty} |k_s(s)|\,\mathrm{d}s \le \delta^{-1} \int_{s_0}^{\infty} |k_s(s)k_\theta(s)|\,r(s)\,\mathrm{d}s \le \delta^{-1} \int_0^{\infty} |K(s)|\,r(s)\,\mathrm{d}s\,.$$

These observations give us a hint how to prove the existence of bound states in the present situation. Even if the mean curvature may decay at infinity, it is not negligible there in the integral sense. On the other hand, K is supposed to be integrable which will make it possible to eliminate the term $(\phi, K\phi)_g$ which was for $\mathcal{K} > 0$ an obstacle in the proof of *Theorem* 4.1. To this end we have to find a family of trial functions supported far from the pole of Σ.

Theorem 4.3 *Let Σ be a C^2-smooth non-planar surface of revolution, with assumptions (i), (ii), and (iv) being satisfied. Then* $\inf\sigma(-\Delta_D^\Omega) < \kappa_1^2$ *, in particular, $\sigma_{\mathrm{disc}}(-\Delta_D^\Omega) \ne \emptyset$ holds under the assumption (iii).*

Proof We have noted that in view of *Theorem* 4.1 one may assume $\mathcal{K} > 0$. We employ trial functions $\psi_{n,\varepsilon}(s,u) := (\varphi_n(s) + \varepsilon\phi_n(s)u)\chi_1(u)$, where ε will be specified later and φ_n, ϕ_n are localized far from the center of coordinates as $n \to \infty$. We define them as follows: consider three integer sequences $\{b_n\}$, $\{c_n\}$, $\{d_n\}$, such that $0 < b_n < c_n < d_n$ and $b_n \to \infty$ as $n \to \infty$, then we set

$$\varphi_n(s) := \frac{\ln(s/b_n)}{\ln(c_n/b_n)}\,\chi_{[b_n,c_n]}(s) + \frac{\ln(s/d_n)}{\ln(c_n/d_n)}\,\chi_{[c_n,d_n]}(s)$$

and $\phi_n(s) := s^{-1}\varphi_n(s)$. They are obviously positive and uniformly bounded, and as before, the corresponding functions $\psi_{n,\varepsilon}$ are not smooth but belong to $Q(\tilde{H})$. Using inequalities (4.3) and Problem 2 we can estimate the longitudinal kinetic parts of $q[\psi_{n,\varepsilon}]$ by means of the integrals

$$t_1[\varphi_n\chi_1] \le C_1 \int_0^{\infty} \dot{\varphi}_n(s)^2 s\,\mathrm{d}s\,, \quad t_1[\phi_n u\chi_1] \le 2a^2 C_1 \int_0^{\infty} \dot{\phi}_n(s)^2 s\,\mathrm{d}s\,,$$

which both converge to zero as $n \to \infty$ provided the three sequences diverge at different rates, i.e. c_n/b_n and d_n/c_n tend to infinity as $n \to \infty$. Moreover, the same is true for the mixed term $t_1(\varphi_n\chi_1, \phi_n u\chi_1)$ by the Schwarz inequality. On the other hand, by an explicit integration in the variable u we find that the remaining part $t_2[\psi_{n,\varepsilon}] - \kappa_1^2\|\psi_{n,\varepsilon}\|_G$ of the form equals

$$(\varphi_n, K\varphi_n)_g - 2\varepsilon(\varphi_n, M\phi_n)_g + \varepsilon^2\left[\|\phi_n\|_g^2 + \frac{\pi^2-6}{3\kappa_1^2}(\phi_n, K\phi_n)_g\right].$$

In the limit $n \to \infty$ the contribution of the terms containing Gauss curvature vanish in view of *(iv)* and the facts that φ_n and ϕ_n are uniformly bounded and their supports move towards infinity as n increases. In this way we arrive at

$$\lim_{n\to\infty} q[\psi_{n,\varepsilon}] = \lim_{n\to\infty}\left[\varepsilon^2\|\phi_n\|_g^2 - 2\varepsilon(\varphi_n, M\phi_n)_g\right] \tag{4.14}$$

on the presumption that the limit on the right-hand side exists. Now we shall make the parameter ε also dependent on n putting $\varepsilon_n := (\varphi_n, M\phi_n)_g^{-1}$ which is a reasonable choice as long as the integral tends to infinity as $n \to \infty$ for particular choices of the sequences; in this way we are trying to achieve

$$\lim_{n\to\infty} \frac{(\phi_n, \phi_n)_g}{(\varphi_n, M\phi_n)_g^2} < 2\,.$$

In the particular case of cylindrically symmetric surfaces when the information concerning the behavior of M at large distances is available, one can show that the limit on the left-hand side is zero. Indeed, since k_s is integrable and ϕ_n is chosen in a way to eliminate the weight r with the help of Problem 2, the contribution from the meridian curvature in the denominator can be neglected as long as the other part diverges. Furthermore, the result of Problem 9 allows us to estimate the factor $|k_\theta r|$ from below by a positive constant at large distances. Using in addition the result of Problem 2 in the numerator, we conclude that it is sufficient to check that the limit of the expression

$$\frac{\int_0^\infty \phi_n(s)^2 s\, \mathrm{d}s}{\left(\int_0^\infty \varphi_n(s)\phi_n(s)\mathrm{d}s\right)^2} = \frac{3}{\ln(d_n/b_n)}$$

vanishes as $n \to \infty$. This happens, e.g., if we choose $b_n = n,\ c_n = n^2,\ d_n = n^3$ with $n \geq 2$, and such sequences also satisfy the other requirements put forth above. In this way the described construction yields a family of trial functions which make the shifted energy form negative for all n large enough. ∎

Requiring asymptotic planarity, we conclude again that $\sigma_{\mathrm{disc}}(-\Delta_D^\Omega)$ is not void. The assumption *(iii)* is necessary, though, because abandoning it we can obtain layers in which the threshold of the essential spectrum is lowered and no bound states exist (Problem 10). The proved result concerns various situations which are not covered by the previous sufficient conditions.

Examples 4.2.2 (a) *Hyperboloid of revolution:* Consider one of the two sheets of the hyperboloid given by $x^2+y^2-z^2\tan^2\vartheta = 1$. It is an asymptotically planar and cylindrically symmetric surface with the total Gauss curvature $\mathcal{K} = 2\pi\cos^2\vartheta(1+\sin\vartheta)^{-1}$ which varies over the interval $(0, 2\pi)$ as a function of ϑ. The assumption *(ii)* is satisfied provided $a < \tan\vartheta$ (Problem 11).
(b) *A nonintegrable* $|\nabla_g M|^2$: There are asymptotically planar surfaces of revolution with curvatures which satisfy *(iv)* but violate the other integral requirements. For an example take $k_s(s) := s^{-2}\sin s^2$ and reconstruct the functions r, z using $b(s) := \int_0^s k_s(\xi)\,\mathrm{d}\xi$ and formulæ analogous to (1.5). It is easy to check that there is a positive c such that $r(s) \geq cs$ for all $s \in \mathbb{R}_+$. Thus $k_\theta = \dot{z}r^{-1} \to 0$ as $s \to \infty$ because $|\dot{z}| = |\sin b| \leq 1$, and the meridian curvature k_s also vanishes in the limit as seen from its definition. It follows that the corresponding surface Σ is asymptotically planar. At the same time, $|K|r = |k_s\dot{z}| \leq |k_s|$ belongs to $L^1(\mathbb{R}_+)$ which gives *(iv)*. On

the other hand, while it is true that $\dot{k}_\theta = k_s r^{-1} \cos b - r^{-2} \sin b \cos b$ is an element of $L^2(\mathbb{R}_+, r(s)\,\mathrm{d}s)$, the same does not hold for the other derivative, $\dot{k}_s$, as is again clear from its definition. Consequently, the vector $\nabla_g M = (\dot{M}, 0)$ does not satisfy the requirement in assumption (c) of *Theorem* 4.2.

In fact, the used construction allows us to make a stronger claim.

Corollary 4.2.2 $\sharp\,\sigma_{\mathrm{disc}}(-\Delta_D^\Omega) = \infty$ *holds if the assumptions of Theorem 4.3 are satisfied and* Σ *is asymptotically planar.*

Proof One has to find an infinite-dimensional subspace on which the shifted energy form is negative. To this end it is sufficient to choose the trial function sequence in the preceding proof in such a way that their supports are disjoint; this happens, e.g., if we set $b_n = 2^{n^2}$, $c_n = 2^{n(n+1)}$, and $d_n = 2^{n(n+2)}$. ∎

Remark 4.2.1 (partial wave decomposition) If the layer Ω has a rotational symmetry, $-\Delta_D^\Omega$ can be naturally decomposed into parts with fixed angular momentum. Consider Cartesian coordinates in $\mathbb{R}^3$ and suppose that the symmetry refers to the third axis. Using polar coordinates $x_1 = \rho\cos\theta$ and $x_2 = \rho\sin\theta$ in the plane $x_3 = 0$, we can decompose $L^2(\Omega)$ and $-\Delta_D^\Omega$ in the same way as one does it for the Dirichlet Laplacian in a planar layer—see relation (5.15) below. Now one can proceed as in Sect. 1.1 introducing

$$\rho := r(s) - u\dot{z}(s)\,, \quad x_3 := z(s) + u\dot{r}(s)\,,$$

where s is the arc length of the curve which is a radial cut of Σ; the meridian curvature $k_s = \dot{r}\ddot{z} - \ddot{r}\dot{z}$ coincides up to the sign with the signed curvature of this curve. On $\Omega_0^+ := \mathbb{R}_+ \times (-a, a)$ we introduce $\tilde{g} := (1 - uk_s)^2$ and the measure $\mathrm{d}\tilde{g} := \tilde{g}^{1/2}\mathrm{d}s\,\mathrm{d}u$. Transforming the partial-wave components into the coordinates (s, u) we get the operators

$$\tilde{H}_m = -\tilde{g}^{-1/2}\partial_s \tilde{g}^{-1/2}\partial_s - \tilde{g}^{-1/2}\partial_u \tilde{g}^{1/2}\partial_u + \frac{4m^2 - 1}{4\rho(s, u)^2}$$

on $L^2(\Omega_0^+, \mathrm{d}\tilde{g})$ with $\{\psi \in H^1(\Omega_0^+, \mathrm{d}\tilde{g}) :\ \tilde{H}_m\psi \in L^2(\Omega_0^+, \mathrm{d}\tilde{g}),\ \psi(\cdot, \pm a) = 0\}$ as the domain for $m \neq 0$, while in the s-wave one has to add the condition $\lim_{s\to 0} \psi(s, \cdot)(s^{1/2} \ln s)^{-1} = 0$ (we take into account that $r(s) = s(1 + \mathcal{O}(s))$ as $s \to 0$ in view of (4.2)). From here one can proceed to operators on $L^2(\Omega_0^+)$ with an effective potential analogous to (1.8) (Problem 12).

To prove the existence of a nonempty discrete spectrum it is more convenient to work with the original operator, however, partial-wave components are useful for a more detailed analysis of the discrete spectrum. Moreover, the decomposition provides an insight which has inspired the proof of *Theorem* 4.3, namely that choosing a family of trial function supported far from the center is a good guess. By (4.13) we have $\dot{r}(\infty) < 1$ for $\mathcal{K} > 0$ and $|\rho(s, u) - r(s)| < a$, hence $\frac{1}{4}(s^{-2} - \rho(s, u)^{-2})$ represents a long-range attractive potential in the s-wave component. It is also clear that in the other partial waves this effect is absent and the existence of bound states with $m \neq 0$ depends on the balance between the curvature-induced attraction and the repulsive centrifugal term.

The fact that any compact neighborhood of the origin can be left out in the construction of the trial function family allows us to extend the results to a much wider class of layers such that the generating surface need not be simply connected and the layer may not have a fixed width or a smooth boundary in a compact region. We have encountered a similar situation in Problems 1.5 and 1.6, however, now we are able to make a stronger claim.

Theorem 4.4 *Let Σ be a C^2-smooth non-planar surface which has at least one cylindrically symmetric end with a positive total Gauss curvature. Suppose that the layer Ω built over Σ satisfies assumptions (i)–(iv), and furthermore, that Ω' is a region in $\mathbb{R}^3$ obtained by a compact deformation of Ω. Then we have* $\inf \sigma_{\text{ess}}(-\Delta_D^{\Omega'}) = \kappa_1^2$ *and* $\sharp\, \sigma_{\text{disc}}(-\Delta_D^{\Omega'}) = \infty$.

Proof Let E be an end with the required properties. If $E = \Sigma$ the result follows from *Corollary* 4.2.2, in the opposite case we use E to construct a new surface E' by attaching smoothly to it a cylindrically symmetric cap C, i.e. a simply connected surface with a compact boundary. Using the Gauss-Bonnet theorem in a way similar to (4.13) we find that $\mathcal{K}_C \geq 0$, and consequently, the total Gauss curvature of E' cannot be smaller than $\mathcal{K}_E$ which means that it is positive. Then the mean curvature of E' behaves at infinity as in the proof of *Corollary* 4.2.2, and the same is true for the mean curvature of E. Thus we may use the sequence of trial functions constructed above because for n large enough their supports are contained in E. Finally, since the essential spectrum is stable under compact deformations and the trial function family can be chosen with supports outside the deformation region, the result extends to Ω'. ∎

Example 4.2.3 Given $\alpha \in \left(0, \frac{1}{2}\pi\right)$, consider a *conical layer* of a width d given in the cylindrical coordinate parametrization by

$$0 < z - \rho \cot \alpha < \frac{d}{\sin \alpha}\,.$$

Since the layers of different thicknesses are related by scaling transformations, we can put $d = \pi$ without loss of generality denoting such a layer as Ω_α. It does not fall into the class of smoothly curved layers discussed so far, of course, but it can be obtained as a compact deformation of the layer built over a smoothed cone. By (4.13) the total Gauss curvature of the latter equals $2\pi(1 - \sin \alpha)$, hence $\inf \sigma_{\text{ess}}(-\Delta_D^{\Omega_\alpha}) = 1$ and $\sharp\, \sigma_{\text{disc}}(-\Delta_D^{\Omega_\alpha}) = \infty$. Notice also that for small α one has a lower bound analogous to that of *Proposition* 1.2.1 (Problem 13). While this is a weaker claim than that following from *Theorem* 4.4, it suggests that a very sharp cylindrical layer may also have numerous bound states supported in the vicinity of its tip.

4.3 Locally Perturbed Layers

Similarly as in the case of strips and tubes, there are other binding mechanisms which can give rise to localized states. Some are of non-geometric type being based, for instance, on an attractive potential or a suitable modification of the kinetic term of the Hamiltonian in a direct analogy with the considerations of Sect. 1.4. We leave this to the reader (Problems 14 and 15).

A geometric binding mechanism is based on suitable local deformations of the layer. For the sake of simplicity we limit ourselves to discussion of a one-sided and strictly local layer protrusion defining Ω_λ^f as

$$\Omega_\lambda^f := \{\vec{x} = (x, y) \in \mathbb{R}^3 : x \in \mathbb{R}^2,\ 0 < y < d + \lambda f(x)\} \tag{4.15}$$

with $\lambda > 0$ and a fixed compactly supported function f; for convenience we may drop indices λ or f from the notation. Then we have the following result.

Theorem 4.5 *Let $f : \mathbb{R}^2 \to [0, \infty)$ be bounded, and suppose that there are $\eta > 0$ and an open set $\mathcal{W} \subset \mathbb{R}^2$ such that $f(x) > \eta$ holds for $x \in \mathcal{W}$. Then $-\Delta_D^{\Omega_\lambda}$ has at least one discrete eigenvalue whenever $\lambda > 0$, and furthermore, there is exactly one such eigenvalue for all λ small enough.*

Proof A standard bracketing argument shows that the essential spectrum of $-\Delta_D^{\Omega_\lambda}$ is $[\kappa_1^2, \infty)$ where $\kappa_1 = \pi/d$ as usual. Due to the eigenvalue monotonicity of the Dirichlet Laplacian with respect to a domain expansion we may suppose without loss of generality that $f \in C_0^\infty(\mathbb{R}^2)$. Indeed, if this is not the case, we can find $g, h \in C_0^\infty(\mathbb{R}^2)$ such that $0 \le g \le f \le h$. Since the essential spectrum threshold is the same for all of them, the existence result for Ω_λ^g implies eigenvalue existence for Ω_λ^f, and *vice versa*, from the eigenvalue uniqueness of the Dirichlet Laplacian in Ω_λ^h the same property for Ω_λ^f will follow.

Now take R large enough to have $\operatorname{supp} f \subset B_R$, where B_R is the disc of radius R centered at the origin. For an arbitrary $\delta \in (0, 1)$ we employ the function $\phi_\delta : \mathbb{R}^2 \to [0, 1]$ which is equal to one in $\overline{B}_R$ and

$$\phi_\delta(x) := |\ln \delta|^{-1} \Big(\ln \frac{R}{\delta\, |x|} \Big)_+ \quad \text{for } |x| > R\,.$$

Choosing $\varepsilon \in (0, \lambda)$ we have $\Omega_\varepsilon \subset \Omega_\lambda$; we define a trial function $u_{\varepsilon,\delta}$ on Ω_λ by

$$u_{\varepsilon,\delta}(x, y) := \phi_\delta(x) \sqrt{\frac{2}{d + \varepsilon f(x)}}\, \sin\Big(\frac{\pi y}{d + \varepsilon f(x)}\Big)$$

for $(x, y) \in \Omega_\varepsilon$ and $u_{\varepsilon,\delta}(x, y) := 0$ elsewhere. It is straightforward to check that $u_{\varepsilon,\delta} \in H_0^1(\Omega_\lambda)$ holds for all $\delta \in (0, 1)$ and $\varepsilon \in (0, \lambda)$; by a direct computation we obtain the inequality

$$\|\nabla u_{\varepsilon,\delta}\|^2 - \kappa_1^2\,\|u_{\varepsilon,\delta}\|^2 \le |\ln\delta|^{-1} + C\,\varepsilon^2 + \int_{\mathbb{R}^2}\Big(\frac{\pi^2}{(d+\varepsilon f(x))^2} - \kappa_1^2\Big)\,\mathrm{d}x\,,$$

where the constant C depends on R only. Taking $\varepsilon = |\ln\delta|^{-1/2}$ and δ small enough we can make the right-hand side of the above inequality negative which proves the existence of an eigenvalue below κ_1^2. For the uniqueness for λ small enough we refer to Remark 4.3.1. ∎

In connection with the second claim of this theorem one naturally expects that the number of such eigenvalues will increase with the growing bulge on the layer. The next result justifies this conjecture and makes it more precise in terms of the deformation function f. As in Sect. 3.1 we use the notation $N(H,\epsilon) := \sharp\sigma(H-\epsilon)_-$ in which we drop ϵ when it is clear from the context.

Theorem 4.6 *Let $f \in C_0^\infty(\mathbb{R}^2)$ be supported in the disc $B_R \subset \mathbb{R}^2$ centered at the origin and $\|f\|_\infty < d$. Then there are constants C_j, $j = 1, 2, 3$, depending on f and the radius R, such that the number of eigenvalues is bounded by*

$$N\big(-\Delta_D^{\Omega_\lambda},\,\kappa_1^2\big) \le N(-\Delta_{\mathbb{R}^2} + 3\,V_{\lambda f})\,, \tag{4.16}$$

where the Schrödinger operators on the right-hand side contains the potential term determined through the formula

$$V_f := -\kappa_1^2\,\frac{f(2d+f)}{(d+f)^2} + C_1\,|\nabla f|^2 + C_2\,|\Delta f|^2 + C_3\,|\nabla f|^4\,.$$

Proof It suffices to prove the claim for $\lambda = 1$. We write Ω_1 instead of Ω_1^f throughout in order to simplify the notation; to distinguish the operators used, we write ∇_x and Δ_x for those acting in $L^2(\mathbb{R}^2)$. Any trial function $\psi \in H_0^1(\Omega_1)$ can be written in the form

$$\psi(x,y) = \varphi(x,y)\,g(x) + h(x,y), \tag{4.17}$$

where

$$\varphi(x,y) = \sqrt{\frac{2}{d+f(x)}}\,\sin\Big(\frac{\pi y}{d+f(x)}\Big)\,, \quad g \in H^1(\mathbb{R}^2)\,,$$

and $h \in H_0^1(\Omega_1)$ is transversally orthogonal to the lowest mode, i.e.

$$\int_0^{d+f(x)} \varphi(x,y)h(x,y)\,\mathrm{d}y = 0 \quad \text{for all } x \in \mathbb{R}^2\,. \tag{4.18}$$

The quadratic form of $-\Delta_D^{\Omega_1} - \kappa_1^2$ can be for such a ψ expressed as

$$\int_{\Omega_1} \left(|\nabla\psi|^2 - \kappa_1^2\,|\psi|^2\right)(x,y)\,\mathrm{d}x\,\mathrm{d}y = \int_{\Omega_1} \Big(|\nabla\varphi|^2|g|^2 + |\nabla_x g|^2|\varphi|^2 + |\nabla h|^2 - \kappa_1^2\,(|\varphi\, g|^2 + |h|^2) + 2g\nabla_x\varphi\cdot\nabla_x h + 2g\varphi\nabla_x\varphi\cdot\nabla_x g + 2g\,\partial_y\varphi\,\partial_y h + 2\varphi\,\nabla_x g\cdot\nabla_x h\Big)(x,y)\,\mathrm{d}x\,\mathrm{d}y\,.$$

To estimate the first two mixed terms in the above identity we note that

$$\begin{aligned} 2\,|g\,\varphi\,\nabla_x\varphi\cdot\nabla_x g| &\le a_1^{-1}\,|\varphi\,\nabla_x g|^2 + a_1\,|g\nabla_x\varphi|^2\,,\\ 2\,|g\,\nabla_x\varphi\cdot\nabla_x h| &\le a_2^{-1}\,|\nabla_x h|^2 + a_2\,|g\nabla_x\varphi|^2\,, \end{aligned} \tag{4.19}$$

where a_1 and a_2 are real positive numbers the values of which will be specified later. As for the last two terms on the right-hand side of the quadratic form expression, integration by parts in combination with (4.18) gives

$$\begin{aligned} \int_{\Omega_1} g\,\partial_y\varphi\,\partial_y h\,\mathrm{d}x\,\mathrm{d}y &= -\int_{\Omega_1} g\,h\,\partial_y^2\varphi\,\mathrm{d}x\,\mathrm{d}y = 0\,,\\ \int_{\Omega_1} \varphi\,\nabla_x g\cdot\nabla_x h\,\mathrm{d}x\,\mathrm{d}y &= \int_{\Omega_1} g(\Delta_x\varphi\, h + \nabla_x\varphi\cdot\nabla_x h)\,\mathrm{d}x\,\mathrm{d}y\,. \end{aligned}$$

The product $g\nabla_x\varphi\cdot\nabla_x h$ is estimated as in (4.19), for the rest we use the inequality

$$|g\,\Delta_x\varphi\, h| \,\le\, a_3\, g^2\,|\Delta_x\varphi|^2 + a_3^{-1}\,h^2\,\chi_f\,,$$

where χ_f denotes the characteristic function of the support of f. Now we put $a_1 = a_2 = 3$ arriving thus at the inequality

$$\begin{aligned} \int_{\Omega_1} \left(|\nabla\psi|^2 - \frac{\pi^2}{d^2}\,|\psi|^2\right)\mathrm{d}x\,\mathrm{d}y &\ge \int_{\mathbb{R}^2} \left(\frac{1}{3}\,|\nabla_x g|^2 + \tilde V_f(x)|g|^2\right)\mathrm{d}x\,\mathrm{d}y\\ &\quad + \int_{\Omega_1} \left(\frac{1}{3}\,|\nabla_x h|^2 + |\partial_y h|^2 - \kappa_1^2\,|h|^2 - a_3^{-1}\,|h|^2\,\chi_f\right)\mathrm{d}x\,\mathrm{d}y \end{aligned} \tag{4.20}$$

with

$$\tilde V_f(x) := \frac{\pi^2}{(d+f(x))^2} - \kappa_1^2 - \int_0^{d+f(x)} \left(8\,|\nabla_x\varphi|^2 + a_3\,|\Delta_x\varphi|^2\right)\mathrm{d}y\,.$$

The function h satisfies Dirichlet boundary conditions at $\partial\Omega_1$, hence it can be extended continuously by zero to $H^1(\mathbb{R}^2\times(0,2d))$; with a slight abuse of notation we will also use the symbol h for this extension. From the assumption $\|f\|_\infty < d$ and from (4.18) we then deduce that

$$\int_{\Omega_1}\left(\frac{1}{3}|\nabla_x h|^2+|\partial_y h|^2-\kappa_1^2|h|^2-a_3^{-1}|h|^2\chi_f\right)\mathrm{d}x\,\mathrm{d}y$$
$$\ge\int_0^d\int_{\mathbb{R}^2}\left(\frac{1}{3}|\nabla_x h|^2+3\kappa_1^2|h|^2-\left(a_3^{-1}+3\kappa_1^2\right)|h|^2\chi_f\right)\mathrm{d}x\,\mathrm{d}y$$
$$+\int_d^{2d}\int_{\operatorname{supp} f}\left(\frac{1}{3}|\nabla_x h|^2-a_3^{-1}|h|^2\right)\mathrm{d}x\,\mathrm{d}y\,.$$

Since $h(\cdot,y)\in H^1(\operatorname{supp} f)$ for every $y\in(d,2d)$ and since $\operatorname{supp} f\subset B_R$, it follows that the last term is non-negative for all $a_3\ge 3\mu(R)^{-1}$, where $\mu(R)$ is the lowest eigenvalue of $-\Delta_x$ on the disc B_R with Dirichlet boundary conditions. Moreover, the expression in the second line can be bounded from below as

$$\int_0^d\int_{\mathbb{R}^2}\left(\frac{1}{3}|\nabla_x h|^2+3\kappa_1^2|h|^2-\left(a_3^{-1}+3\kappa_1^2\right)|h|^2\chi_f\right)\mathrm{d}x\,\mathrm{d}y$$
$$\ge\int_0^d\left(\int_0^\infty\int_0^{2\pi}\left(\frac{1}{3}|\nabla_{r,\theta} h|^2+|h|^2\left(3\kappa_1^2\chi_{[R,\infty)}-\frac{\chi_{[0,R]}}{a_3}\right)\right)r\mathrm{d}r\,\mathrm{d}\theta\right)\mathrm{d}y\,,$$

where we have employed the polar coordinates $(r,\theta)\in(0,\infty)\times[0,2\pi)$. In view of Problem 16, the left-hand side of the last inequality is non-negative for $a_3\ge\max\left\{8R^2,\ \frac{1}{3}\kappa_1^{-2}\right\}$, thus we can choose

$$a_3(R)=\max\left\{\frac{1}{3}\kappa_1^{-2},\ 8R^2,\ 3\mu(R)^{-1}\right\}.$$

Now it remains to estimate the first term on the right-hand side of (4.20). By a direct calculation we arrive at

$$\int_0^{d+f}\left(8|\nabla_x\varphi|^2+a_3(R)\,|\Delta_x\varphi|^2\right)\mathrm{d}y\le C_1\,|\nabla f|^2+C_2\,|\Delta f|^2+C_3\,|\nabla f|^4\,,$$

where C_1, C_2, C_3 are positive numbers which depend only on R. Finally, combining (4.20) with the orthogonality condition (4.18) we obtain

$$\int_{\Omega_1}\left(|\nabla\psi|^2-\kappa_1^2|\psi|^2\right)\mathrm{d}x\,\mathrm{d}y\ge\frac{1}{3}\int_{\mathbb{R}^2}\left(|\nabla_x g|^2+3V_f(x)\,|g|^2\right)\mathrm{d}x\,. \quad (4.21)$$

Let us show that the last inequality implies (4.16). To this end we consider the subspace $\mathcal{M}_0\subset L^2(\mathbb{R}^2)$ spanned by the eigenvectors corresponding to the negative eigenvalues of $-\Delta_{\mathbb{R}^2}+3V_f$ in $L^2(\mathbb{R}^2)$. Its dimension obviously does not exceed $N(-\Delta_{\mathbb{R}^2}+3V_f)$. Next we define $M_0\subset L^2(\Omega_1)$ by $M_0:=\{g\varphi\,:\,g\in\mathcal{M}_0\}$. We assume that $\psi\perp M_0$ and write $\psi=\tilde g\varphi+h$ as in (4.17). Then we have $\tilde g\varphi\perp M_0$, and since $\int_0^{d+f(x)}|\varphi(x,y)|^2\,dy=1$ holds for all $x\in\mathbb{R}^2$, this means that $\tilde g\perp\mathcal{M}_0$.

This means that the right-hand side of (4.21) is non-negative, hence by the variational principle we find

$$N\Big(- \Delta_D^{\Omega_1},\, \kappa_1^2\Big) \,\le\, \dim M_0 = \dim \mathcal{M}_0 \,\le\, N(-\Delta_{\mathbb{R}^2} + 3V_f)\,,$$

which concludes the proof. ∎

Remark 4.3.1 The estimate (4.16) in combination with the uniqueness of the weakly coupled eigenvalue of two-dimensional Schrödinger operators yields the remaining claim of *Theorem* 4.5, namely that $-\Delta_D^{\Omega_\lambda}$ has exactly one simple eigenvalue below κ_1^2 for λ small enough, because the potential is then dominated by the negative term linear in λ. We are going to return to this problem in Chap. 6—see in particular *Theorem* 6.6—where we shall explain that positivity of f is in fact not needed in this case.

4.4 Laterally Coupled Layers

Next we turn to the three-dimensional analogue of the window-coupled waveguides discussed in Sect. 1.5. Consider a pair of adjacent layers of widths $d_j > 0$, $j = 1, 2$, and fix a bounded set $\mathcal{W} \subset \mathbb{R}^2$ which will play the role of the window connecting the two layers. In analogy with the two-dimensional situation we set $\Omega := \mathbb{R}^2 \times (-d_2, d_1) \setminus B_{\mathcal{W}}$, where $B_{\mathcal{W}} := \mathbb{R}^2 \setminus \mathcal{W}$, and define $H(d_1, d_2; \mathcal{W})$ in $L^2(\Omega)$ as the corresponding Dirichlet Laplacian $-\Delta_D^{\Omega}$. The analysis again simplifies in the mirror-symmetric case, $d_1 = d_2$, when the antisymmetric part is trivial and the symmetric one reduces to the investigation of the Laplacian in $L^2(\mathbb{R}^2 \times (0, d))$ with Neumann boundary conditions on $\mathcal{W} \times \{0\}$ and Dirichlet condition on the remaining part of the layer boundary.

Theorem 4.7 *Let $\mathcal{W} \subset \mathbb{R}^2$ be open bounded set and $d := \max\{d_1, d_2\}$. The essential spectrum of $H(d_1, d_2; \mathcal{W})$ coincides with the halfline $[\epsilon_d, \infty)$, where $\epsilon_d := (\frac{\pi}{d})^2$, and $\sigma_{\rm disc}(H(d_1, d_2; \mathcal{W}))$ is non-empty whenever $\mathcal{W} \ne \emptyset$. Moreover, if $\mathcal{W} = aM$ for a fixed simply connected set $M \subset \mathbb{R}^2$ and $a > 0$, then $H(d_1, d_2; \mathcal{W})$ has exactly one eigenvalue below ϵ_d for all a small enough.*

Proof The fact that $\sigma_{\rm ess}(H(d_1, d_2; \mathcal{W})) = [\epsilon_d, \infty)$ follows by a bracketing argument analogous to the one used for window-coupled strips in Sect. 1.5.1. Concerning the existence of the discrete spectrum we first note that as an open set $\mathcal{W}$ contains a disc and that shrinking the window to it raises the eigenvalues. Hence to prove that $\sigma_{\rm disc}(H(d_1, d_2; \mathcal{W}))$ is non-void we may assume without loss of generality that $\mathcal{W} = B_R$, where B_R is a disc of radius $R > 0$. Consider first the symmetric case, $d_1 = d_2 = d$. Here we employ trial functions of the form $\psi_\lambda(x, y) = f_\lambda(x)\, \chi_1(y) + \eta\, g(x, y)$ with

$$f_\lambda(x) := \min\left\{1,\, \frac{K_0(\lambda\,|x|)}{K_0(\lambda R)}\right\}, \qquad g(x,y) = \phi_1(x)\, r(y)\,,$$

where $K_0(\cdot)$ is the Macdonald function, ϕ_1 is the normalized ground-state eigenfunction of the Dirichlet Laplacian $-\Delta_D^{\mathcal{W}}$ on $\mathcal{W} = B_R$ extended by zero to the whole $\mathbb{R}^2$ and corresponding to the eigenvalue $\mu_1 > 0$, and

$$r(y) = \begin{cases} \mathrm{e}^{-\sqrt{\mu_1}\, y} & \dots\ y \in \left(0, \frac{1}{2}\right) \\ 2\left(1 - \frac{y}{d}\right) \mathrm{e}^{-\frac{d}{2}\sqrt{\mu_1}} & \dots\ y \in \left[\frac{1}{2}d, d\right) \end{cases}$$

Using integration by parts in the variable y and the fact that ∇f_λ and $\nabla\phi_1$ have disjoint supports, we can express the quadratic form in question as

$$\|\nabla\psi_\lambda\|^2_{L^2(\mathbb{R}^2)} - \epsilon_d \|\psi_\lambda\|^2_{L^2(\mathbb{R}^2)} = \|\nabla f_\lambda\|^2_{L^2(\mathbb{R}^2)} + C_1\eta^2 - C_2\eta\,,$$

where C_1 and $C_2 > 0$ are constants independent of λ and η. The first term on the right-hand side can be for our choice of f_λ calculated explicitly,

$$\|\nabla f_\lambda\|^2_{L^2(\mathbb{R}^2)} = \frac{\pi}{K_0^2(\lambda R)} \left(\lambda^2 R^2\, K_1'(\lambda R)^2 - (1+\lambda^2 R^2) K_1(\lambda R)^2\right),$$

and using the relation $-K_1'(z) = K_0(z) + z^{-1} K_1(z)$ in combination with the behavior of the modified Bessel functions as $z \to 0$ we arrive at the estimate

$$\|\nabla\psi_\lambda\|^2_{L^2(\mathbb{R}^2)} - \epsilon_d \|\psi_\lambda\|^2_{L^2(\mathbb{R}^2)} \le -\frac{C_3}{\ln \lambda R} + C_1\eta^2 - C_2\eta$$

for some $C_3 > 0$. Hence choosing appropriately small values of λ and η we can make the right-hand side of this inequality negative which implies that the discrete spectrum of $H(d,d;\mathcal{W})$ is nonempty. If $d_1 \neq d_2$ the trial functions have to be modified. The starting point is the product $f_\lambda(x)\,\chi_1(y)$ corresponding to the wider one of the two layers, followed by a suitable modification of this function in the window; we leave this to the reader (Problem 17).

To prove the second part of the statement, consider the family of scaled window sets, $\mathcal{W} = aM$. We impose an additional Neumann boundary condition at $\partial\mathcal{W} \times (-d_2, d_1)$; this yields the operator $H^N(d_1, d_2; \mathcal{W})$ which is a direct sum of three parts. The spectrum of the "outer" parts taken together is obviously $[\epsilon_d, \infty)$, while the "inner" part $H_c^N(d_1, d_2; \mathcal{W})$ is the negative Laplacian in $L^2(\mathcal{W} \times (-d_2, d_1))$ with Neumann boundary conditions at $\partial\mathcal{W} \times (-d_2, d_1)$ and Dirichlet conditions at the remaining part of the boundary; its discrete spectrum clearly coincides with the eigenvalues of $H_c^N(d_1, d_2; \mathcal{W})$ lying below ϵ_d. On the other hand, by separation of variables the eigenvalues of $H_c^N(d_1, d_2; \mathcal{W})$ are equal to

$$\frac{\mu_j(M)}{a^2} + \frac{\pi^2 n^2}{(d_1+d_2)^2} \quad \text{with } j, n \ge 1\,,$$

where $\{\mu_j(M)\}_{j\in\mathbb{N}}$ are the eigenvalues of the Neumann Laplacian in $L^2(M)$. Since $\mu_j(M) > 0$ holds for $j \geq 2$, it follows that for a small enough the second eigenvalue of $H_c^N(d_1, d_2; \mathcal{W})$ equals $4\pi^2/(d_1+d_2)^2 \geq \epsilon_d$, in other words, the operator $H^N(d_1, d_2; \mathcal{W})$ has exactly one eigenvalue below ϵ_d for a sufficiently small. Finally, since $H(d_1, d_2; \mathcal{W}) \geq H^N(d_1, d_2; \mathcal{W})$ holds in the sense of quadratic forms, the same is true for $H(d_1, d_2; \mathcal{W})$. ∎

4.5 Notes

Section 4.1 For geodesic polar coordinates and other geometric notions used here see, e.g., [Kli]. The material of this section is taken from [DEK01], where conditions (b) and (c) of *Theorem* 4.2 were also proved under stronger assumptions, see also [DEK00]. The sufficient condition for invariance of the essential spectrum mentioned in Remark 4.1.1 can be found in [Kr01]. There is some freedom in the choice of the mollifier in the proof of *Theorem* 4.1, but the functions (4.10) are in a sense the most natural ones containing the free Green's function at small negative energies.

Section 4.2 For Huber's lemma see [Hu57], for the Cohn-Vossen inequality [CVo35]. Notice that only the limit $K \to 0$ is needed in order to establish the upper bound in *Proposition* 4.2.1; the latter comes from [CEK04] as well as *Theorem* 4.2. If the layer is not asymptotically planar, the spectral threshold may be lower than κ_1^2 and the spectrum may still be absolutely continuous—cf. [EŠŠ90].

The relation (4.14) derived in the proof of *Theorem* 4.3, which is adopted from [DEK01], can be used even without the surface symmetry, once we are able to show that the limit at its right-hand side is negative. A part of *Theorem* 4.4 stating the existence of bound states in layers with cylindrical ends comes again from [CEK04], a more detailed analysis of the Dirichlet Laplacian spectrum in conical layers mentioned in Example 4.2.3 can be found in [ET10]. Other sufficient conditions for the existence of bound states in terms of the layer geometry have been established in [LR12]; the analogous problem in higher dimensions was discussed in [LL07].

Section 4.3 Weakly coupled eigenvalues of two-dimensional Schrödinger operators are analyzed in [Si76]. The asymptotic behavior of the eigenvalues of $-\Delta_D^{\Omega_\lambda}$ as $\lambda \to 0$ will be discussed in detail in Chap. 6, see especially Sect. 6.2.3. *Theorem* 4.6 is a particular case of a more general result proved in [KV08].

Section 4.4 The proof of existence of the discrete spectrum in window-coupled layers is taken from [EV97b], its extension to the non-symmetric case, $d_1 \neq d_2$, can be done in analogy with the analogous two-dimensional problem [EV96]. The behavior of the lowest eigenvalue of $H(d_1, d_2; aM)$ in the limit $a \to 0$ will be treated in Chap. 6. Dependence of the discrete eigenvalues in window-coupled layers on the shape of the window was studied in [Bo07]. A model of a Dirichlet layer with two concentric Neumann discs of different radii on the opposite boundaries was discussed in [NO11].

4.6 Problems

1. There are surfaces diffeomorphic to $\mathbb{R}^2$ having no pole.
Hint: Consider a sphere and a plane connected smoothly by a long thin tube; destroy the cylindrical symmetry by changing the sphere to an ellipsoid—see also [GM69].
2. Let $K \in L^1(\Sigma_0, d\sigma)$ and denote by $\|K\|_{g,1}$ the corresponding norm. Then the inequality $\int_0^{2\pi} r(s,\theta)\, d\theta \le (2\pi + \|K\|_{g,1})s$ holds for any $s > 0$, and moreover, the same is true, possibly with another constant, if $\mathcal{K}$ exists as a principal value only.
Hint: Integrating (4.2) we get $\dot{r}(s,\theta) \le 1 + \int_0^\infty |K(\zeta,\theta)|r(\zeta,\theta)\, d\theta$ when the initial condition is taken into account, then integrations over θ and s yield the result.
3. Verify relations (4.7) and (4.9).
4. Prove relation (4.11) by computing the integral.
5. Check the claims made in Examples 4.1.2.
Hint: $k_{1,2} = 2g^{-1}\left[2(y^2-x^2) \pm \sqrt{g+4(y^2-x^2)^2}\,\right]$ with $g = 1+4(x^2+y^2)$ holds in the first case and $K = -36g^{-2}(x^2+y^2)$ with $g = 1+9(x^2+y^2)^2$ in the second one.
6. Prove the inequality $\inf \sigma_{\mathrm{ess}}(-\Delta_D^\Omega) \ge \kappa_1^2$ in *Proposition* 4.2.1.
Hint: Adapt the proof from *Proposition* 4.1.1 replacing the geodetical balls used there by a suitable family of compact subsets of Σ.
7. (a) Prove that $M_u := \frac{M-Ku}{1-2Mu+Ku^2}$ in the second one of relations (4.6) is the mean curvature of the parallel surface $\Sigma_u = \mathcal{L}(\Sigma_0 \times \{u\})$; compare to Problem 1.1.
(b) Check the identity $-\Delta\chi_1(u) = 2M_u\, \chi_1'(u) + \kappa_1^2\, \chi_1(u)$.
Hint: Use $|\nabla u| = 1$ and $-\Delta u = 2M_u$.
8. Check the claims made in Example 4.2.1.
Hint: $K = 4(gbc)^{-2}$ with $g = 1+4(x^2b^{-2}+y^2c^{-2})$ and $M = \mathcal{O}(r^{-1})$ as $r \to \infty$.
9. Let $\mathcal{K} > 0$, then there are positive δ, s_0 such that $\delta r(s)^{-1} \le |k_\theta(s)| \le r(s)^{-1}$ holds for all $s > s_0$ and the function k_θ does not change sign there.
Hint: $|k_\theta| = r^{-1}\sqrt{1-\dot{r}^2}$.
10. Find an example of a layer over a surface Σ diffeomorphic to $\mathbb{R}^2$ and equipped with geodesic polar coordinates, which has no bound states.
Hint: Consider a cylinder with a hemispherical "cap" and employ Neumann bracketing, then take a smooth deformation of the connection part and use the domain continuity of Dirichlet eigenvalues [RT75].
11. Find the Gauss curvature and the bound ρ_m in Example 4.2.2a. Check that the surface used in the second part of the example has $\mathcal{K} = 2\pi\left(1-\cos\sqrt{\frac{\pi}{2}}\right) \approx 1.38\,\pi$.
12. Work out the details of the partial-wave decomposition of Remark 4.2.1.
13. Let Ω_α be the conical layer of Example 4.2.3. Check that there is a $C > 0$ such that the bound $\sharp\,\sigma_{\mathrm{disc}}(-\Delta_D^{\Omega_\alpha}) > C\alpha^{-1}$ holds for all α small enough.
Hint: Modify the proof of *Proposition* 1.2.1 inserting a cylinder into the tip region.
14. Consider the operator $H_0 = -\Delta_D^{\Omega_0}$ in $L^2(\Omega_0)$, $d \ge 3$, where $\Omega_0 = \mathbb{R} \times M$ with $M \subset \mathbb{R}^{d-2}$ open and precompact, and its potential perturbation $H = H_0 + V$ defined through the quadratic form analogous to (1.24). Check the following claims:
(a) *Proposition* 1.4.1 remains valid in this case.

(b) Put $\vec{x} = (x, y)$ with $x \in \mathbb{R}^2$ and $y \in M$. Under the same assumptions about the potential V, the condition $\int_{\mathbb{R}^2} \int_M V(x, y)\, \chi_1(y)^2\, d\vec{x} < 0$ implies $\sigma_{\text{disc}}(H) \neq \emptyset$.

15. Extend the claim of *Proposition* 1.4.3 to d-dimensional layers.

16. Prove that for any $u \in H^1(\mathbb{R}_+, r\, dr)$ and any $R > 0$ one has

$$\int_0^R |u(r)|^2\, r\, dr \le \int_R^{2R} |u(r)|^2\, r\, dr + \frac{8}{3} R^2 \int_0^{2R} |u'(r)|^2\, r\, dr\, .$$

Hint: Consider the function $\phi : \mathbb{R}_+ \to \mathbb{R}$ which equals one in $[0, R]$, zero for $r \ge 2R$, and interpolates linearly between the two values in $[R, 2R]$. Use the fact that for any $r \in (0, R)$ we have

$$u(r) = \phi(r)u(r) = -\int_r^{2R} (\phi u)'(t)\, dt = \frac{1}{R} \int_R^{2R} u(t)\, dt - \int_r^{2R} \phi u'(t)\, dt\, ,$$

then apply the Cauchy-Schwarz inequality.

17. Complete the proof of *Theorem* 4.7.

Chapter 5
Point Perturbations

So far we have supposed that the only interaction to which a single particle in a spatial region Ω is exposed comes from its boundaries reflecting their geometry. This is naturally an idealization. Among various sources of additional interactions, material impurities often have to be taken into account. In particular, they are important when we model semiconductor microstructures: we have argued in the introduction that the free motion away of the boundary requires a perfect crystalline structure of the material, i.e. an infinite mean free path.

It is clear that the problem of introducing local perturbations is in general quite complicated. Fortunately we can get a useful insight if we employ another idealization, this time supposing that the impurities have a *point character*. This hypothesis reflects well the fact that the impurities, typically consisting of a single alien atom in the lattice, are much smaller that the system size, and at the same time it simplifies the solution considerably. Point interaction models are used in quantum mechanics from the thirties. They came to use only slowly, however, in part because it was not clear at the beginning what was the proper way to treat them mathematically.

A suitable framework was found in the theory of self-adjoint extensions. If we want to construct a Hamiltonian with point interactions supported by a discrete set $\{\vec{a}_j\}$ of points in the configuration space, we employ the following procedure. Due to the character of the interaction it is natural to require that the sought operator acts as the free one, in our case as $-\Delta_D^\Omega$, outside the interaction support. Restricting $-\Delta_D^\Omega$ to functions which vanish at the points $\vec{a}_j$ we get a symmetric operator with equal deficiency indices; we look for the Hamiltonian among its "local" self-adjoint extensions characterized by appropriate coupling constants (see the notes). The main advantage of such a model is that the resolvent of the full Hamiltonian can be written explicitly through Krein's formula; this makes it possible to turn the spectral analysis into an essentially algebraic problem. Throughout this chapter we suppose that the number of point perturbations is finite.

P. Exner and H. Kovařík, *Quantum Waveguides*,
Theoretical and Mathematical Physics, DOI 10.1007/978-3-319-18576-7_5

5.1 Point Impurities in a Straight Strip

Let us begin with the simplest case of a straight planar strip $\Omega := \mathbb{R} \times (0, d)$ at which the free motion is governed in the chosen units by the Dirichlet Laplacian $-\Delta_D^\Omega$. Its domain and some cores are described in the introduction to Chap. 1, in particular, the condition (1.2) now reads

$$\psi(x, 0) = \psi(x, d) = 0$$

for all $x \in \mathbb{R}$. Since a change of the strip width amounts to a simple scaling transformation (Problem 1) we put $d = \pi$ everywhere in this section.

The abstract scheme of constructing point interaction Hamiltonians simplifies here in view of two facts. First of all, we deal with second-order differential operators where the described restriction and subsequent extension means just a change of boundary conditions at the interaction sites. Furthermore, such a modification of the operator domain bears a local character which means that we may adopt the boundary condition which determine two-dimensional point interactions in $\Omega = \mathbb{R}^2$—cf. [AGHH, Sect. I.5].

On the other hand, the boundary conditions must be introduced with some care, since they involve generalized boundary values at a point $\vec{a} \in \Omega$ defined as

$$L_0(\psi, \vec{a}) := -\lim_{\vec{x}\to\vec{a}} \frac{2\pi\psi(\vec{x})}{\ln|\vec{x}-\vec{a}|}, \quad L_1(\psi, \vec{a}) := \lim_{\vec{x}\to\vec{a}} \left[\psi(\vec{x}) + L_0(\psi, \vec{a}) \frac{\ln|\vec{x}-\vec{a}|}{2\pi}\right], \tag{5.1}$$

which relate ψ to the corresponding fundamental solution of the Laplace equation. Given N-tuples $\alpha := \{\alpha_1, \dots, \alpha_N\} \subset \mathbb{R}$ and $\vec{a} := \{\vec{a}_1, \dots, \vec{a}_N\} \subset \Omega$ we define the ***point-interaction Hamiltonian*** $H_{\alpha,\vec{a}}$ by the boundary conditions

$$L_1(\psi, \vec{a}_j) - \alpha_j L_0(\psi, \vec{a}_j) = 0\,, \quad j = 1, \dots, N\,; \tag{5.2}$$

since Ω is kept fixed we do not indicate it in the symbol. In other words, this operator acts as $(H_{\alpha,\vec{a}}\psi)(\vec{x}) = -(\Delta\psi)(\vec{x})$ for $\vec{x} \neq \vec{a}$, in the sense of distributions, on the domain

$$\mathrm{Dom}\,(H_{\alpha,\vec{a}}) = \left\{\psi \in H_0^1(\Omega \setminus \{\vec{a}\}) :\, -\Delta\psi \in L^2 \text{ and (1.2), (5.2) are satisfied}\right\}.$$

The parameters α_j play the role of coupling constants. The absence of a point interaction at some $\vec{a}_j$ is expressed by the requirement $L_0(\psi, \vec{a}_j) = 0$ for all ψ from the domain which can be formally achieved by putting $\alpha_j = \infty$.

5.1.1 A Single Perturbation

Consider first a single perturbation, $N = 1$, of the free Hamiltonian $H_0 := -\Delta_D^\Omega$ supported by a point $\vec{a} := (a, b)$ with $b \in (0, \pi)$. To find the resolvent of $H_{\alpha,\vec{a}}$ we start with that of H_0. We use the decomposition into transverse modes, $L^2(\Omega) = \bigoplus_{n=1}^\infty L^2(\mathbb{R}) \otimes \{\chi_n\}$, where χ_n is instead of (1.10) a basis modified for the "one-sided" strip, $\chi_n(y) = \sqrt{\frac{2}{\pi}}\,\sin(ny)$. Then the free Hamiltonian can be written as

$$H_0 = \bigoplus_{n=1}^{\infty} h_n \otimes I_n\,, \quad h_n := -\partial_x^2 + n^2 \tag{5.3}$$

with Dom $(h_n) = H^2(\mathbb{R})$. It follows that the free resolvent is an integral operator with the kernel

$$G_0(\vec{x}_1, \vec{x}_2; z) \equiv (H_0 - z)^{-1}(\vec{x}_1, \vec{x}_2) = \frac{i}{\pi} \sum_{n=1}^{\infty} \frac{\mathrm{e}^{ik_n(z)|x_1-x_2|}}{k_n(z)} \sin(ny_1)\, \sin(ny_2)\,,$$

where $\vec{x}_j = (x_j, y_j)$ and $k_n(z) := \sqrt{z-n^2}$ for z in the resolvent set of H_0, i.e. $z \in \mathbb{C}\backslash[1, \infty)$. The function $G_0(\cdot, \cdot; z)$ is defined and smooth except at the diagonal, $\vec{x}_1 = \vec{x}_2$, but the sum on the right-hand side may not converge absolutely if the longitudinal coordinates coincide, $x_1 = x_2$. Moreover, the right-hand side makes also sense for all non-integer $z > 1$, where it gives the boundary value of the kernel at the cut; one has to properly choose the branch of the square root in $k_n(z)$. For $z < 1$ the kernel is strictly positive, $G_0(\vec{x}_1, \vec{x}_2; z) > 0$ for all mutually different $\vec{x}_1, \vec{x}_2 \in \Omega$ – cf. [RS, Appendix to Sect. XIII.12].

Proposition 5.1.1 *The resolvent kernel of $H_{\alpha,\vec{a}}$ equals*

$$(H_{\alpha,\vec{a}}-z)^{-1}(\vec{x}_1, \vec{x}_2) = G_0(\vec{x}_1, \vec{x}_2; z) + \frac{G_0(\vec{x}_1, \vec{a}; z)G_0(\vec{a}, \vec{x}_2; z)}{\alpha - \xi(\vec{a}; z)}\,,$$

where

$$\xi(\vec{a}; z) := \frac{i}{\pi} \sum_{n=1}^{\infty} \left(\frac{\sin^2(nb)}{k_n(z)} - \frac{1}{2in} \right). \tag{5.4}$$

Proof Since $H_{\alpha,\vec{a}}$ and H_0 have a common symmetric restriction with deficiency indices $(1, 1)$, their resolvents differ by a rank-one operator. Its kernel is by Krein's formula equal to $\lambda G_0(\vec{x}_1, \vec{a}; z)G_0(\vec{a}, \vec{x}_2; z)$, so it remains to determine the coefficient λ. We use the fact that by definition $(H_{\alpha,\vec{a}}-z)^{-1}$ maps into Dom $H_{\alpha,\vec{a}}$. Writing $\psi = (H_{\alpha,\vec{a}}-z)^{-1}\phi$ and $\psi_0 = (H_0-z)^{-1}\phi$ for an arbitrary $\phi \in L^2(\Omega)$, we get

$$\psi(x, y) = \psi_0(x, y) + \frac{i\lambda}{\pi} \sum_{n=1}^{\infty} \frac{e^{ik_n(z)|x-a|}}{k_n(z)} \sin(ny) \sin(nb) \psi_0(a, b).$$

Since $\psi_0 \in \text{Dom}(H_0)$ is smooth at $\vec{x} = \vec{a}$, the generalized boundary values can be written as $L_j(\psi, \vec{a}) = \mathcal{L}_j(\vec{a})\psi_0(\vec{a})$. The diagonal singularity of the resolvent kernel for planar regions with a smooth boundary is well known [Ti, Chaps. 11, 14], and it can also be evaluated directly (Problem 2). We find $\mathcal{L}_0(\vec{a}) = -\lambda$ which in turn yields $\mathcal{L}_1(\vec{a}) = 1 + \lambda\xi(\vec{a}; z)$. Using now the boundary conditions (5.2) we get $\lambda = (\alpha - \xi(\vec{a}; z))^{-1}$. ∎

Remark 5.1.1 In the rationalized units with $\hbar^2/2m^* = 1$ and $c = 1$ the energy has the dimension of *length*$^{-2}$; the choice $d = \pi$ then refers to this natural length scale. For another value of d the logarithmic factor in the definition of the regularized boundary values is replaced by $\ln|\kappa_1(\vec{x} - \vec{a})|$, where $\kappa_n := \pi n/d$ as usual. In that case $\mathcal{L}_0(\vec{a})$ does not change, while $\mathcal{L}_1(\vec{a})$ acquires in addition to the scaling the additive factor $(2\pi)^{-1}\ln\kappa_1$. The denominator in the Krein formula is then $\alpha - \xi_d(\vec{a}; z)$, where

$$\xi_d(\vec{a}; z) = \frac{i}{d} \sum_{n=1}^{\infty} \left(\frac{\sin^2(\kappa_n b)}{k_n(z)} - \frac{d}{2\pi i n} \right) + \frac{1}{2\pi} \ln\kappa_1,$$

with $k_n(z) := \sqrt{z - \kappa_n^2}$. Hence changing the width of the strip Ω is equivalent to a shift in the coupling constant for a fixed d, in other words, to replacement of α by $\alpha - (2\pi)^{-1}\ln\kappa_1$.

It is clear from the expression of the resolvent that the function ξ plays a crucial role. Let us review some of its properties.

Proposition 5.1.2 *The function $\xi(\vec{a}; \cdot)$ is for a fixed $\vec{a} \in \Omega$ analytic in $\rho(H_0)$ depending on the transverse component of $\vec{a}$ only. On $(-\infty, 1)$ it is increasing with $\xi(\vec{a}; z) = \pi^{-1}(1-z)^{-1/2}\sin^2 b + \mathcal{O}(1)$ as $z \to 1-$ and*

$$\xi(\vec{a}; z) = -\frac{1}{4\pi} \ln\left(-\frac{z}{4}\right) - \frac{1}{2\pi}\gamma_E + \mathcal{O}\left(e^{-c\sqrt{-z}}\right)$$

as $z \to -\infty$ for any $c < 1$, where $\gamma_E = -\psi(1) = 0.57721\ldots$ is Euler's constant. We also have

$$\xi(\vec{a}; z) > \xi(\vec{a}'; z) \quad \text{if} \quad \left|b - \frac{\pi}{2}\right| < \left|b' - \frac{\pi}{2}\right|.$$

Finally, the series (5.4) also converges in $[1, \infty) \backslash \{n^2\}_{n \in \mathbb{N}}$ giving boundary values of $\xi(\vec{a}; \cdot)$ at the cut which are smooth away from the thresholds; the choice of the branch of the square roots in $\kappa_n(z)$ determines the corresponding sheet of the Riemann surface.

Proof It is useful to introduce $\kappa_n(z) := -ik_n(z) = \sqrt{n^2 - z}$, then

$$\xi(\vec{a}; z) = \frac{1}{\pi} \sum_{n=1}^{\infty} \left(\frac{\sin^2(nb)}{\kappa_n(z)} - \frac{1}{2n} \right) = \xi(\vec{a}; z') + \frac{1}{\pi} \sum_{n=1}^{\infty} \frac{\kappa_n(z') - \kappa_n(z)}{\kappa_n(z)\kappa_n(z')} \sin^2(nb)$$

holds for any $z, z' \in \mathbb{C} \setminus [1, \infty)$. The series on the right-hand side converges because the coefficients at $\sin^2(nb)$ decay like $\mathcal{O}(n^{-3})$ as $n \to \infty$. Moreover, they are analytic for z, z' in any bounded subset of $\rho(H_0)$ and the series converges uniformly there. Since $\xi(\vec{a}; z')$ exists for a particular z' (Problem 4) the first claim follows. The above identity also gives convergence for $z > 1$ away from the thresholds, the smoothness of $\xi(\vec{a}; \cdot)$ there, and the inequality

$$\frac{\partial \xi}{\partial z} = \frac{1}{2\pi} \sum_{n=1}^{\infty} \frac{\sin^2(nb)}{\kappa_n(z)^3} > 0$$

for $z \in (-\infty, 1)$, which yields the monotonicity. The asymptotics for $z \to 1-$ is obvious, for $z \to -\infty$ see Problem 5. To prove the monotonicity across a halfstrip, we use the identity $\sin^2(nb) - \sin^2(nb') = \sin(n(b+b')) \sin(n(b-b'))$ for $0 < b' < b \leq \frac{\pi}{2}$; it shows that

$$\xi(\vec{a}; z) - \xi(\vec{a}'; z) = G_0(0, b + b'; 0, b - b'; z),$$

hence the result follows from the positivity of the free-resolvent kernel. ∎

A single-impurity Hamiltonian then has the following spectral properties.

Theorem 5.1 *For any $\vec{a} \in \Omega$ and $\alpha \in \mathbb{R}$ we have $\sigma_{\mathrm{ess}}(H_{\alpha,\vec{a}}) = \sigma_{\mathrm{ac}}(H_{\alpha,\vec{a}}) = [1, \infty)$ and $\sigma_{\mathrm{sc}}(H_{\alpha,\vec{a}}) = \emptyset$. Moreover, $H_{\alpha,\vec{a}}$ has one eigenvalue $\epsilon_{\alpha,\vec{a}} \in (-\infty, 1)$. The function $\alpha \mapsto \epsilon_{\alpha,\vec{a}}$ is real-analytic and increasing and has the following asymptotic behavior,*

$$\epsilon_{\alpha,\vec{a}} = 1 - \left(\frac{\sin^2 b}{\pi\alpha} \right)^2 + \mathcal{O}(\alpha^{-3}),$$

$$\epsilon_{\alpha,\vec{a}} = -4\,\mathrm{e}^{-4\pi\alpha - 2\gamma_E} \left(1 - \mathcal{O}\left(\exp\left(-2\varrho\, \mathrm{e}^{2\pi\alpha} \right) \right) \right)$$

with any $\varrho < \mathrm{dist}(\vec{a}, \partial\Omega) = \frac{\pi}{2} - \left| b - \frac{\pi}{2} \right|$, for $\alpha \to \pm\infty$, respectively. Furthermore, proximity of the boundary pushes the bound-state energy up,

$$\epsilon_{\alpha,\vec{a}} < \epsilon_{\alpha,\vec{a}'} \quad \text{if} \quad \left| b - \frac{\pi}{2} \right| < \left| b' - \frac{\pi}{2} \right| .$$

Finally, there are no eigenvalues embedded in the continuous spectrum.

Proof The essential and absolutely continuous spectrum are preserved, since we have a rank-one perturbation in the resolvent. Let $\{E_t\}$ be the spectral measure of

$H_{\alpha,\vec{a}}$. To prove the absence of the embedded eigenvalues away from the thresholds, it is sufficient to check that $t \mapsto (\psi, E_t\psi)$ is for all $\psi \in L^2(\Sigma)$ a continuous function in any open interval $I \subset (n^2, (n+1)^2),\ n \in \mathbb{N}$. This follows from Stone's formula and the fact that the boundary values of the full Green's function are smooth by the two preceding propositions; recall that the denominator in the expression of the resolvent cannot be zero in $[1, \infty)$ as the hint to Problem 8 shows. In the same way we get

$$(\psi, (E_t - E_{t_0})\psi) = \frac{1}{\pi} \int_{t_0}^{t} \mathrm{Im}\, \left(\psi, (H_{\alpha,\vec{a}} - u)^{-1}\psi\right) du\,.$$

The left-hand side is thus a smooth function of t away from the thresholds, so the singularly continuous spectrum is absent. To determine the point spectrum, one has to solve the equation

$$\xi(\vec{a}; z) = \alpha\,. \tag{5.5}$$

The existence, uniqueness, and properties of the solution follow easily from *Proposition* 5.1.2; the asymptotics for $\alpha \to -\infty$ is obtained by a bracketing estimate (Problem 5). It remains to check the absence of eigenvalues at thresholds, which can be done directly (Problem 6). ∎

Remark 5.1.2 The residue at the pole given by (5.5) provides us the (non-normalized) wave function of the bound state, $\phi_{\alpha,\vec{a}} = G_0(\cdot, \vec{a}; \epsilon_{\alpha,\vec{a}})$, or more explicitly

$$\phi_{\alpha,\vec{a}}(\vec{x}) = \frac{1}{\pi} \sum_{n=1}^{\infty} \frac{\mathrm{e}^{-\kappa_n(\epsilon_{\alpha,\vec{a}})|x-a|}}{\kappa_n(\epsilon_{\alpha,\vec{a}})} \sin(ny)\, \sin(nb)\,.$$

The limits $\alpha \to \mp\infty$ are naturally associated with the ***strong-*** and ***weak-coupling regime***, respectively. If α decreases, $\phi_{\alpha,\vec{a}}$ becomes well localized and approaches the Hankel eigenfunction of the point interaction in the plane (Problem 7), while in the limit $\alpha \to \infty$ we have

$$\phi_{\alpha,\vec{a}}(\vec{x}) = \alpha\, \frac{\sin y}{\sin b}\, \mathrm{e}^{-|x-a|\sin^2 b/\pi\alpha} + \frac{1}{\pi} \sum_{n=2}^{\infty} \frac{\mathrm{e}^{-|x-a|\sqrt{n^2-1}}}{\sqrt{n^2-1}} \sin(ny)\, \sin(nb) + o(1)\,;$$

the leading term here is the product of $\chi_1(y)$ with the eigenfunction of the one-dimensional attractive point interaction of strength $-(2/\pi\alpha)\sin^2 b$, of course, smoothed at the segment $\{a\} \times ((0, d)\backslash\{b\})$ by the other terms.

Let us turn now to scattering by the point impurity.

Proposition 5.1.3 *The wave operators for the pair* $(H_0, H_{\alpha,\vec{a}})$ *exist and are asymptotically complete. Moreover, the on-shell operator* $S(k)$ *at an energy* $z = k^2 \in (1, \infty)\backslash\{n^2\}_{n\in\mathbb{N}}$ *is a unitary* $2[\sqrt{z}] \times 2[\sqrt{z}]$ *matrix with the blocks*

$$S_{nm} = \sqrt{\frac{k_m}{k_n}} \begin{pmatrix} t_{nm} & r_{nm} \\ \tilde{r}_{nm} & \tilde{t}_{nm} \end{pmatrix},$$

where $k_n = k_n(z) := \sqrt{z-n^2}$, *the indices* $n, m = 1, \dots, [\sqrt{z}]$, *and the transmission and reflection coefficients are given by*

$$(t_{nm}(k) - \delta_{nm})\, e^{i(k_m-k_n)a} = r_{nm}(k)\, e^{-i(k_n+k_m)a} = \frac{i}{\pi} \frac{\sin(nb)\sin(mb)}{k_m(z)(\alpha-\xi(\vec{a};z))},$$

with the tilded (right-to-left) amplitudes obtained by changing the sign of a, *i.e.* $\tilde{r}_{nm} := r_{nm}e^{-2i(k_n+k_m)a}$ *and* $\tilde{t}_{nm} := \delta_{nm} + (t_{nm}-\delta_{nm})e^{2i(k_m-k_n)a}$.

Proof Since the perturbation is rank-one in the resolvent, the first claim follows from the Birman-Kuroda theorem [RS, Sect. XI.3]. To find the scattering matrix, we employ *Proposition* 5.1.1: to any $\psi \in \mathrm{Dom}\,(H_{\alpha,\vec{a}})$ and a non-real z there is a unique decomposition

$$\psi(\vec{x}) = \psi_z(\vec{x}) + \frac{1}{\alpha-\xi(\vec{a};z)}\, G_0(\vec{x},\vec{a};z)\psi_z(\vec{a})$$

with $\psi_z \in \mathrm{Dom}\,(H_0)$ and $(H_{\alpha,\vec{a}}-z)\psi = (H_0-z)\psi_z$. If we choose, for instance, $\psi_z^\varepsilon(\vec{x}) := e^{ik_n(z)x-\varepsilon x^2}\chi_n(y)$ for ψ_z, then the corresponding ψ^ε belongs to $\mathrm{Dom}\,(H_{\alpha,\vec{a}})$ for all $\varepsilon > 0$ and

$$((H_{\alpha,\vec{a}}-z)\psi^\varepsilon)(\vec{x}) = 2\varepsilon\left(2\varepsilon x^2 - 1 - 2ik_n(z)x\right)\psi_z^\varepsilon(\vec{x}).$$

The right-hand side makes sense on the real line, hence $\psi^\varepsilon \in \mathrm{Dom}\,(H_{\alpha,\vec{a}})$ for $z \in [1,\infty)$ and the last relation holds again. We have $\|\psi^\varepsilon\| = \mathcal{O}(\varepsilon^{-1/4})$ as $\varepsilon \to 0+$, while the norm of the right-hand side is $\mathcal{O}(\varepsilon^{1/4})$. Furthermore, the pointwise limit $\psi(\vec{x}) = \lim_{\varepsilon\to 0}\psi^\varepsilon(\vec{x})$ exists and

$$\psi(\vec{x}) = e^{ik_n(z)x}\chi_n(y) + \frac{e^{ik_n(z)a}}{\alpha-\xi(\vec{a};z)}\, G_0(\vec{x},\vec{a};z)\chi_n(b).$$

The function belongs to L^2_{loc}, satisfies the appropriate boundary conditions and solves $(H_{\alpha,\vec{a}} - z)\psi = 0$ as a differential equation, i.e. it is a generalized eigenvector of the operator $H_{\alpha,\vec{a}}$. Substituting for G_0 and comparing the coefficients of the plane waves in different transverse modes as $x \to \pm\infty$, we get $r_{nm}(z)$ and $t_{nm}(z)$ for the incident wave corresponding to the n-th transverse mode; for the right-to-left amplitudes one has to change k_n to $-k_n$ in the above Ansatz. It is clear that an asymptotically non-vanishing quantity is obtained only if both the involved channels are open, i.e. $z > \max\{n^2, m^2\}$. Finally, the S-matrix must be normalized with respect to the relative velocities. Its unitarity follows from the completeness of the wave operators but it can also be checked directly (Problem 8). ∎

Let us finish the single-impurity case with a brief discussion of *scattering resonances*. The square root in each $k_n(z) = \sqrt{z-n^2}$ gives rise to a cut; hence $G_0(\vec{x}, \vec{a}; \cdot)$ as well as other quantities derived from it are in general multi-valued with infinitely sheeted Riemann surfaces. It is then possible that the pole condition (5.5) has solutions on the other sheets.

To fix the branch set $q_n(z) := \sqrt{z-n^2}$ with values in the upper complex halfplane or at the positive real halfline. Given positive integers N, n define $\theta_n^N = -1$ if $n < N$ and $\theta_n^N = 1$ otherwise, then we will use

$$k_n^N(z) := \theta_n^N q_n(z) \tag{5.6}$$

as the channel momenta on the N-th sheet. The condition (5.5) then reads

$$\alpha + \frac{1}{\pi}\sum_{n=1}^{\infty}\left(\theta_n^N \operatorname{Im} q_n^{-1}(z)\sin^2(nb) + \frac{1}{2n}\right) = 0 = -\frac{1}{\pi}\sum_{n=1}^{\infty}\theta_n^N \operatorname{Re} q_n^{-1}(z)\sin^2(nb).$$

The most interesting situation arises in the weak-coupling regime, $\alpha \to \infty$, when there is generically one resonance pole close to each threshold, with the exception of the lowest one; the resonance is absent if Nb/π is an integer so that the incident wave has a node at the impurity. To explain this claim, let us decompose the N-th branch of the function $\xi(\vec{a}; \cdot)$, $N \geq 2$, into the sum $\xi^N(\vec{a}; z) + \tilde{\xi}^N(\vec{a}; z)$, where the two terms include contributions with $n \geq N$ and $1 \leq n \leq N-1$, respectively. Suppose that the latter is switched on with the help of an additional parameter, i.e. that we look for solutions of the equation

$$F(z, \eta) := \alpha - \xi^N(\vec{a}; z) - \eta\tilde{\xi}^N(\vec{a}; z) = 0\,.$$

If $\eta = 0$ we repeat the argument from the proof of *Proposition* 5.1.1 finding that $\xi^N(\vec{a}; \cdot)$ is strictly increasing in the interval $(-\infty, N^2)$ being divergent at the endpoints; hence to a given α there is a unique $z_N^0(\alpha)$ such that $F(z_N^0(\alpha), 0) = 0$. Moreover, the eigenvalues of the "truncated" problem behave as

$$z_N^0(\alpha) = N^2 - \left(\frac{\sin^2(Nb)}{\pi\alpha}\right)^2 + \mathcal{O}(\alpha^{-3})$$

in the limit $\alpha \to \infty$. For a nonzero η the above equation can be solved perturbatively by means of the implicit-function theorem. We have

$$\left.\frac{\partial F}{\partial \eta}\right|_{(z_N^0(\alpha),0)} = -\tilde{\xi}^N(\vec{a}; z_N^0(\alpha)) = \frac{i}{\pi}\sum_{n=1}^{N-1}\left(\pm\frac{\sin^2(nb)}{\sqrt{N^2-n^2}} + \frac{1}{2in}\right) + \mathcal{O}(\alpha^{-2})\,,$$

$$\left.\frac{\partial F}{\partial z}\right|_{(z_N^0(\alpha),0)} = \frac{i}{2\pi}\sum_{n=N}^{\infty}\frac{\sin^2(nb)}{q_N(z_N^0(\alpha))^3} = -\frac{1}{2\pi}\frac{\pi^3\alpha^3}{\sin^4(Nb)} + \mathcal{O}(\alpha^0)\,.$$

The first term is in fact obtained as a limit $z \to z_N^0(\alpha)$; the sign depends on whether it is taken from the upper or lower halfplane. Dividing these quantities we obtain $-(\partial z_N^\eta/\partial\eta)_{\eta=0}$. Moreover, the remainder term coming from $\partial^2 z_N^\eta/\partial\eta^2$ is $\mathcal{O}(\alpha^{-4})$, so for sufficiently large positive α we may use the expansion up to $\eta = 1$. Since we are looking for a solution in the lower halfplane, we get an asymptotic formula for the resonance-pole position,

$$z_N(\alpha) = N^2 - 2i\,\frac{\sin^4(Nb)}{\pi^3\alpha^3}\sum_{n=1}^{N-1}\left(\frac{\sin^2(nb)}{\sqrt{N^2-n^2}} - \frac{1}{2in}\right) + \mathcal{O}(\alpha^{-4})\,.$$

In particular, the resonance width behaves in the weak-coupling limit as

$$\Gamma_N(\alpha) := -2\,\mathrm{Im}\, z_N(\alpha) = 4\,\frac{\sin^4(Nb)}{\pi^3\alpha^3}\sum_{n=1}^{N-1}\frac{\sin^2(nb)}{\sqrt{N^2-n^2}} + \mathcal{O}(\alpha^{-4})\,.$$

Hence a "weak" impurity leads to sharp resonances with poles close to the real axis. Taking into account that the boundary value of $\alpha - \xi(\vec{a}; z)$ is contained in the denominators of the S-matrix elements in *Proposition* 5.1.3 we see that these resonances are manifested in the transmission and reflection probabilities; it is clear that the pole positions depend strongly on the transverse coordinate b.

5.1.2 A Finite Number of Impurities

Let us return now to the operator $H_{\alpha,\vec{a}}$ with N point impurities defined in the beginning of this section. First we find an explicit form of its resolvent.

Proposition 5.1.4 *The resolvent kernel of* $H_{\alpha,\vec{a}}$ *equals*

$$(H_{\alpha,\vec{a}}-z)^{-1}(\vec{x}_1,\vec{x}_2) = G_0(\vec{x}_1,\vec{x}_2;z) + \sum_{j,m=1}^{N}[\Lambda_{\alpha,\vec{a}}(z)]^{-1}_{jm}G_0(\vec{x}_1,\vec{a}_j;z)G_0(\vec{a}_m,\vec{x}_2;z),$$

where $\Lambda \equiv \Lambda_{\alpha,\vec{a}}(z)$ *is the* $N \times N$ *matrix with the elements*

$$\Lambda_{jm} = \big(\alpha_j - \xi(\vec{a}; z)\big)\,\delta_{jm} - G_0(\vec{a}_j,\vec{a}_m;z)(1-\delta_{jm})\,,$$

where $\xi(\vec{a}; z)$ *is given by (5.4).*

Proof Similar to that of *Proposition* 5.1.1 (Problem 9).

This allows us to determine the spectral properties of $H_{\alpha,\vec{a}}$.

Theorem 5.2 *For any* $\vec{a} = \{\vec{a}_j\}$ *with* $\vec{a}_j \in \Omega$, $j = 1,\dots,N$, *and* $\alpha \in \mathbb{R}^N$ *we have* $\sigma_{\mathrm{ess}}(H_{\alpha,\vec{a}}) = \sigma_{\mathrm{ac}}(H_{\alpha,\vec{a}}) = [1,\infty)$ *and* $\sigma_{\mathrm{sc}}(H_{\alpha,\vec{a}}) = \emptyset$. *The discrete spectrum* *consists of k eigenvalues* $\epsilon_i^{\alpha,\vec{a}} \in (-\infty, 1)$, $i = 1, \ldots, k$, *counting multiplicity, arranged in the ascending order, with* $1 \le k \le N$. *They are real-analytic functions of the parameters* α_j. *The ground-state eigenvalue* $\epsilon_1^{\alpha,\vec{a}}$ *is simple while the other eigenvalues may be degenerate. Eigenfunctions corresponding to* $\epsilon_i^{\alpha,\vec{a}}$ *are of the form*

$$\phi_i^{\alpha,\vec{a}}(\vec{x}) = \sum_{j=1}^{N} d_j G_0(\vec{x}, \vec{a}_j; \epsilon_i^{\alpha,\vec{a}}),$$

where the coefficients solve $\sum_{m=1}^{N} \Lambda(\epsilon_i^{\alpha,\vec{a}})_{jm} d_m = 0$. *In particular, for the eigenfunction* $\phi_1^{\alpha,\vec{a}}$ *all of them can chosen positive. Finally,* $z > 1$ *cannot be an eigenvalue with an eigenvector from the subspace* $\bigoplus_{n=1}^{[\sqrt{z}]} L^2(\mathbb{R}) \otimes \{\chi_n\}$.

Proof The continuous spectrum part is analogous to *Theorem* 5.1. The discrete spectrum is again determined by poles of the resolvent coming from the coefficients in the Krein formula; they are given by the condition

$$\det \Lambda_{\alpha,\vec{a}}(z) = 0. \tag{5.7}$$

Comparing to the case $N = 1$, it is now slightly more complicated to determine the eigenfunctions. Suppose that $H \equiv H_{\alpha,\vec{a}}$ satisfies $H\phi = z\phi$ for some $z \in \mathbb{R}$. We pick an arbitrary $z' \in \rho(H)$, then by *Proposition* 5.1.4 there is a $\psi_0 \in \mathrm{Dom}\,(H_0)$ such that the eigenvector ϕ is expressed as

$$\phi = \psi_0 + \sum_{j=1}^{N} d_j G_0(\cdot, \vec{a}_j; z') \tag{5.8}$$

with the coefficients $d_j := \sum_{k=1}^{N} [\Lambda(z')]_{jk}^{-1} \psi_0(\vec{a}_k)$. In addition to that, the relations $(H_0 - z')\psi_0 = (H - z')\phi = (z - z')\phi$ are valid; applying $(H_0 - z')^{-1}$ to this identity we obtain

$$\psi_0 = (z - z') \left[(H_0 - z')^{-1}\psi_0 + \sum_{j=1}^{N} d_j (H_0 - z')^{-1} G_0(\cdot, \vec{a}_j; z') \right],$$

and this in turn yields $(H_0 - z)\psi_0 = (z - z') \sum_{j=1}^{N} d_j G_0(\cdot, \vec{a}_j; z')$. If $z < 1$, the resolvent $(H_0 - z)^{-1}$ exists and may be applied to both sides of the last relation giving

$$\psi_0 = \sum_{j=1}^{N} d_j \left(G_0(\cdot, \vec{a}_j; z) - G_0(\cdot, \vec{a}_j; z') \right)$$

with the help of the first resolvent identity. Substituting into (5.8) and setting $z = \epsilon_i^{\alpha,\vec{a}}$ we arrive at the expression for the (non-normalized) eigenfunction given in the theorem. To determine the coefficients, we use the explicit form of the matrix elements $\Lambda(z)_{jm}$ and the expression of ψ_0 to infer that $\psi_0(\vec{a}_j) = \sum_{m=1}^{N} \left(\Lambda(z')_{jm} - \Lambda(z)_{jm}\right) d_m$. On the other hand, inverting the formula for the coefficients in (5.8) we get $\psi_0(\vec{a}_j) = \sum_{m=1}^{N} \Lambda(z')_{jm} d_m$. Comparing these two expressions for $z = \epsilon_i^{\alpha,\vec{a}}$ we arrive at the sought claim.

The maximum number of eigenvalues equals the deficiency indices of the symmetric operator involved in the construction of $H_{\alpha,\vec{a}}$—cf. [We, Sect. 8.3]. The next question is about the existence of solutions to the Eq. (5.7). We observe that the matrix $\Lambda_{\alpha,\vec{a}}$ has the following asymptotic behavior,

$$\Lambda_{\alpha,\vec{a}}(z) = \frac{1}{4\pi} \ln\left(-\frac{z}{4}\right) I + \mathcal{O}(1), \quad \Lambda_{\alpha,\vec{a}}(z) = -\frac{1}{\pi\sqrt{1-z}} M_1 + \mathcal{O}(1)$$

as $z \to \mp\infty$, respectively, where $M_1 := (\sin b_j \, \sin b_m)_{j,m=1}^{N}$. This matrix has, in particular, the eigenvector $(\sin b_1, \dots, \sin b_N)$ corresponding to the *positive* eigenvalue $\sum_{j=1}^{N} \sin^2 b_j$, and therefore one of the eigenvalues of $\Lambda_{\alpha,\vec{a}}(z)$ tends to $-\infty$ as $z \to 1-$. The matrix elements are real-analytic functions of z, hence the eigenvalues are continuous and at least one of them must cross zero in $(-\infty, 1)$ giving rise to an eigenvalue. Furthermore, $\det \Lambda_{\alpha,\vec{a}}(z)$ is a real-analytic function of α_j and z, so the analytic dependence of $\epsilon_i^{\alpha,\vec{a}}$ on the coupling constants follows by the implicit-function theorem.

To prove the non-degeneracy of the ground state and positivity of the coefficients d_j involved, one has to check that the lowest eigenvalue of $\Lambda(z) \equiv \Lambda_{\alpha,\vec{a}}(z)$ is simple for any $z \in (-\infty, 1)$, which is equivalent to the claim that the matrix semigroup $\{\mathrm{e}^{-t\Lambda(z)} : t \geq 0\}$ is positivity improving [RS, Sect. XIII.12]. The last named property is ensured if all the non-diagonal elements of $\Lambda(z)$ are negative. In our case we have $\Lambda(z)_{jm} = -G_0(\vec{a}_j, \vec{a}_m; z)$ so the desired result follows from the positivity of the free-resolvent kernel.

Let us turn finally to embedded eigenvalues. Suppose that $H\phi = z\phi$ for some $z > 1$. We again employ the formula (5.8) and write ψ_0 as a series, $\psi_0(\vec{x}) = \sum_{n=1}^{\infty} g_n(x)\chi_n(y)$ with the coefficient functions $g_n \in L^2(\mathbb{R})$. Substituting this into the expression for $(H_0 - z)\psi_0$ and using the fact that $\{\chi_n\}$ is an orthonormal basis, we obtain the following system of equations,

$$-g_n''(x) - k_n(z)^2 g_n(x) = \frac{i}{2}\,(z - z') \sum_{j=1}^{N} d_j \chi_n(b_j)\, \frac{\mathrm{e}^{ik_n(z')|x-a_j|}}{k_n(z')}$$

for $n = 1, 2, \dots$. The Fourier-Plancherel operator transforms it into

$$(p^2 - z + n^2)\hat{g}_n(p) = \frac{z - z'}{2\pi} \sum_{j=1}^{N} d_j \chi_n(b_j)\, \frac{\mathrm{e}^{-ipa_j}}{p^2 - z' + n^2}\,. \tag{5.9}$$

If g_n belongs to $L^2(\mathbb{R})$ the same is true for $\hat{g}_n$; this is impossible if $z > n^2$ and the right-hand side of the last equation is nonzero at $\pm\kappa_n(z)$, since $\hat{g}_n^2$ would then have a non-integrable singularity. It is clearly the factor $p^2 - z + n^2$ which matters; recall that $z' \in \rho(H)$ by assumption.

We want to conclude that $\hat{g}_n = 0$ which is obvious for $N = 1$. If $N > 1$ and the a_j's are not the same, it might happen that the right-hand side is not identically zero. However, we can use a longitudinal coordinate shift leading to a common phase factor in front of the sum; then the square integrability requires $\sum_{j=1}^N d_j \chi_n(b_j)\, e^{\mp i\kappa_n(z)(a_j-a)} = 0$ for an arbitrary a. If all the a_j are mutually different (mod $2\pi\kappa_n(z)^{-1}$) it follows that $d_j\chi_n(b_j) = 0$ for each j. On the other hand, if some of them coincide we find $\sum_j d_j\chi_n(b_j) = 0$ where the index runs through the values with the same longitudinal coordinate a_j, and therefore $\hat{g}_n = 0$ again, i.e. $\hat{g}_n$ may be nonzero at most if some a_j differ by multiples of $2\pi p\kappa_n(z)^{-1}$. Consider now an arbitrary $g \in L^2(\mathbb{R})$ and $n < \sqrt{z}$. Using relation (5.8) we find

$$(g\chi_n, \phi) = (\hat{g}, \hat{g}_n)_{L^2(\mathbb{R})} + \frac{i}{2k_n(z')} \sum_{j=1}^N d_j\chi_n(b_j)\left(g, e^{ik_n(z')|\cdot - a_j|}\right)_{L^2(\mathbb{R})},$$

where $\hat{g}_n$ in the first term is given by (5.9). If $d_j\chi_n(b_j) = 0$ for each j, the right-hand side is zero. In the exceptional case mentioned above we use the fact that the left-hand side is independent of z'. The explicit expression for d_j together with the asymptotics of $\Lambda(z)$ show that $d_j \to 0$ as $z' \to -\infty$; the same is true for the inner product in the second term as well as for $(\hat{g}, \hat{g}_n)_{L^2(\mathbb{R})}$; together we find $(g\chi_n, \phi) = 0$ again. This concludes the proof. ∎

The proved result deserves some comments. First of all, the claim about embedded eigenvalues is weaker than in the one-center case. The situation for $N \geq 2$ is indeed different:

Example 5.1.4 The operator $H_{\alpha,\vec{a}}$ can have embedded eigenvalues with eigenfunctions in $\left(\bigoplus_{n=1}^{[\sqrt{z}]} L^2(\mathbb{R}) \otimes \{\chi_n\}\right)^{\perp}$ if $N \geq 2$. Consider a pair of impurities with the *same* coupling constant α placed at $\vec{a}_1 := (0, b)$ and $\vec{a}_2 := (0, \pi - b)$. The eigenvalue problem can be divided into symmetric and antisymmetric parts with respect to the strip axis. In view of Remark 5.1.1 the antisymmetric part is obtained by scaling the single-center problem with $\vec{a} := (0, 2b)$ and coupling constant $\alpha - \frac{1}{2\pi}\ln 2$. The scaled eigenvalue tends to 4 as $\alpha \to \infty$, hence it is embedded in $\sigma_c(H_{\alpha,\vec{a}}) = [1, \infty)$ for all α large enough. In the same way one can construct other examples of embedded eigenvalues. Their common feature is the existence of a symmetry which prevents the (energetically allowed) decay of an eigenstate; a violation of this symmetry turns these eigenvalues into resonances.

Another question concerns possible degeneracy of the discrete spectrum. *Theorem* 5.2 says that the maximum multiplicity is $N - 1$; in particular, the discrete spectrum is always simple for $N = 2$. This may not be true in general:

Example 5.1.5 For brevity put $g_{jm}(z) := -G_0(\vec{a}_j, \vec{a}_m; z)$. Let $N = 3$ with $\vec{a}_{1,3} := \left(\pm a, \frac{\pi}{2}\right)$ and $\vec{a}_2 := (0, b)$, and fix an energy $z < 1$. We have $g_{12}(z) = g_{23}(z)$ for any $b \in (0, \pi)$. If $b = \frac{\pi}{2}$ this value is obviously strictly greater than $g_{13}(z)$; on the other hand, $\lim_{b\to 0} g_{12}(z) = 0$, so there is a $b \in \left(0, \frac{\pi}{2}\right)$ for which all three $g_{jm}(z)$ have the same value. Choosing now the coupling constants α_j in such a way that $\alpha_j - \xi(\vec{a}_j; z) = g_{jm}(z)$, $j = 1, 2, 3$, we find that z is an eigenvalue of multiplicity two.

It should be noted, however, that degenerated eigenvalues occur rather exceptionally. The matrix $\Lambda_{\alpha,\vec{a}}(z)$ has $3N-1$ real parameters, because one of the coordinates a_j may be chosen arbitrarily, while the number of different $(N-1) \times (N-1)$ minors is N^2, and the number of conditions required for a multiplicity larger than two grows even faster with N. If we move the coupling constants we observe more frequently *avoided crossings*. For simplicity fix $N-1$ parameters α_j and let the remaining one run through $\mathbb{R}$. Then the whole discrete spectrum moves up in a peculiar way. If we do not hit a degeneracy point by a chance, we see how the increasing eigenvalue curve in the spectrum graph "exchanges place" with the subsequent constant levels corresponding to the fixed α_j's. The mechanism of this effect can be understood through an example involving a pair of perturbations (Problem 12).

Let us next turn to the asymptotic behavior; for simplicity let us consider only the situation where all the interactions are simultaneously strong or weak. In the first named case, $\max_j \alpha_j \to -\infty$, each perturbation gives rise to a single eigenvalue the behavior of which is in the leading order independent of the other impurities and of the boundary (Problem 13). In the weak-coupling regime the situation is different.

Proposition 5.1.5 *If $\alpha_- := \min_{1\le j\le N} \alpha_j$ is large enough, $H_{\alpha,\vec{a}}$ has a single eigenvalue which behaves as*

$$\epsilon_1^{\alpha,\vec{a}} = 1 - \left(\sum_{j=1}^{N} \frac{\sin^2 b_j}{\pi\alpha_j}\right)^2 + \mathcal{O}(\alpha_-^{-3})$$

as $\alpha_- \to \infty$, with the eigenfunction

$$\phi_1^{\alpha,\vec{a}}(\vec{x}) = \sin y \sum_{j=1}^{N} \exp\left\{-|x-a_j| \sum_{k=1}^{N} \frac{\sin^2 b_k}{\pi\alpha_k}\right\} \sin^2 b_j \left(\sum_{k=1}^{N} \frac{\sin^2 b_k}{\pi\alpha_k}\right)^{-1}$$
$$+ \sum_{n=2}^{\infty} \sin(ny) \sum_{j=1}^{N} \frac{\mathrm{e}^{-\sqrt{n^2-1}|x-a_j|}}{\sqrt{n^2-1}} \sin(nb_j) \sin b_j + o(1)$$

dominated by the product of $\chi_1(y)$ with a linear combination of the eigenfunctions of one-dimensional point interactions placed at a_j, $j = 1, \dots, N$.

Proof We put $\mathcal{A} := \mathrm{diag}(\alpha_1, \dots, \alpha_N)$ and $A(z) := \mathcal{A} - \pi^{-1}(1-z)^{-1/2} M_1$, and employ the decomposition $\Lambda(z) = A(z) + \tilde{\Lambda}(z)$, where $\tilde{\Lambda}(z)$ is a remainder term independent of α. There is a constant $C_{\vec{a}}$ such that $\|\tilde{\Lambda}(z)\| \le C_{\vec{a}}$ for all $z \in (-\infty, 1)$. If we put $\eta := \pi\sqrt{1-z}$, the condition (5.7) can be rewritten as

$$\det\left(M_1 - \eta\mathcal{A} + \eta\tilde{\Lambda}(1-\eta^2/\pi^2)\right) = 0\,.$$

The largest eigenvalue of this matrix satisfies $\mu_N(\eta) \ge \mu_N(0) - (C_{\vec{a}}+\alpha_+)\eta$, where $\alpha_+ := \max \alpha_j$, while for $j = 1, \dots, N-1$ we have $\mu_j(\eta) \le (C_{\vec{a}}-\alpha_-)\eta$. Since $\mu_N(0) > 0$ we see that for α_- large enough the condition has just one solution for with $\eta > 0$. One can check directly (Problem 14) that without $\tilde{\Lambda}(z)$ the condition is satisfied for $\eta = \sum_j \alpha_j^{-1}\sin^2 b_j$. Thus $\eta = \mathcal{O}(\alpha_-^{-1})$ and the eigenvalue expansion follows. To get the eigenfunction we use the eigenvector of M_1 mentioned in the proof of *Theorem* 5.2. ∎

Let us finally mention the scattering problem.

Proposition 5.1.6 *The wave operators for the pair* $(H_0, H_{\alpha,\vec{a}})$ *exist and are asymptotically complete. The on-shell operator* $S(k)$ *at energy* k^2 *with* $k \notin \mathbb{N}$ *is a unitary* $2[k] \times 2[k]$ *matrix with the block structure as in* Proposition *5.1.3 and the transmission and reflection coefficients given by*

$$r_{nm}(k) = \frac{i}{\pi}\sum_{j,l=1}^{N}[\Lambda(z)]_{jl}^{-1}\,\frac{\sin(mb_j)\sin(nb_l)}{k_m(z)}\,\mathrm{e}^{i(k_m a_j + k_n a_l)}\,,$$

$$t_{nm}(k) = \delta_{nm} + \frac{i}{\pi}\sum_{j,l=1}^{N}[\Lambda(z)]_{jl}^{-1}\,\frac{\sin(mb_j)\sin(nb_l)}{k_m(z)}\,\mathrm{e}^{-i(k_m a_j - k_n a_l)}\,,$$

with the right-to-left amplitudes obtained by the mirror transformation which replaces all the a_j *by* $-a_j$.

Proof Analogous to that of *Proposition* 5.1.3. The generalized eigenfunction at a non-integer $z > 1$ is now replaced by

$$\psi(\vec{x}) = \mathrm{e}^{ik_n(z)x}\chi_n(y) + \sum_{j,k=1}^{N}[\Lambda(z)]_{jk}^{-1}G_0(\vec{x}, \vec{a}_j; z)\,\mathrm{e}^{ik_n(z)a_k}\chi_n(b_k)$$

with the incident wave in the n-th channel. Its asymptotic behavior as $x \to \pm\infty$ gives the reflection and transmission amplitudes. ∎

As in the case $N = 1$ scattering resonances are given by complex solutions of the condition (5.7) on other Riemann sheets specified by the choice of signs in (5.6). The explicit form of the S-matrix also makes it possible to find other quantities of physical interest, in the first place the conductance expressed in a way analogous to (2.9), namely

$$G = \frac{2e^2}{h} \sum_{n,m=1}^{[\sqrt{z}]} \frac{k_m}{k_n} |t_{nm}(z)|^2 , \tag{5.10}$$

where z is determined by the Fermi energy and chemical potential of the reservoirs; in the particular case when just two channels are connected to the scattering "target" one usually refers to is as ***Landauer formula***. The presence of impurities again deforms the steplike shape of this function corresponding to an ideal channel, in particular, weak-coupling resonances are manifested by peaks and dips near the thresholds.

5.2 Point Perturbations in a Tube

Now we shall replace the strip of the previous section by a straight tube, $\Omega := \mathbb{R} \times M$ in $\mathbb{R}^3$. The assumptions we adopt are the same as in Sect. 1.4 for $d = 2$, i.e. $M \subset \mathbb{R}^2$ is an open precompact set which is pathwise connected and such that ∂M is piecewise smooth. The free Hamiltonian is the corresponding Dirichlet Laplacian, $H_0 = -\Delta_D^\Omega$ with the form domain $H_0^1(\Omega)$. As above the variables separate and we can write

$$H_0 = -\partial_x^2 \otimes I + I \otimes (-\Delta_D^M) ,$$

where $\vec{x} = (x, y)$ with $y \in M$. Due to the compactness of $\overline{M}$ the last named operator has a purely discrete spectrum; we denote by χ_n, ν_n, $n = 1, 2, \dots$, its eigenfuctions and eigenvalues, respectively. As we have said, the eigenfunctions are supposed to be real-valued. For any $z \in \mathbb{C}\backslash[\nu_1, \infty)$ the free resolvent is thus an integral operator with the kernel

$$G_0(\vec{x}_1, \vec{x}_2; z) \equiv (H_0 - z)^{-1}(\vec{x}_1, \vec{x}_2) = \frac{i}{2} \sum_{n=1}^{\infty} \frac{e^{ik_n(z)|x_1-x_2|}}{k_n(z)} \chi_n(y_1)\chi_n(y_2) ,$$

where $k_n(z) := \sqrt{z-\nu_n}$, which is defined everywhere and smooth except at $\vec{x}_1 = \vec{x}_2$. It is a multi-valued function of z with cuts $[\nu_n, \infty)$, $n = 1, 2, \dots$, the sheets of which are specified by the sign factors in (5.6) with n^2 replaced by ν_n in the definition of $q_n(z)$. Recall that $\nu_1 > 0$ holds by the inequality (1.22).

Fix $\vec{a} := \{a_1, \dots a_N\}$ with $\vec{a}_j = (a_j, b_j)$ and $\alpha := \{\alpha_1, \dots, \alpha_N\} \in \mathbb{R}^N$. The Hamiltonian $H_{\alpha,\vec{a}}$ with N point interactions with the positions and strengths given by the above parameters, respectively, is defined as the self-adjoint extension of the operator $-\Delta_D^M \upharpoonright C_0^\infty(\Omega\backslash\{\vec{a}\})$ specified by the boundary conditions (5.2). Since the underlying space is three-dimensional, however, the generalized boundary values are now different, namely

$$L_0(\psi, \vec{a}) := 4\pi \lim_{\vec{x}\to\vec{a}} \psi(\vec{x})|\vec{x}-\vec{a}| , \quad L_1(\psi, \vec{a}) := \lim_{\vec{x}\to\vec{a}} \left[\psi(\vec{x}) - \frac{L_0(\psi, \vec{a})}{4\pi|\vec{x}-\vec{a}|} \right] . \tag{5.11}$$

The absence of the j-th point interaction means $L_0(\psi, \vec{a}_j) = 0$ which is formally equivalent to putting $\alpha_j = \infty$.

We shall first consider again the case of a single perturbation, $N = 1$. The resolvent kernel of $H_{\alpha,\vec{a}}$ is given by Krein's formula:

$$(H_{\alpha,\vec{a}} - z)^{-1}(\vec{x}_1, \vec{x}_2) = G_0(\vec{x}_1, \vec{x}_2; z) + \frac{G_0(\vec{x}_1, \vec{a}; z)G_0(\vec{a}, \vec{x}_2; z)}{\alpha - \xi(\vec{a}; z)},$$

where $\xi(\vec{a}; z)$ is the regularized Green's function at $\vec{a}$, written with the help of $\kappa_n(z) := \sqrt{\nu_n - z}$ as

$$\xi(\vec{a}; z) = \lim_{u \to 0} \left[\sum_{n=1}^{\infty} \frac{e^{-\kappa_n(z)u}}{2\kappa_n(z)} \chi_n(b)^2 - \frac{1}{4\pi u} \right]. \tag{5.12}$$

Proposition 5.2.1 *Fix $\vec{a} \in \Omega$ and suppose that $|\chi_n(b)| \le C n^{\frac{1}{4}-\varepsilon}$ holds for some positive C, ε and all $n \in \mathbb{N}$. Then the function $\xi(\vec{a}; \cdot)$ is analytic in $\rho(H_0)$. On the interval $(-\infty, \nu_1)$ it is increasing with $\xi(\vec{a}; z) = \frac{1}{2}(\nu_1 - z)^{-1/2}|\chi_n(b)|^2 + \mathcal{O}(1)$ as $z \to \nu_1-$ and*

$$\xi(\vec{a}; z) = -\frac{\sqrt{-z}}{4\pi}\left(1 + \mathcal{O}\left(e^{-c\sqrt{-z}}\right)\right)$$

as $z \to -\infty$ for any $c <$ dist$(b, \partial M)$. Finally, $\xi(\vec{a}; z)$ also makes sense in $[\nu_1, \infty) \backslash \{\nu_n\}_{n \in \mathbb{N}}$ as a boundary value at the cut which is smooth away from the thresholds; the choice of the branch of the square roots in $\kappa_n(z)$ determines the corresponding sheet of the Riemann surface.

Proof The existence follows from the free-resolvent kernel behavior at the singularity [Ti], the other properties are obtained as in *Proposition* 5.1.2 using the identity

$$\xi(\vec{a}; z) - \xi(\vec{a}; z') = \sum_{n=1}^{\infty} \frac{\kappa_n(z') - \kappa_n(z)}{2\kappa_n(z)\kappa_n(z')} \chi_n(b)^2.$$

The behavior of the series is determined by the semiclassical asymptotic properties: the transverse eigenvalues behave as $\nu_n \approx 4\pi|M|^{-1}n$ as $n \to \infty$, where $|M|$ is the area of M, so the coefficient at $\chi_n(b)^2$ is $\mathcal{O}(n^{-3/2})$. Using the assumption about the transverse eigenfunctions, we find that the series on the right-hand side converges uniformly for z, z' in an arbitrary compact set which does not contain any of the thresholds. ∎

Remarks 5.2.1 (a) The assumption about the asymptotic behavior of $|\chi_n(b)|$ was made for convenience and $\xi(\vec{a}; z)$ exists irrespective of it. Moreover, it is not very restrictive and it may happen that even if it is not satisfied the series in the above proof still converges (see the notes).

(b) The limit in (5.12) is not very suitable for practical computation and it is useful to have a prescription to evaluate $\xi(\vec{a}; z)$. A natural idea is to proceed as in the two-dimensional case. The procedure has to be modified, however; we employ the identity

$$\frac{1}{4\pi u} = \frac{1}{4\pi u} \sum_{n=1}^{\infty} \left(\mathrm{e}^{-\gamma u \sqrt{n-1}} - \mathrm{e}^{-\gamma u \sqrt{n}} \right)$$

with a properly chosen γ. The idea is to write the second term on the right-hand side of (5.12) as the series the terms of which have the same asymptotics as the first one with the oscillating factor $\chi_n(b)^2$ replaced by its mean value. If the latter is $|M|^{-1}$ we choose $\gamma := 2\sqrt{\pi}|M|^{-1/2}$ which leads to

$$\xi(\vec{a}; z) = \sum_{n=1}^{\infty} \left[\frac{\chi_n(b)^2}{2\kappa_n(z)} + \frac{\sqrt{n-1} - \sqrt{n}}{2\sqrt{\pi}|M|} \right].$$

However, the convergence in this expression depends strongly on the ergodic properties of the sequence $\{\chi_n(\vec{b})^2\}_{n\in\mathbb{N}}$ and has to be checked separately.

(c) The scaling behavior for $\Omega^\sigma = \mathbb{R} \times M^\sigma$ with $M^\sigma := \sigma M$, $\sigma > 0$, is also more complicated than in the two-dimensional case. One reason is that the definition of $\sigma M := \{\sigma x : x \in M\}$ depends on the choice of the coordinate system origin in the transverse plane, i.e. that a given tube cross section can be scaled in different ways. For a chosen scaling we have $\xi(\vec{a^\sigma}; z\sigma^{-2}) = \sigma^{-1}\xi(\vec{a}; z)$, so any singularity $\epsilon(\alpha, \vec{a})$ of the resolvent kernel transforms as

$$\epsilon^\sigma(\alpha^\sigma, \vec{a^\sigma}) = \sigma^{-2}\epsilon(\alpha, \vec{a}), \quad \alpha^\sigma := \sigma^{-1}\alpha.$$

In particular, the corresponding coupling constant renormalization is multiplicative rather than additive in the three-dimensional case.

Proposition 5.2.2 *For any $\vec{a} \in \Omega$ and $\alpha \in \mathbb{R}$ we have $\sigma_{\mathrm{ess}}(H_{\alpha,\vec{a}}) = \sigma_{\mathrm{ac}}(H_{\alpha,\vec{a}}) = [\nu_1, \infty)$ and $\sigma_{\mathrm{sc}}(H_{\alpha,\vec{a}}) = \emptyset$. Under the assumption of the previous proposition, $H_{\alpha,\vec{a}}$ has one eigenvalue $\epsilon_{\alpha,\vec{a}} \in (-\infty, \nu_1)$ with the eigenfunction*

$$\phi_{\alpha,\vec{a}}(\vec{x}) = \sum_{n=1}^{\infty} \frac{\mathrm{e}^{-\kappa_n(\epsilon_{\alpha,\vec{a}})|x-a|}}{2\kappa_n(\epsilon_{\alpha,\vec{a}})} \chi_n(y)\chi_n(b).$$

The function $\alpha \mapsto \epsilon_{\alpha,\vec{a}}$ is real-analytic and increasing and has the following asymptotic behavior,

$$\begin{aligned} \epsilon_{\alpha,\vec{a}} &= \nu_1 - \frac{\chi_1(b)^4}{4\alpha^2} + \mathcal{O}(\alpha^{-3}), \\ \epsilon_{\alpha,\vec{a}} &= -(-4\pi\alpha)^2 \left(1 - \mathcal{O}\left(\mathrm{e}^{c\alpha}\right)\right) \end{aligned}$$

with any fixed $c < 4\pi\,\mathrm{dist}(\vec{a}, \partial\Omega)$, *for* $\alpha \to \pm\infty$, *respectively. Finally, there are no eigenvalues embedded in the continuous spectrum.*

Proof Analogous to that of *Theorem* 5.1 (Problem 16).

In the case of N point interactions the resolvent kernel is given Krein's formula as in *Proposition* 5.1.4,

$$(H_{\alpha,\vec{a}} - z)^{-1}(\vec{x}_1, \vec{x}_2) = G_0(\vec{x}_1, \vec{x}_2; z) + \sum_{j,m=1}^{N} [\Lambda_{\alpha,\vec{a}}(z)]^{-1}_{jm} G_0(\vec{x}_1, \vec{a}_j; z) G_0(\vec{a}_m, \vec{x}_2; z),$$

where $\Lambda \equiv \Lambda_{\alpha,\vec{a}}(z)$ is the $N \times N$ matrix with the elements

$$\Lambda_{jm} = \big(\alpha_j - \xi(\vec{a}; z)\big)\, \delta_{jm} - G_0(\vec{a}_j, \vec{a}_m; z)(1 - \delta_{jm})$$

and $\xi(\vec{a}; z)$ is now defined by (5.12). The spectral properties of the operator $H_{\alpha,\vec{a}}$ can be then summarized as follows:

Theorem 5.3 *Fix* $\alpha \in \mathbb{R}^N$ *and* $\vec{a} = \{\vec{a}_j\}$ *with* $\vec{a}_j \in \Omega,\ j = 1, \dots, N$, *and suppose that* $|\chi_n(b)| \le C\, n^{\frac{1}{4} - \varepsilon}$ *holds for some positive* C, ε *and all* $n \in \mathbb{N}$. *The spectrum of* $H_{\alpha,\vec{a}}$ *consists of the absolutely continuous part* $[\nu_1, \infty)$ *and eigenvalues* $\epsilon_1^{\alpha,\vec{a}} < \epsilon_2^{\alpha,\vec{a}} \le \cdots \le \epsilon_k^{\alpha,\vec{a}} < \nu_1$ *with* $1 \le k \le N$, *given by the condition*

$$\det \Lambda(\alpha, \vec{a}, z) = 0\,,$$

which are real-analytic functions of the coupling constants. The respective eigenfunctions are $\phi_i^{\alpha,\vec{a}}(\vec{x}) = \sum_{j=1}^{N} d_j G_0(\vec{x}, \vec{a}_j; \epsilon_i^{\alpha,\vec{a}})$, *where the vector* $d \in \mathbb{R}^N$ *is a solution of the linear system* $\sum_{m=1}^{N} \Lambda(z)_{jm} d_m = 0$. *The ground-state eigenfunction can be chosen positive. Furthermore,* $z > \nu_1$ *cannot be an eigenvalue corresponding to an eigenvector from the subspace* $\bigoplus_{\{n:\, \nu_n < z\}} L^2(\mathbb{R}) \otimes \{\chi_n\}$, *while* $H_{\alpha,\vec{a}}$ *can have embedded eigenvalues if the family* $\{\Omega, \vec{a}, \alpha\}$ *has suitable symmetry properties. Finally, in the weak coupling limit,* $\alpha_- := \min_{1 \le j \le N} \alpha_j \to \infty$, *there is a single eigenvalue which behaves as*

$$\epsilon_1^{\alpha,\vec{a}} = \nu_1 - \left(\sum_{j=1}^{N} \frac{\chi_1(b_j)^2}{2\alpha_j} \right)^2 + \mathcal{O}\left(\alpha_-^{-3}\right).$$

Proof For the most part the argument is the same as in the proof of *Theorem* 5.2. The asymptotics of $\Lambda_{\alpha,\vec{a}}(z)$ as $z \to -\infty$ and $z \to \nu_1-$ are now

$$\Lambda_{\alpha,\vec{a}}(z) = \frac{\sqrt{-z}}{4\pi} I + \mathcal{O}(1)\,, \quad \Lambda(\alpha, \vec{a}, z) = -\frac{1}{2\sqrt{\nu_1 - z}} M_1 + \mathcal{O}(1)\,,$$

respectively, where the matrix $M_1 := (\chi_1(b_j)\chi_1(b_m))_{j,m=1}^N$ is rank-one with the positive eigenvalue $\sum_{j=1}^N \chi_1(b_j)^2$ corresponding to $(\chi_1(b_1), \dots, \chi_1(b_N))$. Hence all the eigenvalues of $\Lambda_{\alpha,\vec{a}}(z)$ tend again to $+\infty$ as $z \to -\infty$ while at least one goes to $-\infty$ at the opposite end of the interval $(-\infty, \nu_1)$. Using the continuity we establish that the discrete spectrum is nonempty. The other ingredients of the proof – positivity of $G_0(\vec{a}_j, \vec{a}_m; z)$, orthogonality of the basis $\{\chi_n\}$ and the fact that $\chi_1(b_j)$ is nonzero – also remain valid. The weak-coupling asymptotics is likewise obtained by adapting the argument of *Proposition* 5.1.5. ∎

Remark 5.2.2 We also obtain the weak-coupling asymptotics for the eigenfunction,

$$\phi_1^{\alpha,\vec{a}}(\vec{x}) \approx \chi_1(\vec{y}) \sum_{j=1}^N \exp\left\{-|x-a_j| \sum_{k=1}^N \frac{\chi_1(b_k)^2}{2\alpha_k}\right\} \chi_1(b_j)^2 \left(\sum_{k=1}^N \frac{\chi_1(b_k)^2}{2\alpha_k}\right)^{-1}$$
$$+ \sum_{n=2}^\infty \chi_n(b_j) \sum_{j=1}^N \frac{e^{-\sqrt{\nu_n-\nu_1}|x-a_j|}}{\sqrt{\nu_n-\nu_1}} \chi_n(b_j)\chi_1(b_j)\,,$$

where the leading term combines $\chi_1(y)$ with a linear combination of eigenfunctions of one-dimensional point interactions placed at $a_j,\ j = 1, \dots, N$.

The scattering properties of $H_{\alpha,\vec{a}}$ also adapt easily from the strip case.

Proposition 5.2.3 *The wave operators for the pair $(H_0, H_{\alpha,\vec{a}})$ exist and are asymptotically complete. The on-shell S-matrix at energy $z = k^2$ away from the thresholds is a $2N_z \times 2N_z$ unitary matrix with elementary blocks*

$$S_{nm} = \sqrt{\frac{k_m}{k_n}} \begin{pmatrix} t_{nm} & r_{nm} \\ \tilde{r}_{nm} & \tilde{t}_{nm} \end{pmatrix}, \quad n, m = 1, \dots, N_z\,,$$

where $N_z := \sharp\{\nu_n : \nu_n < z\}$ is the number of open channels at the energy z,

$$r_{nm}(k) = \frac{i}{2} \sum_{j,l=1}^N [\Lambda(z)]_{jl}^{-1} \frac{\chi_m(b_j)\chi_n(b_l)}{k_m(z)}\, e^{i(k_m a_j + k_n a_l)}\,,$$

$$t_{nm}(k) = \delta_{nm} + \frac{i}{2} \sum_{j,l=1}^N [\Lambda(z)]_{jl}^{-1} \frac{\chi_m(b_j)\chi_n(b_l)}{k_m(z)}\, e^{-i(k_m a_j - k_n a_l)},$$

and the tilded quantities are obtained by mirror transformation, $a_j \to -a_j$.

5.3 Point Perturbations in a Layer

Next we will discuss point interactions in another three-dimensional system, namely an infinite planar layer $\Omega := \mathbb{R}^2 \times (0, d)$. We denote the coordinates as $\vec{x} = (x, y)$ with $x = (x_1, x_2) \in \mathbb{R}^2$ and $y \in (0, d)$, and consider the Dirichlet Laplacian $-\Delta_D^\Omega$ as the free Hamiltonian. We shall again put $d = \pi$; a general strip width can be restored using a scaling transformation which is analogous to that of Remark 5.2.1c. Instead of (5.3) we have the decomposition

$$H_0 = \bigoplus_{n=1}^{\infty} h_n \otimes I_n\,, \quad h_n := -\partial_1^2 - \partial_2^2 + n^2, \tag{5.13}$$

where $\partial_i := \partial/\partial x_i$, referring to $L^2(\Sigma) = \bigoplus_{n=1}^{\infty} L^2(\mathbb{R}^2) \otimes \{\chi_n\}$ with the same transverse basis as in Sect. 5.1. Using the properties of the two-dimensional Laplacian we get an expression for the free resolvent kernel,

$$G_0(x, y; x', y'; z) = \frac{i}{4} \sum_{n=1}^{\infty} H_0^{(1)}(k_n|x - x'|)\, \chi_n(y)\, \chi_n(y')\,, \tag{5.14}$$

where $k_n \equiv k_n(z) := \sqrt{z-n^2}$ and $H_0^{(1)}(\cdot)$ is the Hankel function of the first kind. The point-interaction Hamiltonian $H_{\alpha,\vec{a}}$ corresponding to given parameters $\alpha := \{\alpha_1, \dots, \alpha_N\} \in \mathbb{R}^N$ and $\vec{a} := \{a_1, \dots a_N\}$ with $\vec{a}_j = (a_j, b_j) \in \Omega$ is defined by the boundary conditions (5.2) with the "three-dimensional" generalized boundary values introduced at the beginning of the previous section, together with the Dirichlet condition at $\partial\Omega$.

As above it is useful to discuss first the case of a single perturbation. The resolvent kernel of $H_{\alpha,\vec{a}}$ is given by Krein's formula:

$$(H_{\alpha,\vec{a}} - z)^{-1}(\vec{x}_1, \vec{x}_2) = G_0(\vec{x}_1, \vec{x}_2; z) + \frac{G_0(\vec{x}_1, \vec{a}; z) G_0(\vec{a}, \vec{x}_2; z)}{\alpha - \xi(\vec{a}; z)}\,.$$

To find $\xi(\vec{a}; z)$ we employ $\kappa_n := \sqrt{n^2 - z} = -ik_n$ and pass to the Macdonald function $K_0(z) = \frac{\pi i}{2} H_0^{(1)}(iz)$; then it can be written as

$$\xi(\vec{a}; z) = \lim_{\varrho \to 0} \left\{ \frac{1}{\pi^2} \sum_{n=1}^{\infty} K_0(\kappa_n \varrho) \sin^2(nb) - \frac{1}{4\pi\varrho} \right\}.$$

Properties of the regularized Green's function can be summarized as follows.

Proposition 5.3.1 *For fixed $\vec{a} \in \Omega$ and $z \in \rho(H_0)$ we have*

$$\xi(\vec{a}; z) = -\frac{1}{\pi^2} \sum_{n=1}^{\infty} \ln \sqrt{1 - \frac{z}{n^2}} \sin^2(nb) + \frac{1}{4\pi^2} \Big[\gamma_E + \psi\Big(\frac{b}{\pi}\Big) + \frac{\pi}{2} \cot b \Big].$$

The function $\xi(\vec{a}; \cdot)$ is analytic in $\rho(H_0)$. On $(-\infty, 1)$ it is increasing with $\xi(\vec{a}; z) = -\pi^{-2} \ln \sqrt{1-z} \sin^2 b + \mathcal{O}(1)$ as $z \to 1-$, while

$$\xi(\vec{a}; z) = -\frac{\sqrt{-z}}{4\pi} + \mathcal{O}\left(e^{-c\sqrt{-z}}\right) \quad as \quad z \to -\infty$$

holds for any $c < \operatorname{dist}(\vec{a}, \partial\Omega)$. As a function of b it is monotonous across the halflayer,

$$\xi(\vec{a}; z) > \xi(\vec{a}'; z) \quad if \quad \Big|b - \frac{\pi}{2}\Big| < \Big|b' - \frac{\pi}{2}\Big| .$$

Finally, the above expression also makes sense in $[1, \infty)\backslash\{n^2\}_{n\in\mathbb{N}}$ giving boundary values of $\xi(\vec{a}; \cdot)$ at the cut which are smooth away from the thresholds; the Riemann sheet is determined by the branches of the square roots.

Proof To get an expression of $\xi(\vec{a}; z)$ suitable for practical calculations we proceed as in Problem 4 writing $K_0(\kappa_n \varrho) = K_0(n\varrho) + [K_0(\kappa_n \varrho) - K_0(n\varrho)]$ to split the part which can be computed exactly, and at the same time it has the correct asymptotic behavior since $\kappa_n = n(1+\mathcal{O}(n^{-1}))$ as $n \to \infty$. In this way we can write $\xi(\vec{a}; z) = \xi_1 + \xi_2$, where

$$\xi_1 := \lim_{\varrho \to 0} \frac{1}{\pi^2} \sum_{n=1}^{\infty} \Big[K_0(\kappa_n \varrho) - K_0(n\varrho) \Big] \sin^2(nb) ,$$

$$\xi_2 := \lim_{\varrho \to 0} \left\{ \frac{1}{2\pi^2} \sum_{n=1}^{\infty} \Big[K_0(n\varrho) - K_0(n\varrho) \cos(2nb) \Big] - \frac{1}{4\pi\varrho} \right\} .$$

In the first part we use $K_0(\kappa_n \varrho) - K_0(n\varrho) = -\ln \sqrt{1 - zn^{-2}} \left(1 + \mathcal{O}(\varrho^2)\right)$ which shows that the series converges uniformly w.r.t. ϱ and the limit can be interchanged with the sum giving

$$\xi_1 = -\frac{1}{\pi^2} \sum_{n=1}^{\infty} \ln \sqrt{1 - \frac{z}{n^2}} \sin^2(nb) .$$

This series converges for $z \in \mathbb{C}\backslash\{n^2 : n \in \mathbb{N}\}$, because its terms are $\mathcal{O}(n^{-2})$ as $n \to \infty$ as we see using the Taylor expansion of $\ln(1-\zeta)$ to the first order. The series in the second part can be summed and the limit computed explicitly (Problem 17); the resulting ξ_2 is independent of z and finite for any $b \in (0, \pi)$. The rest of the proof is analogous to that of *Proposition* 5.1.2. ∎

Proposition 5.3.2 *Fix $\vec{a} \in \Omega$ and $\alpha \in \mathbb{R}$, then $\sigma_{ess}(H_{\alpha,\vec{a}}) = \sigma_{ac}(H_{\alpha,\vec{a}}) = [1,\infty)$ and $\sigma_{sc}(H_{\alpha,\vec{a}}) = \emptyset$. The operator $H_{\alpha,\vec{a}}$ has a single eigenvalue $\epsilon_{\alpha,\vec{a}} \in (-\infty,1)$ which satisfies*

$$\epsilon_{\alpha,\vec{a}} < \epsilon_{\alpha,\vec{a}'} \quad \textit{if} \quad \left|b - \frac{\pi}{2}\right| < \left|b' - \frac{\pi}{2}\right|,$$

with the eigenfunction

$$\phi_{\alpha,\vec{a}}(\vec{x}) = \frac{1}{4\pi^2} \sum_{n=1}^{\infty} K_0\left(\sqrt{n^2 - \varepsilon_{\alpha,\vec{a}}}\,|x-a|\right) \sin(nb)\sin(ny)\,.$$

The function $\alpha \mapsto \epsilon_{\alpha,\vec{a}}$ is real-analytic and increasing and has the following asymptotic behavior,

$$\begin{aligned} \epsilon_{\alpha,\vec{a}} &= 1 - \exp\left\{-\frac{2\pi^2\alpha}{\sin^2 b}\left(1 + \mathcal{O}(\alpha^{-1})\right)\right\}, \\ \epsilon_{\alpha,\vec{a}} &= -(-4\pi\alpha)^2\left(1 - \mathcal{O}\left(e^{c\alpha}\right)\right) \end{aligned}$$

with any $c < 4\pi\,\mathrm{dist}(\vec{a},\partial\Omega)$, for $\alpha \to \pm\infty$, respectively. Finally, there are no eigenvalues embedded in the continuous spectrum.

Proof Analogous to that of *Theorem* 5.1 (Problem 18).

Let us pass to the scattering problem. The existence of the wave operators is easy to establish due to the finite-rank character of the perturbation. On the other hand, the on-shell scattering operator differs from those of the preceding sections, because we now have more asymptotic directions. In the one-center case one can employ a partial-wave decomposition placing the point perturbation at the origin of polar coordinates in the plane by $T_a : (T_a\phi)(x,y) = \phi(x+a,y)$. We have the tensor-product decomposition $L^2(\Omega) = L^2((0,\infty)\times(0,d); r dr dy) \otimes L^2(S^1)$, where S^1 is the unit circle in $\mathbb{R}^2$ and $r := |x|$. From here we may pass conventionally to

$$L^2(\Omega) = \bigoplus_{m\in\mathbb{Z}} \tilde{U}^{-1} L^2\left((0,\infty)\times(0,d)\right) \otimes \{Y_m\},$$

where the unitary $\tilde{U} : L^2((0,\infty)\times(0,d); r dr dy) \to L^2((0,\infty)\times(0,d))$ is defined by $(\tilde{U}\psi)(r) := r^{1/2}\psi(r)$ and $Y_m(\omega) := (2\pi)^{-1/2} e^{im\theta}$ with $\omega = (\cos\theta, \sin\theta)$. The operator $H_0 = -\Delta_D^\Omega$ then decomposes as $T_a^{-1}\{\bigoplus_{m\in\mathbb{Z}} \tilde{U}^{-1} h_m^{(0)} \tilde{U} \otimes I\} T_a$ with the partial-wave components

$$h_m^{(0)} = -\partial_r^2 - \partial_y^2 + \frac{4m^2-1}{4r^2}, \qquad m \in \mathbb{Z}\,. \tag{5.15}$$

Their domains are given in the usual way. The radial boundary condition at the origin is absent for $m \neq 0$, because the radial part of (5.15) is then a limit-point expression at zero. The point interaction can thus be introduced in the s-wave component only; to this end one has to appropriately modify the free boundary condition, $\lim_{r\to 0}\phi(r,y)\sqrt{r}=0$ for $y\in(0,\pi)$ (Problem 19).

Proposition 5.3.3 *Fix $\vec a\in\Omega$ and $\alpha\in\mathbb{R}$. The on-shell operator $S(k)$ at an energy $z=k^2\in(1,\infty)\backslash\{n^2\}_{n\in\mathbb{N}}$ for the pair $(H_0,H_{\alpha,\vec a})$ is non-trivial in the s-wave subspace only, $m=0$, where it is a unitary $[\sqrt{z}]\times[\sqrt{z}]$ matrix with the elements*

$$S_{nj}(k)=\mathrm{e}^{2i\delta_{nj}(k)}=\delta_{nj}+\frac{i}{\pi}\,\frac{\sin(nb)\sin(jb)}{\alpha-\xi(\vec a;z)}\,.$$

Proof The dimension of $S(k)$ is given by the number of transverse modes in which the particle of energy k^2 can propagate. Using the Krein formula expression of the resolvent it is straightforward to check that

$$\psi_n(\vec x;k)=J_0(k_n(z)r)\chi_n(y)+\frac{i}{4}\sum_{j=1}^{\infty}H_0^{(1)}(k_j(z)r)\frac{\chi_j(b)\chi_n(b)}{\alpha-\xi(\vec a;z)}\chi_j(y)$$

with $r=|x-a|$ is a generalized eigenfunction of $H_{\alpha,\vec a}$ with the eigenvalue k^2 in the s-wave subspace, corresponding to the incident wave in the n-th transverse mode. The S-matrix elements are then obtained using the Bessel function asymptotics for $r\to\infty$ (Problem 20). The unitarity follows from the asymptotic completeness but it can also be checked directly. ∎

In the general case of N point interactions the Krein formula expression for the resolvent kernel reads

$$(H_{\alpha,\vec a}-z)^{-1}(\vec x_1,\vec x_2)=G_0(\vec x_1,\vec x_2;z)+\sum_{j,m=1}^{N}[\Lambda_{\alpha,\vec a}(z)]^{-1}_{jm}G_0(\vec x_1,\vec a_j;z)G_0(\vec a_m,\vec x_2;z),$$

where $\Lambda\equiv\Lambda_{\alpha,\vec a}(z)$ is the $N\times N$ matrix with the elements

$$\Lambda_{jm}=\big(\alpha_j-\xi(\vec a;z)\big)\,\delta_{jm}-G_0(\vec a_j,\vec a_m;z)(1-\delta_{jm})$$

and $\xi(\vec a;z)$ is described in *Proposition* 5.3.1. Notice that while the expression (5.14) for $G_0(\vec a_j,\vec a_m;z)$ makes no sense if the two vectors are vertically arranged, $a_j=a_m$, the Green function still exists and can be computed (Problem 17). The spectral properties of $H_{\alpha,\vec a}$ are now summarized as follows:

Theorem 5.4 *For any $\vec a=\{\vec a_j\}$ with $\vec a_j\in\Omega,\ j=1,\dots,N$, and $\alpha\in\mathbb{R}^N$ we have $\sigma_{\mathrm{ess}}(H_{\alpha,\vec a})=\sigma_{\mathrm{ac}}(H_{\alpha,\vec a})=[1,\infty)$ and $\sigma_{\mathrm{sc}}(H_{\alpha,\vec a})=\emptyset$. Furthermore, $\sigma_{\mathrm{disc}}(H_{\alpha,\vec a})$ consists of the eigenvalues $\epsilon_i^{\alpha,\vec a}\in(-\infty,1)$ with $1\le i\le k\le N$, which are real-analytic functions of the parameters α_j. The ground-state eigenvalue $\epsilon_1^{\alpha,\vec a}$ is simple* *which may not be true for the higher ones. The eigenfunctions corresponding to* $\epsilon_i^{\alpha,\vec a}$ *are of the form*

$$\phi_i^{\alpha,\vec a}(\vec x) = \sum_{j=1}^{N} d_j G_0(\vec x, \vec a_j; \epsilon_i^{\alpha,\vec a})$$

with the coefficients given by solutions to the equation $\sum_{m=1}^{N} \Lambda(\epsilon_i^{\alpha,\vec a})_{jm} d_m = 0$*; those corresponding to* $\phi_1^{\alpha,\vec a}$ *can be chosen positive. The weak-coupling limit,* $\alpha_- := \min_{1\le j\le N} \alpha_j \to \infty$*, of the ground state is*

$$\epsilon_1^{\alpha,\vec a} = 1 - \exp\left\{ -2\pi^2 \left(\sum_{j=1}^{N} \frac{\sin^2 b_j}{\alpha_j} \right)^{-1} (1+\mathcal{O}(\alpha_-^{-1})) \right\}.$$

Finally, $z > 1$ *cannot be an eigenvalue with an eigenvector belonging to the subspace* $\bigoplus_{n=1}^{[\sqrt{z}]} L^2(\mathbb{R}) \otimes \{\chi_n\}$.

Proof The argument is analogous to the proofs of *Theorems* 5.2 and 5.3. The existence of an eigenvalue follows from the asymptotics

$$\Lambda(\alpha, \vec a; z) = \frac{\sqrt{-z}}{4\pi} I + \mathcal{O}(1), \quad \Lambda(\alpha, \vec a; z) = \frac{1}{\pi^2} \ln\sqrt{1-z}\, M_1 + \mathcal{O}(1)$$

as $z \to \mp\infty$, respectively, where M_1 is the matrix $\left(\sin b_j \sin b_m\right)_{j,m=1}^{N}$ as in *Theorem* 5.2. ∎

Remark 5.3.1 The weak-coupling asymptotic expression of the ground-state eigenfunction,

$$\phi_1^{\alpha,\vec a}(\vec x) \approx \sin y \left(\frac{\sum_{j=1}^{N} \sin^2 b_j}{\sum_{j=1}^{N} \frac{\sin^2 b_j}{\alpha_j}} - \frac{1}{\pi^2} \sum_{j=1}^{N} \sin^2 b_j \ln|x - a_j| \right)$$
$$+ \frac{1}{\pi^2} \sum_{n=2}^{\infty} \sin(ny) \sum_{j=1}^{N} \sin b_j \sin(nb_j)\, K_0\left(\sqrt{n^2-1}|x - a_j|\right),$$

is again dominated by the part corresponding to the lowest transverse mode.

Consider finally the scattering by a finite family of point interactions in the layer. With the exception of the case when the impurities are arranged vertically, the Hamiltonian $H_{\alpha,\vec a}$ now loses the invariance with respect to rotations around an axis perpendicular to Σ. Hence we cannot employ the decomposition into angular-momentum eigenspaces and we have to look for an expression of the on-shell scattering operator which mixes different partial waves.

Proposition 5.3.4 *The wave operators for the pair* $(H_0, H_{\alpha,\vec{a}})$ *exist and are asymptotically complete. The on-shell scattering operator* $S(k)$ *at energy* $z = k^2$ *with* $k \notin \mathbb{N}$ *is unitary and acts on vectors* $\phi \in L^2(S^1) \otimes L^2((0,d))$ *as*

$$S(k)\phi = I + \frac{i}{2\pi^2} \sum_{j,l=1}^{N} \sum_{m,n=1}^{[\sqrt{z}]} [\Lambda_{\alpha,\vec{a}}(z)]_{jl}^{-1} \sin(mb_j)\sin(nb_l) \left(\mathrm{e}^{-ik_n(z)(\cdot)a_l}\chi_n, \phi\right) \times \mathrm{e}^{-ik_m(z)(\cdot)a_j}\chi_m \,.$$

Proof We proceed as in the proof of *Proposition* 5.1.3. Starting from $\psi_z^\varepsilon(\vec{x}) = e^{ik_n(z)\omega x - \varepsilon|x|^2}\chi_n(y)$ for ψ_z, where ω is a unit vector in $\mathbb{R}^2$, and using

$$\left((H(\alpha,\vec{a})-z)\psi^\varepsilon\right)(\vec{x}) = 4\varepsilon[1-\varepsilon|x|^2 + ik_n\omega x]\,\psi_z^\varepsilon(\vec{x})\,,$$

we arrive at the generalized eigenvector $\psi_{\alpha,n}(\vec{x}; k_n(z)\omega)$ equal to

$$\mathrm{e}^{ik_n(z)\omega x}\chi_n(y) + \sum_{j,l=1}^{N} [\Lambda_{\alpha,\vec{a}}(z)]_{jl}^{-1}(\alpha,\vec{a};z)\, G_0(\vec{x},\vec{a}_j;z)\, \mathrm{e}^{ik_n(z)\omega a_l}\chi_n(b_l)\,.$$

The components $\left(f_\alpha(k_n(z),\omega',\omega)\right)_{mn}$ of the on-shell scattering amplitude are then given by projections of the following expression,

$$\lim_{|x|\to\infty} |x|^{1/2}\mathrm{e}^{-ik_m|x|}\left[\psi_{\alpha,n}(\vec{x}; k_n(z)\omega) - \mathrm{e}^{ik_n\omega x}\chi_n(y)\right]$$

with $|x|^{-1}x = \omega'$ kept fixed, to the outgoing m-th transverse mode. This yields

$$\left(f_\alpha(k_n(z),\omega',\omega)\right)_{mn} = \frac{\mathrm{e}^{i\pi/4}}{\pi\sqrt{2\pi k_m(z)}} \sum_{j,l=1}^{N} e^{-ik_m(z)\omega' a_j}\,[\Lambda_{\alpha,\vec{a}}(z)]_{jl}^{-1}\,\mathrm{e}^{ik_n(z)\omega a_l}\,\sin(mb_j)\sin(nb_l)$$

and the on-shell scattering operator given above. ∎

As in the previous sections, resonances are determined by poles in the meromorphic continuation of the matrix-valued function $[\Lambda_{\alpha,\vec{a}}(\cdot)]^{-1}$.

5.4 Notes

Point interactions were introduced in the early days of quantum mechanics, first by R. Kronig and W. Penney for one-dimensional systems [KP31], then by E. Fermi in dimension three [Fe36]. A proper mathematical tool to deal with such strongly localized singular interactions was found only three decades later by F. Berezin and L. Faddeev [BF61]. A thorough and extensive discussion of the point-interaction theory, methods and applications can be found in the monograph [AGHH] which we use here as a basic source on the subject, see also [AK]; for a discussion of point interactions from the physicist's point of view the reader can consult [DO].

The theory of *self-adjoint extensions* is one of the important contributions that J. von Neumann made to the newborn quantum mechanics [vN]. It is explained in many mathematical-physics textbooks such as [AG, Chap. VIII], [RS, Sect. X.1], [We, Chap. 8], [BEH, Sect. 4.7], etc. An application of the theory to the description of point interactions was proposed in [BF61]. To get a meaningful physical picture, one has to add an interpretation of the "coupling constants" of such point interactions. A natural way to do that is through approximation of point-interaction Hamiltonians by those with more regular interactions. This idea works well in dimension one, where one can use a family of Schrödinger operators with scaled potentials, convergent in the norm-resolvent topology, which fits into the heuristic concept that a low-energy particle with widely smeared wave packet "sees" the average value of a well localized potential only. If $d = 2, 3$, however, a similar limit must be accompanied by a sophisticated coupling constant renormalization leading an "infinitely weak" interaction—see [AGHH, Sects. I.1, 1.5]. This does not mean that these point interactions are something exceptional: despite their "zero radius" they have a nonzero scattering length and thus represent a natural model of small obstacles [EŠ96]. Note that point interactions as perturbations of second-order operators can be constructed only for $d \leq 3$, since in higher dimensions the construction based on self-adjoint extensions, in the standard quantum mechanical framework at least, leads to a trivial result—cf. [He89, AGHH].

Point interactions as a model of impurity scattering in a two-dimensional electron gas appeared a longtime ago—cf. [Pr81]. However, their nontrivial character is probably the reason why this idea is not used often; for impurities in a hard-wall strip we can mention, e.g., [Ba90, CBC92], where bound and resonance states are discussed in this setting; the first named paper gives a survey of other approaches to this problem. The point scatterers are usually treated in a simplified way, with the model space restricted to a finite number of transverse modes, which makes a comparison of the employed coupling parameters to α_j of (5.2) difficult. Point interactions may also be useful to describe artificial impurities in quantum wires—let us mention [KSH94] as an example—but such a model should be taken *cum grano salis* because these objects are so far much greater than "natural" impurities consisting of isolated alien atoms.

Section 5.1 The material of this section is taken from [EGŠT96] where illustrations of eigenvalue plots, eigenfunctions, transmission probabilities, etc., can also be found. The generalized boundary values were introduced, e.g., in [BG85]. For Krein's formula see, e.g., [AG, Sect. 106], [AGHH, Appendix A], or more generally [KL71]. An alternative expression for the resolvent difference in *Proposition* 5.1.1, namely

$$\lambda^{-1} = \alpha - \xi(\vec{a}; z) = (1+e^{i\theta})^{-1} \left\{ (i-z) \int_\Omega G_0(\vec{x}, \vec{a}; z) G_0(\vec{x}, \vec{a}; i)\, d\vec{x} \right.$$
$$\left. - e^{i\theta}(i+z) \int_\Omega G_0(\vec{x}, \vec{a}; z) G_0(\vec{x}, \vec{a}; -i)\, d\vec{x} \right\},$$

is obtained from an integral formula in [Zo80, Theorem 4.1]. There is a simple relation between the parameters θ and α (Problem 3), however, only the latter has a reasonable physical interpretation. Examples of resonance-pole trajectories are in [EGŠT96] investigated beyond the weak-coupling regime, however, only weak perturbations produce a substantial resonance scattering effect because the pole residue moduli decrease rapidly with the coupling strength.

The argument in the proof of *Theorem* 5.2 is adapted from [AGHH, Sect. II.1]. Monotonicity of the eigenvalues of $\Lambda_{\alpha,\vec{a}}(z)$ w.r.t. z is a general result [KL71], here one can also check it directly (Problem 11). It is easy to see that the coefficients in the ground-state eigenfunction are strictly positive: if some $d_{j_0} = 0$ the eigenfunction would be smooth at $\vec{x} = \vec{a}_{j_0}$ meaning that the corresponding point interaction is absent, formally $\alpha_{j_0} = \infty$. Concerning the number of eigenvalues of the operator $H_{\alpha,\vec{a}}$, one can specify polynomially bounded subsets of the parameter space $\mathbb{R}^N$ corresponding to different fixed numbers—cf. [EGŠT96].

The avoided-crossing effect with respect to a running coupling constant is known for many operators. If the slope of the eigenvalue curves away from the avoided crossings is steep, one also speaks of a *cascading phenomenon*. This typically happens when the involved eigenfunctions are weakly correlated except for a small subset in the parameter space—see, e.g., [GHKSV88]. Problem 12 illustrates this effect for $N = 2$ where the crossing is always avoided and the two eigenvalues follow asymptotically the two branches of the "decoupled" spectrum. The width of the avoided crossing is controlled in this example by $G_0(\vec{a}_1, \vec{a}_2; z)$; since this quantity at a fixed energy decreases exponentially with the distance of the two points, a profound cascading effect may be expected when the impurities are far apart. Moreover, if $N > 2$ and the perturbations produce a multiple eigenvalue or a cluster of almost identical simple eigenvalues, the cascading means that one eigenvalue leaves the cluster and one joins it—cf. [EGŠT96]. In that paper one can also find examples of conductance plots illustrating the resonance effect. Let us remark finally that the explicit formulæ in *Propositions* 5.1.4 and 5.1.6 make it obvious that scattering resonances coincide in the present situation with those defined through poles in the analytically continued resolvent.

Section 5.2 A part of the material of this section is taken from [Ex00]. The assumption about the growth of $|\chi_n(b)|$ in *Theorem* 5.3 and the preceding propositions is not very restrictive. Recall that by [SeS89, Corollary 2.2] there is a constant C such that $|\chi_n(b)| \le Cn^{1/4}$ holds for all n. This is a particular case of the bound $\|\chi\|_\infty \le C\lambda^{(d-1)/4}$ which holds for a d-dimensional M and a normalized eigenfunction χ corresponding to an eigenvalue λ ; it remains true for the Laplace-Beltrami operator on a Riemann manifold M with a smooth boundary—see [Gr02]. The bound may be saturated if there are points in M where many classical trajectories intersect:

a prime example is a circular M. However, even in such cases it may happen that the series in the proof of *Proposition* 5.2.1 converges (Problem 15).

Section 5.3 The material of this section is taken from [EN02].

5.5 Problems

1. Check the properties of H_0 and $H_{\alpha,\vec{a}}$ with respect to a transverse scaling of the strip Ω—cf. Remark 5.1.1.

2. Check that $\mathcal{L}_0(\vec{a}) := -\lim_{\vec{x}\to\vec{a}} \frac{2\pi}{\ln|\vec{x}-\vec{a}|} \left\{1+\frac{i\lambda}{\pi}\sum_{n=1}^{\infty} \frac{e^{ik_n|x-a|}}{k_n} \sin(ny)\sin(nb)\right\}$ is equal to $-\lambda$ by a direct computation.

Hint: $k_n^{-1}-(in)^{-1}$ and $e^{ik_n+n-z/2n}-1$ are $\mathcal{O}(n^{-3})$ as $n\to\infty$.

3. The resolvent formula of *Proposition* 5.1.1 and Zorbas' parametrization given in the notes to Sect. 5.1 are related by $\alpha(\theta) = F(b)_+ + \frac{\sin\theta}{1+\cos\theta} F_-(b)$, where

$$F_\pm(b) := \frac{1}{\pi}\sum_{n=1}^{\infty}\left(\sqrt{\frac{\sqrt{n^4+1}\pm n^2}{2(n^4+1)}}\,\sin^2(nb) - \frac{1\pm 1}{4n}\right).$$

Hint: Use $i\mp z = \pm k_n(\pm i)^2 \mp k_n(z)^2$.

4. Check directly that $\xi(\vec{a};0) = (2\pi)^{-1}\ln(2\sin b)$ holds for the function defined in *Proposition* 5.1.1 and use this result to express $\xi(\vec{a};z)$.

5. Prove the asymptotics of $\xi(\vec{a};z)$ as $z\to-\infty$ in *Proposition* 5.1.2.

Hint: By Dirichlet bracketing the eigenvalue of $H_{\alpha,\vec{a}}$ satisfies $\epsilon_\alpha \le \epsilon_{\alpha,\vec{a}} \le \epsilon_\alpha^R$, where ϵ_α refers to a single point interaction with coupling constant α in $\mathbb{R}^2$ and ϵ_α^R to such an interaction in the center of a Dirichlet circle of radius R.

6. In the vicinity of m^2, $m=1,2,\dots$, the resolvent kernel of $H_{\alpha,\vec{a}}$ behaves as

$$\frac{i\tilde{\alpha}}{\pi}\,\frac{\sin(my_1)\sin(my_2)}{\sqrt{z-m^2}}\left(\tilde{\alpha}-\frac{i}{\pi}\frac{\sin^2(mb)}{\sqrt{z-m^2}}\right)^{-1}\left(1+\mathcal{O}(\sqrt{z-m^2})\right),$$

where $\tilde{\alpha} := \alpha + \frac{1}{2\pi m} - \frac{i}{\pi}\sum_{n\ne m}\left(\frac{\sin^2(nb)}{k_n(m^2)} - \frac{1}{2in}\right)$, so it has no pole there.

7. Let $\psi_{\alpha,\vec{a}} := \phi_{\alpha,\vec{a}}\|\phi_{\alpha,\vec{a}}\|^{-1}$ be the normalized eigenfunction of $H_{\alpha,\vec{a}}$ according to Remark 5.1.2 and $\kappa_\alpha := 2\,e^{-2\pi\alpha-\gamma_E}$, then

$$\lim_{\alpha\to-\infty}\left\|\psi_{\alpha,\vec{a}} - \frac{\kappa_\alpha}{\sqrt{\pi}}K_0(\kappa_\alpha|\cdot-\vec{a}|)\right\| = 0$$

holds in $L^2(\Omega)$.

Hint: By Remark 5.1.1 one can reformulate the problem to keep α fixed and to consider a family $\{H^d_{\alpha,\vec{a}_d}\}$ of operators corresponding to transversally scaled strips. The latter converges to the Hamiltonian with a point interaction in $L^2(\mathbb{R})$ by *Proposition* 5.1.1, [AGHH, Theorem I.5.2], and the corresponding result for Dirichlet Laplacians—see, e.g., [RT75].

8. Check the unitarity of the S-matrix in *Proposition* 5.1.3, in particular, the conservation of probability flow, $\sum_{m=1}^{[\sqrt{z}]} k_m(|t_{nm}|^2 + |r_{nm}|^2) = k_n$.
Hint: Use $t_{nn} = 1 + r_{nn}\mathrm{e}^{-2ik_n a}$ and $\operatorname{Im}\xi(\vec{a}; z) = \pi^{-1}\sum_{m=1}^{[\sqrt{z}]} \frac{\sin^2(mb)}{k_m(z)}$, $z \neq n^2$.
9. Prove *Proposition* 5.1.4.
10. Any solution of the linear system $\sum_{m=1}^{N} \Lambda(z)_{jm} d_m = 0$, where $\Lambda \equiv \Lambda_{\alpha,\vec{a}}(z)$ is the matrix from *Proposition* 5.1.4, determines an eigenstate of the operator $H_{\alpha,\vec{a}}$ by $\sum_{j=1}^{N} d_j G_0(\cdot, \vec{a}_j; z)$.
Hint: Invert the argument from *Theorem* 5.2—cf. [AGHH, Sect. II.1].
11. The eigenvalues of $\Lambda_{\alpha,\vec{a}}(z)$ are decreasing w.r.t. the variable z.
Hint: Check that $\frac{d}{dz}(\eta, \Lambda(z)\eta) < 0$ holds for any $\eta \in \mathbb{C}^N$ using the fact that the function $f : f(x) = \mathrm{e}^{-\kappa|x|}(1+\kappa|x|)$ is of the positive type as follows from Bochner's theorem [RS, Sect. IX.2].
12. Consider $\alpha := (\tilde{\alpha}, \alpha_2)$, where α_2 is kept fixed and $\tilde{\alpha}$ runs through $\mathbb{R}$. If one of the perturbations is absent, the other one gives rise to a single eigenvalue which we denote by $e_1(\tilde{\alpha})$ and $e_2(\alpha) \equiv e_2$, respectively. Find the the eigenvalues $\epsilon_j(\tilde{\alpha}) := \epsilon_j^{\alpha,\vec{a}}$ of $H_{\alpha,\vec{a}}$ and show that

$$\epsilon_2(\tilde{\alpha}) = e_2 + \frac{c_2(e_2)}{\alpha_0 - \tilde{\alpha}} + \mathcal{O}(\tilde{\alpha}^{-2}),$$

where $c_j := g_{12}^2/\xi_j' > 0$ and $\alpha_0 := \xi_1(e_2) - 2(g_{12}g_{12}')(e_2)/\xi_2'(e_2)$ with $g_{12}(z) := G_0(\vec{a}_1, \vec{a}_2; z)$ and $\xi_j := \xi(\vec{a}_j, \cdot)$, as $\tilde{\alpha} \to -\infty$. Similarly, the ground state behaves in the strong-coupling limit as

$$\epsilon_1(\tilde{\alpha}) \approx e_1(\tilde{\alpha}) + \frac{4\pi}{\tilde{\alpha}}\,(g_{12}(e_1(\tilde{\alpha})))^2\,\mathrm{e}^{-2(2\pi\tilde{\alpha}+\gamma_E)},$$

where $g_{12}(e_1(\cdot))$ is exponentially decaying with the same rate. Find the behavior above the avoided crossing under the assumption that $2g_{12}g_{12}'$ remains small over a wide range of energies.
13. Find the eigenvalue asymptotics of $H_{\alpha,\vec{a}}$ with $N > 1$ as $\max_j \alpha_j \to -\infty$.
Hint: Write $\Lambda(z) = \operatorname{diag}\left(\alpha_j + \frac{1}{4\pi}\ln\left(-\frac{z}{4}\right) + \frac{1}{2\pi}\gamma_E\right) + \tilde{\Lambda}(z)$, where $\tilde{\Lambda}(z)$ is a remainder matrix, independent of α, with the norm vanishing exponentially fast as $z \to -\infty$.
14. Complete the proof of *Proposition* 5.1.5. Show that $M_1 - \eta\mathcal{A}$ has a zero eigenvalue if $\eta = \sum_j \alpha_j^{-1}\sin^2 b_j$.
15. Let M be a circle of radius $R > 0$ and denote by χ_{jm} the corresponding normalized eigenfunctions of the Dirichlet Laplacian $-\Delta_D^M$, i.e. $\chi_{jm}(r,\varphi) = c_j J_m(\kappa_{jm} r)\,\mathrm{e}^{im\varphi}$ with appropriate parameters. Show that

$$\chi_{jm}(0,0)^2 = \frac{\pi j}{2R^2} + \mathcal{O}(1)$$

as $j \to \infty$. What does it mean for the sequence $\{|\chi_n(\vec{0})|^2\}_{n\in\mathbb{N}}$ obtained by ordering all the eigenfunctions in the order of their eigenvalues? Check that the corresponding series in the proof of *Proposition* 5.2.1 converges.
Hint: $\chi_{jm}(0,0) = 0$ for $m \neq 0$.
16. Prove *Proposition* 5.2.2.
17. Show that the quantity $\xi_2 \equiv \xi_2(b)$ in the proof of *Proposition* 5.3.1 equals

$$\xi_2 = \frac{1}{4\pi^2}\left\{-\frac{1}{2\beta} - \sum_{n=1}^{\infty}\frac{\beta^2}{n(n^2-\beta^2)}\right\} = \frac{\gamma_E}{4\pi^2} + \frac{1}{8\pi^2}\Big(2\psi(\beta) + \pi\cot b\Big),$$

where $\beta := b/\pi$. Furthermore, $G_0(\vec{a}, \vec{a}'; z)$ for a pair of vertically arranged vectors $\vec{a} = (a, b)$ and $\vec{a}' = (a, b')$ can be expressed as

$$-\frac{1}{\pi^2}\sum_{n=1}^{\infty}\ln\sqrt{1-\frac{z}{n^2}}\,\sin(nb)\sin(nb') + \xi_2\left(\frac{b+b'}{2}\right) - \xi_2\left(\frac{|b-b'|}{2}\right).$$

Hint: Express $\sum_{n=1}^{\infty} K_0(n\varrho)\cos(2nb)$ using [PBM, Sects. II.5.9.1.4 and I.5.1.15.2].
18. Prove *Proposition* 5.3.2.
19. Let $H_{\alpha,\vec{a}}$ describe a layer with a single point interaction. Its partial-wave components in the appropriate system of coordinates act as those of the free Hamiltonian (5.15). Write the corresponding boundary conditions.
Hint: Use the generalized boundary values $\ell_0(\phi)(y) := 4\pi \lim_{r\to 0}\phi(r, y)\sqrt{r}$ and $\ell_1(\phi)(y) := \lim_{r\to 0} r^{-1/2}[\phi(r, b) - 4\pi\ell_0(\phi)(y)r^{-1/2}]$ at the symmetry axis.
20. Fill in the details into the proof of *Proposition* 5.3.3. Check the unitarity relation, $\sum_{j=1}^{[\sqrt{z}]} S_{nj}\bar{S}_{sj} = \delta_{ns}$, in the same way as in Problem 8.
Hint: $\psi_n(\vec{x}; k) \approx e^{i\delta_{nj}(k)}\cos(k_n(z)r - \pi/4 + \delta_{nj}(k))$ as $r \to \infty$.

Chapter 6
Weakly Coupled Bound States

Properties of the discrete spectrum in a tube or layer induced either by a local change of geometry or by a potential depend, of course, on the perturbation. Situations where the latter is weak are of special interest. We have already encountered such problems in the previous chapter in the particular context of point interactions. Now we are going to discuss systematically the weak coupling behavior of the bound states treated above; our aim is to demonstrate several different methods which can be used to this purpose.

6.1 Birman-Schwinger Analysis

One of the most useful tricks in the theory of Schrödinger operators was invented more than forty years ago simultaneously and independently by M.S. Birman, a mathematician, and J. Schwinger, a physicist. It consists of transforming the original differential-operator problem to solution of an integral equation which is in many respects an easier task.

In this section we will describe how the ***Birman-Schwinger method*** (or BS-method) works for Schrödinger operators in straight Dirichlet tubes and layers. As above we start from the Cartesian product $\Omega_0 = \mathbb{R}^c \times M \subset \mathbb{R}^d$, $d \geq 2$, where $M \subset \mathbb{R}^{d-c}$ is an open, precompact, pathwise connected set, with ∂M piecewise smooth if $d - c \geq 2$. The free Hamiltonian is the Dirichlet Laplacian, $H_0 = -\Delta_D^{\Omega_0}$, defined through its quadratic form on $H_0^1(\Omega_0)$. The spectrum is obtained by separation of variables $\vec{x} = (x, y)$ as in (1.23) and Problem 4.14. We shall again use the symbols χ_n, ν_n with $n = 1, 2, \dots$ for the eigenfuctions and eigenvalues, respectively, of the Dirichlet Laplacian $-\Delta_D^M$ in $L^2(M)$ assuming without loss of generality that the eigenfunctions are real-valued. The ground-state eigenvalue ν_1 is positive in view of the inequality (1.22) which extends naturally to higher dimensions.

P. Exner and H. Kovařík, *Quantum Waveguides*,
Theoretical and Mathematical Physics, DOI 10.1007/978-3-319-18576-7_6

The full Hamiltonian is a perturbation of H_0. In this section we shall be concerned with potential perturbations which can be defined conveniently through the respective quadratic forms analogous to (1.24). It is useful, however, to include from the beginning more general interaction terms. Let $U,\ \tilde U$ be operators on $L^2(\Omega_0)$ which allow us to write the Hamiltonian as

$$H_\lambda = H_0 + \lambda \tilde U^* U\,; \tag{6.1}$$

with the needs of this chapter in mind we introduce the explicit coupling constant $\lambda \in \mathbb{R}$. Naturally, as a quantum observable H_λ has to be self-adjoint. For unbounded $U,\ \tilde U$ it is guaranteed, e.g., if the operator $\lambda \tilde U^* U$ is H_0-bounded with relative bound less than one. We shall adopt this assumption throughout, of course, checking its validity in particular situations.

For a real $z < \nu_1 = \inf \sigma(H_0)$ the resolvent $(H_0 - z)^{-1}$ is bounded, hence one can define the operator

$$K_\lambda^z := \lambda U (H_0 - z)^{-1} \tilde U^* \tag{6.2}$$

provided $\mathrm{Dom}\,(H_0) \subset \mathrm{Dom}\,(U)$. We have the following important relation between the spectral properties of K_λ^z and those of the original operator H_λ.

Proposition 6.1.1 (BS principle) *Suppose that* $\mathrm{Dom}\,(H_\lambda) \subset \mathrm{Dom}\,(U)$*, then* $z \in \sigma_{\mathrm{disc}}(H_\lambda)$ *holds if and only if* $-1 \in \sigma_{\mathrm{disc}}(K_\lambda^z)$.

Proof If $K_\lambda^z \psi = -\psi$, the vector $\phi := -\lambda (H_0 - z)^{-1} \tilde U^* \psi$ is easily checked to satisfy $H_\lambda \phi = z\phi$. Conversely, if $H_\lambda \phi = z\phi$ we have $\varphi \in \mathrm{Dom}\,(H_\lambda) \subset \mathrm{Dom}\,(U)$ by assumption, so $\psi := U\phi$ belongs to $L^2(\Omega_0)$ and $K_\lambda^z \psi = -\psi$. ∎

Remarks 6.1.1 (a) If V is a potential on Ω_0, one usually takes multiplication by the functions $|V(\cdot)|^{1/2}$ and $V(\cdot)^{1/2} := |V(\cdot)|^{1/2}\,\mathrm{sgn}\, V(\cdot)$, respectively, for the pair of factorizing operators $U,\ \tilde U$.
(b) The result also remains valid in the case when $U,\ \tilde U$ map into another Hilbert space $\mathcal{G}$, then K_λ^z is, of course, an operator on this space.
(c) Another generalization concerns the situation where one or both of the operators $U,\ \tilde U$ depend on the parameter λ. The BS-principle holds again, but caution is needed when the weak-coupling analysis is performed by expansion in λ (see the notes).

Let us now apply the BS-principle to Schrödinger operators in tubes and layers, $H_\lambda = -\Delta_D^{\Omega_0} + \lambda V$. An important role will be played again by the matrix representation of the potential similar to that appearing in (2.12), namely

$$V_{mn} :\ V_{mn}(x) = \int_M V(x,y)\,\chi_m(y)\chi_n(y)\,\mathrm{d}y\,, \tag{6.3}$$

by functions which are well defined and measurable in the variable $x \in \mathbb{R}^c$. The last claim depends, of course, on regularity properties of the function V. Although we will deal mostly with bounded potentials, it is useful to consider a wider class borrowed from the usual Schrödinger operator theory: we shall suppose generally that $V \in (L^p + L^\infty)(\Omega_0)$, where $p \geq \max\{2, \frac{d}{2}\}$ for $d \neq 4$ and $p > 2$ for $d = 4$. This requirement ensures, in particular, that H_λ is well defined as an operator sum (Problem 1).

Since the free Green's function depends on d, we shall treat the different codimension cases separately starting with the tubes, $c = 1$. Under slightly stronger assumptions about the potential, we can prove the following claim.

Theorem 6.1 *Suppose that* $V \in (L^p + L^\infty_\varepsilon)(\Omega_0)$, *where* $p = 2$ *for* $d = 2, 3$ *and* $p > \frac{d}{2}$ *for* $d \geq 4$, *nonzero, and such that* $|V|_{11} \in L^1(\mathbb{R}, |x|\,dx)$; *then* H_λ *has for small enough* $|\lambda|$ *at most one simple eigenvalue* $\epsilon(\lambda) < \nu_1$, *and this happens if and only if* $\int_{\mathbb{R}} \lambda V_{11}(x)\,dx \leq 0$. *Moreover, if this condition holds the following expansion is valid,*

$$\sqrt{\nu_1 - \epsilon(\lambda)} = -\frac{\lambda}{2}\int_{\mathbb{R}} V_{11}(x)\,dx - \frac{\lambda^2}{4}\left\{\int_{\mathbb{R}^2} V_{11}(x)|x-x'|\,V_{11}(x')\,dx\,dx'\right.$$
$$\left. - \sum_{n=2}^{\infty}\int_{\mathbb{R}^2} V_{1n}(x)\,\frac{e^{-\sqrt{\nu_n-\nu_1}\,|x-x'|}}{\sqrt{\nu_n-\nu_1}}\,V_{n1}(x')\,dx\,dx'\right\} + \mathcal{O}(\lambda^3)\,. \tag{6.4}$$

Proof Recall that $\inf \sigma_{ess}(H_\lambda) = \nu_1$ by *Proposition* 1.4.1. Since the BS-principle applies to H_λ in view of Problem 1, we have to find the spectrum of the operator $K_\lambda^z = \lambda|V|^{1/2}(-\Delta_D^{\Omega_0} - z)^{-1}V^{1/2}$. The free resolvent is an integral operator with the kernel we know from Sect. 5.1.2. Its part coming from the lowest transverse mode has a singularity as $z \to \nu_1-$ which we shall single out introducing the decomposition $K_\lambda^z = \lambda Q_z + \lambda P_z$ with

$$Q_z(\vec{x}, \vec{x}') = \frac{e^{-\kappa_1(z)\,|x|}}{2\kappa_1(z)}\,|V(x,y)|^{1/2}\chi_1(y)\,e^{-\kappa_1(z)\,|x'|}\,\chi_1(y')V(x',y')^{1/2},$$

where we use again $\kappa_n(z) := \sqrt{\nu_n - z}$, and $P_z := A_z + |V|^{1/2}B_z V^{1/2}$ with

$$A_z(\vec{x}, \vec{x}') := |V(x,y)|^{1/2}\chi_1(y)\,\frac{e^{-\kappa_1(z)\,|x|_>}\,\sinh(\kappa_1(z)\,|x|_<)}{\kappa_1(z)}\,\chi_1(y')V(x',y')^{1/2}$$

and $B_z(\vec{x}, \vec{x}')$ representing the sum of the higher-mode contributions to the Green function $G_0(\vec{x}, \vec{x}'; z)$; in the last formula we use the notation

$$|x|_< := \max\{0, \min(|x|, |x'|)\,\mathrm{sgn}\,(xx')\} \tag{6.5}$$

and $|x|_> := \max\{|x|, |x'|\}$. It is obvious that $\|B_z\|$ has for any $z \leq z_0 < \nu_2$ a bound independent of z and, under our hypothesis on the potential, the same is true for $\||V|^{1/2}B_z V^{1/2}\|$ (Problem 2). In the first term $\|A_z\|^2$ can be estimated by the Hilbert-Schmidt norm,

$$\|A_z\|_{\mathrm{HS}}^2 \leq \int_{\Omega_0\times\Omega_0} |V(x,y)\chi_1(y)|^2\, |x|_<^2\, |V(x',y')\chi_1(y')|^2\, \mathrm{d}\vec{x}\, \mathrm{d}\vec{x}'$$
$$\leq \left(\int_{\mathbb{R}} |x|\, |V|_{11}(x)\, \mathrm{d}x\right)^2,$$

with the right-hand side finite by assumption, where we have employed the inequalities $\mathrm{e}^{-z'}\sinh z \leq z$ for $z' \geq z$ and $|x|_<^2 \leq |xx'|$. Moreover, the same estimate shows by dominated convergence that A_z tends to a bounded operator in the limit $z \to \nu_1-$. Hence $\|P_z\|$ has in $(-\infty, \nu_1]$ a bound independent of z and $\|\lambda P_z\| < 1$ holds for $|\lambda|$ small enough.

In such a case $I + \lambda P_z$ is invertible and we may rewrite the operator, the singularities of which we are seeking, using the identity

$$(I + K_\lambda^z)^{-1} = \left[I + \lambda(I+\lambda P_z)^{-1}Q_z\right]^{-1}(I+\lambda P_z)^{-1}. \tag{6.6}$$

It follows that for $|\lambda|$ small enough the operator K_λ^z has eigenvalue -1 *iff* the same is true for $\lambda(I+\lambda P_z)^{-1}Q_z$. Since λQ_z is a rank-one operator, K_λ^z acts as $(\psi,\cdot)\phi$ with $\psi := \frac{\lambda}{2\kappa_1(z)}\,\mathrm{e}^{-\kappa_1(z)|\cdot|}V^{1/2}\chi_1$ and $\phi := (I+\lambda P_z)^{-1}\mathrm{e}^{-\kappa_1(z)|\cdot|}|V|^{1/2}\chi_1$, and consequently, it has just one eigenvalue which is (ψ,ϕ). Putting it equal to -1 we get for $\kappa_1(z) =: \zeta$ the following equation

$$\zeta = \mathcal{G}(\lambda,\zeta)\,, \tag{6.7}$$

where the right-hand side is defined as

$$-\frac{\lambda}{2}\int_{\Omega_0} \mathrm{e}^{-z|x|}V(x,y)^{1/2}\chi_1(y)\left[(I+\lambda P_{z(\zeta)})^{-1}\mathrm{e}^{-z|\cdot|}|V|^{1/2}\chi_1\right](x,y)\,\mathrm{d}x\,\mathrm{d}y\,.$$

If V decays sufficiently fast, $\mathcal{G}$ is analytic around $(0,0)$ and the assertion follows by the implicit-function theorem; it also gives us a prescription of how to compute higher-order terms in the expansion (6.4). For a general V one has to modify to the present situation the standard argument used for weakly coupled one-dimensional Schrödinger operators (Problem 3).

It remains to check that the second-order term in the expansion is positive provided $\int_{\mathbb{R}} \lambda V_{11}(x)\,\mathrm{d}x = 0$, so the bound state exists in this case too. If the corresponding functions V_{mn} belong to $L^2(\mathbb{R})$, the corresponding coefficient in the Taylor expansion is proportional to

$$\lim_{\varepsilon\to 0+} 2\int_{\mathbb{R}^2} V_{11}(x)\,\frac{\mathrm{e}^{-\varepsilon|x-x'|}-1}{2\varepsilon}\,V_{11}(x')\,\mathrm{d}x\,\mathrm{d}x'$$
$$+\sum_{n=2}^{\infty}\int_{\mathbb{R}^2} V_{1n}(x)\,\frac{\mathrm{e}^{-\sqrt{\nu_n-\nu_1}\,|x-x'|}}{\sqrt{\nu_n-\nu_1}}\,V_{n1}(x')\,\mathrm{d}x\,\mathrm{d}x'$$
$$=\lim_{\varepsilon\to 0+} 2\int_{\mathbb{R}} |\hat V_{11}(k)|^2\frac{\mathrm{d}k}{k^2+\varepsilon^2}+\sum_{n=2}^{\infty}\int_{\mathbb{R}} |\hat V_{1n}(k)|^2\,\frac{\mathrm{d}k}{k^2+\nu_n-\nu_1}>0\,,$$

otherwise we use L^2 approximations and check that the sign is preserved in the limit when the regularization is removed. This concludes the proof. ∎

The obtained result shows, in particular, that in case of a weak coupling the condition (1.25) allows a zero mean value of the potential. For a layer, $c=2$, the analogous criterion and asymptotic expansion are as follows.

Theorem 6.2 *Let $V\in(L^p+L^\infty_\varepsilon)(\Omega_0)$, where $p=2$ for $d=2,3$ and $p>\frac{d}{2}$ for $d\ge 4$; suppose that V is nonzero and satisfies the condition*

$$|V|_{11}\in L^{1+\delta}(\mathbb{R}^2)\cap L^1(\mathbb{R}^2,(1+|x|^\delta)\,\mathrm{d}x)$$

for some $\delta>0$. Then H_λ has for small enough $|\lambda|$ at most one simple eigenvalue $\epsilon(\lambda)<\nu_1$, which happens iff $\int_{\mathbb{R}^2}\lambda V_{11}(x)\,\mathrm{d}x\le 0$. If this is the case we have $\epsilon(\lambda)=\nu_1-\mathrm{e}^{2w(\lambda)^{-1}}$, where $w(\lambda)$ has the following asymptotic expansion,

$$\begin{aligned} w(\lambda)=&\frac{\lambda}{2\pi}\int_{\mathbb{R}^2} V_{11}(x)\,\mathrm{d}x \\ &+\left(\frac{\lambda}{2\pi}\right)^2\left\{\int_{\mathbb{R}^2\times\mathbb{R}^2} V_{11}(x)\left(\gamma_{\mathrm{E}}+\ln\frac{|x-x'|}{2}\right)V_{11}(x')\,\mathrm{d}x\,d\mathrm{d}x'\right.\\ &\left.-\sum_{n=2}^{\infty}\int_{\mathbb{R}^2\times\mathbb{R}^2} V_{1n}(x)\,K_0\left(\sqrt{\nu_n-\nu_1}|x-x'|\right)\,V_{n1}(x')\,\mathrm{d}x\,\mathrm{d}x'\right\}+\mathcal{O}(\lambda^{2+\eta})\end{aligned} \tag{6.8}$$

with γ_{E} being Euler's constant and $\eta:=\min\{1,\delta/2\}$.

Proof As above the essential spectrum starts at ν_1 and one can apply the BS-principle. The free Green's function is given by (5.14) with the χ_n's being now eigenfunctions of $-\Delta_D^M$ for the general cross section considered here. The kernel of $K_\lambda^z=\lambda|V|^{1/2}(-\Delta_D^{\Omega_0}-z)^{-1}V^{1/2}$ has in the layer case a logarithmic singularity as $z\to\nu_1-$. We split it using the decomposition $K_\lambda^z=\lambda L_z+\lambda M_z$, where

$$L_z(\vec x,\vec x')=-\frac{1}{2\pi}\,|V(x,y)|^{1/2}\,\chi_1(y)\,\ln\kappa_1(z)\,\chi_1(y')\,V(x',y')^{1/2}$$

and the regular part written as $M_z=A_z+|V|^{1/2}B_zV^{1/2}$ with the last term having for any $z\le z_0<\nu_2$ a bound independent of z, which can be checked as in the preceding proof (Problem 2). The first term has the kernel

$$\frac{1}{2\pi}\,|V(x,y)|^{1/2}\,\chi_1(y)\,\Big(K_0\left(\kappa_1(z)|x-x'|\right)+\ln\kappa_1(z)\Big)\,\chi_1(y')\,V(x',y')^{1/2},$$

which shows that M_z is well defined at $z=\nu_1$, in particular, that we have

$$A_{\nu_1}(\vec{x},\vec{x}')=-\frac{1}{2\pi}\,|V(x,y)|^{1/2}\,\chi_1(y)\left(\gamma_{\mathrm{E}}+\ln\frac{|x-x'|}{2}\right)\,\chi_1(y')V(x',y')^{1/2}.$$

To proceed we need several bounds the proof of which is left to the reader (Problem 5); in all of them C is an unspecified positive number.

Lemma 6.1.1 *(a)* $\|M_z\|\le C$ *for any* $z<\nu_1$,
(b) $\|M_z-M_{\nu_1}\|\le C\kappa_1(z)^{2\eta}$ *with* η *defined in the theorem,*
(c) and finally, $\left\|\frac{\mathrm{d}M_{z(w)}}{\mathrm{d}w}\right\|\le C|w|^{-1}$ *for small enough* $w:=(\ln\kappa_1(z))^{-1}$.

By the first claim of the lemma $\|\lambda M_z\|<1$ holds for all sufficiently small $|\lambda|$, hence one can employ the identity

$$\left(I+K_\lambda^z\right)^{-1}=\left[I+\lambda(I+\lambda M_z)^{-1}L_z\right]^{-1}(I+\lambda M_z)^{-1}$$

and reduce the problem to the question of whether the rank one-operator $(\psi,\cdot)\phi$ with $\psi:=-\frac{\lambda}{2\pi}\ln\kappa_1(z)\,V^{1/2}\chi_1$ and $\phi:=(I+\lambda M_z)^{-1}|V|^{1/2}\chi_1$ has an eigenvalue equal to -1. This requirement yields the equation

$$w=\mathcal{G}(\lambda,w)\,, \tag{6.9}$$

where

$$\mathcal{G}(\lambda,w):=\frac{\lambda}{2\pi}\left(V^{1/2}\chi_1,\left(I+\lambda M_{z(w)}\right)^{-1}|V|^{1/2}\chi_1\right)$$

and the auxiliary variable w determines the energy via $z=\nu_1-\mathrm{e}^{2w^{-1}}$. To solve (6.9), we rewrite the last expression using the identity

$$(I+\lambda M_z)^{-1}=I-\lambda M_{\nu_1}-\lambda(M_z-M_{\nu_1})+\lambda^2M_z^2(I+\lambda M_z)^{-1}\,;$$

this leads to the asymptotic expansion (6.8) in view of claims (a), (b) of *Lemma* 6.1.1 and the coefficients are well defined owing to our assumptions about $|V|_{11}$. By an elementary perturbation argument $\inf\sigma(H_\lambda)\ge\nu_1-c\lambda$ holds for some $c>0$, which is possible only if w approaches zero assuming negative values. This yields the sign definiteness as a necessary and sufficient condition for the existence of a weakly bound state as above (Problem 6). The same argument gives $\kappa_1(z)^2\le c\lambda$, and thus the error term in combination with *Lemma* 6.1.1*b*.

It remains to check that (6.8) is the only solution of (6.9) for small $|\lambda|$. By minimax we can suppose without loss of generality that V is non-positive, $V = -|V|$, and $\lambda > 0$. The Schwarz and triangle inequalities together with claims (a), (c) of *Lemma* 6.1.1 allow us to estimate $\left|\frac{\partial \mathcal{G}}{\partial w}(\lambda, w)\right|$ for small λ by

$$\begin{aligned}
&\frac{\lambda}{2\pi}\left|\left(|V|^{1/2}\chi_1, (I+\lambda M_{z(w)})^{-1}\lambda \frac{\mathrm{d}M_{\alpha(w)}}{\mathrm{d}w}(I+\lambda M_{z(w)})^{-1}|V|^{1/2}\chi_1\right)\right| \\
&\quad \le \frac{\lambda^2}{2\pi}\left\||V|^{1/2}\chi_1\right\|_2^2 \left\|\left(I+\lambda M_{z(w)}\right)^{-1}\right\|^2 \left\|\frac{\mathrm{d}M_{\alpha(w)}}{\mathrm{d}w}\right\| \\
&\quad \le \frac{\lambda^2}{2\pi}(1-\lambda C_1)^{-2}\frac{C_2}{|w|}\,\| |V|_{11} \|_1^2 ,
\end{aligned}$$

with some $C_1, C_2 > 0$, the last norm being finite and nonzero by assumption. Hence there is a $c' > 0$ such that the inequality $|w|^{-1} \le c'\lambda^{-1}$ is valid for any solution w of the implicit equation (6.9) and λ small enough. Consequently, $\left|\frac{\partial \mathcal{G}}{\partial w}\right|$ is bounded by $C_3\lambda$ for λ small enough. Any two solutions w_1, w_2 of the equation $w = \mathcal{G}(\lambda, w)$ have to fulfill

$$|w_2 - w_1| = \left|\int_{w_1}^{w_2} \frac{\partial \mathcal{G}}{\partial w}\,\mathrm{d}w\right| \le \left|\int_{w_1}^{w_2} \left|\frac{\partial \mathcal{G}}{\partial w}\right|\,\mathrm{d}w\right| \le C_3\lambda\,|w_2 - w_1|,$$

thus the uniqueness is ensured as long as $\lambda < C_3^{-1}$. ∎

6.2 Applications to Tubes and Layers

Our next aim is to apply the Birman-Schwinger technique to different types of weakly bound states in systems which we considered in Chaps. 1 and 4.

6.2.1 Mildly Bent Tubes

The most straightforward way is to use the above results to estimate the operator in question from below and from above. If the two bounds squeeze in the limit we can determine the true asymptotics. We shall illustrate this method on Hamiltonians of bent tubes considered in Sects. 1.1 and 1.3.

To compare different shapes we need to introduce a parametrization which allows us to say precisely what a mildly curved tube is. In the two-dimensional case a natural possibility is suggested by relation (1.4): we shall consider families of strips with the generating curves Γ_β characterized by the curvature

$$\gamma_\beta(s) := \beta\gamma(s) \tag{6.10}$$

for a fixed function γ and $\beta > 0$. If $\int_{\mathbb{R}} \gamma(s)\,\mathrm{d}s \neq 0$ we may put it equal to one without loss of generality; then β controls the total bending. If the integral is zero, diminishing of β again means straightening of the curve. Its nature is then illustrated by the example of γ with a compact support: the length of the curved part remains preserved and the curvature radius at each point grows proportionally to β^{-1}. In the three-dimensional case the situation is not so simple but we know from Sect. 1.3 that it is the curvature again which is responsible for the existence of bound states. Hence we consider families of tubes with the curvature scaled according to (6.10) and the torsion fixed; the interpretation of the tube straightening as $\beta \to 0$ remains the same.

Theorem 6.3 *Let $\{\Omega_\beta\} \subset \mathbb{R}^d$, $d = 2, 3$, be a family of tubes with a fixed cross section built over the curves Γ_β described above. If $d = 2$ we adopt the assumptions (i)–(v) with $k = 2$ of Sect. 1.1. If $d = 3$ we assume (i)–(iii) with $k = 2$ of Sect. 1.3 and, in addition, we suppose that $\gamma^{(k)} \in L^1(\mathbb{R}, |s|\,\mathrm{d}s)$ for $k \leq 2$. Then if Γ_1 is not straight, the operator $-\Delta_D^{\Omega_\beta}$ has for small enough nonzero β exactly one isolated eigenvalue $\epsilon(\beta)$ in $(0, \nu_1)$. Moreover,*
(a) in the two-dimensional case the asymptotic behavior is given by

$$\sqrt{\kappa_1^2 - \epsilon(\beta)} = \frac{\beta^2}{8}\left\{ \|\gamma\|^2 - \frac{1}{2}\sum_{n=2}^{\infty} (\chi_n, u\chi_1)^2 \right.$$
$$\left. \times\, \varrho_n \int_{\mathbb{R}^2} \dot\gamma(s)\,\mathrm{e}^{-\varrho_n|s-s'|}\dot\gamma(s')\,\mathrm{d}s\,\mathrm{d}s' \right\} + \mathcal{O}(\beta^3)\,, \tag{6.11}$$

where $\nu_1 = \kappa_1^2$ and $\varrho_n := \kappa_1\sqrt{n^2-1}$; in fact the sum only runs over even n.
(b) Similarly, for $d = 3$ we have

$$\sqrt{\nu_1 - \epsilon(\beta)} = \frac{\beta^2}{8}\|\gamma\|^2 - \frac{\beta^2}{16}\sum_{n=2}^{\infty}\int_{M\times M} \mathrm{d}y\,\mathrm{d}y'\,\chi_1(y)\chi_1(y')\chi_n(y)\chi_n(y')$$
$$\times\, \varrho_n \int_{\mathbb{R}^2} \mathrm{d}s\,\mathrm{d}s'\, h_s(s,y)\,\mathrm{e}^{-\varrho_n|s-s'|}\, h_s(s',y') + \mathcal{O}(\beta^3)\,, \tag{6.12}$$

where h_s is the expression contained in (1.21) and $\varrho_n := \sqrt{\nu_n - \nu_1}$.

Proof We shall consider the two-dimensional situation only, the case $d = 3$ is similar (Problem 7). We pass to the operator (1.7) and squeeze it between

$$H_\pm := -(1 \mp a\beta\|\gamma\|_\infty)^{-2}\partial_s^2 - \partial_u^2 + V_\beta(s,u)\,,$$

where V_β is the potential (1.8) referring to γ_β. By *Theorem* 6.1 and the minimax principle there is a single eigenvalue for small nonzero β which is squeezed between the eigenvalues $\epsilon_\pm(\beta)$ of the estimating operators $H_\pm$; if we rescale the longitudinal variable to $s_\pm := (1 \mp a\beta\|\gamma\|_\infty)s$ and appropriately change the integrations, we obtain

$$\begin{aligned}\sqrt{\kappa_1^2 - \epsilon_\pm(\beta)^2} = &- \frac{1}{2}(1 \mp a\beta\|\gamma\|_\infty) \int_{\mathbb{R}} (V_\beta)_{11}(s)\, \mathrm{d}s \\ &- \frac{1}{4}(1 \mp a\beta\|\gamma\|_\infty)^2 \bigg\{ \int_{\mathbb{R}^2} (V_\beta)_{11}(s)\, |s - s'| (V_\beta)_{11}(s')\, \mathrm{d}s\, \mathrm{d}s' \\ &- \sum_{n=2}^{\infty} \varrho_n^{-1} \int_{\mathbb{R}^2} (V_\beta)_{1n}(s)\, \mathrm{e}^{-\varrho_n |s-s'|}\, (V_\beta)_{n1}(s')\, \mathrm{d}s\, \mathrm{d}s' \bigg\} + \mathcal{O}(\beta^3)\,.\end{aligned}$$

Since V_β itself expands in terms of β and we are interested in the leading term only, the scaling factors $1 \mp a\beta\|\gamma\|_\infty$ play no role. Let us denote the explicit part of the right-hand side by $I_1 + \sum_{n=1}^{\infty} I_{2,n}$. The first term can be calculated through integration by parts which leads to

$$\begin{aligned} I_1 :=& \frac{\beta^2}{8} \int_{-a}^{a} \mathrm{d}u \int_{\mathbb{R}} \mathrm{d}s\, \chi_1(u)^2 \left[\frac{\gamma^2}{(1+u\beta\gamma)^2} - \frac{u^2\dot{\gamma}^2}{(1+u\beta\gamma)^4} \right] \\ =& \frac{\beta^2}{8} \left\{ \|\gamma\|^2 - \|u\chi_1\|^2 \|\dot{\gamma}\|^2 \right\} + \mathcal{O}(\beta^3)\,, \end{aligned}$$

where we have used the fact that $\dot{\gamma}(s) \to 0$ as $|s| \to \infty$ due to *(iv)*. The important thing is that in the integration by parts the term linear in β is canceled. Hence we need to compute $\sum_{n=1}^{\infty} I_{2,n}$ too because it contains parts of order β^2 coming from the second term in (1.8). In particular, $I_{2,1}$ equals

$$-\frac{\beta^2}{16} \int_{-a}^{a} \int_{-a}^{a} \mathrm{d}u\, \mathrm{d}u'\, uu'\, \chi_1(u)^2 \chi_1(u')^2 \int_{\mathbb{R}^2} \mathrm{d}s\, \mathrm{d}s'\, \ddot{\gamma}(s)\, |s - s'|\, \ddot{\gamma}(s') + \mathcal{O}(\beta^3)\,,$$

where the inner integral is $-2 \int_{\mathbb{R}} \dot{\gamma}(s)^2 \mathrm{d}s$ by a double integration by parts, so

$$I_{2,1} = \frac{\beta^2}{8} (\chi_1, u\chi_1)^2 \|\dot{\gamma}\|^2 + \mathcal{O}(\beta^3)\,.$$

Splitting the leading terms in the other expressions we get

$$\begin{aligned} I_{2,n} = \frac{\beta^2}{16} \int_{-a}^{a} \int_{-a}^{a} & \mathrm{d}u\, \mathrm{d}u'\, uu'\, \chi_1(u) \chi_N(u) \chi_1(u') \chi_N(u') \\ & \times \int_{\mathbb{R}^2} \mathrm{d}s\, \mathrm{d}s'\, \ddot{\gamma}(s)\, \frac{\mathrm{e}^{-\varrho_n |s-s'|}}{\varrho_n}\, \ddot{\gamma}(s') + \mathcal{O}(\beta^3)\,. \end{aligned}$$

The inner integral can be rewritten by a repeated integration by parts as

$$2\int_{\mathbb{R}} \dot\gamma(s)^2 \mathrm{d}s - \varrho_n \int_{\mathbb{R}^2} \mathrm{d}s\,\mathrm{d}s'\,\dot\gamma(s)\,\mathrm{e}^{-\varrho_n|s-s'|}\,\dot\gamma(s')\,, \tag{6.13}$$

hence putting all the contributions together and using the Parseval identity, $\|u\chi_1\|^2 = \sum_{n=1}^\infty (\chi_n, u\chi_1)^2$, we find that the terms containing $\|\dot\gamma\|^2$ cancel and we arrive at the relation (6.11). ∎

Remark 6.2.1 In thin tubes the leading-term coefficient is dominated by $\frac{1}{8}\|\gamma\|^2$ as one expects from (1.9) and its three-dimensional analogue in combination with the weak-coupling asymptotics for one-dimensional Schrödinger operators; it is not difficult to check that the other contributions are of order $\mathcal{O}(a^2)$. On the other hand, in the general case the coefficient has a more complicated structure with competing contributions from the first and the higher transverse modes. While we know from the existence results that the coefficient is non-negative, it is not obvious from the above asymptotic formulæ. Nevertheless, one can check this fact directly as we shall illustrate in the case $d = 2$. To this end, one has to integrate the second term in (6.13) by parts again twice which gives $2\,\varrho_n^2\|\gamma\|^2\mathrm{d}s - \varrho_n^3 \int_{\mathbb{R}^2} \gamma(s)\,\mathrm{e}^{-\varrho_n|s-s'|}\,\gamma(s')\,\mathrm{d}s\,\mathrm{d}s'$ and to sum the series referring to the first term (Problem 8). This leads to cancelation of the terms containing $\|\gamma\|^2$. The remaining part contains the Green function and can thus be rewritten using Fourier transformation as in the proof of *Theorem* 6.1; in this way we arrive at an equivalent form of the asymptotic expansion,

$$\sqrt{\kappa_1^2 - \epsilon(\beta)} = \frac{\beta^2}{8}\sqrt{2\pi}\sum_{n=2}^\infty (\chi_1, u\chi_j)_u^2\,\varrho_n^4 \int_{\mathbb{R}} \frac{|\hat\gamma(p)|^2}{p^2+\varrho_n^2}\,\mathrm{d}p + \mathcal{O}(\beta^3)\,, \tag{6.14}$$

which shows that the leading-order coefficient is always positive.

6.2.2 Gently Curved Layers

Two-sided bounds can also be used to determine the weak-coupling behavior of curved layers. The first question is how to choose a family of layers to make it planar in the limit. One way is to multiply the principal curvatures of a fixed Σ by a scaling parameter in analogy with (6.10). However, since such an Ansatz lacks a straightforward interpretation as we had in the tube case, we adopt another approach and consider a class of layers built over surfaces which are graphs of a function $\mathbb{R}^2 \to \mathbb{R}^3$ containing a scaling parameter,

$$\Sigma_\beta := p\,(\mathbb{R}^2)\,, \quad p\,(x^1, x^2; \beta) = \left(x^1, x^2, \beta f(x^1, x^2)\right)\,, \tag{6.15}$$

where f is a C^4-smooth function and $\beta > 0$; the layer $\Omega_\beta := \mathcal{L}(\Omega)$ is defined by (4.1). While this choice has an illustrative meaning, it naturally requires us to first express the geometric characteristics of such a Σ_β and to assess how they behave for small β which makes the analysis more complicated.

The function f has to satisfy additional requirements. First of all Σ_β must be asymptotically planar, i.e. the Gauss and mean curvatures must vanish at large distances from the origin. This will hold if we require

(i) $f_{,\mu} \in L^\infty(\mathbb{R}^2)$ and $f_{,\mu\nu} \to 0$ as $|x| \to \infty$

for $\mu, \nu = 1, 2$ as we shall see a little below. Since the injectivity hypotheses *(i)*, *(ii)* of Sect. 4.1.1 are satisfied for small enough β, *Proposition* 4.1.1 tells us that $\inf \sigma_{\mathrm{ess}}(-\Delta_D^{\Omega_\beta}) \ge \kappa_1^2$. To get an opposite inequality, we assume

(ii) $f_{,\mu\nu\rho} \to 0$ as $|x| \to \infty$, and $f_{,\mu\nu\rho\sigma} \in L^\infty(\mathbb{R}^2)$;

we also add the following integrability hypotheses,

(iii) $f_{,\mu\nu}\,,\ f_{,\mu\nu\rho}\,,\ |f_{,\mu\nu\rho\sigma}|^{1/2} \in L^2\left(\mathbb{R}^2, (1+|x|^\delta)\,\mathrm{d}x\right)$ for some $\delta > 0$.

A consequence of these assumptions is that the total Gauss curvature of Σ_β tends to zero faster than the natural scaling by the factor β^2 would suggest (Problem 9). This fact is reflected in the asymptotic expansion given below.

Theorem 6.4 *Let $\{\Omega_\beta\}$ be a family of layers of halfwidth a built over the surfaces Σ_β described above. Suppose that the function $f \in C^4(\mathbb{R}^2)$ satisfies the assumptions (i)–(iii). If Σ_1 is not planar, then for all β small enough $-\Delta_D^{\Omega_\beta}$ has exactly one eigenvalue $\epsilon(\beta)$ below the threshold of the essential spectrum. Moreover, it can be expressed as $\epsilon(\beta) = \kappa_1^2 - \mathrm{e}^{2w(\beta)^{-1}}$, where $w(\beta)$ has the following asymptotic expansion,*

$$w(\beta) = -\beta^2 \sum_{n=2}^{\infty} (\chi_1, u\chi_n)^2 \left(\kappa_n^2 - \kappa_1^2\right)^2 \int_{\mathbb{R}^2} \frac{|\hat{m}_0(\omega)|^2}{|\omega|^2 + \kappa_n^2 - \kappa_1^2}\,\mathrm{d}\omega + \mathcal{O}(\beta^{2+\eta})$$

with $\eta := \min\{1, \delta/2\}$, where in fact the sum only runs over even n. The function $\hat{m}_0$ in the integral is the Fourier image of the leading term in the expansion of the mean curvature of Σ_β given by $m_0 = \frac{1}{2}(f_{,11} + f_{,22})$.

Proof First we have to express the needed geometric quantities. Consider the 2×2 matrices $(\eta_{\mu\nu})$, $(\tilde{\eta}^{\mu\nu})$, $(\theta_{\mu\nu})$ defined by

$$\eta := \begin{pmatrix} f_{,1}^2 & f_{,1}f_{,2} \\ f_{,1}f_{,2} & f_{,2}^2 \end{pmatrix}, \quad \tilde{\eta} := \begin{pmatrix} f_{,2}^2 & -f_{,1}f_{,2} \\ -f_{,1}f_{,2} & f_{,1}^2 \end{pmatrix}, \quad \theta := \begin{pmatrix} f_{,11} & f_{,12} \\ f_{,21} & f_{,22} \end{pmatrix},$$

respectively. The metric tensor of Σ_β is of the form $g_{\mu\nu}(\beta) = \delta_{\mu\nu} + \beta^2\eta_{\mu\nu}$ and

$$g(\beta) := \det(g_{\mu\nu}) = 1 + \beta^2({f_{,1}}^2 + {f_{,2}}^2)$$

together with $g^{\mu\nu}(\beta) = g(\beta)^{-1}(\delta^{\mu\nu}+\beta^2\tilde{\eta}^{\mu\nu})$. The Jacobian $g^{1/2}$ defines the invariant surface element $\mathrm{d}\sigma_\beta := g^{1/2}\mathrm{d}x$. Under the assumption *(i)* the matrix function η is bounded with the norm $\|\eta\|_\infty =: \eta_\infty$, and thus $g_{\mu\nu}(\beta)$ is uniformly elliptic for small β, because

$$c_-\delta_{\mu\nu} \le g_{\mu\nu}(\beta) \le c_+\delta_{\mu\nu}\,, \quad c_\pm := 1 \pm \beta^2\eta_\infty\,.$$

The second fundamental form is also easily computed, $h_{\mu\nu}(\beta) = \beta g(\beta)^{-1/2}\theta_{\mu\nu}$, which in turn yields the Weingarten tensor $h_\mu^{\ \nu} := h_{\mu\rho}g^{\rho\nu}$, and subsequently, the Gauss and mean curvatures of the surface Σ_β,

$$\begin{aligned} K(\beta) &= \beta^2 g(\beta)^{-2}k_0\,, \quad k_0 := \det(\theta_{\mu\nu}) = f_{,11}f_{,22} - f_{,12}{}^2\,, \\ M(\beta) &= \beta g(\beta)^{-3/2}\left(m_0 + \beta^2 m_1\right)\,, \quad m_0 := \frac{1}{2}\,\mathrm{tr}(\theta_{\mu\nu}) = \frac{1}{2}\left(f_{,11} + f_{,22}\right)\,, \\ m_1 &:= \frac{1}{2}\mathrm{tr}\,(\theta_{\mu\rho}\tilde{\eta}^{\rho\nu}) = \frac{1}{2}\left(f_{,1}{}^2 f_{,22} + f_{,2}{}^2 f_{,11} - 2f_{,1}f_{,2}f_{,12}\right). \end{aligned}$$

In view of the assumption *(i)*, Σ_β is asymptotically planar. As we have said, the injectivity hypotheses of Sect. 4.1.1 are satisfied for small β because $\rho_m^{-1} := \max\{\|k_1\|_\infty, \|k_2\|_\infty\} = \mathcal{O}(\beta)$. This also means that the constants $C_\pm := (1\pm a\rho_m^{-1})^2$ in the bounds (4.3) approach one as $\beta \to 0$.

Since Σ_β is C^4-smooth by assumption we can cast the operator $-\Delta_D^{\Omega_\beta}$ into the form (4.7) with the transverse part H_2 given by (4.8), while for the longitudinal one we have the estimate

$$\frac{C_-}{C_+^2}\big(-\Delta_g + v_1\big) \le H_1 \le \frac{C_+}{C_-^2}\big(-\Delta_g + v_1\big)$$

which holds in $L^2(\mathbb{R}^2 \times (-a,a),\, g^{1/2}\,\mathrm{d}x\,\mathrm{d}u)$ in the form sense. For brevity we use the shorthand $-\Delta_g$ for the Laplace-Beltrami operator $-g^{-1/2}\partial_\mu g^{1/2}g^{\mu\nu}\partial_\nu$. The potential v_1 in the above inequalities is obtained replacing $G^{\mu\nu}$ by $g^{\mu\nu}$ in $V_1 = g^{-1/2}(g^{1/2}G^{\mu\nu}J_{,\nu})_{,\mu} + J_{,\mu}G^{\mu\nu}J_{,\nu}$; a direct computation gives

$$v_1 = -\frac{|u^2\nabla_g K - 2u\nabla_g M|_g^2}{4(1-2Mu+Ku^2)^2} + \frac{u^2\Delta_g K - 2u\Delta_g M}{2(1-2Mu+Ku^2)}\,,$$

where $|\cdot|_g$ and ∇_g are the norm and gradient operator induced by the metric $g_{\mu\nu}$, respectively. Furthermore, one can get rid of the metric using

$$-\frac{c_-^2}{c_+^3}\,\Delta \le -\Delta_g \le -\frac{c_+^2}{c_-^3}\,\Delta$$

with the constants introduced above. The Hilbert space can be identified with $L^2(\mathbb{R}^2 \times (-a, a))$ as a set in view of the uniform ellipticity, and using finally the scaling $x \mapsto \zeta_\pm x$ with $\zeta_\pm^2 := c_\mp^3 C_\mp^2/(c_\pm^2 C_\pm)$, we get the bounds

$$H_- \le H \le H_+ , \quad H_\pm := -\Delta - \partial_3^2 + \beta V_\pm ,$$

where

$$V_\pm(x, u) := \frac{1}{\beta}\left(\frac{C_\pm}{C_\mp^2}\, v_1 + V_2\right)\left(\frac{x}{\zeta_\pm}, u\right).$$

Now we are able to apply *Theorem* 6.2 to the operators $H_\pm$; notice that as in the tube case the estimating potentials are β-dependent.

In view of *(i), (ii)* we have $K,\ M,\ |\nabla_g K|_g,\ |\nabla_g M|_g,\ \Delta_g K \to 0$ as $|x| \to \infty$, so the same is true for $(V_\pm)_{11}$; recall that $u\Delta_g M$ vanishes when projected onto the first transverse mode. This means that $\sigma_{\mathrm{ess}}(-\Delta_D^{\Omega_\beta}) = [\kappa_1^2, \infty)$. Since the potentials $V_\pm$ are bounded and *(iii)* ensures that $\sup\{V_\pm(\cdot, u) : |u| < a\}$ belongs to $L^1(\mathbb{R}^2, (1 + |x|^\delta)\,\mathrm{d}x)$ the assumptions of *Theorem* 6.2 are satisfied and $-\Delta_D^{\Omega_\delta}$ has by the minimax principle for small $\beta > 0$ exactly one eigenvalue which is squeezed between the eigenvalues $\epsilon_\pm(\beta) = \kappa_1^2 - \mathrm{e}^{2w_\pm(\beta)^{-1}}$ of the estimating operators, where $w_\pm(\beta)$ are given by (6.8) with λV replaced by $\beta\zeta_\pm^2 V_\pm(\zeta_\pm \cdot)$. The matrix elements of the potential V_2 from (4.8) satisfy

$$\beta^{-1}(V_2)_{1j}(x; \beta) = \delta_{1j}\left[\beta\left(k_0(x) - m_0(x)^2\right) + \mathcal{O}(\beta^2)\right], \quad j \in \mathbb{N},$$

where the error term is an integrable function of x. The expansion of $\beta^{-1}v_1$ is more involved because $\beta^{-1}\Delta_g M$ appearing in the second term is of order one. We begin by writing $v_1(x, u; \beta)$ as

$$-u(\Delta_g M)(x; \beta) + u^2\left(\frac{1}{2}\Delta_g K - |\nabla_g M|_g^2 - 2M\Delta_g M\right)(x; \beta) + r_1(x, u; \beta),$$

where r_1 is an integrable function of order $\mathcal{O}(\beta^3)$, then we expand $|\cdot|_g$, ∇_g and Δ_g. The first term above is an odd function of u, so it does not contribute to $(V_\pm)_{11}$, while it plays an important role in the higher modes,

$$(v_1)_{11}(x; \beta) = \|u\chi_1\|^2\left(\frac{1}{2}\Delta_g K - |\nabla_g M|_g^2 - 2M\Delta_g M\right)(x; \beta) + \mathcal{O}(\beta^3),$$

$$(v_1)_{1n}(x; \beta) = -(\chi_1, u\chi_j)\,(\Delta_g M)(x; \beta) + \mathcal{O}(\beta^2), \quad n = 2, 3, \ldots .$$

In the next step we employ the relations $|\nabla_g M|_g^2 = |\nabla M|^2 + \beta^2 M_{,\mu}\tilde{\eta}^{\mu\nu} M_{,\nu}$ and $-\Delta_g = -\Delta + \beta^2 L(\beta)$, where $L(\beta)$ is a second-order differential operator with coefficients which expand as $\mathcal{O}(1)$. Further we use the expressions of the curvatures

K, M and the fact that $C_\pm/C_\mp^2$ and $\zeta_\pm$ expand as $1 + \mathcal{O}(\beta)$; in this way we arrive at the formula for $w_\pm(\beta)$, which is up to $\mathcal{O}(\beta^{2+\eta})$ equal to

$$\frac{\beta^2}{2\pi}\left\{\int_{\mathbb{R}^2}\left(k_0 - m_0^2\right)(x)\,\mathrm{d}x + \|u\chi_1\|^2\int_{\mathbb{R}^2}\left(\frac{1}{2}\Delta k_0 - |\nabla m_0| - 2m_0\Delta m_0\right)(x)\,\mathrm{d}x\right.$$
$$\left.-\frac{1}{2\pi}\sum_{n=2}^{\infty}(\chi_1, u\chi_n)^2\int_{\mathbb{R}^2\times\mathbb{R}^2}(\Delta m_0)(x)\,K_0\left(\varrho_n\zeta_\pm|x-x'|\right)\,(\Delta m_0)(x')\,\mathrm{d}x\,\mathrm{d}x'\right\},$$

where $\varrho_n = \sqrt{\kappa_n^2 - \kappa_1^2}$ and the sum runs in fact over even n only. This result can be further simplified. By a double integration by parts using the fact that $f_{,\mu}f_{,\nu\rho} \to 0$ as$|x| \to \infty$ by *(i)*, we find that the integral of k_0 is zero; we also have to employ *(iii)* here to ensure that the involved integrals exist. Moreover, Gauss' theorem together with *(i)*, *(ii)* by which $\nabla K \to 0$ as $|x| \to \infty$ imply that the integral of Δk_0 vanishes as well. A similar argument employing Green's formula gives $\int_{\mathbb{R}^2}(m_0\Delta m_0)(x)\,\mathrm{d}x = -\int_{\mathbb{R}^2}|\nabla m_0|^2(x)\,\mathrm{d}x$. It is useful to rewrite the integral in the last part of the above expression for $w_\pm(\beta)$ as

$$\frac{1}{2\pi}\int_{\mathbb{R}^2\times\mathbb{R}^2}(\Delta m_0)(x)\,K_0(\varrho_n\sigma_\pm|x-x'|)\,(\Delta m_0)(x')\,\mathrm{d}x\,\mathrm{d}x' = (\Delta m_0, G_k * \Delta m_0)\,,$$

where $G_k(\cdot) := (2\pi)^{-1}\,K_0(k|\cdot|)$ and k stands for $\varrho_n\sigma_\pm$. Since G_k is the fundamental solution of the distributional equation $(-\Delta + k^2)G_k = \delta$, we get

$$\begin{aligned}(\Delta m_0, G_k * \Delta m_0) &= (\Delta m_0, \Delta G_k * m_0) = \left(\Delta m_0, (k^2G_k - \delta) * m_0\right)\\ &= k^2\,(\Delta m_0, G_k * m_0) - (\Delta m_0, m_0) = k^2\,(G_k * \Delta m_0, m_0) + \|\nabla m_0\|^2\\ &= k^2\,(\Delta G_k * m_0, m_0) + \|\nabla m_0\|^2 = k^2\left((k^2G_k - \delta) * m_0, m_0\right) + \|\nabla m_0\|^2\\ &= k^4\,(m_0, G_k * m_0) - k^2\|m_0\|^2 + \|\nabla m_0\|^2.\end{aligned}$$

Using the Parseval identity, $\|u\chi_1\|^2 = \sum_{n=1}^{\infty}(\chi_1, u\chi_n)^2$, we find that the terms containing $\|\nabla m_0\|$ in the expression for $w_\pm(\beta)$ cancel giving

$$-\frac{\beta^2}{2\pi}\left\{\|m_0\|^2 + \sum_{n=2}^{\infty}(\chi_1, u\chi_j)^2\,k^2\Big[k^2\,(m_0, G_k * m_0) - \|m_0\|^2\Big]\right\} + \mathcal{O}(\beta^{2+\eta})\,,$$

where $k = \varrho_n + \mathcal{O}(\beta)$, and by Problem 8 the terms containing $\|m_0\|$ cancel too. Finally we rewrite $(m_0, G_k * m_0)$ using Fourier transformation and expand the remaining k from G_k with respect to β including the higher orders in the error term. Since the obtained leading order is the same for both $w_\pm(\beta)$, the same is true for $w(\beta)$ and we arrive at the sought formula. ∎

Remarks 6.2.1 (a) The leading-order coefficient given in the theorem is strictly negative. If this was not the case, m_0 would be zero, so $f_{,11}(x) = -f_{,22}(x)$ everywhere. This yields $k_0 = -(f_{,11}{}^2 + f_{,12}{}^2)$ and thus $k_0 = 0$; recall that $\int_{\mathbb{R}^2} k_0(x)\,\mathrm{d}x = 0$ holds by Problem 8 under the assumptions *(i)–(iii)*. Hence $\mathcal{K}(\Sigma_\beta) = 0$, and at the same time, $f_{,\mu\nu} = 0$ gives $m_1 = 0$, and therefore $\mathcal{M}(\Sigma_\beta) = 0$, so the coefficients vanish only in the trivial case of a plane.
(b) We have formulated the asymptotic expansion in a way analogous to (6.14). If we are interested in the thin-layer situation, we can rewrite the formula for w_1 in the expansion $w(\beta) = \beta^2 w_1 + \mathcal{O}(\beta^{2+\eta})$ in a different way,

$$w_1 = -\frac{1}{2\pi}\,\|m_0\|^2 + \frac{\pi^2-6}{24\pi^3}\,\|\nabla m_0\|^2 d^2 + \mathcal{O}(d^4) \tag{6.16}$$

(Problem 10). To interpret this recall that $\int_{\mathbb{R}^2} k_0(x)\,\mathrm{d}x = 0$ which means

$$\int_{\Sigma_\beta} \left(K - M^2\right) d\sigma_\beta = -\beta^2\|m_0\|^2 + \mathcal{O}(\beta^4)\,,$$

and thus the first term in (6.16) comes from the surface attractive potential $K - M^2$ which dominates the picture for thin layers—cf. (4.9).

6.2.3 A Direct Estimate: Local Deformations

The preceding discussion has shown that an application of the BS method based on two-sided estimates by Schrödinger operators in tubes and layers is not always easy. It would become even more complicated when the map $\Omega \to \Omega_0$ used to transform the Hamiltonian lacks the properties such as local orthogonality. This often happens when Ω is obtained from Ω_0 by a cross-section variation. In such a case it is more reasonable to avoid the intermediate step and to apply *Proposition* 6.1.1 directly. We will illustrate this approach on asymptotic properties of weakly coupled bound states in locally deformed tubes and layers.

Let us begin with the two-dimensional case. For simplicity we consider only a one-sided deformation of $\Omega_0 := \mathbb{R} \times (0, d)$, and moreover, we suppose that the deformation is compactly supported with the boundary $\partial\Omega$ infinitely smooth. In other words, we define $\Omega_\lambda^f \equiv \Omega_\lambda$ by

$$\Omega_\lambda := \{\vec{x} \in \mathbb{R}^2 :\ 0 < y < d + \lambda f(x)\} \tag{6.17}$$

for $\lambda \geq 0$ and a fixed $f \in C_0^\infty(\mathbb{R})$. From the general discussion in Sect. 1.4 we know only that $-\Delta_D^{\Omega_\lambda}$ has at least one eigenvalue below κ_1^2 if f is nonzero and $f \geq 0$. In the weak-coupling case we can prove a lot more.

Theorem 6.5 *Let $\{\Omega_\lambda\}$ be a family of strips (6.17) with a nonzero function $f \in C_0^\infty(\mathbb{R})$. Then $-\Delta_D^{\Omega_\lambda}$ has for small enough $\lambda > 0$ at most one simple eigenvalue $\epsilon(\lambda) \in (0, \kappa_1^2)$. This is the case if $\langle f \rangle := \int_{\mathbb{R}} f(x)\,\mathrm{d}x > 0$, when the eigenvalue is a real-analytic function at $\lambda = 0$, and*

$$\epsilon(\lambda) = \kappa_1^2 - \lambda^2\kappa_1^4\langle f\rangle^2 + \mathcal{O}(\lambda^3).$$

Proof We employ the simplest possible coordinate change passing to the operator $H_\lambda := U_\lambda(-\Delta_D^{\Omega_\lambda})U_\lambda^{-1}$ on $L^2(\Omega_0)$ defined by means of the unitary map $U_\lambda : (U_\lambda\psi)(x, y) = \sqrt{1+\lambda f(x)}\psi(x, (1+\lambda f(x))y)$. To find the explicit form of H_λ is a matter of straightforward computation which yields

$$H_\lambda = H_0 + \lambda\sum_{j=1}^{3} A_j^* B_j + \lambda^2\sum_{j=4}^{7} A_j^* B_j\,, \tag{6.18}$$

where the A_j's and B_j's are multiplication or first-order differential operators given explicitly in Problem 11. To cast it into a more compact form of the type (6.1), we define a pair of operators $C_\lambda,\ D : L^2(\Omega_0) \to L^2(\Omega_0)\otimes\mathbb{C}^7$ by

$$(C_\lambda\phi)_j := \begin{cases} A_j\phi\,, & j = 1, 2, 3 \\ \lambda A_j\phi\,, & j = 4, 5, 6, 7 \end{cases} \qquad (D\phi)_j := B_j\phi\,, \quad j = 1, \dots, 7,$$

and rewrite the operator (6.18) as $H_\lambda = H_0 + \lambda C_\lambda^* D$. In view of *Proposition* 6.1.1 we have to analyze then the BS operator $K_\lambda^z := \lambda D(H_0 - z)^{-1}C^*$.

The argument follows the same lines as the proof of *Theorem* 6.1 so we concentrate on the differences. We employ a simpler decomposition of the free resolvent writing $K_\lambda^z = \lambda L_z + \lambda M_z$, where $L_z := D\mathcal{L}_z C^*$ with the middle part kernel $\mathcal{L}_z(\vec{x}, \vec{x}') = (2\kappa_1(z))^{-1}\chi_1(y)\chi_1(y')$, and $M_z := D(N_z + R_0^\perp(z)C^*$, where $N_z(\vec{x}, \vec{x}') = (2\kappa_1(z))^{-1}\chi_1(y)(\mathrm{e}^{-\kappa_1|x-x'|} - 1)\chi_1(y')$ and $R_0^\perp(z)$ is the free-resolvent part in $L^2(\mathbb{R})\otimes\{\chi_1\}^\perp =: \mathcal{H}_1^\perp$. It is important that M_z is regular, more exactly, that in the present case the function $\zeta \mapsto M_{\kappa_1^2-\zeta}$ is a bounded operator-operator valued function which is analytic in a neighborhood of the point $\zeta = 0$ (Problem 12). This makes it possible to use a factorization analogous to (6.6) and to ask when the eigenvalue of the rank-one operator $\lambda(I + \lambda M_z)^{-1}L_z$ is equal to -1. This yields the implicit equation $\zeta = \mathcal{G}(\lambda, \zeta)$ with

$$\mathcal{G}(\lambda, \zeta) := -\frac{\lambda}{2}\int_{\Omega_0} \chi_1(y)\left(C_\lambda^*(I + \lambda M_{\kappa_1^2-\zeta})^{-1}D\chi_1\right)(\vec{x})\,\mathrm{d}\vec{x}\,.$$

The mentioned analyticity means that $\mathcal{G}$ is analytic around $(0, 0)$, so the solution $\zeta(\lambda)$ is easily found by the implicit function theorem to be

$$\zeta(\lambda) = \frac{\lambda}{2}\,(C_\lambda\chi_1, D\chi_1) + \mathcal{O}(\lambda^2) = \frac{\lambda}{2}\sum_{j=1}^{3}(A_j\chi_1, B_j\chi_1) + \mathcal{O}(\lambda^2)$$

with the contributions from $\lambda A_j^* B_j$, $j = 4, 5, 6, 7$, included in the error term, and the function ζ is analytic around $\lambda = 0$. Moreover, the sum is reduced to the first term only, because $B_3\chi_1 = 0$ and $(A_2\chi_1, B_2\chi_1) = 0$ follows from $\int_{\mathbb{R}} f''(x)\,\mathrm{d}x = 0$. This gives the expression $\lambda \int_{\Omega_0} f(x)\,\chi_1'(y)^2\,\mathrm{d}\vec{x} = \lambda\kappa_1^2\langle f\rangle$ for the leading-order coefficient. The solution gives rise to an eigenvalue $\epsilon(\lambda)$ by $\zeta(\lambda) = \sqrt{\kappa_1^2 - \epsilon(\lambda)}$ provided the function takes non-negative values for small λ which is ensured if $\langle f\rangle > 0$. Taking then $\zeta(\lambda)$ to the square we get the asymptotic expansion of the theorem. ∎

The same technique can be applied to weakly deformed layers. Given a function $f : \mathbb{R}^2 \to [0, \infty)$ of compact support and a number $\lambda > 0$ we define

$$\Omega_\lambda := \{\vec{x} \in \mathbb{R}^3 : 0 < y < d + \lambda f(x)\},$$

as in (4.15). Similarly to the curved layer case, the inherent two-dimensional nature of the problem means that the coupling caused by the deformation is exponentially weak as $\lambda \to 0$.

Theorem 6.6 *Let $f \in C_0^\infty(\mathbb{R}^2)$ be a nonzero function determining a family $\{\Omega_\lambda\}$ of layers by (4.15). Then $-\Delta_D^{\Omega_\lambda}$ has for sufficiently small $\lambda > 0$ at most one simple eigenvalue in $(0, \kappa_1^2)$, which happens if $\langle f\rangle := \int_{\mathbb{R}^2} f(x)\,\mathrm{d}x > 0$. The eigenvalue can then be expressed as $\epsilon(\lambda) = \kappa_1^2 - \mathrm{e}^{2w(\lambda)^{-1}}$, where w is an analytic function in the vicinity of $\lambda = 0$ with the expansion*

$$w(\lambda) = -\lambda\frac{\kappa_1^2}{\pi}\,\langle f\rangle^2 + \mathcal{O}(\lambda^2)\,.$$

Proof We begin as in the deformed-strip case with the "straightening" map $U_\lambda : L^2(\Omega_\lambda) \to L(\Omega_0)$, $(U_\lambda\psi)(x, y) = \sqrt{1 + \lambda f(x)}\psi(x, (1 + \lambda f(x))y)$, by which we pass from $-\Delta_D^{\Omega_\lambda}$ to the unitarily equivalent operator H_λ which can be written in the form (6.18) by Problem 11, or more concisely as $H_\lambda = H_0 + \lambda C_\lambda^* D$ if we adopt the notation from the preceding proof. We shall focus on the aspects in which the subsequent analysis differs from that of *Theorems* 6.2 and 6.5.

We employ the decomposition $K_\lambda^z = \lambda L_z + \lambda M_z$, where the singular rank-one part is $L_z := D\mathcal{L}_z C^*$ with the kernel of the middle factor now being $\mathcal{L}_z(\vec{x}, \vec{x}') = -(2\pi)^{-1}\chi_1(y)\ln\kappa_1(z)\chi_1(y')$. The operator appearing in the remaining part is $M_z := D(N_z + R_0^\perp(z)C^*$, where $R_0^\perp(z)$ is again the resolvent restricted to the subspace spanned by the higher transverse modes, and

$$N_z(\vec{x}, \vec{x}') = \frac{1}{2\pi}\,\chi_1(y)\left(K_0(\kappa_1(z)|x - x'|) + \ln\kappa_1(z)\right)\chi_1(y')\,.$$

In the next step we have to check that M_z is bounded and even analytic with respect to the variable $w := (\ln \kappa_1(z))^{-1}$ around $w = 0$. The part coming from $R_0^\perp(z)$ causes no problems and can be dealt with as in Problem 12. Also for the lowest-mode remainder we use the same route and rewrite the term in question as $DhN_{\kappa_1^2-\zeta}hC^*$ with a suitable $h \in C_0^\infty(\mathbb{R})$; using integration by parts and the explicit form of C_λ, D we conclude that it is sufficient to check the boundedness and analyticity of the integral operators hn_zh and $hn_{z,\mu}h$, $\mu = 1, 2$, on $L^2(\mathbb{R}^2)$ with the kernels

$$n_z(x,x') := \frac{1}{2\pi}\left(K_0(k_1(z)|x-x'|) + \ln k_1(z)\right),$$
$$n_{z,\mu}(x,x') = -\frac{1}{2\pi}\frac{x_\mu - x'_\mu}{|x-x'|}\,k_1(z)K_1(k_1(z)|x-x'|);$$

recall that $K_0' = -K_1$. At this stage comes a difference, however. The singularity of the second term prevents us from using the Hilbert-Schmidt norm to estimate these operators. Instead we employ the Schur-Holmgren bound (Problem 14). Defining $\rho_h := \operatorname{diam}(\operatorname{supp} h)$ and using further the fact that $|(K_0(z) + \ln z)e^{-z}| \le c_1$ for some $c_1 > 0$ we arrive at the following estimates,

$$\|hn_zh\|_{\mathrm{SH}} = \sup_{x\in\mathbb{R}^2} |h(x)| \int_{\mathbb{R}^2} |n_z(x,x')h(x')|\,\mathrm{d}x' \le c_1\|h\|_\infty^2\,\rho_h c_h(z)$$

with $c_h(z) := \rho_h e^{\kappa_1(z)\rho_h} + \max\{e^{-1}, \rho_h \ln \rho_h\}$, and $\|hn_{z,\mu}h\|_{\mathrm{SH}} \le \|h\|_\infty^2\,\rho_h$. To check the analyticity one has to inspect properties of the (complex) derivatives of operator-valued functions $w \mapsto hn_{z(w)}h$ and $w \mapsto hn_{z(w),\mu}h$, where we have introduced $z(w) := \kappa_1^2 - e^{2w(\lambda)^{-1}}$. Using $K_1' = -(K_0 + K_2)$ and defining $\zeta := \kappa_1(z(w))$ we have $\frac{dn_{z(w)}}{dw}(x,x') = (2\pi)^{-1}\zeta w^{-2}(K_1(\zeta) - \zeta^{-1})$ and

$$\frac{dn_{z(w),\mu}}{dw}(x,x') = \frac{1}{2\pi}\frac{x_\mu - x'_\mu}{|x-x'|}\frac{e^{w^{-1}}}{w^2}\left[K_1(\zeta) - \frac{\zeta}{2}\Big(K_0(\zeta) + K_2(\zeta)\Big)\right].$$

Now we notice that $|\zeta K_1(\zeta)| \le 1$ and that the combinations $K_1(\zeta) - \zeta^{-1}$ and $2K_1(\zeta) - \zeta(K_0(\zeta) + K_2(\zeta))$ are also bounded in $\mathbb{R}_+$. Furthermore, $w^{-2}e^{w^{-1}}$ is bounded in $(-\infty, 0)$. Taken together these inequalities allow us to check the finiteness of the Schur-Holmgren bounds,

$$\max\left\{\left\|h\frac{dn_{z(w)}}{dw}h\right\|_{\mathrm{SH}},\ \left\|h\frac{dn_{z(w),\mu}}{dw}h\right\|_{\mathrm{SH}}\right\} \le c_2\|h\|_\infty^2\,\rho_h^2$$

for some $c_2 > 0$ and all $w \in (-\infty, 0)$. Finally, since the limits as $w \to 0-$ in these bounds make sense, the operators hn_zh and $hn_{z,\mu}h$ can be analytically continued to a neighborhood of $w = 0$.

The rest of the proof is the same as in the preceding theorem. Factorizing out the singular part we get the implicit equation $w = \mathcal{G}(\lambda, w)$ with

$$\mathcal{G}(\lambda, w) := \frac{\lambda}{2} \int_{\Omega_0} \chi_1(y) \left(C_\lambda^*(I + \lambda M_{z(w)})^{-1} D\chi_1 \right) (\vec{x})\, \mathrm{d}\vec{x} \,,$$

which has an analytic solution $w(\lambda)$ in the vicinity of $\lambda = 0$. Computing the derivative of this implicitly given function we get

$$\frac{\mathrm{d}w}{\mathrm{d}\lambda}(0) = \frac{1}{2\pi} (A_1\chi_1, B_1\chi_1) = -\frac{1}{\pi} \|\chi_1'\|^2 \int_{\mathbb{R}^2} f(x)\, \mathrm{d}x = -\frac{\kappa_1^2}{\pi} \langle f \rangle \,,$$

because $B_3\chi_1 = 0$ and $\int_{\mathbb{R}^2} (\Delta f)(x)\, \mathrm{d}x = 0$ gives $(A_2\chi_1, B_2\chi_1) = 0$, while the contributions of the remaining operators in (6.18) can be included in the error term. The obtained expression gives rise to an eigenvalue if the derivative is negative, which is the case if $\langle f \rangle > 0$. ∎

An attentive reader has surely noticed that in contrast to the results of Sect. 6.1, *Theorems* 6.5 and 6.6 provided only a sufficient condition for the existence of a weakly coupled bound state, because the critical case, $\langle f \rangle = 0$, was left out. A naive guess would suggest that such a state exists with a higher power of the deformation parameter in the asymptotic expansion. However, since the mode coupling given by the operators C_λ and D differs from the potential case, the answer is not *a priori* clear. Later in this chapter we shall see that this caution is justified: depending on the particular shape of the function f critically deformed strips may or may not have weakly coupled bound states.

6.3 A Generalized BS Technique

The basic idea of the Birman-Schwinger method which is to replace the original operator, typically partial differential and unbounded, by a mathematically better manageable object can be used even in situations when the assumptions of *Proposition* 6.1.1 are not satisfied. This applies, for instance, to strongly singular interactions supported by a set of zero measure for which the formula (6.2) defining the BS operator makes no sense. To be specific we shall discuss such a generalization in the example of a double waveguide with a semitransparent barrier described in Sect. 1.5.2. However, since the resolvent formula which is a core of the argument is useful in different contexts, we shall derive it in a more general form than needed for this particular example.

6.3.1 A Resolvent Formula

Let Ω be an open set in $\mathbb{R}^d$. Consider a positive Radon measure m on Ω and a Borel function $\alpha : \mathbb{R}^d \to \mathbb{R}$ such that the inequality

$$\int_\Omega \left(1+\alpha(x)^2\right) |\psi(x)|^2 \,\mathrm{d}m(x) \le a \int_\Omega |\nabla\psi(x)|^2 \,\mathrm{d}x + b \int_\Omega |\psi(x)|^2 \,\mathrm{d}x \tag{6.19}$$

is satisfied for all $\psi \in C_0^\infty(\Omega)$ with some positive $a < 1$ and b. Since $C_0^\infty(\Omega)$ is dense in $H_0^1(\Omega)$ by definition, there is a unique bounded linear operator $I_m : H_0^1(\Omega) \to L^2(m) := L^2(\Omega, \mathrm{d}m)$ such that $I_m\psi = \psi$ holds for an arbitrary $\psi \in C_0^\infty(\Omega)$; abusing notation and identifying a continuous function ψ with the corresponding equivalence classes in $H_0^1(\Omega)$ and $L^2(m)$ we can also write the last relation as $(I_m\psi)(x) = \psi(x)$ for $x \in \operatorname{supp} m$. By density, (6.19) holds for all $\psi \in H_0^1(\Omega)$ if ψ is replaced by $I_m\psi$ on the left-hand side. A sufficient condition under which (6.19) is valid is, e.g., that α is bounded and m belongs to the generalized Kato class (see the notes).

The class of operators we are interested in here can be formally written in the form $-\Delta_D^\Omega + \alpha(x)m(x)$. The most natural way to define such measure-type perturbations is through the corresponding quadratic form,

$$t_{\alpha m}[\psi] := \int_\Omega |\nabla\psi(x)|^2 \,\mathrm{d}x + \int_\Omega \alpha(x)\, |(I_m\psi)(x)|^2 \,\mathrm{d}m(x) \tag{6.20}$$

with the domain $\mathrm{Dom}\,(t_{\alpha m}) = H_0^1(\Omega)$. In view of (6.19) and the KLMN theorem [RS, Sect. X.1], the above form is closed and bounded below on $H_0^1(\Omega)$, hence there is a unique self-adjoint operator $H_{\alpha m}$ associated with $t_{\alpha m}$. Of course, this definition includes the particular case (1.39) which we will discuss below, but it is also worth mentioning that regular potential perturbations fit into the scheme as well, and moreover in different ways: one can put, e.g., $\mathrm{d}m(x) = |V(x)|\,\mathrm{d}x$ and $\alpha(x) = \operatorname{sgn} V(x)$, or $V = \alpha$ with m being the Lebesgue measure restricted to $\operatorname{supp} V$, etc.

To derive a formula for the resolvent of $H_{\alpha m}$ we need some more notation. Put $\mathbb{C}_\Omega^+ := \left\{k :\, \operatorname{Im} k > 0 \text{ or } k^2 \in [0, \inf\sigma(-\Delta_D^\Omega))\right\} \subset \mathbb{C}$. For any $z = k^2$ the free resolvent is an integral operator with the kernel $G_0(\cdot,\cdot;k^2)$. Let further μ, ν be positive Radon measures without a discrete component, i.e. $\mu(\{x\}) = \nu(\{x\}) = 0$ for any $x \in \Omega$, then we denote by $R_{\mu,\nu}^k$ the integral operator from $L^2(\mu) := L^2(\Omega, \mathrm{d}\mu)$ to $L^2(\nu)$ with the kernel $G_0(\cdot,\cdot;k^2)$. In other words, the operator acts at a vector $\psi \in \mathrm{Dom}\,(R_{\mu,\nu}^k) \subset L^2(\mu)$ as

$$(R_{\mu,\nu}^k\psi)(x) = \int_\Omega G_0(x,y;k^2)\psi(y)\,\mathrm{d}\mu(y)\,,$$

where the right-hand side is understood with respect to x as an element of $L^2(\nu)$. We are interested primarily in the situations when μ, ν are the measure m appearing in (6.20) and the Lebesgue measure on Ω in various combinations.

Proposition 6.3.1 *Let $k \in \mathbb{C}^+_\Omega$ and $\psi \in L^2(m)$, then $R^k_{m,\mathrm{d}x}\psi \in H^1_0(\Omega)$ and*

$$t_0(R^k_{m,\mathrm{d}x}\psi, \phi) - (k^2 R^k_{m,\mathrm{d}x}\psi, \phi) = \int_\Omega \bar{\psi}(y)(I_m\phi)(y)\,\mathrm{d}m(y)$$

holds for all $\phi \in H^1_0(\Omega)$. In particular, $R^k_{m,\mathrm{d}x}$ is injective.

Proof If we prove the first claim, the injectivity will follow by density of Ran I_m in $L^2(m)$. Assume first $k^2 < 0$, so k belongs to the positive imaginary axis. Then $\langle\psi, \phi\rangle_k := t_0(\psi, \phi) - k^2(\psi, \phi)$ defines an inner product on $H^1_0(\Omega)$ and the corresponding norm is equivalent to the Sobolev norm. Fix a $\psi \in L^2(m)$. Using the Schwarz inequality and the fact that I_m is bounded we infer that

$$\left|\int_\Omega \bar{\psi}(y)(I_m\phi)(y)\,\mathrm{d}m(y)\right|^2 \le \int_\Omega |\psi(y)|^2\mathrm{d}m(y)\int_\Omega |(I_m\phi)(y)|^2\mathrm{d}m(y) \le c\,\langle\phi, \phi\rangle_k$$

holds for any $\phi \in H^1_0(\Omega)$ and a constant c depending on ψ. Hence the linear functional $\phi \mapsto \int_\Omega \bar{\psi}(y)(I_m\varphi)(y)\,\mathrm{d}m(y)$ on the Hilbert space $(H^1_0(\Omega), \langle\cdot\,,\cdot\rangle_k)$ is bounded, and by Riesz's lemma, there is a unique $\psi^k_m \in H^1_0(\Omega)$ such that

$$\langle\psi^k_m, \phi\rangle_k = \int_\Omega \bar{\psi}(y)(I_m\phi)(y)\,\mathrm{d}m(y) \tag{6.21}$$

for all $\phi \in H^1_0(\Omega)$. Thus it is sufficient to show that $(R^k_{m,\mathrm{d}x}\psi)(x) = \psi^k_m(x)$ holds a.e. with respect to the Lebesgue measure $\mathrm{d}x$.

Recall that the free Green's function is positive for all $x, y \in \Omega$, $x \neq y$, if $k^2 < 0$ and $\mathrm{d}x$-integrable if the other variable is fixed, in fact exponentially decaying for a non-compact Ω. Moreover, we have

$$\int_\Omega G_0(x, y; k^2)(-\Delta^\Omega_D - k^2)\phi(x)\,\mathrm{d}x = \phi(y)$$

for all $y \in \Omega$, $\phi \in H^1_0(\Omega)$ and the functions $\phi := (-\Delta^\Omega_D - k^2)^{-1}\eta$ with some $\eta \in C^\infty_0(\Omega)$ are C^∞, bounded, and have the same decay as the Green's function for a non-compact Ω. Suppose first that $\psi \in L^1(m) \cap L^2(m)$. For ϕ of the indicated form we may then employ the Fubini's theorem which in combination with the above relation and the fact that G_0 is real-valued gives

$$\int_\Omega \overline{\left(\int_\Omega G_0(x, y; k^2)\psi(y)\,\mathrm{d}m(y)\right)}(-\Delta^\Omega_D - k^2)\phi(x)\,\mathrm{d}x = \int_\Omega \bar{\psi}(y)\phi(y)\,\mathrm{d}m(y),$$

so by (6.21) and the second representation theorem we have

$$\int_\Omega \left(\overline{\psi_m^k}(-\Delta_D^\Omega - k^2)\phi\right)(x)\,\mathrm{d}x = \langle \psi_m^k, \phi\rangle_k = \int_\Omega \bar{\psi}(y)\phi(y)\,\mathrm{d}m(y)$$

for all $\phi \in (-\Delta_D^\Omega - k^2)^{-1}C_0^\infty(\Omega)$. The last identity uses the fact that $I_m\phi = \phi$ holds m-a.e. This requires a comment because in general $\phi \notin C_0^\infty(\Omega)$, however, if this is the case one can approximate ϕ by $C_0^\infty(\Omega)$ functions using a standard family of mollifiers and use dominated convergence to check that the relation in question is valid. Since such ϕ are dense in $L^1(\Omega)$, comparing the last two displayed equations we find that $\psi_m^k = R_{m,\mathrm{d}x}^k\psi$ holds dx-a.e. By standard approximation arguments one can check that the same is true for any $\psi \in L^2(m)$ and $k^2 \in \mathbb{C}_\Omega^+$ (Problems 15 and 16). ∎

This result allows us to derive an explicit formula for the resolvent of $H_{\alpha m}$ as a perturbation of the free resolvent which we denote for brevity by R_0^k.

Proposition 6.3.2 *Let $k \in \mathbb{C}_\Omega^+$. Suppose that $I + \alpha I_m R_{m,\mathrm{d}x}^k$ is invertible on $L^2(m)$ and the operator*

$$R^k := R_0^k - R_{m,\mathrm{d}x}^k(I + \alpha I_m R_{m,\mathrm{d}x}^k)^{-1}\alpha I_m R_0^k$$

is defined everywhere in $L^2(\Omega)$. Then $k^2 \in \rho(H_{\alpha m})$ and $(H_{\alpha m} - k^2)^{-1} = R^k$.

Proof The free resolvent maps $L^2(\Omega)$ onto $H_0^1(\Omega)$ and the same is true for the second term in view of the assumed invertibility and *Proposition* 6.3.1. Hence $R^k\psi \in H_0^1(\Omega)$ and by Problem 15 we have to check that

$$t_{\alpha m}(R^k\psi, \phi) - (k^2 R^k\psi, \phi) = (\psi, \phi)$$

holds for all $\psi \in L^2(\Omega)$ and $\phi \in H_0^1(\Omega) = \mathrm{Dom}\,(t_{\alpha m})$. Substituting for R^k and defining $\chi := (I + \alpha I_m R_{m,\mathrm{d}x}^k)^{-1}\alpha I_m R_0^k\psi$ we rewrite the left-hand side as

$$\begin{aligned} t_0(R_0^k\psi, \phi) - (k^2 R_0^k\psi, \phi) - t_0(R_{m,\mathrm{d}x}^k\chi, \phi) + (k^2 R_{m,\mathrm{d}x}^k\chi, \phi) \\ + \int_\Omega \alpha(x)(I_m\overline{R^k\psi})(x)(I_m\phi)(x)\,\mathrm{d}m(x)\,. \end{aligned}$$

The first two pairs of terms equal (ψ, ϕ) and $-\int_\Omega \bar{\chi}(x)(I_m\phi)(x)\,\mathrm{d}m(x)$ by Problem 15 and *Proposition* 6.3.1, respectively. Some simple algebra then gives

$$\alpha I_m R^k\psi = (I + \alpha I_m R_{m,\mathrm{d}x}^k\psi)(I + \alpha I_m R_{m,\mathrm{d}x}^k\psi)^{-1}\alpha I_m R_0^k\psi - \alpha I_m R_{m,dx}^k\chi = \chi$$

for any $\psi \in L^2(\Omega)$, thus the sought relation holds for all $\phi \in H_0^1(\Omega)$. ∎

In view of the relative boundedness (6.19) the invertibility assumption is satisfied if κ corresponds to a large enough negative energy (Problem 17). Since our aim is to extend the BS method to the present case of measure-type perturbations, we have to ask about the relation between the kernels of the operators $H_{\alpha m}$ and $I + \alpha I_m R_{m,\mathrm{d}x}^k$ for $k \in \mathbb{C}_\Omega^+$. It is not difficult to check that they have the same dimension (Problem 18), and thus there is again a close relation between the solutions of the two spectral problems. To make use of it we cast the above derived resolvent expression into a more convenient form.

Theorem 6.7 *(a)* $I_m R^k_{m,\mathrm{d}x} = R^k_{m,m}$ *and* $I_m R^k_0 = R^k_{\mathrm{d}x,m}$ *for all* $k \in \mathbb{C}^+_{\Omega,0}$.
(b) $I + \alpha R^{i\kappa}_{m,m}$ *has a bounded inverse in* $L^2(m)$ *for all* $\kappa > 0$ *large enough.*
(c) Assume that $I + \alpha R^k_{m,m}$ *is invertible for* $k \in \mathbb{C}^+_\Omega$ *and the operator*

$$R^k := R^k_0 - R^k_{m,\mathrm{d}x}(I + \alpha R^k_{m,m})^{-1}\alpha R^k_{\mathrm{d}x,m}$$

is defined on $L^2(\Omega)$. *Then* $k^2 \in \rho(H_{\alpha m})$ *and* $(H_{\alpha m} - k^2)^{-1} = R^k$.
(d) $\dim \operatorname{Ker}(H_{\alpha m} - k^2) = \dim \operatorname{Ker}(I + \alpha R^k_{m,m})$ *for all* $k \in \mathbb{C}^+_{\Omega,0}$.

Proof By *Proposition* 6.3.2 and Problems 17, 18 it is sufficient to establish the claim (a). We begin with the free Green's function $(x, x') \mapsto G_0(x, x'; -\kappa^2)$ which exists for all $\kappa > 0$, is positive, smooth in each argument, and decays exponentially as $|x - x'| \to \infty$ if Ω is non-compact. Moreover, for $d \geq 2$ it has a singularity at $x = x'$ which forces us to employ a smooth approximation. To this end, choose a function $\eta \in C^\infty(\mathbb{R}_+)$ which is monotonous, $\eta(r) = 1$ for $r \geq 1$, and behaves around the origin as $\eta(r) = c(\ln r)^{-1} + o((\ln r)^{-1})$ and $\eta(r) = c\, r^{d-2} + o(r^{d-2})$ with a nonzero c for $d = 2$ and $d \geq 3$, respectively. Define $\eta_n(r) := \eta(nr)$. Furthermore, take an increasing sequence $\zeta_n \in C_0^\infty(\Omega)$ such that $\lim_{n\to\infty} \zeta_n(x) = 1$ for a fixed $x \in \Omega$, and put

$$G_n(x, x'; -\kappa^2) := \eta_n(|x - x'|)\, G_0(x, x'; -\kappa^2)\, \zeta_n(x)\,.$$

In this way we obtain a sequence which is non-decreasing for fixed $x, x' \in \Omega$ with $G_n(\cdot, x'; -\kappa^2) \in C_0^\infty(\Omega)$. Moreover, $|\nabla_x G_n(x, x'; -\kappa^2)| \leq c_1|x - x'|^{1-d}$ for $x \neq x'$ and $|G_n(x, x'; -\kappa^2)| + |\nabla_x G_n(x, x'; -\kappa^2)| \leq c_2\, \mathrm{e}^{-c_3|x-x'|}$ for $|x - x'|$ large enough holds with suitable constants.

For brevity we use the symbol μ for m, $\mathrm{d}x$. With an arbitrary $\psi \in L^2(\mu)$ we associate the vector $\phi_n := \int_\Omega G_n(\cdot, x'; -\kappa^2)\psi(x')\, \mathrm{d}\mu(x') \in C_0^\infty(\Omega)$. Its Sobolev norm can be estimated using the above listed properties. We have

$$\int_\Omega |\nabla_x \phi_n(x)|^2\, \mathrm{d}x \leq \int_{\mathbb{R}^2} \left| \int_{\mathbb{R}^2} \frac{c_1}{|x - x'|^{d-1}}\, \psi(x')\, \mathrm{d}\mu(x') \right|^2 \mathrm{d}x\,,$$

where ψ on the right-hand side means the trivial extension from Ω to $\mathbb{R}^2$; the integral is easily seen to be finite using Fourier transformation. Furthermore, the non-derivative contribution to the norm has a bound independent of n because $R^{i\kappa}_{\mu,\mathrm{d}x}\psi \in H_0^1(\Omega) \subset L^2(\Omega)$. Hence $\{\phi_n\} \subset H_0^1(\Omega)$ is a bounded sequence.

Now we put $\chi_n := \frac{1}{n}\sum_{j=1}^n \phi_j$ and use the diagonal trick to show (choosing a subsequence if necessary) that $\{\chi_n\}$ converges strongly in $H_0^1(\Omega)$. By the monotone convergence theorem we then get $\chi_n \to R^{i\kappa}_{\mu,\mathrm{d}x}\psi$ in $H_0^1(\Omega)$ as $n \to \infty$. From the definition of I_m we have $I_m\chi_n \to I_m R^{i\kappa}_{\mu,\mathrm{d}x}\psi$ in $L^2(m)$, and since $I_m\chi_n \in C_0^\infty(\Omega)$ holds by construction we conclude that $I_m \chi_n = \chi_n$ is valid m-a.e. which proves the result for k purely imaginary. It extends to any $k \in \mathbb{C}_\Omega^+$ by means of the Hilbert identity as in Problem 16. ∎

6.3.2 A Semitransparent Barrier

If we want to apply the above result to weak-coupling analysis of systems with a locally modified leaky barrier, it is useful to cast the resolvent formula of *Theorem* 6.7 into still another form that would allow us to compare resolvents corresponding to two different measure perturbations. With this aim in mind, we assume that the measure m is the same for both perturbations and that one of the functions is constant, $\alpha_0(x) = \alpha_0$ for some $\alpha_0 \in \mathbb{R}$ and all $x \in \Omega$. Notice that this is not a strong restriction, because the possible x-dependence of the perturbation can be included in the measure m.

Denote the corresponding resolvents by $R^k(\alpha)$ and $R^k(\alpha_0)$, respectively. For any $k \in \mathbb{C}_\Omega^+$ with $k^2 \in \rho(H_{\alpha_0 m}) \cap \rho(H_{\alpha m})$ *Theorem* 6.7 gives

$$\begin{aligned} R^k(\alpha) - R^k(\alpha_0) &= R^k_{m,\mathrm{d}x}\left[(I + \alpha_0 R^k_{m,m})^{-1}\alpha_0 - (I + \alpha R^k_{m,m})^{-1}\alpha\right] R^k_{dx,m} \\ &= R^k_{m,\mathrm{d}x}(I + \alpha R^k_{m,m})^{-1}(\alpha_0 - \alpha)(I + \alpha_0 R^k_{m,m})^{-1} R^k_{\mathrm{d}x,m}\,, \end{aligned}$$

where we have used the fact that α_0 commutes with $I + \alpha R^k_{m,m}$; the resolvent traces on the right-hand side correspond to the function α. Furthermore,

$$R^k_{\mathrm{d}x,m}(\alpha_0) = R^k_{\mathrm{d}x,m} - R^k_{m,m}(I + \alpha_0 R^k_{m,m})^{-1}\alpha_0 R^k_{\mathrm{d}x,m} = (I + \alpha_0 R^k_{m,m})^{-1} R^k_{\mathrm{d}x,m}\,,$$

and using $(R^k_{\mu,\nu})^* = R^{\bar{k}}_{\nu,\mu}$ we get $R^k_{m,\mathrm{d}x}(\alpha_0) = R^k_{m,\mathrm{d}x}(I + \alpha_0 R^k_{m,m})^{-1}$. Finally, applying *Theorem* 6.7*a* once more we get the relations

$$R^k_{m,m}(\alpha_0) = (I + \alpha_0 R^k_{m,m})^{-1} R^k_{m,m} = R^k_{m,m}(I + \alpha_0 R^k_{m,m})^{-1}\,.$$

In particular, the "mixed-term" expressions allow us to write the resolvent difference as $-R^k_{m,\mathrm{d}x}(\alpha_0)(I + \alpha_0 R^k_{m,m})(I + \alpha R^k_{m,m})^{-1}(\alpha - \alpha_0) R^k_{\mathrm{d}x,m}(\alpha_0)$. It is useful to rewrite it further in a form symmetric with respect to the perturbation. With the usual BS convention, $(\alpha - \alpha_0)^{1/2} := |\alpha - \alpha_0|^{1/2}\mathrm{sgn}(\alpha - \alpha_0)$, the central part of the last expression equals

$$\begin{aligned} (I + \alpha_0 R^k_{m,m})(I + \alpha R^k_{m,m})^{-1}(\alpha - \alpha_0) &= \left[I + (\alpha - \alpha_0) R^k_{m,m}(\alpha_0)\right]^{-1}(\alpha - \alpha_0) \\ = \left[I + (\alpha - \alpha_0) R^k_{m,m}(\alpha_0)\right]^{-1} &\left[I + (\alpha - \alpha_0) R^k_{m,m}(\alpha_0)\right](\alpha - \alpha_0)^{1/2} \\ \times \left[I + |\alpha - \alpha_0|^{1/2} R^k_{m,m}(\alpha_0)(\alpha - \alpha_0)^{1/2}\right]^{-1} &|\alpha - \alpha_0|^{1/2} \\ = (\alpha - \alpha_0)^{1/2}\left[I + |\alpha - \alpha_0|^{1/2} R^k_{m,m}(\alpha_0)(\alpha - \alpha_0)^{1/2}\right]^{-1} &|\alpha - \alpha_0|^{1/2}\,. \end{aligned}$$

The obtained expression for the resolvent no longer contains R_0^k, hence we can extend its validity to $\mathbb{C}^+_{\Omega,\alpha_0} := \{k : \operatorname{Im} k > 0 \text{ or } k^2 \in [0, \inf \sigma(H_{\alpha_0 m}))\}$ by means of the Hilbert identity. Summing up the argument, we arrive at the following conclusion.

Proposition 6.3.3 *Under the stated assumptions, the resolvents of $H_{\alpha m}$ and $H_{\alpha_0 m}$ are related for an arbitrary $k \in \mathbb{C}^+_{\Omega,\alpha_0}$ by the formula*

$$R^k(\alpha) = R^k(\alpha_0) - R^k_{m,\mathrm{d}x}(\alpha_0)\,(\alpha-\alpha_0)^{1/2}$$
$$\times \Big[I + |\alpha-\alpha_0|^{1/2}\,R^k_{m,m}(\alpha_0)\,(\alpha-\alpha_0)^{1/2}\Big]^{-1}\,|\alpha-\alpha_0|^{1/2} R^k_{\mathrm{d}x,m}(\alpha_0)\,.$$

Let us return now to the weak-coupling analysis of the example from Sect. 1.5.2 in which m is the Dirac measure supported by the x-axis in the strip $\Omega := \mathbb{R}\times(-d_2, d_1)$. It is convenient in the present situation to regard α as a Borel function $\mathbb{R}\to\mathbb{R}$ in correspondence with the definition of the Hamiltonian by (1.39). In view of *Proposition* 6.3.3 the question about bound states of $H_{\alpha m} \equiv H_\alpha$ is equivalent to spectral analysis of the integral operator

$$K^k_\alpha := |\alpha-\alpha_0|^{1/2}\,R^k_{m,m}(\alpha_0)\,(\alpha-\alpha_0)^{1/2} \tag{6.22}$$

on $L^2(m) \equiv L^2(\mathbb{R})$. For simplicity we adopt slightly stronger assumptions about the coupling than those of *Proposition* 1.5.1.

Proposition 6.3.4 *Suppose that $\alpha \in L^{1+\eta}_{\mathrm{loc}}(\mathbb{R})$ and $\alpha(x)-\alpha_0 = \mathcal{O}(|x|^{-2-\varepsilon})$ holds as $|x|\to\infty$ for some $\eta,\varepsilon>0$. Then*
(a) K^k_α is Hilbert-Schmidt for any $k \in \mathbb{C}^+_{\Omega,\alpha_0}$,
(b) $I + K^{i\kappa}_\alpha$ has a bounded inverse for all $\kappa>0$ large enough,
(c) $\dim\operatorname{Ker}(H_\alpha - k^2) = \dim\operatorname{Ker}(I+K^k_\alpha)$ holds for all $k\in\mathbb{C}^+_{\Omega,\alpha_0}$ and the generalized Birman-Schwinger principle is valid: $k^2\in\sigma_{\mathrm{disc}}(H_\alpha)$ if and only if -1 is an isolated eigenvalue of the operator K^k_α.

Proof We postpone verification of the claims (a), (b) to *Proposition* 6.3.5 below. In the last part it is sufficient to assume $\alpha\in L^\infty$. Indeed, define $\alpha_N(x) := \operatorname{sgn}\alpha(x)\,\min\{|\alpha(x)|, N\}$. Then $K^k_{\alpha_N}\to K^k_\alpha$ in the operator norm as $N\to\infty$, again by the argument of *Proposition* 6.3.5. On the other hand, due to absolute continuity of the Lebesgue integral the values of the quadratic form $t_{\alpha_N m}$ converge to those of (6.20) so $H_{\alpha_N}\to H_\alpha$ in the strong resolvent sense by [Ka, Theorem VIII.3.6]. Hence the discrete spectrum of H_α is approximated by that of H_{α_N}; both are finite in each interval $(-\infty, k_1^2]$ with $k_1^2 < \nu_1(\alpha_0)$.

If $\|\alpha-\alpha_0\|_\infty < \infty$ then $R^k_{m,\mathrm{d}x}(\alpha_0)\,(\alpha-\alpha_0)^{1/2}$ and $|\alpha-\alpha_0|^{1/2}R^k_{\mathrm{d}x,m}(\alpha_0)$ are obviously bounded. The operator $I+K^k_\alpha$ has by part (a) a purely discrete spectrum, every non-unit eigenvalue being of finite multiplicity. Thus if -1 is an eigenvalue of K^k_α, the number k^2 belongs to $\sigma(H_{\alpha m})$ with the same multiplicity. On the other hand, if there is no ψ solving $K^k_\alpha\psi = -\psi$, then $(I+K^k_\alpha)^{-1}$ is bounded, and so is $R^k(\alpha)$, hence $k^2\in\rho(H_{\alpha m})$. ∎

To be able to control the local variation of the barrier which plays here the role of an effective potential, we will consider $\alpha(x) = \alpha_0 + \lambda\beta(x)$, where $\lambda>0$ is the parameter by which we tune the coupling and the function $\beta : \mathbb{R} \to \mathbb{R}$ has the properties stated in the last proposition for $\lambda = 1$.

To be consistent with the notation of the previous sections, we denote the Birman-Schwinger operator (6.22) corresponding to this α by K_λ^k. One finds easily that it is an integral operator on $L^2(\mathbb{R})$ with the kernel

$$K_\lambda^k(x,x') = \lambda|\beta(x)|^{1/2} \sum_{n=1}^{\infty} \frac{|\chi_n(0;\alpha_0)|^2}{2\kappa_n(z)}\, \mathrm{e}^{-\kappa_n(z)|x-x'|}\, \beta(x')^{1/2}\,,$$

where $\kappa_n(z) := \sqrt{\nu_n(\alpha_0) - z}$ corresponds to the energy $z = k^2$ and $\{\chi_n\}$ are the transverse eigenfunctions from *Lemma* 1.5.1 . As in the case of the regular BS method, the idea is to split the rank-one part of the above operator which diverges as $z \to \nu_1(\alpha_0)-$. We write $K_\lambda^k = \lambda Q^k + \lambda P^k$, where

$$Q^k(x,x') = |\beta(x)|^{1/2}\, \mathrm{e}^{-\kappa_1|x|}\, \frac{|\chi_1(0;\alpha_0)|^2}{2\kappa_1}\, \mathrm{e}^{-\kappa_1|x'|}\, \beta(x')^{1/2},$$

and furthermore, we again split off the lowest-mode contribution to the remaining part, $P^k = A^k + C^k$. In particular, the first-term kernel at the singularity equals $A^{k_0}(x,x') = |\beta(x)|^{1/2}\, |\chi_1(0;\alpha_0)|^2\, |x|_<\, \beta(x')^{1/2}$, where we use the notation (6.5). The function $k \mapsto P^k$ is regular around the threshold:

Proposition 6.3.5 *Let β satisfy the same assumptions as $\alpha - \alpha_0$ in the previous proposition. Then in a neighborhood of the point $k_0 := \nu_1(\alpha_0)^{1/2}$ the operators A^k, C^k are Hilbert-Schmidt and*

$$\lim_{k\to k_0} \|A^k - A^{k_0}\|_{\mathrm{HS}} = \lim_{k\to k_0} \|C^k - C^{k_0}\|_{\mathrm{HS}} = 0\,.$$

Proof We have $\|A^{k_0}\|_{\mathrm{HS}}^2 = |\chi_1(0;\alpha_0)|^4 \int_{\mathbb{R}^2} |\beta(x)|\, |x|_<^2\, |\beta(x')|\, \mathrm{d}x\, \mathrm{d}x' < \infty$, since the last integral is estimated by the square of $\int_{\mathbb{R}} |x|\, |\beta(x)|\, \mathrm{d}x$ which is finite by assumption. Using $|A^k(x,x')| \le |A^{k_0}(x,x')|$ and the dominated convergence we establish the first claim. The squared Hilbert-Schmidt norm of C^{k_0} is

$$\sum_{m,n=2}^{\infty} \frac{|\chi_n(0;\alpha_0)\, \chi_m(0;\alpha_0)|^2}{4\sqrt{\nu_n - \nu_1}\sqrt{\nu_m - \nu_1}} \int_{\mathbb{R}^2} |\beta(x)|\, \mathrm{e}^{-(\sqrt{\nu_n-\nu_1}+\sqrt{\nu_m-\nu_1})|x-x'|}\, |\beta(x')|\, \mathrm{d}x\, \mathrm{d}x'\,,$$

where the interchange of summation and integration is justified by the monotone convergence theorem. Using Hölder's inequality we estimate the integral by

$$2(\eta')^{\eta'}\, \|\beta\|_{1+\eta} \|\beta\|_1 \left(\sqrt{\nu_n - \nu_1} + \sqrt{\nu_m - \nu_1}\right)^{-\eta'}$$

with $\eta' := \frac{\eta}{1+\eta} > 0$ where the norms are again finite by assumption. Since the sequence $\{\chi_n(0;\alpha_0)\}$ is bounded by Problem 1.22, it is sufficient to check convergence of the double series

$$\sum_{m,n=2}^{\infty} \frac{1}{\sqrt{\nu_n - \nu_1}\sqrt{\nu_m - \nu_1}\,(\sqrt{\nu_n - \nu_1} + \sqrt{\nu_m - \nu_1})^{\eta'}}$$

which follows from the fact that $\nu_n^{-1/2} = o(n^{-1})$ as $n \to \infty$ – cf. *Lemma* 1.5.1 . As in the case of operators A^k we complete the proof using the pointwise limit of the kernel as $k \to k_0$ in combination with the dominated convergence.

Notice finally that an analogous argument yields the claims (a), (b) of *Proposition* 6.3.4. The summation includes in this case also the first transverse mode and $\sqrt{\nu_n - \nu_1}$ is replaced by $\sqrt{\nu_n + \kappa^2}$, so the modification of the above series makes sense and its sum tends to zero as $\kappa \to \infty$. ∎

Having established the decomposition of the Birman-Schwinger operator (6.22) into the parts which are singular and regular at $k = k_0$, we can proceed with the analysis following the argument which led to *Theorem* 6.1.

Theorem 6.8 *Let $H_{\alpha_0+\lambda\beta}$ correspond to a nonzero function β which obeys the assumptions of Proposition 6.3.5. Then an eigenvalue $\epsilon(\lambda) < \nu_1(\alpha_0)$ exists* iff $\int_{\mathbb{R}} \beta(x)\,\mathrm{d}x \le 0$. *In such a case, it is unique, simple, and satisfies the asymptotic expansion*

$$\sqrt{\nu_1(\alpha_0) - \epsilon(\lambda)} = -\frac{\lambda}{2}\,|\chi_1(0)|^2 \int_{\mathbb{R}} \beta(x)\,\mathrm{d}x$$
$$-\frac{\lambda^2}{4}\Bigg\{|\chi_1(0)|^4 \int_{\mathbb{R}^2} \beta(x)\,|x-x'|\,\beta(x')\,\mathrm{d}x\,\mathrm{d}x'$$
$$-|\chi_1(0)|^2 \sum_{n=2}^{\infty} |\chi_n(0)|^2 \int_{\mathbb{R}^2} \beta(x)\,\frac{\mathrm{e}^{-\sqrt{\nu_n-\nu_1}|x-x'|}}{\sqrt{\nu_n - \nu_1}}\,\beta(x')\,\mathrm{d}x\,\mathrm{d}x'\Bigg\} + \mathcal{O}(\lambda^3),$$

where $\chi_n(0) := \chi_n(0;\alpha_0)$ and $\nu_n := \nu_n(\alpha_0)$.

Proof is left to the reader (Problem 19).

Remark 6.3.1 Making the barrier variation small is just one way to achieve a weak-coupling regime in the model we are discussing here. Another possibility is to shrink the support of the effective potential while keeping $\|\alpha - \alpha_0\|_\infty$ fixed. Consider a longitudinal scaling of the coupling function,

$$\alpha_\sigma(x) := \alpha\left(\frac{x}{\sigma}\right),$$

with the parameter $\sigma \in (0, 1]$ and ask about the behavior in the limit $\sigma \to 0+$. A modification of the above analysis to this case is not difficult (Problem 19). Since it is convenient to employ a less sophisticated kernel decomposition similar to that in proof of *Theorem* 6.5 one has to require a stronger decay, e.g. $\alpha(x) - \alpha_0 = \mathcal{O}(|x|^{-3-\varepsilon})$ as $|x| \to \infty$. Under this assumption H_{α_σ} has a weakly coupled bound state with energy $\epsilon(\sigma)$ for all σ small enough if and only if $\int_{\mathbb{R}}(\alpha(x) - \alpha_0)\,\mathrm{d}x \le 0$, and one has the asymptotic expansion

$$\sqrt{\nu_1(\alpha_0) - \epsilon(\sigma)} = -\frac{\sigma}{2}\, |\chi_1(0)|^2 \int_{\mathbb{R}} (\alpha(x) - \alpha_0)\,\mathrm{d}x - \frac{\sigma^2}{4}\, |\chi_1(0)|^2 \sum_{n=2}^{\infty} |\chi_n(0)|^2$$
$$\times \int_{\mathbb{R}^2} (\alpha(x) - \alpha_0)\, \frac{\mathrm{e}^{-\sigma\sqrt{\nu_n-\nu_1}|x-x'|}}{\sqrt{\nu_n - \nu_1}}\, (\alpha(x') - \alpha_0)\,\mathrm{d}x\,\mathrm{d}x' + \mathcal{O}(\sigma^3)\,.$$

The formula differs from that of *Theorem* 6.8 mostly by the absence of the contribution from the remainder operator in the lowest mode which is in the present case of order of σ^4 and thus absorbed in the error term.

6.4 Variational Estimates

The previous sections illustrated the power of the Birman-Schwinger technique in analyzing the weak-coupling behavior of tube and layer systems. At the same time, the method is not universal. Sometimes it is just too complicated as, for instance, in situations when the leading-order term vanishes and the next one is difficult to compute. There are cases, however, where the problem goes deeper. If the perturbation is not additive in the sense of an operator or quadratic-form sum, an analogue of the BS principle is missing. An alternative method is based on variational estimates. It typically does not yield the exact asymptotics but it can provide upper and lower bounds which capture the correct power behavior with respect to the coupling parameter. In this section we are going to illustrate this claim on the examples discussed above.

6.4.1 A Critically Deformed Strip

The analysis of a weakly deformed strip (6.17) given in *Theorem* 6.5 is restricted to the leading term and gives no answer in the critical case when $\langle f \rangle = 0$. The variational technique allows us to make the following conclusion.

Theorem 6.9 *Let $\{\Omega_\lambda\}$ be the family (6.17) with a nonzero $f \in C_0^\infty(\mathbb{R})$ such that* $\operatorname{supp} f \subset [-b, b]$ *and $\int_{-b}^{b} f(x)\,\mathrm{d}x = 0$. Then the discrete spectrum of $-\Delta_D^{\Omega_\lambda}$ is empty for small enough $|\lambda|$ provided $8b < d\sqrt{3}$. On the other hand, if*

$$\frac{\|f'\|^2}{\|f\|^2} < \frac{24\kappa_1^2}{9 + \sqrt{117 + 48\pi^2}}\,, \tag{6.23}$$

then the same operator has for small nonzero $|\lambda|$ exactly one isolated eigenvalue $\epsilon(\lambda)$ and in that case there are positive constants $c_1,\ c_2$ such that

$$c_1\lambda^4 \le \kappa_1^2 - \epsilon(\lambda) \le c_2\lambda^4\,. \tag{6.24}$$

Proof We employ the unitary equivalence of $-\Delta_D^{\Omega_\lambda}$ with the operator (6.18) which we will for the present purpose write as $H_\lambda = -\Delta_D^{\Omega_0} + \lambda W_1 - \lambda^2 W_2$. Consider first the non-existence condition. In view of *Theorem* 1.4 we have to check that $(\psi, H_\lambda \psi) \geq \kappa_1^2 \|\psi\|^2$ holds for all ψ from a core of the operator, say, from $C_0^2(\Omega_0)$. Moreover, since H_λ commutes with complex conjugation, it is sufficient to consider real-valued ψ only. Any such function can be written as $\psi = h + r$, where $h(x, y) = \phi(x)\chi_1(y)$ and $r(x, \cdot) \perp \chi_1$ for all $x \in \mathbb{R}$, which yields for the value $(\psi, H_\lambda \psi)$ of the quadratic form in question the expression

$$-(h, \Delta_D^{\Omega_0} h) - (r, \Delta_D^{\Omega_0} r) - \sum_{j=1}^{2} (-\lambda)^j \left[(h, W_j h) + (r, W_j r) + 2(r, W_j h) \right].$$

Let us begin with the terms linear in λ. We put $\alpha := \phi(-b)$ and $g(x) := \phi(x) - \alpha$; without loss of generality we may suppose that $\alpha \geq 0$. Then with an abuse of notation we can write

$$(h, W_1 h) = \alpha^2 (\chi_1, W_1 \chi_1) + 2\alpha (g\chi_1, W_1 \chi_1) + (g\chi_1, W_1 g\chi_1).$$

The first term on the right-hand side vanishes in view of $\langle f \rangle = \langle f'' \rangle = 0$, the second one is easily calculated to be $-4\alpha\kappa_1^2 (g, f)_{L^2(-b,b)}$ using Problem 11 and the relation $-\chi_1'' = \kappa_1^2 \chi_1$. Since $f \in C_0^\infty$, the last term can be estimated by a linear combination of the L^2 norms of the functions g, g' in $(-b, b)$. However, $\|g\|_{L^2(-b,b)} \leq (4b/\pi)\|g'\|_{L^2(-b,b)}$ because $g(-b) = 0$ by construction, so we get

$$\lambda(h, W_1 h) \geq -4\lambda\alpha\kappa_1^2 (g, f)_{L^2(-b,b)} - \lambda C \|g'\|_{L^2(-b,b)}^2,$$

where C here and in the following is an unspecified positive constant. In a similar way properties of f together with the Schwarz inequality yield $|(r, W_1 r)| \leq C \|R\|_{H^1(\Omega_b)}^2$, where $\Omega_b := (-b, b) \times (0, d)$, and an estimate for the α-independent part of the mixed term,

$$2|(r, W_1 g\chi_1)| \leq C \left(\|g'\|_{L^2(-b,b)}^2 + \|r\|_{L^2(\Omega_b)}^2 \right).$$

The remaining term linear in λ requires more attention. Since r is smooth by assumption, it expands pointwise as $r(x, y) = \sum_{n=2}^\infty r_n(x)\chi_n(y)$ giving

$$2\alpha(r, W_1 \chi_1) = -2\alpha \sum_{n=2}^\infty (r_n', f')_{L^2(-b,b)} (\chi_n, y\chi_1')_{L^2(0,a)}.$$

The last inner product equals $(-1)^n 2n/(n^2 - 1)$. We introduce the quantity

$$K := \left(\sum_{n=2}^\infty \left(\frac{2n}{n^2 - 1} \right)^2 \right)^{1/2} = \sqrt{\frac{\pi^2}{3} + \frac{1}{4}} \approx 1.881 \qquad (6.25)$$

(Problem 20a), which allows us to estimate $2\alpha|(r, W_1\chi_1)|$ by

$$
\begin{aligned}
2\alpha\left(|f'|, \sum_{n=2}^{\infty}|r_n'|\,\frac{2n}{n^2-1}\right)_{L^2(-b,b)} &\le 2\alpha K\left(|f'|, \left(\sum_{n=2}^{\infty}|r_n'|^2\right)^{1/2}\right)_{L^2(-b,b)} \\
&\le \alpha^2K^2\lambda(1+\tilde{c}\lambda)\|f'\|^2_{L^2(-b,b)} + \frac{1}{\lambda(1+\tilde{c}\lambda)}\|r_x\|^2_{L^2(\Omega_b)},
\end{aligned}
$$

where $r_x := \partial r/\partial x$ and $\tilde{c}$ is a fixed positive number.

Among the quadratic terms we observe that $\lambda^2|(r, W_2r)| \le C\lambda^2\|r\|^2_{H^1(\Omega_b)}$ because $f \in C_0^\infty$. In a similar way we find

$$\lambda^2|(r, W_2h)| \le C\lambda^3\left(\alpha^2 + \|g'\|^2_{L^2(-b,b)}\right) + \lambda\|r\|^2_{L^2(\Omega_b)},$$

where we have also employed the Schwarz inequality and the fact the L^2 norm of g is dominated by that of g', and furthermore,

$$\lambda^2|2\alpha(\chi_1, W_2g\chi_1) + (g\chi_1, W_2g\chi_1)| \le C\lambda^2\left(\alpha^2\lambda + \|g'\|^2_{L^2(-b,b)}\right) + \lambda\|g\|^2_{L^2(-b,b)}.$$

The remaining part of the mixed term is found explicitly in the leading term,

$$-\lambda^2\alpha^2(\chi_1, W_2\chi_1) = \lambda^2\alpha^2\left(3\kappa_1^2\|f\|^2_{L^2(-b,b)} + K^2\|f'\|^2_{L^2(-b,b)}\right) + \mathcal{O}(\lambda^3) \quad (6.26)$$

with K given by (6.25) – cf. Problem 20a. Now putting all the estimates together, we find that $(\psi, H_\lambda\psi)$ is bounded from below by

$$
\begin{aligned}
&\|\nabla h\|^2 + \|\nabla r\|^2 + 3\lambda^2\alpha^2\kappa_1^2\|f\|^2_{L^2(-b,b)} - 4\lambda\alpha\kappa_1^2(g, f)_{L^2(-b,b)} \\
&\quad -(1+\tilde{c}\lambda)^{-1}\|r_x\|^2_{L^2(\Omega_b)} - C\left(\lambda\|g'\|^2_{L^2(-b,b)} + \lambda\|r\|^2_{H^1(\Omega_b)} + \alpha^2\lambda^3\right)
\end{aligned}
$$

for all $|\lambda|$ small enough. To estimate the kinetic part of the quadratic form, we employ the lower bound $\|\nabla r\|^2 \ge \|r_x\|^2_{L^2(\Omega_b)} + 4\kappa_1^2\|r\|^2_{L^2(\Omega_b)}$ together with the explicit expression of the Sobolev norm; this shows that the inequality

$$\|\nabla r\|^2 - C\lambda\|r\|^2_{H^1(\Omega_b)} - (1+\tilde{c}\lambda)^{-1}\|r_x\|^2_{L^2(\Omega_b)} \ge \kappa_1^2\|r\|^2_{L^2(\Omega_b)}$$

holds if $(1-C\lambda)(1+\tilde{c}\lambda) \ge 0$ and $3 - 4\lambda C \ge 0$, i.e. for $|\lambda|$ small enough. In a similar way, the first term is estimated by $\|\nabla h\|^2 \ge \|g'\|^2_{L^2(-b,b)} + \kappa_1^2\|h\|^2_{L^2(\Omega_b)}$, and since $\|g'\|^2_{L^2(-b,b)} \ge (\pi/4b)^2\|g\|^2_{L^2(-b,b)}$ we arrive at the lower bound

$$
\begin{aligned}
(\psi, H_\lambda\psi) - \kappa_1^2\|\psi\|^2 &\ge \pi^2\,\frac{1-C\lambda}{16b^2}\,\|g\|^2_{L^2(-b,b)} - 4\lambda\alpha\kappa_1^2(g, f)_{L^2(-b,b)} \\
&\quad + \lambda^2\alpha^2\|f\|^2_{L^2(-b,b)}\left(3\kappa_1^2 - C\lambda\|f\|^{-2}_{L^2(-b,b)}\right),
\end{aligned}
$$

where we may suppose without loss of generality that f is nonzero. It is straightforward to see that the right-hand side of the last inequality is positive for small $|\lambda|$ as long as $\kappa_1^2 < 3\pi^2/64b^2$.

The same argument yields the upper bound in (6.24). Computing the minimum of the right-hand side in the above relation and taking into account the contribution to the kinetic energy from $\Omega_b^c := \Omega_0 \setminus \Omega_b$ which we have previously neglected we see that the shifted energy form is bounded from below by

$$(\psi, H_\lambda\psi) - \kappa_1^2\|\psi\|^2 \geq \|h_x\|^2_{L^2(\Omega_b^c)} - \lambda^2\alpha^2 c_0\|f\|^2 + \mathcal{O}(\lambda^3)\,,$$

where $c_0 := (2\pi b a^{-2})^2 - 3\kappa_1^2$. For a $\psi \in C_0^2(\Omega_0)$ such that the left-hand side of the last inequality is negative we may use $\|\psi\|^2 \geq \|h\|^2_{L^2(\Omega_b^c)} \geq \|\phi\|_{L^2(-\infty,-b)}$ and an analogous lower bound for $\|h_x\|^2_{L^2(\Omega_b^c)}$; minimizing the corresponding functional we find that the second inequality in (6.24) holds with $c_2 > c_0^2\|f\|^4$.

To get the lower bound we employ a family of trial functions defined by $\psi_{\eta,\rho}(x,y) := (1+\lambda\eta f(x))\chi_1(y)$ in Ω_b and $\psi_{\eta,\rho}(x,y) := \mathrm{e}^{-\rho|x\mp b|}\chi_1(y)$ in the tail parts. Using again the above estimates, this time with $r = 0$, $\alpha = 1$, and $g = \lambda\eta f$, we find that $\|H_\lambda^{1/2}\psi_{\eta,\rho}\|^2 - \kappa_1^2\|\psi_{\eta,\rho}\|^2$ is bounded from above by

$$\rho + \lambda^2\left\{(3-4\eta)\kappa_1^2\|f\|^2 + (\eta^2 + K^2)\|f'\|^2\right\} + C\lambda^3\,.$$

Since $\rho > 0$ can be chosen arbitrarily small, a bound state exists as long as the curly bracket is negative for all η which gives the condition (6.23). Finally, optimizing the ratio of the above quantity to $2b - \rho^{-1} - C\lambda$ which estimated $\|\psi_{\eta,\rho}\|^2$ from below, we get the first inequality in (6.24). ∎

The most important consequence of this result is that in the critical case weakly bound states may not exist if the strip width is too large relative to the support of the perturbation. Of course, the method does not give an exact coefficient in the asymptotics, and thus there are situations in which the existence question cannot be decided in this way—see Problem 20b and the notes.

6.4.2 Window-Coupled Strips

Now we want to show that variational estimates can be useful in situation where the BS method does not work. We shall illustrate this claim on the example of lateral coupling through the window in a Dirichlet boundary. Consider first a pair of coupled strips in the plane discussed in Sect. 1.5.1. By *Theorem* 1.5 the respective operator $-\Delta_D^\Omega$ has for small enough $a > 0$ exactly one simple eigenvalue $\epsilon(a)$ below the essential-spectrum threshold ϵ_d.

Theorem 6.10 *In the described situation there are positive constants* $c_1,\ c_2$ *such that*

$$c_1 a^4 \le \epsilon_d - \epsilon(a) \le c_2 a^4 \tag{6.27}$$

holds for all sufficiently small positive window halfwidths a.

Proof The lower bound to the gap width is again easier requiring the selection of a good enough trial function. Consider first the symmetric case, $\varrho = 1$, and put $\psi = h + g$ in one guide, where $h(x,y) := \phi(x)\chi_1(y)$ with $\phi(x) := \alpha \min\{1, \mathrm{e}^{-\kappa||x|-a|}\}$ and $g(x,y) := \eta\,\chi_{(-a,a)}(x)\,r(y)\,\cos\sqrt{\epsilon_\ell}\,x$ with

$$r(y) := \chi_{(0,d/2]}(y)\,\mathrm{e}^{-\pi y/2a} + 2\chi_{(d/2,d)}(y)\left(1 - \frac{y}{d}\right)\mathrm{e}^{-\pi d/4a}\,;$$

recall that $\epsilon_\ell := (\pi/2a)^2$. It is obvious that such a ψ belongs to $Q(-\Delta_D^\Omega)$ being thus an admissible trial function. To evaluate the shifted energy form $q[\psi] := \|\nabla\psi\|^2 - \epsilon_d\|\psi\|^2$ we employ the identity $-\chi_1'' = \epsilon_d\chi_1$ and a simple integration by parts together with the explicit value $\chi_1'(0) = \pi\sqrt{2/d^3}$ obtaining

$$q[\psi] = \|\psi_x\|^2 + \|g_y\|^2 - \epsilon_d\|g\|^2 - 2\alpha\,\epsilon_d^{1/2}\sqrt{\frac{2}{d}}\int_{-a}^{a} g(x,0)\,\mathrm{d}x\,. \tag{6.28}$$

Since $h_x,\ g_x$ have disjoint supports, we have $\|\psi_x\|^2 = \|h_x\|^2 + \|g_x\|^2$. The second term in this expression equals $\eta^2\,a\epsilon_a\,\|r\|^2_{L^2(0,d)}$, where the squared norm can be estimated by $(a/\pi)(1+\varepsilon_1)$ for any $\varepsilon_1 > 0$ and all a small enough. In the same way, $\int_{-a}^{a} g(x,0)\,\mathrm{d}x = 4a\eta/\pi$. Finally, we have $\|r'\|^2_{L^2(0,d)} < \pi/4a$ for $a < \pi d/8$, which means that $\|g_y\|^2 < \eta^2\pi/4$. Putting these estimates together, using $\|h_x\|^2 = \alpha^2\kappa$, and neglecting the negative term $-\epsilon_d\|w\|^2$, we arrive at the inequality

$$q[\psi] < \alpha^2\kappa - \frac{8\sqrt{2}\,\alpha a}{d^{3/2}}\,\eta + \frac{\pi}{4}\,(2+\varepsilon_1)\eta^2\,.$$

Taking the minimum of the right-hand side with respect to η and using a simple bound, $\|\psi\|^2 > \alpha^2(1-\varepsilon_2)\kappa^{-1}$ for any $\varepsilon_2 > 0$ and all a small enough, we get

$$q[\psi] < (1-\varepsilon_2)^{-1}\left(\kappa^2 - \frac{2^7 a^2\kappa}{\pi d^3(1+\varepsilon_1)}\right)\|\psi\|^2\,;$$

the sought inequality follows by minimizing the right-hand side with respect to κ. In the asymmetric case, $\varrho < 1$, we use an analogous argument; the part of the trial function in the wider channel is as above and the other one is just transversally rescaled, $\tilde\psi(x,y) := \psi(x, -\varrho y)$.

The other bound is more difficult, however, by Neumann bracketing we are allowed to consider the symmetric case only, i.e. a single guide with the Neumann segment $(-a,a)$ at the boundary. As in the previous proof we have to estimate the form $q[\psi] := \|\nabla\psi\|^2 - \epsilon_d\|\psi\|^2$ from below for all real-valued C^2-smooth ψ with $\|\psi\| = 1$ satisfying the needed boundary conditions. It is also clear that we may restrict ourselves to functions even with respect to x because the operator is reduced by the parity eigenspaces and a lower bound to the symmetric part will estimate the antisymmetric one as well. Any such function can be written through the Fourier expansion

$$\psi(x,y)=\sum_{n=1}^{\infty} c_n(x)\chi_n(y)\,, \tag{6.29}$$

where the real-valued coefficients $c_n(x)=(\chi_n,\psi(x,\cdot))$ are smooth and even with respect to x and the series is uniformly convergent for $|x|\ge a$.

It is useful to split off the part of the coefficient c_1 which is constant in the doubled window. We put $\alpha := c_1(2a)$ and $\hat f_1(x) := (c_1(x)-\alpha)\chi_{[-2a,2a]}(x)$ so that, in particular, $\hat f_1(\pm 2a)=0$. Then we define $f_1 := c_1-\hat f_1$ and use the decomposition $\psi = f+g$ with $f(x,y) := f_1(x)\chi_1(y)$ for which the relation (6.28) holds again. We shall estimate separately contributions to its terms from different regions indicating the respective parts of the norms by subscripts. We begin with the part exterior to the window and for a later purpose we put aside one half of the longitudinal kinetic energy, considering

$$\frac12\|\psi_x\|^2_{|x|\ge a}+\|g_y\|^2_{|x|\ge a}-\epsilon_d\|g\|^2_{|x|\ge a}=\sum_{n=1}^{\infty}\int_a^{\infty}\left[(c_n')^2+\left(n^2-1\right)\epsilon_d c_n^2\right](x)\,\mathrm{d}x\,.$$

To estimate the right-hand side we employ a bound obtained easily by solving the appropriate Euler equation: a function $\phi\in C^2(\mathbb{R}_+)$ with $\phi(0)=\alpha$ satisfies

$$\int_0^{\infty}\left[\phi'(t)^2+m^2\phi^2(t)\right](t)\,\mathrm{d}t\ge m\alpha^2 \tag{6.30}$$

for a fixed $m\ge 0$. Consequently, the expression in question is bounded from below by $\epsilon_d^{1/2}\sum_{n=2}^{\infty} nc_n(a)^2$ if we neglect the non-negative term $\int_a^{\infty} c_1'(x)^2\,\mathrm{d}x$. Using further the fact that $\psi_x=g_x$ for $|x|\le 2a$ we get the lower bound

$$\begin{aligned} q[\psi] >& \frac12\|\psi_x\|^2_{|x|\ge a}+\|g_x\|^2_{|x|\le a}+\|g_y\|^2_{|x|\le a}-\epsilon_d\|g\|^2_{|x|\le a}\\ &+\epsilon_d^{1/2}\sum_{n=2}^{\infty} nc_n(a)^2-2\alpha\,\epsilon_d^{1/2}\sqrt{\frac{2}{d}}\int_{-a}^{a} G(x,0)\,\mathrm{d}x\,. \end{aligned}$$

Next we want to estimate the sum of the first two terms on the right-hand side. To this end we use the decomposition $g=g_1+g_2$, where $g_1(x,y)=\hat f_1(x)\chi_1(y)$ and g_2 contains the contributions from the higher transverse modes. Since $\hat f_1(\pm 2a)=0$ by definition, we have $\|g_{1,x}\|^2_{|x|\le 2a}\ge \frac14\epsilon_\ell\|g\|^2_{|x|\le 2a}$. Moreover,

$$\frac12\|\psi_x\|^2_{a\le|x|\le 2a}+\|g_x\|^2_{|x|\le a}\ge\frac12\|g_x\|^2_{|x|\le 2a}=\frac12\|g_{1,x}\|^2_{|x|\le 2a}+\frac12\|g_{2,x}\|^2_{|x|\le 2a}\,,$$

where the subscript in the last norm may be dropped because $g_1(x, y) = 0$ for $|x| \geq 2a$. Combining these estimates, we find

$$q[\psi] > \frac{1}{2}\,\|\psi_x\|^2_{|x|\geq 2a} + \frac{1}{2}\,\|g_{2,x}\|^2_{|x|\leq 2a} + \|g_y\|^2_{|x|\leq a} - \epsilon_d\|g\|^2_{|x|\leq a}$$
$$+\,\epsilon_d^{1/2}\sum_{n=2}^{\infty} n c_n(a)^2 - 2\alpha\,\epsilon_d^{1/2}\sqrt{\frac{2}{d}}\int_{-a}^{a} g(x,0)\,\mathrm{d}x + \frac{1}{8}\,\epsilon_\ell\,\|g_1\|^2\,.$$

The function g_2 will also be split into parts, $g_2(x, y) = \hat g_2(x, y) + \tilde g_2(y)$, with $\tilde g_2(x, y) := \sum_{n=2}^{\infty} c_n(2a)\chi_n(y)$ independent of x. The last property means that $g_{2,x} = \hat g_{2,x}$, and since $\hat g_2(\pm 2a) = 0$ holds by construction, we have

$$\|g_{2,x}\|^2_{|x|\leq 2a} \geq \frac{1}{4}\,\epsilon_\ell\,\|\hat g_2\|^2_{|x|\leq 2a} \geq \frac{1}{8}\,\epsilon_\ell\,\|g_2\|^2_{|x|\leq 2a} - \frac{1}{4}\,\epsilon_\ell\,\|\tilde g_2\|^2_{|x|\leq 2a}\,,$$

where in the second step we have used the Schwarz inequality. Since it is not easy to find a lower bound for the last term, we restrict ourselves to the vicinity of the window introducing $\Omega_a := (-2a, 2a) \times (0, a)$ which gives

$$\|g_{2,x}\|^2_{|x|\leq 2a} \geq \frac{1}{4}\,\epsilon_\ell\,\|\hat g_2\|^2_{\Omega_a} \geq \frac{1}{8}\,\epsilon_\ell\,\|g_2\|^2_{\Omega_a} - \frac{1}{4}\,\epsilon_\ell\,\|\tilde g_2\|^2_{\Omega_a}\,. \tag{6.31}$$

To find an upper bound to $\|\tilde g_2\|^2_{\Omega_a}$, notice that we may restrict ourselves to ψ with the higher Fourier coefficients exponentially dominated outside of the window, $|c_n(x)| \leq |e_n(x)|$, where $e_n(x) := c_n(a)\,\exp\{-\epsilon_d^{1/2}\sqrt{n^2-1}\,(|x| - a)\}$ for $|x| \geq a$ and $n \geq 2$. Indeed, write $\psi(x, y) = \tilde\psi(x, y) + \chi_{|x|\geq a}(x)c_n(x)\chi_n(y)$ to isolate the contribution from the exterior part of the n-th mode contribution, then the normalized quadratic form $q[\psi]\|\psi\|^{-2}$ can be expressed as

$$\frac{\|\nabla\tilde\psi\|^2 - \epsilon_d\|\tilde\psi\|^2 + 2\int_a^\infty \left[c_n'(x)^2 + (n^2-1)\,\epsilon_d\,c_n(x)^2\right]\mathrm{d}x}{\|\tilde\psi\|^2 + 2\int_a^\infty c_n(x)^2\mathrm{d}x}\,.$$

Without loss of generality we may consider only those ψ for which the numerator is negative. As in (6.30) its last term is minimized by the function e_n, hence replacing $c_n(x)^2$ by $\min\{c_n(x)^2, e_n(x)^2\}$ we can only get a more negative number, and at the same time, the positive denominator can only be diminished, which justifies the claim about the exponential dominance.

To get the sought norm estimate we divide the series expressing $\tilde g_2$ into two parts referring to small and large values, relative to a^{-1}. For the former we employ the smallness of the χ_n norm restricted to $(0, a)$, while the latter will be estimated by means of the subexponential decay specified above,

$$\|\tilde g_2\|^2_{\Omega_a} = \int_{-2a}^{2a} \mathrm{d}x \int_0^a \mathrm{d}y \left(\sum_{n=2}^{\infty} c_n(2a)\chi_n(y)\right)^2$$

$$\le 8a \int_0^a \left(\sum_{n=2}^{[a^{-1}]+1} c_n(2a)\chi_n(y)\right)^2 \mathrm{d}y + 8a \int_0^a \left(\sum_{n=[a^{-1}]+2}^{\infty} c_n(2a)\chi_n(y)\right)^2 \mathrm{d}y$$

$$\le 8a \left(\sum_{n=2}^{[a^{-1}]+1} n^{-1}c_n(a)^2 \int_0^a \chi_n(y)^2\mathrm{d}y\right)\left(\sum_{n=2}^{[a^{-1}]+1} n\right)$$

$$+8a \left(\sum_{n=[a^{-1}]+2}^{\infty} nc_n(a)^2 \int_0^a \chi_n(y)^2\mathrm{d}y\right)\left(\sum_{n=[a^{-1}]+2}^{\infty} n^{-1}e_n(a)^2\right),$$

by Hölder's inequality and the rough bound $c_n(2a) < c_n(a)$. Evaluating the integral $\int_0^a \chi_n(y)^2\mathrm{d}y = \frac{a}{d}\left[1-\left(\frac{d}{2\pi na}\right)\sin\left(\frac{2\pi na}{d}\right)\right] \le \frac{a}{d}\min\left\{\frac{1}{6}\left(\frac{2\pi na}{d}\right)^2, 2\right\}$, and using the two bounds in the first and the second term, respectively, we get

$$\|\tilde g_2\|^2_{\Omega_a} \le \frac{16a^2}{d}\left\{\frac{2}{3}\epsilon_d + \sum_{n=[a^{-1}]+1}^{\infty} n^{-1}\mathrm{e}^{-2\pi an/d}\right\}\sum_{n=2}^{\infty} nc_n(a)^2,$$

where the estimates $\sum_{n=2}^{[a^{-1}]+1} n \le 2a^{-1}$ and $\sqrt{n^2-1} < n-1$ have also been employed. The sum in the curly bracket on the right-hand side has a bound independent of a, being the Darboux sum of the integral $\int_1^\infty x^{-1}\,\mathrm{e}^{-2\pi x/d}\,\mathrm{d}x = -\mathrm{Ei}(-2\pi/d)$. Hence there is a positive C such that

$$\|\tilde g_2\|^2_{\Omega_a} \le Ca^2 \sum_{n=2}^{\infty} nc_n(a)^2,$$

so using (6.31) we arrive at the inequality

$$\frac{1}{2}\|g_{2,x}\|^2_{|x|\le 2a} + \epsilon_d^{1/2}\sum_{n=2}^{\infty} nc_n(a)^2 \ge \frac{\delta}{16}\epsilon_\ell\|g_2\|^2_{\Omega_a} - \frac{\delta}{2}\|\tilde g_2\|^2_{\Omega_a} + \epsilon_d^{1/2}\sum_{n=2}^{\infty} nc_n(a)^2$$

valid for any $\delta \in (0,1]$; the above estimate shows that the sum of the last two terms can be made non-negative by choosing δ small enough. Putting $m := \frac{\pi}{8}\sqrt{\delta}$, we can estimate the shifted energy form by

$$q[\psi] > \frac{1}{2}\|\psi_x\|^2_{|x|\ge 2a} + \|g_y\|^2_{|x|\le a} - \epsilon_d\|g\|^2_{|x|\le a} + \frac{1}{8}\epsilon_\ell\,\|g_1\|^2$$
$$- 2\alpha\,\epsilon_d^{1/2}\sqrt{\frac{2}{d}}\int_{-a}^{a} g(x,0)\,\mathrm{d}x + \frac{m^2}{a^2}\|g_2\|^2_{\Omega_a}.$$

Next we express the window contribution to the norm of g_y using the decomposition $g = g_1 + g_2$, an integration by parts, the relation $g_2(x, 0) = g(x, 0)$, and the fact that $g_2(x, \cdot)$ is orthogonal to χ_1. This yields the relation

$$\|g_y\|^2_{|x|\le a} = \|g_{1,y}\|^2_{|x|\le a} + \|g_{2,y}\|^2_{|x|\le a} - 2\,\epsilon_d^{1/2}\sqrt{\frac{2}{d}}\int_{-a}^{a} \hat{f}_1(x) g(x,0)\,\mathrm{d}x\,,$$

in which the last term does not exceed $(\pi/d^2)\left(2\|g_1\|^2_{|x|\le a} + d\|g(\cdot,0)\|^2_{|x|\le a}\right)$ by the Schwarz inequality. We use this conclusion in the last bound of $q[x]$ in combination with the identity $\|g\|^2_{|x|\le a} = \|g_1\|^2_{|x|\le a} + \|g_2\|^2_{|x|\le a}$. We may also neglect $\|g_{1,y}\|^2_{|x|\le a}$ as well as the term $(\pi^2/32a^2)\|g_1\|^2 - \pi(\pi+2)d^{-2}\|g_1\|^2_{|x|\le a}$ which is positive for a small enough, obtaining

$$\begin{aligned} q[\psi] >& \frac{1}{2}\,\|\psi_x\|^2_{|x|\ge 2a} + \|g_{2,y}\|^2_{|x|\le a} + \frac{m^2}{a^2}\,\|g_2\|^2_{\Omega_a} - \epsilon_d\,\|g_2\|^2_{|x|\le a} \\ & - \epsilon_d^{1/2}\,\|g_2(\cdot,0)\|^2_{|x|\le a} - 2\alpha\,\epsilon_d^{1/2}\sqrt{\frac{2}{d}}\int_{-a}^{a} g_2(x,0)\,\mathrm{d}x\,. \end{aligned}$$

Now we notice that the function $g_2(x, \cdot)$ satisfies for a fixed $x \in [-a, a]$ the assumptions of Problem 21b, thus the sum of the second, third, and fourth term on the right-hand side is bounded from below by $(c_0/a)\|g_2(\cdot, 0)\|^2_{|x|\le a}$. Furthermore, $\frac{c_0}{2a} > \epsilon_d^{1/2}$ holds for small a, which yields

$$q[\psi] > \frac{1}{2}\,\|\psi_x\|^2_{|x|\ge 2a} + \frac{c_0}{2a}\,\|g_2(\cdot,0)\|^2_{|x|\le a} - 2\alpha\,\epsilon_d^{1/2}\sqrt{\frac{2}{d}}\,\|g_2(\cdot,0)\|_{|x|\le a}\sqrt{2a}\,,$$

where we have employed the Schwarz inequality again; taking the minimum of the right-hand side over $\|g_2(\cdot, 0)\|_{|x|\le a}$, we arrive finally at the estimate

$$q[\psi] > \frac{1}{2}\,\|\psi_x\|^2_{|x|\ge 2a} - \frac{8\pi^2\alpha^2}{c_0 d^3}\,a^2\,.$$

The remaining part of the argument is simple. We have $\|\psi\| \ge \|\psi_x\|^2_{|x|\ge 2a}$, and estimating the norms of ψ and ψ_x outside of the doubled window from below by the contribution from the lowest transverse mode, we get

$$\frac{q[\psi]}{\|\psi\|^2} > \frac{\int_{2a}^{\infty} c_1'(x)^2\mathrm{d}x - \frac{8\pi^2\alpha^2}{c_0 d^3}\,a^2}{2\int_{2a}^{\infty} c_1(x)^2\mathrm{d}x} > -2\left(\frac{4\pi^2}{c_0 d^3}\right)^2 a^4\,.$$

The second inequality, which is the final result, follows from the fact that the extremum of the functional in question over functions with a fixed value at $x = 2a$ is achieved with $c_1(2a)\,\mathrm{e}^{-\kappa(x-2a)}$ and equals $(\kappa^2/2) - (8\pi^2/c_0 d^3)\kappa a^2$; it is then sufficient to take the minimum over κ. ∎

6.4.3 Window-Coupled Layers

In the three-dimensional case the setting is similar. Using the notation from Sect. 6.2.3 we consider a pair of adjacent layers of widths $d_j > 0,\ j = 1, 2$, i.e. $\Omega = \mathbb{R}^2 \times (-d_2, d_1) \backslash B_{\mathcal{W}}$ coupled by a boundary window $\mathcal{W}$ which is supposed to be an open bounded subset of $\mathbb{R}^2$. As before we denote by $H(d_1, d_2; \mathcal{W})$ the Dirichlet Laplacian in $L^2(\Omega)$. By *Theorem* 4.7 the discrete spectrum of $H(d_1, d_2; \mathcal{W})$ is nonempty whenever $\mathcal{W} \neq \emptyset$, while the essential spectrum coincides with the half-line $[\epsilon_d, \infty)$, where $d := \max\{d_1, d_2\}$ and $\epsilon_d := \frac{\pi^2}{d^2}$. Moreover, if the window is sufficiently small there is just one eigenvalue of $H(d_1, d_2; \mathcal{W})$; now we are going to analyze its asymptotic behavior.

Theorem 6.11 *Suppose that $\mathcal{W} = aM$, where $a > 0$ and $M \subset \mathbb{R}^2$ is open and nonempty. Then for all a small enough the operator $H(d_1, d_2; \mathcal{W})$ has exactly one eigenvalue $\epsilon(a)$ and there are positive constants c_1, c_2 such that*

$$\exp(-c_2 a^3) \le \epsilon_d - \epsilon(a) \le \exp(-c_1 a^3). \tag{6.32}$$

Proof The existence of a single eigenvalue for small a was established in *Theorem* 4.7. As in the previous section, we shall treat in detail only the symmetric case, $d_1 = d_2 = d$, leaving the general situation to the reader (see also the notes); in Sect. 4.4 we mentioned that the problem then reduces to an analysis of the operator $H(d, aM)$ on $L^2(\mathbb{R}^2 \times (0, d))$ which acts as Laplacian with combined boundary conditions, Neumann on $\mathcal{W}$ and Dirichlet on the rest of the boundary. Moreover, by assumption $\mathcal{W}$ as a set can be sandwiched between two open discs, and since by Neumann bracketing the eigenvalue is monotonous with respect to a window enlargement, it is sufficient to prove (6.32) for $M = B_1$ with $B_1 \subset \mathbb{R}^2$ being open unit disc centered at the origin. We shall again use the notation $\vec{x} = (x, y) \in \Omega$ with $x \in \mathbb{R}^2$ and $y \in (0, d)$, as well as $\kappa_1 = \frac{\pi}{d} = \epsilon_d^{1/2}$.

To prove the lower bound in (6.32) we employ a trial function argument; we mimic the choice used in Sect. 4.4 and put $\psi_\kappa(x, y) = f_\kappa(x)\chi_1(y) + \eta G(x, y)$ with

$$f_\kappa(x) := \min\left\{1, \frac{K_0(\kappa|x|)}{K_0(\kappa a)}\right\}, \quad G(x, y) = \phi_1^{(a)}(x)\, R(y; a)\,,$$

where $\phi_1^{(a)}$ is the normalized ground-state eigenfunction of the Dirichlet Laplacian $-\Delta_D^{aM}$ on aM extended by zero to the whole $\mathbb{R}^2$ and corresponding to the eigenvalue $\mu_1(a) = \mu_1(1)a^{-2}$, and

$$R(y; a) = \begin{cases} \mathrm{e}^{-\sqrt{\mu_1(a)}\, y} & \dots\ y \in \left(0, \frac{d}{2}\right) \\ 2\left(1 - \frac{y}{d}\right) \mathrm{e}^{-\frac{d}{2}\sqrt{\mu_1(a)}} & \dots\ y \in \left[\frac{d}{2}, d\right) \end{cases}$$

Using integration by parts with respect to the variable y and the fact that ∇f_κ and $\nabla\phi_1^{(a)}$ have disjoint supports, we obtain

$$\begin{aligned} L(\psi_\kappa) :=& \|\nabla\psi_\kappa\|^2 - \kappa_1^2\, \|\psi_\kappa\|^2 = \|\nabla f_\kappa\|^2_{L^2(\mathbb{R}^2)} + \eta^2 \left(\mu_1(a) - \kappa_1^2\right) \|R\|^2_{L^2(0,d)} \\ &+ \eta^2\, \|R'\|^2_{L^2(0,d)} - 2\kappa_1 \eta a \int_M \phi_1^{(1)}(x)\, \mathrm{d}x\,. \end{aligned} \tag{6.33}$$

By a direct calculation one can check that there is an $\varepsilon_1 > 0$, independent of a, such that

$$\mu_1(a)\,\|R\|^2_{L^2(0,d)} + \|R'\|^2_{L^2(0,d)} \le \frac{\sqrt{\mu_1(1)}}{a}\,\varepsilon_1$$

holds for all a sufficiently small. The first term on the right-hand side of (6.33) has been calculated explicitly as in the proof of *Theorem* 4.7; we have just to replace R by a in the formula derived there. Using again the identity $-K_1'(z) = K_0(z) + z^{-1}K_1(z)$ in combination with the asymptotic expansions $K_0(z) = -\ln z + \mathcal{O}(1)$ and $K_1(z) = z^{-1} + \mathcal{O}(\ln z)$ we arrive then at the bound

$$\|\nabla f_\kappa\|^2_{L^2(\mathbb{R}^2)} \le -\frac{\varepsilon_2}{\ln \kappa a}$$

valid for some $\varepsilon_2 > 0$ and all a small enough. Now we set $\eta = \beta a^2$ with $\beta > 0$ and insert all these estimates into (6.33). Since $\phi_1^{(1)}$ can be chosen positive in M, it is then easy to see that for β small enough but fixed there exists a constant $C > 0$, independent of a, such that

$$L(\psi_\kappa) \le -\frac{\varepsilon_2}{\ln \kappa a} - Ca^3\,. \tag{6.34}$$

We also need a lower bound on $\|\psi_\kappa\|^2$. Estimating the contribution from the window region by $2\|f_\kappa\chi_1\|_{|x|\le a} + 2\eta^2\,\|G\|_{|x|\le a}$ we get

$$\|\psi_\kappa\|^2 \ge \|\psi_\kappa\|^2_{|x|\ge a} - 2\pi\,a^2 - 2\eta^2\,\|R\|^2_{L^2(0,d)}\,,$$

where the last term is of order $\mathcal{O}(a^4)$ by assumption, and the first term can be calculated similarly as above,

$$\|\psi_\kappa\|^2_{|x|\ge a} = \frac{\pi a^2}{K_0^2(\kappa a)}\left(K_1(\kappa a)^2 - K_0(\kappa a)^2\right).$$

From the mentioned asymptotic expansions we get $\|\psi_\kappa\|^2 \ge \kappa^{-2}(\ln \kappa a)^{-2}\,\varepsilon_3$ with a fixed $\varepsilon_3 > 0$ for all a small enough; this inequality in combination with (6.34) gives the estimate

$$\frac{L(\psi_\kappa)}{\|\psi_\kappa\|^2} \le -\frac{\kappa^2\,\ln \kappa a}{\varepsilon_3}\left(Da^3\ln \kappa a + E\right),$$

where D, E are positive constants independent of a. The lower-bound part in (6.32) then follows by minimizing the right-hand side of the above inequality with respect to the parameter κ.

Passing to the other inequality we follow the argument scheme of the previous section: to find a lower bound on $\epsilon(a)$ we have to estimate $L(\psi)/\|\psi\|^2$ from below for all real $\psi \in L^2(\Omega_+)$ which belong to $H^1(\Omega_+)$, are radially symmetric and vanish at the boundary except for the window. We can write such a ψ in the form of a series (6.29) where the coefficients depend only on $r := |x|$. As in the proof of *Theorem* 6.10 we assume that

$$|c_n(r)| \le |c_n(a)| \, \frac{K_0(\kappa_1\sqrt{n^2-1}\,r)}{K_0(\kappa_1\sqrt{n^2-1}\,a)}, \quad n \ge 2. \tag{6.35}$$

In analogy with the waveguide case we write the lowest-mode component as

$$F(x,y) := \begin{cases} \alpha\,\chi_1(y) & \dots \quad r \le 2a \\ c_1(r)\,\chi_1(y) & \dots \quad r > 2a \end{cases}$$

with $\alpha := c_1(2a)$ and we divide $G(x,y) := \psi(x,y) - F(x,y)$ into $G_1(x,y) := (c_1(r)-\alpha)\chi_1(y)$ supported in the extended window region of radius $2a$ and $G_2(x,y) := \hat{G}(x,y)+\Gamma(x,y)$ with $\Gamma(x,y) := \sum_{n=2}^{\infty} c_n(2a)\chi_n(y)$. Using these definitions we divide the corresponding quadratic form,

$$L[\psi] = \|\nabla_x\psi\|^2 + \|\partial_y G\|^2 - \epsilon_d\|G\|^2 - 2\alpha\chi_1'(0)\int_{B_a} G(x,0)\,\mathrm{d}x\,, \tag{6.36}$$

into several parts which we estimate separately. We start with one half of the first term combined with the second and the third one,

$$\begin{aligned} L_1[\psi] &:= \frac{1}{2}\,\|\nabla_x\psi\|^2_{r\ge a} + \|\partial_y G\|^2_{r\ge a} - \epsilon_d\|G\|^2_{r\ge a} \\ &= \pi\sum_{n=1}^{\infty}\int_a^{\infty}\left(c_n'(r)^2 + 2\epsilon_d(n^2-1)c_n(r)^2\right) r\,\mathrm{d}r \\ &\ge \pi\sum_{n=2}^{\infty}\int_a^{\infty}\left(c_n'(r)^2 + \epsilon_d n^2 c_n(r)^2\right) r\,\mathrm{d}r \\ &\ge \pi\sum_{n=2}^{\infty} c_n(a)^2\kappa_1 na\,\frac{K_1(\kappa_1 na)}{K_0(\kappa_1 na)} \ge \pi\kappa_1 a\sum_{n=2}^{\infty} nc_n(a)^2, \end{aligned} \tag{6.37}$$

where the estimate in the last line employs the fact that the functional $u \mapsto \int_a^{\infty}(|u'(t)|^2 + m^2|u(t)|^2)\,t\,\mathrm{d}t$ on $H^1(a,\infty)$ with the condition $u(a) = \alpha$ is minimized by the function

$$u_0(t) = \alpha\,\frac{K_0(mt)}{K_0(ma)},$$

as can be verified by inspecting the corresponding Euler equation, in combination with the inequality $K_1(z) \ge K_0(z)$, see [AS]. Next we consider

$$L_2[\psi] := \|\nabla_x\psi\|^2_{r\le 2a} = \|\nabla_x G_1\|^2_{r\le 2a} + \|\nabla_x G_2\|^2_{r\le 2a}.$$

Since G_1 vanishes at $r = 2a$ by assumption, the Friedrichs inequality implies

$$\|\nabla_x G_1\|^2_{r\le 2a} \ge C_1\, a^{-2}\, \|G_1\|^2 \tag{6.38}$$

for some $C_1 > 0$. Using now the notation $\Omega_a := B_{2a} \times (0, a)$ we get

$$\begin{aligned}\|\nabla_x G_2\|^2_{r\le 2a} &= \|\nabla_x \hat{G}\|^2_{r\le 2a} \ge \frac{C_1}{a^2}\,\|\hat{G}\|^2_{r\le 2a} \ge \frac{C_1}{a^2}\,\|\hat{G}\|^2_{\Omega_a} \\ &\ge \frac{\delta C_1}{2a^2}\,\|G_2\|^2_{\Omega_a} - \frac{\delta C_1}{a^2}\,\|\Gamma\|^2_{\Omega_a}\end{aligned} \tag{6.39}$$

for all $a \le d$ and $0 < \delta \le 1$. The last term on the right-hand side can be estimated by combining the dominated decay (6.35) with the smallness of the norm of χ_n restricted to $[0, a]$ in complete analogy with the lower-dimensional case of the previous section; we obtain

$$\|\Gamma\|^2_{\Omega_a} \le C_2\, a^3 \Big(C_3 + \sum_{n=[a^{-1}]+2}^{\infty} \frac{K_0^2(2\kappa_1\sqrt{n^2-1}\,a)}{n\, K_0^2(\kappa_1\sqrt{n^2-1}\,a)} \Big) \sum_{n=2}^{\infty} n c_n(a)^2$$

with some C_2, C_3. The sum can be estimated by the corresponding integral,

$$\sum_{n=[a^{-1}]+2}^{\infty} \frac{K_0^2(2\kappa_1\sqrt{n^2-1}\,a)}{n\, K_0^2(\kappa_1\sqrt{n^2-1}\,a)} \le \int_1^{\infty} \frac{K_0^2(\kappa_1 z)}{n\, K_0^2(\frac{1}{2}\kappa_2 z)}\, \mathrm{d}z$$

which is convergent since $K_0(z) \sim \sqrt{\frac{\pi}{2z}}\, \mathrm{e}^{-z}$ as $z \to \infty$, hence there is a $C_4 > 0$, independent of a, such that

$$C_1 \|\Gamma\|^2_{\Omega_a} \le C_4 a^3 \sum_{n=2}^{\infty} n c_n(a)^2. \tag{6.40}$$

Combining now the estimates (6.37)–(6.40) we obtain

$$L_1(\psi) + L_2(\psi) \ge \frac{C_1}{a^2}\,\|G_1\|^2 + \frac{m^2}{a^2}\,\|G_2\|^2_{\Omega_a}$$

for some $m > 0$ and all a small enough. Using next the claim of Problem 21b and the estimate

$$\|\partial_y G\|^2_{\{x:r\le a\}} \ge \|\partial_y G_2\|^2_{r\le a} - \frac{2\pi}{d^2}\left(2\|G_1\|^2_{r\le a} + d\, \|G_2(\cdot,0)\|^2_{r\le a}\right)$$

obtained again as in the two-dimensional case, we arrive at the inequality

$$L_1[\psi] + L_2[\psi] + \|\partial_y G\|^2_{\{x:r\le a\}} - \epsilon_d \|G\|^2_{\{x:r\le a\}} \ge \frac{C_5}{a}\,\|G_2(\cdot,0)\|^2_{r\le a}$$

valid for a positive C_5 and a small enough. Inserting this into (6.36) and applying the Cauchy-Schwarz inequality we get

$$L[\psi] \geq \frac{1}{2}\,\|\nabla_x\psi\|^2_{r\geq 2a} - C_6 a^3$$

with some $C_6 > 0$. The first term on the right-hand side can be estimated from below by the first transverse-mode contribution. Estimating in the same way $\|\psi\|^2$ we get

$$\frac{L[\psi]}{\|\psi\|^2} \geq \frac{\int_{2a}^{\infty} c_1'(r)^2 r dr - C_7\, a^3 c_1(2a)^2}{2\int_{2a}^{\infty} c_1(r)^2 r dr} \tag{6.41}$$

with a $C_7 > 0$. Solving the corresponding Euler equation one can check that the right-hand side of the last inequality is minimized by

$$c_1(r) = c_1(2a)\,\frac{K_0(\kappa r)}{K_0(2\kappa)}$$

for some $\kappa > 0$. If we insert this into (6.41) and evaluate the resulting integrals, we arrive at

$$\frac{L[\psi]}{\|\psi\|^2} \geq -\kappa^2 \,\ln(2\kappa a)\left(D' a^3 \ln(2\kappa a) + E'\right)$$

for some $D', E' > 0$; minimization of the right-hand side with respect to κ then leads to second inequality in (6.32) concluding thus the proof. ∎

6.5 Distant Perturbations: Matching Methods

In the last section of this chapter we describe one more type of weak-coupling problem in waveguides together with a method to analyze it. If a perturbation is supported by two regions which are very far from each other, one can regard the spectrum coming from each of them as the unperturbed one and examine how it is changed by the presence of the remote component of the perturbation.

We shall illustrate this situation on the example of a planar strip with Dirichlet boundaries one of which has two Neumann segments; as we know it describes the nontrivial part of the spectral problem for a symmetric pair of waveguides coupled by two windows in the common boundary. We have encountered already, in Sect. 1.2 and elsewhere, matching of solutions as a tool of spectral analysis, based on finding Fourier coefficients from a comparison of solutions in adjacent regions. Here we will deal with a different type of matching which consists of a smooth interpolation between solutions; we are going to show that it can also be used to derive spectral results.

We consider thus a horizontal strip $\Omega = \mathbb{R}\times(0,d)$ in the lower boundary of which we fix two segments $\gamma_l^{\pm}(a)$ of the same lengths $2a$. Specifically, we put $\gamma_l^{\pm}(a) = \{(x,y) : |x \mp l| < a,\ y = 0\}$ and $\gamma_l(a) = \gamma_l^{+}(a) \cup \gamma_l^{-}(a)$. We suppose that the behavior of the system is governed by the Laplacian on $L^2(\Omega)$ subject to Dirichlet boundary conditions on $\partial\Omega \setminus \gamma_l(a)$ and to Neumann condition on $\gamma_l(a)$; we denote this operator by $H_l(a)$. We are particularly interested in the asymptotic behavior of the discrete spectrum of $H_l(a)$ in the limit $l \to \infty$.

To begin with, let us briefly recall some basic facts about the single-window operators discussed in Sect. 1.5.1 which we shall denote here by $H(a)$. By *Theorem* 1.5 this operator has for any $a > 0$ a finite number $N \geq 1$ of discrete eigenvalues $\epsilon_j(a),\ j = 1, \dots, N$ associated with normalized eigenfunctions ψ_j having a definite parity, $\psi_j(-x, y) = (-1)^{j-1}\psi_j(x, y)$. There are critical values $0 = a_1 < a_1 < a_2 < \dots$ at which new eigenvalues emerge from the continuum; in such a case the equation $(H(a_n) - \kappa_1^2)\psi = 0$ has a bounded solution ψ^n describing a threshold resonance, again of a definite parity in the variable x. This solution and the eigenfunctions ψ_j behave in the limit $x \to +\infty$ as

$$\psi^n(x, y) = \sqrt{\frac{2}{d}}\, \sin \kappa_1 y + \beta_n\, \mathrm{e}^{-\kappa_1\sqrt{3}x} \sin 2\kappa_1 y + \mathcal{O}\big(\mathrm{e}^{-\kappa_1\sqrt{8}x}\big) \tag{6.42}$$

and

$$\psi_j(x, y) = \alpha_j\, \mathrm{e}^{-\sqrt{\kappa_1^2 - \epsilon_j(a)}\, x} \sin \kappa_1 y + \mathcal{O}\big(\mathrm{e}^{-\sqrt{4\kappa_1^2 - \epsilon_j(a)}\, x}\big), \tag{6.43}$$

respectively, with some constants α_j, β_n, see also Problem 1.22.

The spectral behavior is different in those two situations. Let us first describe what it looks like if a single window does not have a threshold resonance.

Theorem 6.12 *Let $a \in (a_n, a_{n+1})$ for some $n \in \mathbb{N}$. Then for the window distance l large enough the operator $H_l(a)$ has exactly $2n$ eigenvalues $\lambda_j^{\mp}(l, a),\ j = 1, \dots, n$, situated in the interval $(\frac{1}{4}\kappa_1^2, \kappa_1^2)$. Each of them is simple and satisfies the asymptotic expansion*

$$\lambda_j^{\pm}(l, a) = \epsilon_j(a) \mp \mu_j(a)\, e^{-2l\sqrt{\kappa_1^2 - \epsilon_j(a)}} + \mathcal{O}\Big(e^{-\left(4\sqrt{\kappa_1^2 - \epsilon_j(a)} - \sigma\right) l}\Big), \tag{6.44}$$

as $l \to +\infty$, where σ is a positive number and the coefficients $\mu_j(a)$ are given by

$$\mu_j(a) = \alpha_j(a)^2\, d\, \sqrt{\kappa_1^2 - \epsilon_j(a)}\,.$$

Before turning to a proof sketch we note that in view of the natural scaling behavior it suffices to consider the case $d = \pi$ putting $\kappa_1 = 1$ in the above formulæ. We shall need a couple of auxiliary results which we state without proofs referring to the literature given in the notes.

Lemma 6.5.2 *The discrete spectrum of the operator $H_l(a)$ is simple and nonempty for any $a > 0$. The eigenvalues depend continuously on l and a. Those corresponding to even and odd eigenfunctions are increasing and decreasing, respectively, as functions of l. Moreover, all the eigenvalues of $H_l(a)$ converge for a fixed $a > 0$ to the eigenvalues of $H(a)$ as $l \to +\infty$.*

This lemma shows, among other things, that $H(a)$ is in a sense the limiting operator of our system. This leads us to the analysis of the following problem,

$$-(\Delta+\lambda)u=f\,,\qquad u=0\quad\text{on}\;\Gamma(a)\,,\qquad \partial_y u=0\quad\text{on}\;\gamma(a)\,,\tag{6.45}$$

where $\Gamma(a):=\partial\Omega\setminus\gamma(a)$ and $f\in L^2(\Omega)$ is assumed to be supported in the rectangle $\Omega_b:=\Omega\cap\{(x,y):\,|x|<b\}$ for some $b>0$. The structure of the solution depends on whether λ is close to one or not. We first pick $\delta\in(\epsilon_n(a),1)$ and assume that $\lambda\in D_\delta$, where $D_\delta:=\{\lambda\in\mathbb{C}:\,\mathrm{Re}\,\lambda<\delta\}$. It is then sufficient to consider solutions to problem (6.45) which behave as $\mathcal{O}\big(\mathrm{e}^{-\sqrt{1-\lambda}\,|x|}\big)$ in the limit $|x|\to\infty$. Let $\Omega^\pm=\{(x,y)\in\Omega:\,\pm x>0\}$ and consider the following pair of boundary value problems,

$$-(\Delta+\lambda)v^\pm=g\quad\text{in}\;\Omega^\pm\,,\qquad v^\pm=0\quad\text{on}\;\partial\Omega^\pm\,,\tag{6.46}$$

where g is an arbitrary function from $L^2(\Omega)$ supported in Ω_A for some number $A\geq\max\{a,b-1\}$. Solutions of problems (6.46) can easily be found by separation of variables,

$$v^\pm(x,y)=\int_{\Omega^\pm}G^\pm(x,y,x',y';\lambda)\,g(x',y')\,\mathrm{d}x'\,\mathrm{d}y'\,,\tag{6.47}$$

$$G^\pm(x,y,x',y';\lambda)=\sum_{j=1}^\infty\frac{\mathrm{e}^{-\kappa_j(\lambda)|x-x'|}-\mathrm{e}^{\mp\kappa_j(\lambda)(x+x')}}{\pi\kappa_j(\lambda)}\,\sin jy\,\sin jy'$$

with $\kappa_j(\lambda):=\sqrt{j^2-\lambda}$. We can also write the function as $v^\pm=T_1^\pm(\lambda)g$, where $T_1^\pm(\lambda):L^2(\Omega_A^\pm)\to H^2(\Omega^\pm)$ with $\Omega_A^\pm=\Omega^\pm\cap\{(x,y):\,|x|<A\}$ are bounded linear operators which form a holomorphic family with respect to the variable $\lambda\in D_\delta$. We also introduce the "glued" function v equal to v^+ if $x\geq0$ and to v^- if $x<0$. Then we consider the problem

$$\Delta w=\Delta v\quad\text{in}\;\Omega_A\,,\quad \partial_y w=0\quad\text{on}\;\gamma(a)\,,\quad w=v\quad\text{on}\;\partial\Omega_A\setminus\gamma(a)\,,\tag{6.48}$$

which is posed in a bounded domain, hence by the standard theory of elliptic boundary-value problems the function w exists, is unique and belongs to $H^1(\Omega_A)$. Consequently, there is a bounded linear operator $T_2(\lambda):\,L^2(\Omega_A)\to H^1(\Omega_A)\cap H^2(\Omega_A\setminus S_r)$ with $S_r:=\{(x,y):\,(x\pm a)^2+y^2<r^2\}$ for any $r>0$, and such that $w=T_2(\lambda)g$. Now we introduce a C^∞ interpolation function χ such that $\chi(t)=1$ holds if $|t|<A-1$ and $\chi(t)=1$ for $|t|>A$, and construct a solution to the problem (6.45) in the following way,

$$u(x,y)=\chi(x)w(x,y)+(1-\chi(x))v(x,y)\,.\tag{6.49}$$

Since $w=T_2(\lambda)g$ and $v^\pm=T_1^\pm(\lambda)g$, we can further express the solution as $u=T_3(\lambda)g$, where $T_3(\lambda)$ is the appropriate bounded linear operator from $L^2(\Omega_A)$ into $H^1(\Omega_A)\cap H^2(\Omega_A\setminus S_r)$, holomorphic in the variable $\lambda\in D_\delta$. In view of the definition of w and v the interpolated function (6.49) satisfies by construction the boundary conditions of (6.45), hence it represents a solution to this boundary-value problem if and only if it satisfies the differential equation in question. Substituting u into it, we arrive at the equation

$$g + T_4(\lambda)g = f\,, \tag{6.50}$$

where $T_4(\lambda) : L^2(\Omega_A) \to L^2(\Omega_A)$ is the bounded linear operator defined by

$$T_4(\lambda)\, g := -2\nabla_x\chi \cdot \nabla_x(w - v) - (w - v)(\Delta + \lambda)\,\chi\,.$$

Since Ω_A is a bounded region, it is possible to verify that $T_4(\lambda)$ defined in this way is a compact operator from $L^2(\Omega_A)$ to $L^2(\Omega_A)$ (see the notes). This allows us to apply to (6.50) the standard Fredholm technique. In this way one is able to derive another pair of auxiliary results.

Lemma 6.5.3 *To any solution g of (6.50) there exists a unique solution $u = T_3(\lambda)g$ of (6.45). Conversely, to any solution u of (6.45) there exists a unique solution to (6.50) such that $u = T_3(\lambda)g$. The equivalence holds for any $\lambda \in D_\delta$.*

Thus the resolvent family $(T_4(\lambda) + I)^{-1}$ is meromorphic and its only poles are the eigenvalues of $H(a)$. To proceed we shall also need the following result concerning the behavior of $(T_4(\lambda) + I)^{-1}$ in the vicinity of the poles.

Lemma 6.5.4 *If $\epsilon_0 < 1$ is an eigenvalue of $H(a)$, then for any λ close to ϵ_0 we have*

$$(T_4(\lambda) + I)^{-1} = \frac{\phi}{\epsilon_0 - \lambda}\,(\cdot\,,\,\psi)_{L^2(\Omega)} + T_5(\lambda), \tag{6.51}$$

where ψ is an eigenfunction of $H(a)$ associated with ϵ_0 and normalized in $L^2(\Omega)$, the function ϕ is such that $\psi = T_3(\epsilon_0)\phi$, and $T_5 : L^2(\Omega_A) \to L^2(\Omega_A)$ is a bounded linear operator holomorphic in the variable λ.

After dealing with the single window case let us turn to the perturbed operator. Keeping the window in the center and putting the axis of mirror symmetry at the point $x = -l$, we have to solve the boundary-value problem

$$\begin{aligned} &-(\Delta + \lambda)u = f \quad \text{on } \{(x, y) \in \Omega : x > -l\} \\ &u = 0 \quad \text{on } \Gamma(a)\,, \quad \partial_y u = 0 \quad \text{on } \gamma(a)\,, \quad hu = 0 \quad \text{at } x = 0\,, \end{aligned} \tag{6.52}$$

where $hu := u$ in the odd case and $hu := \partial_x u$ in the even one. Let us consider for simplicity the former situation only, the latter can be treated in the same way. In analogy with (6.46) we have the problems

$$\begin{aligned} -(\Delta + \lambda)\, v_l^+ &= g \quad \text{on } \Omega^+\,, \quad v_l^+ = 0 \quad \text{on } \partial\Omega^+\,, \\ -(\Delta + \lambda)\, v_l^- &= g \quad \text{on } \Omega_l^+\,, \quad v_l^- = 0 \quad \text{on } \partial\Omega_l^-\,. \end{aligned}$$

While $v_l^+ = v^+$, where v^+ is the solution to (6.46), the equation for v_l^- takes into account the perturbation; by separation of variables we get

$$v_l^-(x,y) = v^-(x,y) + \int_{\Omega^-} G_l^-(x,y,x',y';\lambda)\, g(x',y')\, \mathrm{d}x'\mathrm{d}y',$$
$$G^-(x,y,x',y';\lambda) = -\sum_{j=1}^{\infty} \frac{2\,\mathrm{e}^{-\kappa_j(\lambda)l}}{\pi\kappa_j(\lambda)\sinh\kappa_j(\lambda)l} \sinh\kappa_j(\lambda)x\, \sinh\kappa_j(\lambda)x'$$
$$\times \sin jy \sin jy'\,. \tag{6.53}$$

Since the rest of the analysis of (6.52) follows the same line as in the case of the limiting problem (6.45), we skip the details and conclude that (6.52) is equivalent to the second-kind Fredholm operator equation

$$g + T_4(\lambda)g + T_6(\lambda,l) = f\,, \tag{6.54}$$

where $T_6 : L^2(\Omega_A) \to L^2(\Omega_A)$ is a compact linear operator which is holomorphic in the variable λ and jointly continuous in (λ, l) provided $\lambda \in D_\delta$ and l is large enough. Moreover,

$$\|T_6\| = \mathcal{O}\big(\mathrm{e}^{-2l\sqrt{1-\lambda}}\big) \quad \text{as } l \to \infty\,. \tag{6.55}$$

With these prerequisites we can *sketch the proof of Theorem* 6.12: If we consider the odd case, the eigenvalues of $H_l(a)$ and the corresponding eigenfunctions are given by solutions to the problem (6.52) with $hu = u$ and $f = 0$, hence we should look for λ such that the equation

$$\Phi + T_4(\lambda)\Phi + T_6(\lambda,l) = 0$$

has a non-trivial solution. We are interested in eigenvalues which are close to a fixed eigenvalue ϵ_0 of $H(a)$, thus we consider λ's which lie in a neighborhood of ϵ_0 containing neither any other eigenvalue of $H(a)$ nor the point 1. In view of *Lemma* 6.5.4 we can rewrite the last equation in the form

$$\Phi - \frac{\phi}{\lambda-\epsilon_0}\,(\psi, T_6(\lambda,l)\Phi)_{L^2(\Omega)} + T_5(\lambda)\,T_6(\lambda,l)\Phi = 0$$

recalling that $\psi = T_3(\lambda)\phi$. Since $I + T_5(\lambda)\,T_6(\lambda,l)$ is invertible for l large enough, by (6.55), the last equation implies

$$\Phi - \frac{1}{\lambda-\epsilon_0}\,(\psi, T_6(\lambda,l)\Phi)_{L^2(\Omega)}(I + T_5(\lambda)\,T_6(\lambda,l))^{-1}\phi = 0\,.$$

Moreover, the inner product in the second term cannot vanish, otherwise Φ would be trivial. We are thus able to express Φ from the above equation and to compute the inner product $(\psi, T_6(\lambda,l)\Phi)_{L^2(\Omega)}$. This yields the equation

$$\lambda - \epsilon_0 - \big(\psi, T_6(\lambda, l)(I + T_5(\lambda)\, T_6(\lambda, l))^{-1}\phi\big)_{L^2(\Omega)} = 0$$

which determines eigenvalues of the problem (6.52) and hence of the operator $H_l(a)$. In view of the asymptotics (6.55) we get from here

$$\lambda - \epsilon_0 - \big(\psi, T_6(\lambda, l)\phi\big)_{L^2(\Omega)} + \mathcal{O}\big(\mathrm{e}^{-2(2\sqrt{1-\epsilon_0}-\sigma)l}\big) = 0 \tag{6.56}$$

with some $\sigma > 0$. Next we introduce the function V on Ω by

$$\begin{aligned} V(x, y) := &-\frac{4\,\mathrm{e}^{-2\kappa_1(\epsilon_0)l}}{\pi\,\kappa_1(\epsilon_0)} \sinh\kappa_1(\epsilon_0)x\, \sinh\kappa_1(\epsilon_0)y \\ &\times \int_{\Omega^-} \sinh\kappa_1(\epsilon_0)x'\, \sinh\kappa_1(\epsilon_0)y'\, \phi(x', y')\,\mathrm{d}x'\mathrm{d}y' \end{aligned}$$

for $x < 0$ and $V(x, y) = 0$ elsewhere; using it we can rewrite the leading term of the operator $T_6(\lambda, l)$ which comes from the lowest contribution to the sum in (6.53). Assuming that W solves (6.48) with $v = V$ we arrive at

$$T_6(\lambda, l)\phi = -(\Delta + \epsilon_0)(V + \chi(W - V)) + \mathcal{O}\big(\mathrm{e}^{-2(2\sqrt{1-\epsilon_0}-\sigma)l}\big)$$

in $L^2(\Omega_A)$. With the help of this expansion we can calculate the leading term of the second summand in (6.56). Using integration by parts we obtain

$$(\psi, T_6(\lambda, l)\phi)_{L^2(\Omega)} = \lim_{R\to\infty} \int_0^\pi \big(V\partial_x\psi - \psi\partial_x V\big)(-R, y)\,\mathrm{d}y + \mathcal{O}\big(\mathrm{e}^{-2(2\sqrt{1-\epsilon_0}-\sigma)l}\big).$$

To evaluate the above integral we use the fact that in view of $\psi = T_3(\epsilon_0)\phi$ and of the definition of T_3 the constant $\alpha_j = \alpha$ in (6.43) is given by

$$\alpha = \frac{2}{\pi\kappa_1(\epsilon_0)} \int_{\Omega^-} \sinh\kappa_1(\epsilon_0)x'\, \sinh\kappa_1(\epsilon_0)y'\, \phi(x', y')\,\mathrm{d}x'\mathrm{d}y'.$$

In combination with (6.43) and the definition of V we thus get

$$(\psi, T_6(\lambda, l)\phi)_{L^2(\Omega)} = -\pi\alpha^2\kappa_1(\epsilon_0)\, e^{-2\kappa_1(\epsilon_0)l} + \mathcal{O}\big(\mathrm{e}^{-2(2\sqrt{1-\epsilon_0}-\sigma)l}\big),$$

which together with (6.56) implies the validity of the expansion (6.44) for $\lambda_j^-(l, a)$. A similar reasoning in the even case, when $hu = \partial_x u$, leads to the respective expansion for the eigenvalues $\lambda_j^+(l, a)$. ∎

The threshold case when the Neumann segments have a critical length can be treated using the same technique, however, the analysis is more subtle since one also has to take into account the threshold resonances of the operator $H(a)$, therefore we limit ourselves to stating the asymptotic expansion result, referring to the notes for further reading.

Theorem 6.13 *Let $a = a_n$ for some $n \geq 2$. Then the operator $H_l(a)$ has for all l large enough $(2n-1)$ eigenvalues. The first $2n-2$ of them obey the asymptotic expansion given in Theorem 6.12, while the last eigenvalue, which we denote by $\lambda_{2n-1}^{+}(l, a_n)$, corresponds to an even eigenfunction and satisfies*

$$\lambda_{2n-1}^{+}(l, a_n) = \kappa_1^2 - 3\beta_n^4 d^2 \,\mathrm{e}^{-4\sqrt{3}\kappa_1 l} + \mathcal{O}\big(\mathrm{e}^{-2(\sqrt{8}+\sqrt{3})\kappa_1 l}\big)$$

as $l \to \infty$, where β_n are the coefficients in (6.42).

6.6 Notes

Section 6.1 The Birman-Schwinger principle was discovered simultaneously and independently in [Bi61] and [Sch61]. It is discussed in many places; the present formulation is adapted from [BGRS97, Lemma 2.1]. Subtle features of weak-coupling expansion for interactions depending on λ in a nonlinear way are usually manifested in situations when the leading term vanishes as we noted at the end of Sect. 6.2.3; a different, physically interesting example of such an effect is given in [BCEZ99]. The slight difference in the assumptions of Problems 1 and 2 does not concern the cases $d = 2, 3$ which are the most interesting. Notice also that for $d = 3$ we can replace L^2 by Rollnik's class provided $H_\lambda = H_0 + V$ is defined through the quadratic form as in (1.24). *Theorems* 6.1 and 6.2 are taken from [DE95] and [EKr01a], respectively.
Section 6.2 The applications of the above results to weakly-coupled states in tubes and layers in *Theorems* 6.3 and 6.4 come again from [DE95] and [EKr01a], respectively. *Theorem* 6.5 was proved in [BGRS97] where the authors also conjectured various extensions of the result; we leave the reader to work them out (Problem 13).
Section 6.3 Radon measure is an abstraction of Lebesgue's outer measure for general topological spaces – see, e.g., [Rao]. A measure m is said to belong to the generalized Kato class if for every open set Ω we have

$$\lim_{\varepsilon\to 0+} \sup_{x\in\Omega} \int_{B_\varepsilon(x)\cap\Omega} g(x-y)\,\mathrm{d}m(y) = 0\,,$$

where $g(x-y) := |x-y|^{2-d}$ for $d \geq 3$ and $g(x-y) := |\ln|x-y||$ for $d = 2$, while for $d = 1$ the condition reads $\sup_{x\in\mathbb{R}} m([x, x+1]) < \infty$. The inequality (6.19) for such m and a bounded α was established in [SV96], for other examples of singular measures with this property see [He89]. *Theorem* 6.7 was proved in [BEKŠ94], *Proposition* 6.3.3 and its application to weak-coupling analysis of a double waveguide with a leaky interface come from [EKr01b] where slightly weaker assumptions were used.
Section 6.4 *Theorem* 6.9 is taken (with a slight correction) from [EV97a]. As mentioned in Problem 20b, it leaves some situations undecided; this gap can be covered by another method, cf. [BEGK01]. Note that the nontrivial part of *Theorem* 6.9 is the non-existence condition; the $\mathcal{O}(\lambda^4)$ asymptotic behavior of the gap can be expected from the BS analysis of Sect. 6.2.3. However, the variational technique allows us to determine the correct power

of leading term even in situations when the Hamiltonian cannot be expressed through a quadratic form sum as is the case of windows in Dirichlet barriers. Concerning such results, *Theorem* 6.10 comes from [EV96], *Theorem* 6.11 is essentially due to [EV97b]; we refer to those papers for details of the proofs including extensions to the asymmetric cases. Weakly coupled bound states also appear in waveguides composed of two semi-strips of slightly different widths and a small bulge near their junction, see [CNR13].
Section 6.5 For the proofs of *Lemmata* 6.5.2–6.5.4 and of *Theorem* 6.13 see [BE04] where asymptotic expansions of the eigenfunctions of $H_l(a)$ are also calculated. The reduction procedure which transforms the eigenvalue problem for the operator $H_l(a)$ to an appropriate Fredhom operator equation follows a general scheme proposed by E. Sánchez-Palencia [SP]. The application of this reduction technique in the proof of *Theorem* 6.12 is simplified by the mirror symmetry of the model, i.e. by the fact that the two Neumann segments have the same lengths. The method also works without this assumption, however, the argument is more complicated [BE07]. Moreover, by the same technique one can also treat resonances: an example of a guide in which there are two long but finite Dirichlet barriers separating the Neumann window from semi-infinite Neumann boundary segments has been worked out in [BEG13].

6.7 Problems

1. Let $V \in (L^p + L^\infty_\varepsilon)(\Omega_0)$, where $p \geq \max\left\{2, \frac{d}{2}\right\}$ for $d \neq 4$ and $p > 2$ for $d = 4$, then the operator $H_\lambda := -\Delta_D^{\Omega_0} + \lambda V$ defined on $\mathrm{Dom}\,(-\Delta_D^{\Omega_0})$ is self-adjoint and the BS-principle holds for the factorization of Remark 6.1.1a.
Hint: Using Problem 1.18 show that V is $-\Delta_D^{\Omega_0}$-bounded with relative bound zero, and employ the inclusion $\mathrm{Dom}\,(V) \subset \mathrm{Dom}\,(|V|^{1/2})$.

2. Let $V \in (L^p + L^\infty)(\Omega_0)$, where $p = 2$ for $d = 2, 3$ and $p > \frac{d}{2}$ for $d \geq 4$, then the norm $\||V|^{1/2} B_z V^{1/2}\|$ in the proof of *Theorem* 6.1 has for any $z \leq z_0 < \nu_2$ a bound independent of z.
Hint: Employ the inequality $\|f(x)g(-i\nabla)\|_q \leq (2\pi)^{-d/q}\|f\|_q\|g\|_q$, where $\|\cdot\|_q$ on the left-hand side is the Schatten norm with $q \geq 2$, for appropriate functions f, g – cf. [RS, Theorem XI.20].

3. Complete the proof of *Theorem* 6.1.
Hint: Solve the equation (6.7) using the implicit-function theorem. Show that the solution is valid for any V satisfying the hypotheses up to $\mathcal{O}(\lambda^3)$ modifying the argument of [Si76], see also [DE95].

4. There are real-analytic functions f, g which factorize the Macdonald function, $K_0(\zeta) = f(\zeta)\ln\zeta + g(\zeta)$ for any $\zeta \in (0, \infty)$, such that
(a) $f(\zeta) = -1 + \mathcal{O}(\zeta^2)$ and $g(\zeta) = (\ln 2 - \gamma_E) + \mathcal{O}(\zeta^2)$ as $\zeta \to 0+$,
(b) $\max\{f(\zeta), g(\zeta)\} \leq C\mathrm{e}^{-\zeta}$ for some $C > 0$ and all $\zeta > 0$.
Hint: Use an interpolation, e.g., $f(\zeta) = -\mathrm{e}^{-\zeta^2} I_0(\zeta) - (1 - \mathrm{e}^{-\zeta^2})\, K_0(\zeta)$.

5. Prove *Lemma* 6.1.1.
Hint: To check (a), compute the HS-norm of A_z using the assumption on $|V|_{11}$, employ Hölder and Young inequalities. In the A_z part of (b) compute the HS-norm again using

the previous problem; for the B_z part employ the resolvent identity. The derivative of A_z in (c) can be expressed by Cauchy's formula, that of B_z is computed directly. See [EKr01a] for details.

6. Using Fourier transformation and an approximation argument, check that the second-order term in (6.8) is negative when the first one vanishes.

7. Prove *Theorem* 6.3*b*.

8. Check that the identity $\sum_{n=2}^{\infty}(\chi_1, u\chi_n)^2(\kappa_n^2-\kappa_1^2) = \|u\chi_1'\|^2 - \kappa_1^2\|u\chi_1\|^2 = 1$ is valid for the elements of the orthonormal basis (1.10).

9. Let Σ_β be the surfaces (6.15) satisfying the assumptions *(i)–(iii)* of Sect. 6.2.2. Show that $\lim_{\beta\to 0}\beta^{-\mu}\mathcal{K}(\Sigma_\beta) = 0$ holds for any $\mu < 4$.

10. Prove the relation (6.16).
Hint: At the intermediate stage in the proof of *Theorem* 6.4 apply the Fourier transformation directly to $(\Delta m_0, G_k * \Delta m_0)$ and expand the result in powers of β, compute $\|u\chi_1\|^2$ and notice that $\|\Delta m_0\| < \infty$ by *(iii)*.

11. Show that the formulæ (6.18) are valid for strips and layers with

$$\begin{aligned}
A_1^* &:= 2f\partial_y\,, & B_1 &:= g\partial_y\,,\\
A_2^* &:= \Delta f\,, & B_2 &:= g\left(y\partial_y+\frac{1}{2}\right),\\
A_3^* &:= \left(2y\partial_y+1\right)g\,, & B_3 &:= (\nabla f)\cdot\nabla\,,\\
A_4^* &:= -\frac{3f^2+|\nabla f|^2y^2+2\lambda f^3}{(1+\lambda f)^2}\,\partial_y\,, & B_4 &:= g\partial_y\,,\\
A_5^* &:= -\frac{f\Delta f}{1+\lambda f}\,, & B_5 &:= g\left(y\partial_y+\frac{1}{2}\right),\\
A_6^* &:= -\frac{3|\nabla f|^2}{(1+\lambda f)^2}\,, & B_6 &:= g\left(y\partial_y+\frac{1}{4}\right),\\
A_7^* &:= -\frac{2y\partial_y+1}{1+\lambda f}\,f\,, & B_7 &:= (\nabla f)\cdot\nabla\,,
\end{aligned}$$

where $g \in C_0^\infty$ is any function such that $g(x) = 1$ holds on supp f; in the case of a strip ∇, ∇f, Δf mean simply $\frac{\mathrm{d}}{\mathrm{d}x}$, f', and f'', respectively.

12. Consider the operators M_z introduced in the proof of *Theorem* 6.5. Show that the map $\zeta \mapsto M_{\kappa_1^2-\zeta}$ is a bounded operator-valued function for $\mathrm{Re}\,\zeta < \kappa_1^2$ which can be analytically continued to a region containing the point $\zeta = 0$ (in fact, a circle of radius smaller than $\sqrt{3}\kappa_1$).
Hint: In the part containing $R_0^\perp(z)$ use the fact that the operators $C_\lambda P_1^\perp R_0^\perp(z)^{-1/2}$ and $DP_1^\perp R_0^\perp(z)^{-1/2}$ are bounded, and write the rest as $DhN_{\kappa_1^2-\zeta}hC^*$, where $h \in C_0^\infty(\mathbb{R})$ equals one on supp f. It is an operator on $\mathcal{H}_1$ the kernel of which is found using integration by parts, then one can estimate it and its derivative w.r.t. ζ in the Hilbert-Schmidt norm.

13. Discuss the extension of the result in *Theorem* 6.5 to the following situations:
(a) less regularity, e.g., f piecewise C^2 with derivatives bounded to the second order, and vanishing at large distances, $\lim_{|x|\to\infty} f(x) = 0$,
(b) two-sided deformation, $a - \lambda f_-(x) < y < a + \lambda f_+(x)$, with a pair of function $f_\pm$

having suitable regularity and decay properties,
(c) more generally, a weak local deformation of $\Omega_0 := \mathbb{R} \times M$ with $M \subset \mathbb{R}^{d-1}$ satisfying the assumptions of Sect. 1.4.
Consider the same problem for layers with a weak local deformation.

14. Let $K : L^2(M) \to L^2(M)$, where $M \subset \mathbb{R}^d$ is an open set, be an integral operator with the kernel $K(\cdot,\cdot)$. Prove the *Schur-Holmgren bound*,

$$\|K\| \le \|K\|_{\mathrm{SH}} := \left(\sup_{x\in M} \int_M |K(x,x')|\mathrm{d}x' \sup_{x'\in M} \int_M |K(x,x')|\mathrm{d}x \right)^{1/2}.$$

Check that $\|\cdot\|_{\mathrm{SH}}$ is not a norm, and express it in case of a symmetric kernel.
Hint: More generally, $\|K\|_{p,p} \le \|K\|_{1,1}^{1/p} \|K\|_{\infty,\infty}^{1/q}$, where K is an operator on $L^p(M)$, $p^{-1}+q^{-1} = 1$, and $\|K\|_{\infty,\infty}$, $\|K\|_{1,1}$ are the two suprema appearing in the bound. If they are finite, the result follows by interpolation, see [BEGK01] and [Mad, Theorem 7.1.9] for the better known discrete case.

15. Let t be a densely defined form on a Hilbert space $\mathcal{H}$ which is closed and bounded from below, and denote by H the self-adjoint operator associated with t. Then for $z \in \mathbb{C}$ and a map $R : \mathcal{H} \to \mathrm{Dom}\,(t)$ the following claims are equivalent:
(a) $z \in \rho(H)$ and $(H-z)^{-1} = R$,
(b) $t(R\phi,\psi) = (zR\phi + \phi, \psi)$ for all $\phi \in \mathcal{H}$ and $\psi \in \mathrm{Dom}\,(t)$.

16. Complete the proof of *Proposition* 6.3.1.
Hint: To prove $\psi_m^k = R_{m,\mathrm{d}x}^k \psi$ for a general $\psi \in L^2(m)$, approximate it by a non-decreasing sequence from $\psi \in L^1(m) \cap L^2(m)$. Use the previous problem together with the first resolvent identity to extend the validity of the result to any $k^2 \in \mathbb{C}_\Omega^+$ – cf. [BEKŠ94].

17. There is a $\kappa_0 > 0$ such that $\|\alpha I_m R_{m,\mathrm{d}x}^{i\kappa}\| < 1$ holds for any $\kappa \ge \kappa_0$.
Hint: Rewrite (6.19) as $\int_\Omega |I_m\phi(x)|^2(1+\alpha(x)^2)\,\mathrm{d}m(x) \le a\,\langle\phi,\phi\rangle_{i\kappa_0}$ with $\kappa_0 := \sqrt{a/b}$, estimate the *r.h.s.* by means of *Proposition* 6.3.1 and the Schwarz inequality.

18. $\dim \mathrm{Ker}\,(H_{\alpha m} - k^2) = \dim \mathrm{Ker}\,(I + \alpha I_m R_{m,\mathrm{d}x}^k)$ holds for any $k \in \mathbb{C}_\Omega^+$.
Hint: Use *Proposition* 6.3.1—cf. [BEKŠ94].

19. Prove *Theorem* 6.8 and its modification from Remark 6.3.1.
Hint: Cf. [EKr01b].

20. (a) Prove the relations (6.25) and (6.26).
(b) Find an example of the shape function f for which *Theorem* 6.9 says nothing about the existence of a weakly bound state.
Hint: Write the sum in terms of di- and tri-gamma functions [PBM, Sect. 5.1]. As for (b), notice that the left-hand side of (6.23) can be made as small as $(\pi/2b)^2$ leaving for d/b open approximately the gap $(1.697, 4.619)$.

21. Let $\phi \in C^2[0,d]$ be real-valued with $\phi(d) = 0$. Then
(a) there are $\varepsilon_1, \varepsilon_1 > 0$ such that $\left|\int_0^d \phi(t)\chi_1(t)\,Dt\right| < \varepsilon_1\|\phi\|$ implies

$$\int_{-d}^0 \phi'(t)^2\mathrm{d}t > (1+\varepsilon_2)\varepsilon_d\|\phi\|^2,$$

(b) if $(\phi, \chi_1) = 0$ and $\phi(0) = \beta$, then to each $m > 0$ there is a $c_0 > 0$ such that

$$\int_0^d \phi'(t)^2 \mathrm{d}t + \left(\frac{m}{a}\right)^2 \int_0^a \phi(t)^2 \mathrm{d}t - \varepsilon_d \int_0^d \phi(t)^2 \mathrm{d}t \geq \frac{c_0 \beta^2}{a}$$

holds for all a small enough.
Hint: Consider an even extension of ϕ to $[-d, d]$ and estimate the projection onto the lowest Dirichlet eigenfunction in terms of ε_1. Part (b) is obtained by solving Euler's equation on $[0, a]$ if $\|\phi\|^2_{L^2(0,a)} > a^2 \phi(a)^2 + \|\phi\|^2_{L^2(a,d)}$ while in the opposite case part (a) has to be applied – cf. [EV96].

Chapter 7
External Fields and Magnetic Transport

Next we are going to discuss how the behavior of guided quantum particles is influenced by electric or magnetic fields. Two questions are of a particular interest. First we shall discuss the influence of such external fields on the curvature-induced discrete spectrum in waveguides. Among other things, we will derive conditions under which an electric or magnetic field prevents the existence of bound states below the threshold of the essential spectrum. The second main topic addressed here concerns transport properties of a two-dimensional electron gas subject to a perpendicular magnetic field; we shall analyze the link between the properties of such a field and the existence of states carrying electric current.

7.1 External Fields

For simplicity we shall discuss the influence of electric and magnetic fields separately focusing on a few particular situations. The guided particles are supposed to be two-dimensional and charged; we neglect possible finer properties such as their magnetic moment. As the charge value plays no significant role in the following considerations and a change of its sign is equivalent to a switch in potential orientation we suppose everywhere that $|e| = 1$.

7.1.1 Homogeneous Electric Fields

We restrict our attention to the situation when the electric field is homogeneous corresponding to a potential linear in a coordinate in the plane and the particle is confined to a curved strip $\Omega \subset \mathbb{R}^2$. It is evident the behavior of the system depends on the interplay between the field and the geometry of Ω and the influence of the field cannot be neglected anywhere. We shall discuss a particular case where Ω is curved only locally and the effect outside the bent region is limited to a shift in the

P. Exner and H. Kovařík, *Quantum Waveguides*,
Theoretical and Mathematical Physics, DOI 10.1007/978-3-319-18576-7_7

essential spectrum threshold as the field is perpendicular to the axis of Ω outside a compact set.

We thus choose Cartesian coordinates in the plane in such a way that the straight parts of Ω are parallel to the x-axis and the Hamiltonian has the form

$$-\Delta_D^\Omega + Fy \quad \text{in } L^2(\Omega)$$

with the domain $H^2(\Omega) \cap H_0^1(\Omega)$, where $F > 0$ is the strength of the electric field. As usual we introduce the curvilinear coordinates (s, u), but in contrast to other considerations it is convenient to let the coordinate u run through the interval $(0, d)$. Without loss of generality we may suppose that $\gamma(s) = 0$ for $s < 0$; applying then the straightening transformation of Sect. 1.1 we find that the above Hamiltonian is unitarily equivalent to the operator

$$H_\Omega(F) = -\partial_s\, g^{-1}(s,u)\, \partial_s - \partial_u^2 + V_F(s,u) \quad \text{in } L^2(\mathbb{R} \times (0,d))$$

with $g(s,u) = (1 + u\gamma(s))^2$ and

$$V_F(s,u) = V_0(s,u) - F \int_0^s \sin\beta(s_1)\, \mathrm{d}s_1 + Fu\cos\beta(s)\,, \tag{7.1}$$

where $V_0(s,u)$ is given by the effective-potential formula (1.8) and $\beta(s) = \beta(s,0)$.

Suppose now that the bend causes a one-sided tilt of the guide. In such a case there is a potential difference between the two straight parts of Ω which is proportional to F, and it is thus not surprising that for a sufficiently large field strength the discrete spectrum of $H_\Omega(F)$ can be empty.

Theorem 7.1 *Adopt the assumptions (i) and (ii)$_2$ of Sect. 1.1 together with $d\|\gamma\|_\infty < 1$. Moreover, suppose that γ is non-vanishing with* $\operatorname{supp}\gamma \in [0, s_0]$ *for some $s_0 > 0$ and zero mean, $\int_\mathbb{R} \gamma(s)\,\mathrm{d}s = 0$, giving rise to a one-sided tilt, $\beta(s) = \int_0^s \gamma(s')\,\mathrm{d}s' \in [-\pi, 0]$ for all $s \in [0, s_0]$. Then there is an $F_0 > 0$ such that $\sigma_{\mathrm{disc}}(H_\Omega(F)) = \emptyset$ holds for all $F > F_0$.*

Proof We start with the essential spectrum. In view of (7.1) and the assumptions imposed on γ the variables separate outside the central part of Ω and we have

$$\lambda_1(F) := \inf \sigma_{\mathrm{ess}}(H_\Omega(F)) = \inf \sigma(h(F))\,,$$

where $h(F) := -\partial_u^2 + Fu$ on $L^2(0,d)$ with the Dirichlet condition at the endpoints of the interval. To find it we note that the Airy functions

$$w_\lambda(y) := \mathrm{Ai}\Big(F^{1/3}\Big(y - \frac{\lambda}{F}\Big)\Big)\,, \quad v_\lambda(y) := \mathrm{Bi}\Big(F^{1/3}\Big(y - \frac{\lambda}{F}\Big)\Big)$$

are fundamental solutions of the equation $(-\partial_y^2 + Fy)\psi = \lambda\psi$ and $\lambda_1(F)$ coincides with the first root of the implicit equation $w_\lambda(0)v_\lambda(d) - w_\lambda(d)v_\lambda(0) = 0$. The asymptotic behavior of the Airy functions for large arguments then implies

$$\lambda_1(F) = F^{2/3}\Big[c_1 + c_2\,\mathrm{e}^{-(4/3)d^{3/2}\sqrt{F}}\big(1 + \mathcal{O}(F^{-1/2})\big)\Big] \quad \text{as} \quad F \to \infty,$$

where $-c_1 \approx -2.33$ is the first zero of Ai and c_2 is a positive constant.

To study the discrete spectrum, it is convenient to work with the original operator $-\Delta_D^\Omega + Fy$ in $L^2(\Omega)$; we want to estimate it from below by an operator $\tilde H_\Omega(F)$ such that $\inf\sigma(\tilde H_\Omega(F)) \ge \lambda_1(F)$. First we note that by assumption there is an $s_1 \in (0, s_0)$ such that $-\frac{1}{2}\pi < \beta(s) \le 0$ holds in $(0, s_1)$, in other words, the lower boundary of Ω is the graph of a function smoothly increasing from zero at $x = 0$ to $b := y(s_1, 0)$ at $a := x(s_1, 0)$. Using the domain monotonicity of $-\Delta_D^\Omega$ together with the monotonicity of the potential with respect to y we can estimate the operator in question from below by $H_{\Omega_1}(F) := -\Delta_D^{\Omega_1} + Fy$ where Ω_1 is the protruded strip with the indicated part of the boundary replaced by the segments $[0, a]$ on the x-axis followed by $[0, b]$ perpendicular to it.

In the next step we use a bracketing argument estimating $H_{\Omega_1}(F)$ from below by the operator $\tilde H_\Omega(F) := H_1(F) \oplus H_2(F) \oplus H_3(F)$, the three parts of which correspond to a dissection of the protruded strip Ω_1 by additional Neumann conditions imposed at the segment $(0, d)$ of the y-axis and the segment connecting the points $(0, d)$ and (a, d). The spectrum of $H_1(F)$ corresponding to the halfstrip is $[\lambda_1(F), \infty)$ and $H_3(F) \ge Fb > \lambda_1(F)$ holds for F large enough. The remaining part corresponds to a rectangle and one can find its purely discrete spectrum by separation of variables, in particular, the lowest eigenvalue is found using the first root of the equation $w_\lambda(0)v'_\lambda(\beta) - w'_\lambda(\beta)v_\lambda(0) = 0$ in combination with the asymptotic relation

$$\mu_1(F) = \frac{\pi^2}{4a^2} + F^{2/3}\Big[c_1 - c_2\,\mathrm{e}^{-(4/3)b^{3/2}\sqrt{F}}\big(1 + \mathcal{O}(F^{-1/2})\big)\Big]$$

holds as $F \to \infty$. From the two asymptotic formulæ we see that $\mu_1(F) > \lambda_1(F)$ holds for all F large enough, which concludes the argument. ∎

7.1.2 Local Magnetic Fields

Let us pass next to two-dimensional quantum waveguides exposed to a magnetic field perpendicular to the guide plane. First we will be concerned with the effect of a local, compactly supported field; we are going to demonstrate that it gives rise to an effective repulsive interaction in many respects similar to that coming from the twisting of three-dimensional tubes analyzed in Sect. 1.7. A mathematical expression of this claim is a ***Hardy-type inequality*** induced by the magnetic field which we shall now derive.

Without loss of generality we may put $d = \pi$, the general case then follows by scaling. Let $\Omega_0 = \mathbb{R} \times (0, \pi)$ and let $B : \Omega_0 \to \mathbb{R}$ be a bounded magnetic field—speaking of such a field here we always have in mind its intensity, the direction being fixed as mentioned above. As before we shall describe points of Ω_0 by their Cartesian coordinates $\vec{x} = (x, y)$. Choose a point $p \in \Omega_0$ such that there is a ball $B_R(p) \subset \Omega_0$ of radius R, centered in p and such that the flux

$$\Phi(r) := \frac{1}{2\pi} \int_{B_r(p)} B(\vec{x})\, \mathrm{d}x\, \mathrm{d}y$$

through $B_r(p)$ is not identically zero for any $r \in (0, R)$; for simplicity we may suppose that $p = (0, y_0)$ for some $y_0 \in (0, \pi)$. We associate with B a magnetic vector potential $A(\vec{x}) = (a_1(\vec{x}), a_2(\vec{x}))$ defined on $\mathbb{R}^2$, for instance, by

$$a_1(\vec{x}) = -(y - y_0) \int_0^1 B(tx, t(y - y_0) + y_0)\, \mathrm{d}t\,,$$

$$a_2(\vec{x}) = x \int_0^1 B(tx, t(y - y_0) + y_0)\, \mathrm{d}t\,;$$

it is obvious that $\partial_x a_2(\vec{x}) - \partial_y a_1(\vec{x}) = B(\vec{x})$ and that the transverse gauge condition $A(\vec{x}) \cdot (x, y - y_0) = 0$ holds. From the ***diamagnetic inequality***,

$$|\nabla |v|(\vec{x})| \le |(i\nabla - A)v(\vec{x})| \quad \text{for almost all } \vec{x}\,, \tag{7.2}$$

we easily conclude that the estimate

$$\int_{\Omega_0} |(i\nabla - A)u(\vec{x})|^2\, \mathrm{d}x\, \mathrm{d}y \ge \int_{\Omega_0} |u(\vec{x})|^2\, \mathrm{d}x\, \mathrm{d}y$$

holds for all $u \in H_0^1(\Omega_0)$. One can make, however, a stronger claim.

Theorem 7.2 *Let $B \in C_0^1(\mathbb{R}^2)$ be a real-valued magnetic field which does not vanish in Ω, then the inequality*

$$\int_{\Omega_0} \Big(|(i\nabla - A)u(\vec{x})|^2 - |u(\vec{x})|^2\Big)\, \mathrm{d}x\, \mathrm{d}y \ge c_B \int_{\Omega_0} \frac{|u(\vec{x})|^2}{1 + x^2}\, \mathrm{d}x\, \mathrm{d}y \tag{7.3}$$

holds for all $u \in H_0^1(\Omega_0)$, where A is a magnetic vector potential associated with B and c_B is a positive constant.

Proof For convenience we denote by $c, c', \ldots$ generic constants which may vary from line to line, but do not depend on the test function u. It is straightforward to check that the inequality (7.3) is gauge-invariant, hence we can without loss of generality assume that the components of A are given by the above formulæ. Using polar coordinates (r, θ) centered at the point p, we will check that

$$c\int_{B_R(p)} |u|^2 r\,\mathrm{d}r\,\mathrm{d}\theta \le \int_{B_R(p)} \left(|\partial_r u|^2 + \frac{1}{r^2}\,|i\partial_\theta u - ra(r,\theta)u|^2\right) r\,\mathrm{d}r\,\mathrm{d}\theta$$

holds for all $u \in H_0^1(\Omega_0)$, where $a(r,\theta) := A\cdot(-\sin\theta, \cos\theta)$ is the (anticlockwise) tangent component of the vector potential. For a fixed r we consider the operator $K_r = i\partial_\theta - ra$ in $L^2(0,2\pi)$ which is self-adjoint on the domain consisting of functions from $H^1(0,2\pi)$ with periodic boundary conditions. The spectrum of K_r is discrete consisting of the eigenvalues $\{\lambda_k\}_{k=-\infty}^{\infty}$ given by

$$\lambda_k = \lambda_k(r) = k + \frac{r}{2\pi}\int_0^{2\pi} a(r,\theta)\,\mathrm{d}\theta = k + \Phi(r)$$

which correspond to orthonormal eigenfunctions

$$\varphi_k(r,\theta) = \frac{1}{\sqrt{2\pi}}\,\mathrm{e}^{-i\lambda_k\theta + ir\int_0^\theta a(r,s)\,\mathrm{d}s}\,.$$

This implies, in particular, that the quadratic form associated with K_r^2 satisfies for all $u(r,\cdot) \in H^1(0,2\pi)$ with periodic boundary conditions the inequality

$$\mu(r)^2\int_0^{2\pi} |u|^2\,\mathrm{d}\theta \le \int_0^{2\pi} |i\partial_\theta u - rau|^2\,\mathrm{d}\theta\,,$$

where $\mu(r) := \mathrm{dist}(\Phi(r),\mathbb{Z})$. Integrating over the radial variable we get

$$\int_{B_R(p)} \frac{\mu^2}{r^2}\,|u|^2\,r\,\mathrm{d}r\,\mathrm{d}\theta \le \int_{B_R(p)} \frac{1}{r^2}\,|i\partial_\theta u - rau|^2\,r\,\mathrm{d}r\,\mathrm{d}\theta$$

for all $u \in H^1(\Omega_0)$. Define next the function $\chi : [0,R] \to [0,1]$ by

$$\chi(r) := \frac{\mu_0^2\,\mu(r)^2}{r^2} \quad\text{with}\quad \mu_0 := \left(\max_{r\in[0,R]} \frac{\mu(r)}{r}\right)^{-1}.$$

Since Φ is by our assumption about B piecewise continuously differentiable and $\Phi(0) = 0$ we see that χ is well defined; by definition there is at least one $r_0 \in (0,R]$ where it reaches its maximum, $\chi(r_0) = 1$. Suppose that $v \in H^1(0,R)$ satisfies $v(r_0) = 0$, then we have the inequalities

$$\int_{r_0}^{R} |v(r)|^2\,r\,\mathrm{d}r \le \frac{2R^3 - 3R^2r_0 + r_0^3}{6r_0}\int_{r_0}^{R} |v'(r)|^2\,r\,\mathrm{d}r$$

and

$$\int_0^{r_0} |v(r)|^2\,r\,\mathrm{d}r \le \frac{r_0^2}{j_{0,1}^2}\int_0^{r_0} |v'(r)|^2\,r\,\mathrm{d}r\,,$$

where $j_{0,1} \approx 2.40$ is the first zero of Bessel function J_0. The latter comes from the lowest eigenvalue of $-\Delta_D$ in a circle of radius r_0, the former follows by

$$|v(r)|^2 = \left| \int_{r_0}^{r} v'(t)\,\mathrm{d}t \right|^2 \leq (r - r_0) \int_{r_0}^{r} |v'(r)|^2\,\mathrm{d}r \leq \frac{r - r_0}{r_0} \int_{r_0}^{R} |v'(r)|^2\, r\,\mathrm{d}r\,.$$

Using the the above estimates we infer that

$$\begin{aligned}
\int_{B_R(p)} |u|^2\, r\,\mathrm{d}r\,\mathrm{d}\theta &\leq 2 \int_{B_R(p)} \left(|\chi u|^2 + |(1-\chi)u|^2 \right) r\,\mathrm{d}r\,\mathrm{d}\theta \\
&\leq 2\mu_0^2 \int_{B_R(p)} \frac{1}{r^2}\, |i\partial_\theta u - rau|^2\, r\,\mathrm{d}r\,\mathrm{d}\theta + 2 \int_0^{2\pi} \Big(\frac{r_0^2}{j_{0,1}^2} \int_0^{r_0} |((1-\chi)u)'|^2\, r\,\mathrm{d}r \\
&\quad + \frac{2R^3 - 3R^2 r_0 + r_0^3}{6r_0} \int_{r_0}^{R} |((1-\chi)u)'|^2\, r\,\mathrm{d}r \Big) \mathrm{d}\theta \qquad (7.4) \\
&\leq 2\mu_0^2 \int_{B_R(p)} \frac{1}{r^2}\, |i\partial_\theta u - rau|^2\, r\,\mathrm{d}r\,\mathrm{d}\theta + c \int_{B_R(p)} \left(|\chi' u|^2 + |\partial_r u|^2 \right) r\,\mathrm{d}r\,\mathrm{d}\theta \\
&\leq c' \int_{B_R(p)} \left(|\partial_r u|^2 + \frac{1}{r^2}\, |i\partial_\theta u - rau|^2 \right) r\,\mathrm{d}r\,\mathrm{d}\theta\,.
\end{aligned}$$

The operator $-\frac{\mathrm{d}^2}{\mathrm{d}y^2} - 1$ on the domain $H^2(0,\pi) \cap H_0^1(0,\pi)$ with the additional Dirichlet condition imposed at $y = y_0$ has the spectrum bounded from below by $\pi^2 \min\left\{ y_0^{-2}, (\pi - y_0)^{-2} \right\} - 1$, which in terms of quadratic forms means that for $v \in H^1(0,\pi)$ satisfying $v(y_0) = 0$ we have

$$\left(\pi^2 \min\left\{ y_0^{-2}, (\pi - y_0)^{-2} \right\} - 1\right) \int_0^{\pi} |v(y)|^2 \sin^2 y\,\mathrm{d}y \leq \int_0^{\pi} |v'(y)|^2 \sin^2 y\,\mathrm{d}y\,.$$

Let $u \in H^1(\Omega_0)$ and define $\psi : \Omega_0 \to [0,1]$ by

$$\psi(\vec{x}) = \begin{cases} \frac{|y - y_0|}{\sqrt{R^2 - x^2}} & \text{if } |x| < R \text{ and } h_-(x) < y < h_+(x) \\ 1 & \text{otherwise} \end{cases}$$

where $h_\pm(x) := y_0 \pm \sqrt{R^2 - x^2}$; then we write $u = (1-\psi)u + \psi u$ and use the above estimate to obtain for a fixed $x \in (-R, R)$ the relation

$$\begin{aligned}
\int_0^{\pi} |u(\vec{x})|^2 \sin^2 y\,\mathrm{d}y &\leq 2 \int_{h_-(x)}^{h_+(x)} |(1-\psi)u|^2(\vec{x}) \sin^2 y\,\mathrm{d}y \\
&\quad + c \left(\int_0^{\pi} |\psi\, \partial_y u|^2(\vec{x}) \sin^2 y\,\mathrm{d}y + \int_{h_-(x)}^{h_+(x)} \frac{|u(\vec{x})|^2}{R^2 - x^2} \sin^2 y\,\mathrm{d}y \right);
\end{aligned}$$

combining it with (7.4) and defining $\Omega_R := (-R, R) \times (0, \pi)$ we get

$$\int_{\Omega_R} (R^2 - x^2)|u|^2 \sin^2 y \, dy \, dx \le c \int_{B_R(p)} |(i\nabla - A)u|^2 \sin^2 y \, dx \, dy + c \int_{\Omega_R} |\partial_y u|^2 \sin^2 y \, dy \, dx$$

for any $u \in H^1(\Omega_0)$. The diamagnetic inequality (7.2) then implies

$$\int_{\Omega_R} (R^2 - x^2)|u|^2 \sin^2 y \, dx \, dy \le c \int_{\Omega_R} |(i\nabla - A)u|^2 \sin^2 y \, dx \, dy$$

for all $u \in C^\infty(\overline{\Omega}_0)$. Recall next the classical one-dimensional Hardy inequality

$$\int_{-\infty}^{\infty} \frac{|v(t)|^2}{t^2} \, dt \le 4 \int_{-\infty}^{\infty} |v'(t)|^2 \, dt \tag{7.5}$$

that holds for any $v \in H^1(\mathbb{R})$ satisfying $v(0) = 0$. Take finally $m = \frac{R}{\sqrt{2}}$ and define the map $\phi : \mathbb{R} \to [0, 1]$ by

$$\phi(x) := \begin{cases} 1 & \text{if } |x| > m \\ \frac{|x|}{m} & \text{if } |x| < m \end{cases}$$

Decomposing $u \in C^\infty(\overline{\Omega}_0) \cap L^2(\Omega)$ in a similar way as above, $u = u\phi + u(1 - \phi)$, and using the last two inequalities in combination with the diamagnetic one, we arrive at

$$\begin{aligned} \int_{\Omega_0} \frac{|u|^2 \sin^2 y}{1 + x^2} \, dx \, dy &\le 2 \int_{\Omega_0} \frac{|u\phi|^2 + |u(1 - \phi)|^2}{1 + x^2} \sin^2 y \, dx \, dy \\ &\le 16 \int_{\Omega_0} \left(|\phi \, \partial_x u|^2 + |u\phi'|^2\right) \sin^2 y \, dx \, dy + 2 \int_{\Omega_m} \frac{|u|^2 \sin^2 y}{1 + x^2} \, dx \, dy \\ &\le 16 \int_{\Omega_0} |\partial_x u|^2 \sin^2 y \, dx \, dy + c \int_{\Omega_R} (R^2 - x^2)|u|^2 \sin^2 y \, dx \, dy \\ &\le c \int_{\Omega_0} |(i\nabla - A)u|^2 \sin^2 y \, dx \, dy \, ; \end{aligned}$$

it is now sufficient to substitute $v(\vec{x}) = u(\vec{x}) \sin y$ to obtain (7.3). ∎

The proved theorem has consequences for the existence of geometrically induced bound states. Consider, for instance, a bent strip Ω the axis of which has signed curvature γ and which is exposed to such a magnetic field. The corresponding magnetic Hamiltonian H_B^Ω is the unique self-adjoint operator associated with the closed quadratic form

$$Q[\varphi] = \int_\Omega |(i\nabla - A)\,\varphi(\vec{x})|^2\,\mathrm{d}x\,\mathrm{d}y \tag{7.6}$$

defined on $H_0^1(\Omega)$. It is not difficult to locate the essential spectrum of H_B^Ω.

Proposition 7.1.1 *Adopt the assumptions (i) and (ii)$_2$ of Sect. 1.1 and suppose, in addition, that $\|\gamma\|_\infty < 2/\pi$ and that γ has a compact support. If the magnetic field satisfies the hypotheses of Theorem 7.2, then $\sigma_{\mathrm{ess}}(H_B^\Omega) = [1, \infty)$.*

The proof is left to the reader (Problem 1). The inequality (7.3) can be used to demonstrate that the discrete spectrum of H_B^Ω is empty provided both γ and $\dot{\gamma}$ are sufficiently small.

Theorem 7.3 *Let B be as in Theorem 7.2. If γ satisfies the assumptions of Proposition 7.1.1 and $\|\gamma\|_\infty + \|\dot{\gamma}\|_\infty$ is small enough, then $\sigma_{\mathrm{disc}}(H_B^\Omega) = \emptyset$.*

Proof We apply again the straightening transformation $U : L^2(\Omega) \to L^2(\Omega_0)$ introduced in Sect. 1.1. Adopting the notation used there we define

$$\tilde{A}(s,u) := (\tilde{a}_1(s,u), \tilde{a}_2(s,u)) = A(\xi(s) - u\dot{\eta}(s), \eta(s) + u\dot{\xi}(s))\,;$$

the operator H_B^Ω is then unitarily equivalent to $\tilde{H}_B^\Omega = U H_B^\Omega U^{-1}$ in $L^2(\Omega_0)$ generated by the quadratic form $q_\gamma[\psi] := Q[U^{-1}\psi]$ defined on $H_0^1(\Omega_0)$. Using the hypotheses about γ, we find after a straightforward calculation that

$$q_\gamma[\psi] - \|\psi\|^2 \ge q_0[\psi] - \|\psi\|^2 - c(\|\gamma\|_\infty + \|\dot{\gamma}\|_\infty) \int_{\Omega_0} \chi(|\nabla\psi|^2 + |\psi|^2)\,\mathrm{d}s\,du\,,$$

where χ is the characteristic function of $\operatorname{supp}\gamma$, c is a positive constant, and

$$q_0[\psi] := \int_{\Omega_0} |i\partial_s\psi - (\dot{\xi}\tilde{a}_1 + \dot{\eta}\tilde{a}_2)\psi|^2 + |i\partial_u\psi - (\dot{\xi}\tilde{a}_2 - \dot{\eta}\tilde{a}_1)\psi|^2\,\mathrm{d}s\,\mathrm{d}u\,.$$

Since the magnetic field generated by the potential $(\dot{\xi}\tilde{a}_1 + \dot{\eta}\tilde{a}_2,\ \dot{\xi}\tilde{a}_2 - \dot{\eta}\tilde{a}_1)$ satisfies the assumptions of *Theorem* 7.2, we can apply the Hardy-type inequality (7.3). The latter in combination with the pointwise estimate

$$|\nabla\psi|^2 \le 2\big(|i\nabla\psi - \tilde{A}\,\psi|^2 + |\tilde{A}|^2|\psi|^2\big)$$

and the above derived lower bound on $q_\gamma[\psi] - \|\psi\|^2$ gives

$$\begin{aligned} q_\gamma[\psi] - \|\psi\|^2 \ge{}& \Big(\frac{1}{2} - c_1\,M_\gamma\Big)(q_0[\psi] - \|\psi\|^2) \\ &+ \Big(\frac{1}{2}c_B - c_2\,(1 + |\operatorname{supp}\gamma|^2)\,M_\gamma\Big) \int_{\Omega_0} \frac{|\psi(s,u)|^2}{1+s^2}\,\mathrm{d}s\,du \end{aligned}$$

for any $\psi \in H_0^1(\Omega)$ (Problem 2), where $M_\gamma := \|\gamma\|_\infty + \|\dot{\gamma}\|_\infty$ and c_1, c_2 are constants independent of γ. Using the diamagnetic inequality (7.2) once again we see that the right-hand side of the last estimate is positive for M_γ small enough, which concludes the proof. ∎

Remarks 7.1.1 (a) While both *Theorem* 7.1 and the above result provide conditions under which an external field destroys curvature-induced bound states, they refer to rather different situations. The perturbation in the former case is global and, in the latter, local with with appropriate consequences for the essential spectrum. In addition, the claim of *Theorem* 7.1 depends on a particular type of bending which need not be weak. On the other hand, a local magnetic field has the effect only if the curvature is sufficiently small.
(b) A similar behavior can be observed for other geometrically-induced bound states; the example of a locally bulged strip is proposed as Problem 3. The reason behind this effect is that, in contrast to the non-magnetic case, the corresponding Hamiltonians are in view of *Theorem* 7.2 and perturbative arguments ***subcritical***, in other words, they have no threshold resonances.
(c) The effect of local magnetic fields is robust, present even if the hypotheses of *Theorem* 7.2 are not valid. A case of particular interest concerns ***Aharonov-Bohm fields***, i.e. a singular magnetic field vanishing everywhere except at a flux line, for which a modified version of inequality (7.3) holds (Problem 4).

7.1.3 Nöckel's Model Revisited

In the rest of this chapter we shall deal with a homogeneous magnetic field of intensity B, again perpendicular to the plane to which the particles are confined. First we look how such a field can influence spectral properties using for this purpose the model of an open quantum dot introduced in Sect. 2.3.

The vector potential corresponding to the field can be chosen, of course, in different ways; for the present purpose the best option is to use the ***Landau gauge***, $A(\vec{x}) = (-By, 0)$. The split of the potential in (2.10) into an x-dependent part and a perturbation controlled by the coupling constant is not essential here, hence we use a single local potential writing the Hamiltonian as

$$H_U(B) = H_0(B) + U\,, \quad H_0(B) = (-i\partial_x - By)^2 - \partial_y^2$$

with Dirichlet boundary conditions at $|y| = a$; the operator acts on $L^2(\Omega_0)$, where $\Omega_0 = \mathbb{R} \times (-a, a)$ as before, and its domain coincides with that of $-\Delta_D^{\Omega_0}$. In the absence of the magnetic field the potential U gives rise to a nonempty discrete spectrum as long as $\int_{\Omega_0} U(\vec{x})\,\chi_1(y)^2\,dy\,dx < 0$, in particular, if U is purely attractive. Our first aim is to find out what happens with the discrete eigenvalues of $H_U(0)$ when the magnetic field is switched on.

The free operator $H_0(B)$ exhibits a translational invariance in the longitudinal variable x. This makes it possible to simplify its spectral analysis since by a partial Fourier transformation, or more exactly, by means of $F \otimes I$, where F is the one-dimensional Fourier-Plancherel operator, one finds that $H_0(B)$ is unitarily equivalent to a direct integral,

$$H_0(B) \simeq \int_{\mathbb{R}}^{\oplus} h_B(p)\,\mathrm{d}p\,, \quad h_B(p) := -\partial_y^2 + (p - yB)^2 \;\text{ in }\; L^2(-a, a)$$

with Dirichlet boundary conditions at $|y| = a$. Each fiber operator $h_B(p)$ has a purely discrete spectrum; we denote their eigenvalues by $\nu_j(p)$ and the corresponding normalized eigenfunctions by $\chi_j^B(p, \cdot)$ so we can write

$$h_B(p) = \sum_{j=1}^{\infty} \nu_j(p)\,\pi_j^B(p)\,, \quad \pi_j^B(p) := (\chi_j^B(p, \cdot), \cdot)_{L^2(-a,a)}\,\chi_j^B(p, \cdot)\,.$$

A closer inspection shows that $\nu_j(p)$ are real even analytic functions of p and that $p\nu'(p) > 0$ holds if $p \neq 0$ (see the notes), hence we get

$$\inf \sigma_{\mathrm{ess}}(H_0(B)) = \min_{p\in\mathbb{R}} \nu_1(p) = \nu_1(0)\,.$$

To derive a sufficient condition under which $H_U(B)$ has at least one eigenvalue below $\nu_1(0)$ we introduce as in Sect. 2.3 the potential projection

$$U_{11}(x) := \int_{-a}^{a} U(x, y)\,|\chi_1^B(0, y)|^2\,\mathrm{d}y\,,$$

now with respect to the magnetic transverse eigenfunctions.

Theorem 7.4 *Assume that* $U \in L^\infty(\Omega_0)$ *and* $\lim_{|x|\to\infty} \|U(x, \cdot)\|_\infty = 0$. *Then the discrete spectrum of* $H_U(B)$ *is nonempty provided* $\int_{\mathbb{R}} U_{11}(x)\,\mathrm{d}x < 0$.

Proof In a standard way one checks that the decaying potential does not alternate the essential spectrum, which means that the relation

$$\inf \sigma_{\mathrm{ess}}(H_U(B)) = \inf \sigma_{\mathrm{ess}}(H_0(B)) = \nu_1(0)$$

holds and it is sufficient to find a suitable trial function $\Phi \in L^2(\Omega_0)$ which would make the quadratic form

$$q[\Phi] := \|(-i\partial_x - By)\Phi\|^2 + \|\partial_y \Phi\|^2 + (\Phi, U\Phi) - \nu_1(0)\|\Phi\|^2$$

negative. We again employ the criticality of the free operator $H_0(B)$ and choose $\Phi(x, y) := \phi(x)\,\chi_1^B(0, y)$, where the function ϕ is to be specified, obtaining

$$q[\phi\chi_1^B] = \|\phi'\|^2_{L^2(\mathbb{R})} - 2B\,\mathrm{Im}\,\left(\phi, \langle y\rangle^B_{11}\,\phi'\right)_{L^2(\mathbb{R})} + (\phi, U_{11}\,\phi)_{L^2(\mathbb{R})}\,, \qquad (7.7)$$

where $\langle y\rangle^B_{11} := \int_{-a}^a y\,|\chi_1^B(0, y)|^2 \mathrm{d}y$. Note that the middle term on the right-hand side of (7.7) vanishes if ϕ is real-valued. We choose a real function $g \in \mathcal{S}(\mathbb{R})$ such that $g(x) = 1$ in $[-d, d]$ for some $d > 0$ and define

$$\phi_\varepsilon(x) := \begin{cases} g(x) & \ldots\ |x| \le d \\ g(\pm d + \varepsilon(x \mp d)) & \ldots\ \pm x > d \end{cases}$$

The exterior scaling allows us to make the kinetic term in (7.7) arbitrarily small by the choice of ε because

$$q[\phi_\varepsilon\chi_1^B] = \varepsilon\|g'\|^2_{L^2(\mathbb{R})} + (\phi, U_{11}\,\phi)_{L^2(\mathbb{R})}\;;$$

on the other hand, the potential term tends to $\int_{\mathbb{R}} U_{11}(x)\,\mathrm{d}x$ as $\varepsilon \to 0$ by dominated convergence, so the right-hand side is negative for ε small enough. ∎

Remark 7.1.2 We see that, in contrast to the previous section, even a strong magnetic field cannot remove the discrete spectrum arising from the presence of the potential well U. Note that the question of whether the same could be true for geometrically induced bound states remains open. Note also that in view of the magnetic field homogeneity the coordinate choice does not matter; we can always choose a coordinate system in which the guide axis is given by $y = 0$.

The present model can also be used for a different purpose. In Sect. 2.3 we have shown how embedded eigenvalues which exist due to a symmetry can turn into resonances under a perturbation that breaks the symmetry. Now we are going to show that magnetic field can have the same effect. To this end replace the first term in (2.10) by its magnetic counterpart and we leave out the potential perturbation, in other words, we consider the Hamiltonian

$$H(B) := (-i\partial_x - By)^2 + V(x) - \partial_y^2$$

on $L^2(\Omega_0)$ with the Dirichlet condition imposed at $|y| = a$. We shall regard the magnetic field as a perturbation of the operator $H(0)$ the spectral properties of which were discussed earlier. We use the basis of transverse-mode eigenfunctions to write $H(B)$ as a matrix differential operator analogous to (2.10), the potential perturbation $\lambda U_{jk}(x)$ now being replaced by

$$U_{jk}(B) := 2i\,B\,m^{(1)}_{jk}\partial_x + B^2 m^{(2)}_{jk}\,,$$

where the involved momenta are defined by $m^{(r)}_{jk} := \int_{-a}^a y^r\overline{\chi}_j(y)\chi_k(y)\,\mathrm{d}y$. Next we use the complex scaling described by the operator S_θ, see (2.13), and perform the perturbation expansion for the eigenvalues of the obtained non-selfadjoint operator revealed by the rotation of the essential spectrum (Problem 5). This leads us to the following conclusion.

Theorem 7.5 *Adopt the assumption (i) of Sect. 2.3, and moreover, suppose that $\epsilon_0 = \mu_n + \nu_j$ is a simple eigenvalue of $H(0)$ satisfying the conditions (2.11). Then a weak magnetic field turns it into a resonance pole $\epsilon(B)$ of $H(B)$ with*

$$\operatorname{Im}\epsilon(B) = -2B^2 \sum_{k=1}^{k(\epsilon_0)} \sum_{\sigma=\pm} \frac{\pi}{\sqrt{\epsilon_0-\nu_k}} |m_{jk}^{(1)}|^2 \left|\tau_\sigma(\epsilon_0-\nu_k)\omega(\epsilon_0-\nu_k+i0)\phi_n'\right|^2 + \mathcal{O}(B^3),$$

where the involved symbols are as in Theorem 2.4.

As in the non-magnetic case the conclusions of this section, both concerning the discrete spectrum and resonances, can be extended to the case of soft channels, where the free operator is $H_0(B) := (-i\partial_x - By)^2 - \partial_y^2 + W(y)$ on $L^2(\mathbb{R}^2)$ with a suitable confining potential W (Problem 6).

7.2 Magnetic Transport in Electron Gas

The results of the previous section are not the only examples of the influence that a homogeneous magnetic field has on two-dimensional charged particles. Even more important ones concern the absolutely continuous spectrum of the one-particle magnetic Hamiltonian, often also referred to as the ***Landau Hamiltonian***, which, written in Cartesian coordinates, acts as

$$(-i\partial_x + A_x)^2 + (-i\partial_y + A_y)^2\,,$$

where $A = (A_x, A_y)$ is a vector potential corresponding to the magnetic field $B = \partial_x A_y - \partial_y A_x$ parallel to the z-axis. Our interest stems from the fact that electron transport is associated with states carrying current. To achieve a macroscopic electric current, a large number of electrons is naturally needed, however, if the electron gas is dilute and the mutual interactions can be neglected, the one-particle Hamiltonian provides useful information.

The existence of an absolutely continuous spectrum in this situation is a nontrivial effect. Recall that if the particle motion in the plane is not restricted, then in the absence of a magnetic field the energy spectrum is purely absolutely continuous and covers the positive half-line. If a magnetic field is added, the situation changes completely. It is well known that in the presence of a homogeneous magnetic field its character changes to pure point. This effect has a classical counterpart in the fact that charged particles in a homogeneous magnetic field are fully localized circling on the appropriate cyclotron orbits, which means that transport is absent in both the classical and quantum case. It appears, however, that it can result from a suitable perturbation of the system.

We are going to discuss two mechanisms which make such a transport possible. One consists of restricting the particle motion by a boundary which is infinitely extended but otherwise it can be of various nature. The other involves a local variation of the magnetic field which is localized in one direction and infinitely extended in the perpendicular one. Since our main motivation here comes from description of magnetic transport in solids and real-life materials are never absolutely pure, it is also useful to ask whether the transport remains preserved in the presence of a potential modeling impurities.

7.2.1 Edge States

Bertrand Halperin showed in the seminal paper [Ha82] that the presence of boundaries in two-dimensional systems with homogeneous magnetic field induces the existence of current-carrying states. He also suggested that these ***edge states***, localized in the vicinity of the boundary but extended along it, play an important role in the understanding of the ***quantum Hall effect*** contributing to the so-called Hall current. We will analyze this phenomenon in a couple of particular situations with different types of boundaries.

Consider first a *straight hard wall*, in other words, suppose that the particle motion is confined to a halfplane with Dirichlet condition imposed at its boundary. In such a system an unrestricted motion is possible also classically provided the particle is close enough to the boundary. Indeed, if its cyclotron orbit crosses the boundary, the particle gets reflected from it, and since the incidence and reflection angles coincide, it bounces from the boundary periodically moving thus in a hopping fashion in the direction determined by the magnetic field orientation. It is easy to see that the speed of such a propagation does not exceed the initial velocity of the particle and that it decreases with the distance of the cyclotron-orbit center from the boundary; once it matches or exceeds its radius, the particle does not hit the boundary and remains therefore localized. This, as we are going to show, is not the case in the quantum situation unless, of course, an additional potential is introduced.

We choose the coordinates in such a way that the particle is confined to the right halfplane, $x > 0$, and assume for definiteness that $B > 0$. The spectrum is gauge invariant and it is again convenient to work in the Landau gauge, hence the Hamiltonian in the absence of perturbations is

$$H_0 = -\partial_x^2 + (-i\partial_y + Bx)^2 \qquad \text{in } L^2(\mathbb{R}_+ \times \mathbb{R})$$

with Dirichlet boundary condition at $x = 0$. In view of its translational invariance in the y-direction, we can use the same argument as in the previous section and to pass by means of $I \otimes F$ to the unitarily equivalent operator

$$\hat{H}_0 = \int_{\mathbb{R}}^{\oplus} h(p)\,\mathrm{d}p \quad \text{with} \quad h(p) := -\partial_x^2 + (p - Bx)^2\,, \tag{7.8}$$

where the fiber operator is acting in $L^2(\mathbb{R}_+)$ with Dirichlet condition at $x = 0$. The parameter $p \in \mathbb{R}$ is the momentum component canonically conjugated with the coordinate y. The spectrum of H_0 is then given by

$$\sigma(H_0) = \sigma(\hat{H}_0) = \overline{\bigcup_{p\in\mathbb{R}} \sigma(h(p))}\,.$$

As the potential term of $h(p)$ diverges to $+\infty$ when $x \to +\infty$, the spectrum of each $h(p)$ is purely discrete and consists of simple eigenvalues $\epsilon_n(p)$, $n \in \mathbb{N}_0$, which are usually referred to as ***band functions***.

Proposition 7.2.1 *(a) The functions $\epsilon_n(\cdot)$ are real analytic on $\mathbb{R}$.*
(b) For any $n \in \mathbb{N}_0$ and all $p \in \mathbb{R}$ we have $\partial_p\epsilon_n(p) = -\frac{1}{B}|\partial_x u_n(0; p)|^2 < 0$, where $u_n(\cdot; p)$ is the normalized real-valued eigenfunction of $h(p)$ corresponding to the eigenvalue $\epsilon_n(p)$.
(c) $\epsilon_n(p) \ge \epsilon_n(0) + p^2$ holds for all $p \le 0$ and any $n \in \mathbb{N}_0$.
(d) $\lim_{p\to\infty} \epsilon_n(p) = (2n + 1)B$ for any $n \in \mathbb{N}_0$.

Proof The eigenvalue problem for $h(p)$ is explicitly solvable; one gets

$$u_n(x; p) = c_n\, D_{\epsilon_n(p)-1}\big(B^{1/2}x - B^{-1/2}p\big)\,,$$

where c_n a normalization constant, D_ν is the Whittaker function [AS, 19.3], and the eigenvalue $\epsilon_n(p)$ is determined by the condition

$$D_{\epsilon_n(p)-1}\big(-B^{-1/2}p\big) = 0\,.$$

One can check easily that $\operatorname{Dom} h(p)$ is independent of p, then (d) follows from the above condition in combination with properties of Whittaker functions (see also Problem 9). To prove (a) we observe that for any $u \in H_0^1(\mathbb{R}_+)$ and any $\varepsilon > 0$ one has the estimate

$$\begin{aligned}(u, Bxu)_{L^2(\mathbb{R}_+)} &\le B\|xu\|_2\,\|u\|_2 \le B(\|xu\|_2^2 + \|u\|_2^2)\\ &\le (h(0)u, u)_{L^2(\mathbb{R}_+)} + B\|u\|_2^2 \le \varepsilon\|h(0)u\|_2^2 + (\varepsilon^{-1} + B)\|u\|_2^2\,.\end{aligned}$$

Hence the operator Bx is relatively bounded with respect to $h(0)$ with relative bound zero, and the assertion (a) follows from analytic perturbation theory, cf. [Ka, Theorem VII.2.6]. Furthermore, using the eigenvalue equation

$$-\partial_x^2 u_n(x; p) + (p - Bx)^2 u_n(x; p) = \epsilon_n(p)u_n(x; p),$$

together with the Feynman-Hellmann formula and integration by parts we get

$$\begin{aligned}\partial_p \epsilon_n(p) &= 2\int_0^\infty (p - Bx)\, u_n^2(x;\, p)\, \mathrm{d}x = \frac{2}{B}\int_0^\infty (p - Bx)^2\, u_n(x;\, p)\, \partial_x u_n(x;\, p)\, \mathrm{d}x \\ &= \frac{2}{B}\int_0^\infty \big(\epsilon_n(p)\, u_n(x;\, p)\, \partial_x u_n(x;\, p) + \partial_x^2 u_n(x;\, p)\, \partial_x u_n(x;\, p)\big)\, \mathrm{d}x \\ &= -\frac{1}{B}\, |\partial_x u_n(0;\, p)|^2 < 0\,,\end{aligned}$$

because the eigenfunction $u_n(\cdot;\, p)$ satisfying the Dirichlet boundary condition cannot have at the same time vanishing derivative at $x = 0$. The proof of (c) is left to the reader (Problem 8). ∎

A consequence of the claim (b) is that the hard-wall magnetic half-plane exhibits transport at all energies, without any localized states.

Corollary 7.2.1 $\sigma(H_0) = [B, \infty)$ *is purely absolutely continuous.*

This result demonstrates universality of free magnetic transport in the half-plane. Let us next look at what happens in the presence of an additional impurity potential, more specifically, whether and for which energies the absolutely continuous spectrum survives when the Hamiltonian H_0 is replaced by

$$H = H_0 + W$$

with a bounded potential W. A suitable tool toward this goal is the Mourre theory of positive commutators which has been described in Sect. 2.1. Its central point is to find a conjugate operator Π such that the commutator $E(\Delta)[i\Pi, H]E(\Delta)$, where $E(\Delta)$ denotes the spectral projection of H corresponding to a fixed interval Δ, is positive. It is convenient to establish the corresponding estimate first for the free Hamiltonian H_0. We denote by

$$L_n = ((2n+1)B, (2n+3)B]$$

the interval between neighboring Landau levels and define

$$\theta_n(p, n', n'') := \begin{cases} |\epsilon_{n'}(p) - \epsilon_{n''}(p)| & \text{if both } \epsilon_{n'}(p) \text{ and } \epsilon_{n''}(p) \text{ are in } L_n \\ 2B & \text{otherwise} \end{cases}$$

which obviously satisfies $\theta_n(p, n', n'') = \theta_n(p, n'', n')$. Furthermore, put

$$\delta_n := \inf_{n' < n'' \le n} \inf_p \theta_n(p, n', n'')\,.$$

It follows from *Proposition* 7.2.1 that $\theta_n(p, n', n'') > cB$ with $c > 0$ outside a compact interval, and since $|\epsilon_{n'}(p) - \epsilon_{n''}(p)| > 0$ for all p we conclude that $\delta_n > 0$

holds for all n. Let $E_0(\Delta)$ be the spectral projection of H_0 corresponding to a set Δ, then we have the following estimate.

Lemma 7.2.1 *Let $\Delta \subset L_n$ be a closed interval with $|\Delta| < \delta_n B$. Define*

$$\nu_-(\Delta) := \inf_{\{n',p:\,\epsilon_{n'}(p)\in\Delta\}} |\partial_p \epsilon_{n'}(p)| > 0 \tag{7.9}$$

and $\nu_+(\Delta)$ similarly with inf *replaced by* sup. *Then*

$$\sqrt{B}\,\nu_-(B^{-1}\Delta)E_0(\Delta) \le E_0(\Delta)\,[iy, H_0]\,E_0(\Delta) \le \sqrt{B}\,\nu_+(B^{-1}\Delta)\,E_0(\Delta)$$

holds true in the sense of quadratic forms on $L^2(\mathbb{R}_+ \times \mathbb{R})$.

Proof The positivity of $\nu_-(\Delta)$ follows from *Proposition* 7.2.1b. Suppose first that $B = 1$. Denote by $\pi_{n'}(p)$ the eigenprojection associated with $\epsilon_{n'}(p)$ and by $\mathcal{F} = I \otimes F$ the Fourier-Plancherel operator acting in the y-direction. Since $|\Delta| < \delta_n$ by assumption, we have $\epsilon_{n'}^{-1}(\Delta) \cap \epsilon_{n''}^{-1}(\Delta) = \emptyset$ if $n' \neq n''$ which allows us to use the Feynman-Hellmann formula to infer

$$\begin{aligned}\mathcal{F}E_0(\Delta)\,[iy, H_0]\,E_0(\Delta)\mathcal{F}^{-1} &= \sum_{n'} \int_{\epsilon_{n'}^{-1}(\Delta)} \pi_{n'}(p)[-\partial_p,\, h(p)]\pi_{n'}(p)\,\mathrm{d}p \\ &= \sum_{n'} \int_{\epsilon_{n'}^{-1}(\Delta)} \pi_{n'}(p)\big(-\partial_p\epsilon_{n'}(p)\big)\pi_{n'}(p)\,\mathrm{d}p \\ &\ge \nu_-(\Delta)\sum_{n'} \int_{\epsilon_{n'}^{-1}(\Delta)} \pi_{n'}(p)\,dp = \nu_-(\Delta)\hat{P}_0(\Delta)\,,\end{aligned}$$

where $\hat{P}_0(\Delta)$ is the corresponding spectral projection of $\hat{H}_0 = \mathcal{F}H_0\mathcal{F}^{-1}$; in the same way we obtain the upper bound with $\nu_-(\Delta)$ replaced by $\nu_+(\Delta)$. Hence

$$\nu_-(\Delta)E_0(\Delta) \le E_0(\Delta)\,[iy, H_0]\,E_0(\Delta) \le \nu_+(\Delta)\,E_0(\Delta)$$

and the claim for any $B > 0$ follows by scaling, $(x, y) \mapsto \sqrt{B}\,(x, y)$. ∎

This result allows us to demonstrate that if the perturbation W is small enough, the absolutely continuous spectrum of H in certain intervals of the positive real axis remains preserved.

Theorem 7.6 *Fix $n \in \mathbb{N}_0$ and let $\lambda, \lambda' > 0$ be such that $\lambda + \lambda' < 2$. Then there exists $\eta(n, \lambda, \lambda') > 0$ such that for $\|W\|_\infty \le \eta(n, \lambda, \lambda')\,B$ we have*

$$\sigma_{\mathrm{sing}}(H) \cap \big(B\,(2n + 1 + \lambda),\, B\,(2n + 3 - \lambda')\big) = \emptyset\,.$$

Proof As in the previous proof it suffices to check the claim for $B = 1$, the general case will then follow by scaling. Let $\alpha \in (2n + 1 + \lambda, 2n + 3 - \lambda')$ and set

$$\sigma := \frac{1}{4}\min(\delta_n, \lambda, \lambda').$$

Our aim is to show that if $W_0 := \|W\|_\infty$ and $\varepsilon \in (0, \sigma]$ are small enough, then the inequality

$$E(\Delta_\varepsilon)\,[iy, H]\,E(\Delta_\varepsilon) \geq c(\varepsilon, W_0)\,E(\Delta_\varepsilon) \tag{7.10}$$

holds in the sense of quadratic forms on the interval $\Delta_\varepsilon = [\alpha-\varepsilon, \alpha+\varepsilon]$ and with some constant $c(\varepsilon, W_0) > 0$. To this end we consider the interval $\Delta_\sigma = [\alpha-\sigma, \alpha+\sigma]$, which contains Δ_ε since $\varepsilon \leq \sigma$, and its complement $\Delta_\sigma^c = \mathbb{R}\setminus\Delta_\sigma$. By construction we have $\Delta_\sigma \subset L_n$. For any vector $\psi \in E(\Delta_\varepsilon)L^2(\mathbb{R}_+\times\mathbb{R})$ we can then write the following inequality,

$$\|(H_0-\alpha)\,\psi\| \leq \|(H-\alpha)E(\Delta_\varepsilon)\,\psi\| + W_0\,\|\psi\| \leq (\varepsilon + W_0)\,\|\psi\|,$$

and moreover, since $\min\{|s-\alpha| : s \in \Delta_\sigma^c\} \geq \sigma$, we also have

$$\|E_0(\Delta_\sigma^c)\,\psi\| \leq \|(H_0-\alpha)^{-1}E_0(\Delta_\sigma^c)\|\,\|(H_0-\alpha)\psi\| \leq \sigma^{-1}(\varepsilon+W_0)\,\|\psi\|. \tag{7.11}$$

Using next the commutator identity $[iy, H] = [iy, H_0] = 2(i\partial_y + Bx)$ in combination with the fact that $E(\Delta_\sigma^c)\psi = 0$ we obtain

$$(\psi, [iy, H]\psi) \geq \big(E_0(\Delta_\sigma)\psi, [iy, H_0]E_0(\Delta_\sigma)\psi\big) - 2\,\|[iy, H_0]\,E_0(\Delta_\sigma^c)\psi\|\,\|\psi\|.$$

In order to control the second term on the right-hand side we note that

$$\begin{aligned}\|[iy, H_0]\,E_0(\Delta_\sigma^c)\psi\| &\leq 2\big(E_0(\Delta_\sigma^c)\psi, H_0E_0(\Delta_\sigma^c)\psi\big)^{1/2}\\ &\leq 2\,\|E_0(\Delta_\sigma^c)\psi\|^{1/2}\,\|H_0E_0(\Delta_\sigma^c)\psi\|^{1/2}.\end{aligned}$$

If $W_0 \leq 1$, then the last norm can be estimated by

$$\|H_0\,E_0(\Delta_\sigma^c)\psi\|^{1/2} \leq (\|H\psi\| + W_0\|\psi\|)^{1/2} \leq (2n+4)^{1/2}\,\|\psi\|^{1/2}$$

and putting the above estimates together we arrive at

$$(\psi, [iy, H]\psi) \geq (E_0(\Delta_\sigma)\psi, [iy, H_0]E_0(\Delta_\sigma)\psi) - 4(2n+4)^{1/2}\sigma^{-1/2}(\varepsilon+W_0)^{1/2}\|\psi\|.$$

On the other hand, since $|\Delta_\sigma| = 2\sigma < \delta_n$ holds by assumption, we can apply *Lemma* 7.2.1 with $B = 1$ to the first term on the right-hand side of the last inequality. This yields the estimate

$$(E_0(\Delta_\sigma)\psi,\,[iy, H_0]\,E_0(\Delta_\sigma)\psi) \geq \nu(n, \lambda)\|E_0(\Delta_\sigma)\psi\|^2,$$

where $\nu(n,\lambda) := \nu_-\big((2n+1+\lambda, 2n+3)\big)$ with $\nu_-(\cdot)$ being defined by (7.9). This in combination with (7.11), the above lower bound to $(\psi, [iy, H]\psi)$, and the identity $\|\psi\|^2 = \|E_0(\Delta_\sigma)\psi\|^2 + \|E_0(\Delta_\sigma^c)\psi\|^2$ finally implies

$$(\psi, [iy, H]\,\psi) \geq \nu(n,\lambda)\Big[1 - \Big(\frac{(\varepsilon+W_0)^2}{\sigma^2} + \frac{4(2n+4)^{1/2}(\varepsilon+W_0)^{1/2}}{\sigma^{1/2}\,\nu(n,\lambda)}\Big)\Big]\,\|\psi\|^2 .$$

Hence taking W_0 and ε small enough, the choice being dependent on n, λ, λ', we conclude that

$$(\psi, [iy, H]\,\psi) \geq c(\varepsilon, W_0)\,\|\psi\|^2$$

holds for some $c(\varepsilon, W_0) > 0$ and all $\psi \in E(\Delta_\varepsilon)L^2(\mathbb{R}_+ \times \mathbb{R})$ giving (7.10).

Now it remains to apply the Mourre theory. It can easily be checked that the unitary group $\{e^{ity} : t \in \mathbb{R}\}$ preserves the domain of H. Moreover, $[iy, H_0] = [iy, H]$ is relatively bounded with respect to H_0 and the double commutator $[iy, [iy, H_0]]$ is bounded (Problem 10). Thus the spectrum of H in the interval Δ_ε is purely absolutely continuous, and since $\alpha \in (2n+1+\lambda, 2n+3-\lambda')$ is arbitrary, the claim of the theorem follows. ∎

It is important to realize that the choice of the conjugate operator Π as multiplication by y in *Lemma* 7.2.1 and *Theorem* 7.6 is essential for the argument. We have observed in the notes to Sect. 2.1 that the classical counterpart of a positive commutator is the existence of an observable which increases in time with a positive lower bound to its derivative. Looking at the classical trajectories near the boundary described at the opening of this section, it is clear that the position coordinate parallel to the boundary is a suitable candidate.

It is thus natural to ask whether the result on the stability of a part of the absolutely continuous spectrum can be extended to domains with more complicated boundaries by choosing a suitable conjugate operator replacing $\Pi = y$. It turns out that the domain in question must satisfy some geometrical assumptions. The most important is that it has to contain a conical subset. This requirement is related to interference coming from scattering on the impurity potential. Note that the transport associated with the edge states has a unique direction determined by the orientation of the magnetic field. If two parts of the boundary face each other, the motion along them occurs in opposite directions. Even in this situation the spectrum can be absolutely continuous as the case of a straight magnetic strip without a potential discussed in Sect. 7.1.3 shows, however, it appears to be more sensitive to the presence of impurities.

Since the application of Mourre's theory is more involved in the case of a general boundary, we limit ourselves to stating the result, referring to the notes for reference to the proof and a guide for further reading.

Theorem 7.7 *Let $\Omega \subset \mathbb{R}^2$ be an open, simply connected region containing a wedge-shaped subset, the boundary of which is C^3-smooth apart from a finite number of* *points where its segments meet at nonzero angles. Suppose that $\partial\Omega$ is parametrized by its arc length and there is a function $s \in C^2(\overline{\Omega})$ such that $s(\gamma(s')) = s'$ for $s' \in \mathbb{R}$ having uniformly bounded derivatives, $\|\partial_x s\|_\infty + \|\partial_y s\|_\infty < \infty$ and $\|\partial_i \partial_j s\|_\infty < \infty$ for $i, j = x, y$. Fix $E \notin \{(2n+1)B : n \in \mathbb{N}_0\}$ and suppose that $\|W\|_\infty < \infty$. Then the spectrum of the operator $H = H_0 + W$ is absolutely continuous in the vicinity of E provided B is large enough.*

7.2.2 Edge States Without a Classical Analogue

Magnetic transport along a boundary considered above has a classical counterpart in trajectories of particles moving on segments of cyclotron orbits and bouncing from the wall. Our next aim is to demonstrate that the transport is also possible in the absence of such a hopping propagation, more exactly, in situations where such trajectories refer to a zero-measure family of initial conditions. The magnetic transport then becomes a purely quantum effect in the same sense as the curvature-induced bound states discussed in Chap. 1. The model we are going to analyze describes the free magnetic Hamiltonian H_0 in the plane perturbed by a periodic array of point interactions.

Without loss of generality we may suppose that the point interactions are situated along the x-axis. Using again a Landau gauge and choosing $A(\vec{x}) = (-By, 0)$ we can write the Hamiltonian formally as

$$H_{\alpha,\ell} = (-i\partial_x - By)^2 - \partial_y^2 + \sum_{j\in\mathbb{Z}} \tilde{\alpha}\,\delta(x - x_0 - j\ell),$$

where ℓ is the period of the array. The interaction term with "coupling constant" $\tilde{\alpha}$ naturally has a symbolic meaning only; the proper way to introduce the perturbation is to impose the boundary conditions (5.2) with a parameter $\alpha \in \mathbb{R}$ at the points $\vec{a}_j := (x_0 + j\ell, 0),\ j \in \mathbb{Z}$. We denote the resulting operator as $H_{\alpha,\ell}$. The free operator H_0 is identified with $H_{\infty,\ell}$ and as we have already indicated, its spectrum consists of eigenvalues of infinite multiplicity, the ***Landau levels***, being

$$\sigma(H_{\infty,\ell}) = \{(2n+1)\,B \,:\, n \in \mathbb{N}_0\}\,.$$

Our aim is to show that whenever the perturbation is present, i.e. for *any* finite value of α there is an unbounded set of energies for which the transport occurs.

Theorem 7.8 *The spectrum of $H_{\alpha,\ell}$ consists for any $\alpha \in \mathbb{R}$ of the Landau levels $B(2n+1)$, $n \in \mathbb{N}_0$, and an infinite family of absolutely continuous spectral bands situated between any two adjacent Landau levels and below B.*

Proof Without loss of generality we may suppose that $0 < x_0 < \ell$. Using the periodicity, we can perform the Bloch decomposition in the x direction and write

$$H_{\alpha,\ell} = \frac{\ell}{2\pi} \int_{|\theta\ell|\le\pi}^{\oplus} H_{\alpha,\ell}(\theta)\, \mathrm{d}\theta\,,$$

where the fiber operator $H_{\alpha,\ell}(\theta)$ on the period cell acts on $L^2(0,\ell)\otimes L^2(\mathbb{R})$ with the domain being specified by the boundary conditions

$$\partial_x^i \psi(\ell-, y) = \mathrm{e}^{i\theta\ell} \partial_x^i \psi(0+, y)\,, \quad i = 0, 1\,.$$

Since the strip $(0,\ell)\times\mathbb{R}$ contains a single point perturbation, the Green function of the operator $H_{\alpha,\ell}(\theta)$ can, in analogy with *Proposition* 5.1, be expressed by means of Krein's formula

$$(H_{\alpha,\ell}(\theta)-z)^{-1}(\vec{x},\vec{x}\,') = G_0(\vec{x},\vec{x}\,';\theta,z) + \frac{G_0(\vec{x},\vec{a}_0;\theta,z)G_0(\vec{a}_0,\vec{x}\,';\theta,z)}{\alpha-\xi(\vec{a}_0;\theta,z)}\,, \quad (7.12)$$

where G_0 is the corresponding free Green's function and

$$\xi(\vec{a};\theta,z) := \lim_{|\vec{x}-\vec{a}|\to 0} \left(G_0(\vec{a},\vec{x};\theta,z) - \frac{1}{2\pi} \ln|\vec{x}-\vec{a}| \right)$$

is its regularized value at the point $\vec{a}$. To find G_0 we employ a transverse mode decomposition similarly as we did in Sect. 5.1.1, just the longitudinal part is a bit more involved. The above boundary conditions determine eigenvalues and eigenfunctions of the transverse part of the free operator, namely

$$\mu_m(\theta) = \left(\frac{2\pi m}{\ell} + \theta\right)^2, \quad \eta_m^\theta(x) = \frac{1}{\sqrt{\ell}}\, \mathrm{e}^{i(2\pi m+\theta\ell)x/\ell}\,,$$

where the index m runs through integers. Then we have

$$G_0(\vec{x},\vec{x}\,';\theta,z) = -\sum_{m=-\infty}^{\infty} \frac{u_m^\theta(y_<)v_m^\theta(y_>)}{W(u_m^\theta, v_m^\theta)}\, \eta_m^\theta(x)\, \overline{\eta_m^\theta}(x')\,,$$

where $y_<, y_>$ are as usual the smaller and the larger value of y, y', respectively, and u_m^θ, v_m^θ are solutions to the equation

$$-u''(y) + \left(By + \frac{2\pi m}{\ell} + \theta\right)^2 u(y) = zu(y)$$

such that u_m^θ is L^2 at $-\infty$ and v_m^θ is L^2 at $+\infty$; in the denominator we have their Wronskian. Making the appropriate argument shift we can write

$$u_m^\theta(y) = u\left(y + \frac{2\pi m + \theta\ell}{B\ell}\right)$$

and a similar relation for v_m^θ, where u, v are the corresponding solutions of the equation with harmonic-oscillator potential. We have $W(u_m^\theta, v_m^\theta) = W(u, v)$, of course, and u, v express in terms of confluent hypergeometric functions,

$$v(y) = \mathrm{e}^{-By^2/2}U\left(\frac{B-z}{4B}, \frac{1}{2}; By^2\right)$$

away from zero, and u is obtained by analytical continuation in the y^2 variable; one can express both of them through the formula

$$\begin{Bmatrix} u \\ v \end{Bmatrix}(y) = \sqrt{\pi}\, e^{-By^2/2}\left[\frac{M\left(\frac{B-z}{4B}, \frac{1}{2}; By^2\right)}{\Gamma\left(\frac{3B-z}{4B}\right)} \pm 2\sqrt{B}y\, \frac{M\left(\frac{3B-z}{4B}, \frac{3}{2}; By^2\right)}{\Gamma\left(\frac{B-z}{4B}\right)}\right].$$

Computing the Wronskian and using the explicit form of the transverse eigenvalues and eigenfunctions we then get

$$G_0(\vec{x}, \vec{x}'; \theta, z) = -\frac{2^{(z/2B)-(3/2)}}{\sqrt{\pi B}\ell}\,\Gamma\left(\frac{B-z}{2B}\right) \mathrm{e}^{i\theta(x-x')} \tag{7.13}$$
$$\times \sum_{m=-\infty}^{\infty} u\left(y_< + \frac{2\pi m + \theta\ell}{B\ell}\right) v\left(y_> + \frac{2\pi m + \theta\ell}{B\ell}\right) \mathrm{e}^{2\pi i m(x-x')/\ell}.$$

As expected the function has singularities independent of θ which coincide with the Landau levels, $z_n = B(2n+1),\ n = 0, 1, 2, \ldots$. Inspecting the functions $w^k \sin\left(\frac{\pi w}{\ell}\right) \mathrm{e}^{-B|w|^2/4}$ with $k \in \mathbb{N}_0$ and $w := x - x_0 + iy$ which span the corresponding eigenspaces and noting that the point-interaction condition is satisfied for them automatically, it is not difficult to see that each z_n remains to be an infinitely degenerate eigenvalue of the perturbed fiber operator $H_{\alpha,\ell}(\theta)$.

On the other hand, $H_{\alpha,\ell}(\theta)$ also has eigenvalues different from z_n which we denote as $\epsilon_n^{(\alpha,\ell)}(\theta)$. In view of (7.12) they are given by the implicit equation

$$\alpha = \xi(\vec{a}_0; \theta, \epsilon) \tag{7.14}$$

referring to non-normalized eigenfunctions $\psi_n^{(\alpha,\ell)}(\cdot;\theta) = G_0(\cdot, \vec{a}_0; \theta, \epsilon_n(\theta))$. In order to evaluate them, we have to assess the convergence of the series in (7.13). Using the asymptotic behavior

$$\begin{Bmatrix} u \\ v \end{Bmatrix}(y) = e^{\mp\{\pm\}By^2/2}\left(\mp\sqrt{B}y\right)^{\frac{z-B}{2B}}\left(1 + \mathcal{O}(|y|^{-2})\right)$$

for $y \to \mp\infty$, we find that the product

$$s_m := u\left(y_< + \frac{2\pi m + \theta\ell}{B\ell}\right) v\left(y_> + \frac{2\pi m + \theta\ell}{B\ell}\right)$$

is for $y \neq y'$ governed by the exponential term, namely

$$s_m = \exp\left\{\frac{B}{2}\left(y_<^2 - y_>^2\right) + \left(\theta - \frac{2\pi|m|}{\ell}\right)(y_> - y_<)\right\}\left(|m|^{-1} + \mathcal{O}(|m|^{-2})\right)$$

as $|m| \to \infty$, while for $y = y'$ we have $s_m = -\frac{1}{4\pi}|m|^{-1} + \mathcal{O}(|m|^{-2})$, hence the series (7.13) is not absolutely convergent. Summing now the contributions from $\pm m$ we see that in the limit $x' \to x$ it diverges at the same rate as the Taylor series of $-(1/2\pi)\ln\zeta$ does for $\zeta \to 0+$, and consequently, we get

$$\xi(\vec{x};\theta,z) = \sum_{m=-\infty}^{\infty}\left\{\frac{1-\delta_{m,0}}{4\pi|m|} - \frac{2^{-2\zeta-1}}{\sqrt{\pi B}\ell}\Gamma(2\zeta)\,(uv)\left(y + \frac{2\pi m + \theta\ell}{B\ell}\right)\right\}, \tag{7.15}$$

where $\zeta := \frac{B-z}{4B}$. Note that the expression is independent of x as it should be since the free operator H_0 is translationally invariant. We can write it by means of the first hypergeometric function alone, since

$$(uv)\left(\frac{\xi}{\sqrt{B}}\right) = \pi\,\mathrm{e}^{-\xi^2}\left[\frac{M(\zeta,\frac{1}{2};\xi^2)^2}{\Gamma(\zeta+\frac{1}{2})^2} - 4\xi^2\frac{M(\zeta+\frac{1}{2},\frac{3}{2};\xi^2)^2}{\Gamma(\zeta)^2}\right],$$

where we have used the shorthand $\xi := \sqrt{B}\left(y + \frac{2\pi m+\theta\ell}{B\ell}\right)$.

The spectral bands of $H_{\alpha,\ell}$ are in view of the direct-integral decomposition given by the ranges of the functions $\epsilon_n^{(\alpha,\ell)}(\cdot)$. Solutions of the condition (7.14) do not cross the Landau levels, because $\xi(\vec{a}_0;\theta,\cdot)$ is increasing in the spectral gaps of H_0, i.e. the intervals $(-\infty, B)$ and $(B(2n-1), B(2n+1))$, and diverges at their endpoints. The spectrum will be continuous away from z_n if the band functions are nowhere constant. In view of the spectral condition (7.14) one has to check that $\xi(\vec{x};\theta,z)$ is nowhere constant as a function of θ. Notice that each term in (7.15) is real-analytic for $z \in \mathbb{R}$ and the series has a convergent majorant independent of θ, hence $\xi(\vec{x};\cdot,z)$ is real-analytic as well and one has to check that it is non-constant in the whole Brillouin zone $[-\pi/\ell, \pi/\ell)$.

Suppose that the opposite is true. Then the Fourier coefficients of the function, $c_k := \int_{-\pi/\ell}^{\pi/\ell}\xi(\vec{x};\theta,z)\,\mathrm{e}^{ik\ell\theta}\,\mathrm{d}\theta$, must vanish for any nonzero integer k. Since the summand in (7.15) behaves as $\mathcal{O}(|m|^{-2})$ as $|m| \to \infty$, we may interchange the summation and integration; a simple change of variables then gives

$$c_k = -\frac{2^{-2\zeta-1}}{\sqrt{\pi B}\ell}\Gamma(2\zeta)\lim_{M\to\infty}\int_{-\pi(2M+1)}^{\pi(2M+1)}(uv)\left(y + \frac{\vartheta}{B\ell}\right)e^{ik\vartheta}\,d\vartheta,$$

hence $\hat{F}_y(k) := \int_{-\infty}^{\infty}F_y(\vartheta)\,\mathrm{e}^{ik\vartheta}\,\mathrm{d}\vartheta = 0$ holds for $F_y(\vartheta) := (uv)\left(y + \frac{\vartheta}{B\ell}\right)$. The same reasoning applies to any finitely periodic extension of $\xi(\vec{x};\theta,z)$ suggesting that $\hat{F}_y(k) = 0$ is also valid for any nonzero rational k. Some caution is needed, however, because the function decays as $\mathcal{O}(|\vartheta|^{-1})$ and the integral makes sense only as the principal value. To avoid this problem, we use the mentioned asymptotic behavior which implies, in particular, $F_y(\vartheta) = -\frac{1}{4\pi}(1+\vartheta^2)^{-1/2} + f_y(\vartheta)$, where $f_y(\vartheta) = \mathcal{O}(|\vartheta|^{-2})$ uniformly in $y \in [0, \ell]$. The Fourier transform of the first term can be computed explicitly giving

$$\hat{F}_y(k) = -\frac{1}{2\pi} K_0(k) + \hat{f}_y(k)\,,$$

and since $f_y \in L^1$, the second term on the right-hand side is continuous with respect to k. The same is then true for $\hat{F}_y$; this means that we have $\hat{F}_y(k) = 0$ for any nonzero k. Furthermore, $\hat{f}_y$ is bounded and the Macdonald function K_0 diverges logarithmically at $k = 0$, hence $\int_{-N}^{N} F_y(\vartheta)\, \mathrm{e}^{ik\vartheta}\, \mathrm{d}\vartheta$ can be majorized by an integrable function independent of N. This yields

$$\int_{-\infty}^{\infty} \hat{F}_y(k)\phi(k)\, \mathrm{d}k = \int_{-\infty}^{\infty} \mathrm{d}k\, \phi(k) \lim_{N\to\infty} \int_{-N}^{N} F_y(\vartheta)\, \mathrm{e}^{ik\vartheta}\, \mathrm{d}\vartheta = \int_{-\infty}^{\infty} F_y(\vartheta)\, \hat{\phi}(\vartheta)\, \mathrm{d}\vartheta$$

for any $\phi \in \mathcal{S}(\mathbb{R})$, in other words, $\hat{F}_y(k)$ is the Fourier transform of $F_y(\vartheta)$ in the sense of tempered distributions. Since this is a one-to-one correspondence we arrive at the conclusion that $F_y = 0$, which is a contradiction. ∎

7.2.3 The Iwatsuka Model

Let us turn now to the second mechanism mentioned in the opening of this section. Our aim is to demonstrate that transport can also be induced by variations of a homogeneous magnetic field, in other words, that the operator

$$H_A = (-i\partial_x + A_x)^2 + (-i\partial_y + A_y)^2$$

on $L^2(\mathbb{R}^2)$ can have an absolutely continuous spectrum if the vector potential A generates a magnetic field of nonconstant intensity. Looking for such an effect one naturally considers variations which are infinitely extended in some direction. A. Iwatsuka noticed that the question simplifies considerably if the field exhibits a translational symmetry proving the following result.

Theorem 7.9 *Assume that $B(x, y) = B(x)$ depends on x only and that there are constants $M_\pm$ such that $0 < M_- \le B(x) \le M_+ < \infty$ holds for all $x \in \mathbb{R}$. Suppose, in addition, that either* $\limsup_{x\to-\infty} B(x) < \liminf_{x\to+\infty} B(x)$ *or* $\limsup_{x\to+\infty} B(x) < \liminf_{x\to-\infty} B(x)$. *Then the spectrum of H_A with a vector potential A satisfying* $\operatorname{rot} A = B$ *is purely absolutely continuous.*

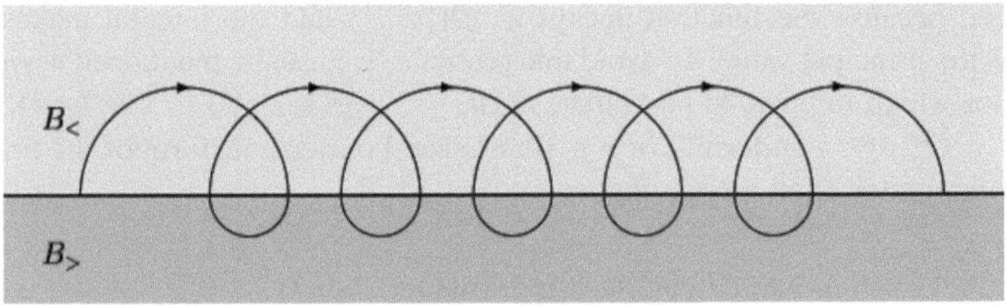

Fig. 7.1 Classical propagation along a magnetic-field step

Note that, similarly to the hard-wall case discussed above, such a non-constant field can also force a classical charged particle to propagate, naturally only if its trajectory passes through regions where the field varies. The simplest example is a step-shaped field, $B(x) = B_<$ for $x \le 0$ and $B(x) = B_> > B_<$ for $x > 0$. Unless the trajectory is confined to one of the half-planes, the difference of the cyclotron radii causes a motion along the step—cf. Fig. 7.1. The method used by Iwatsuka to prove *Theorem* 7.9 can also be applied under different assumptions, and in fact, one expects that *any* nonconstant and translationally invariant field gives rise to an absolutely continuous spectrum (see the notes). The validity of this conjecture remains, however, an open question; what is known are various sufficient conditions.

We limit ourselves here to discussing in more detail a particular situation where the magnetic field variation is *local* in the sense that it is restricted to a straight strip in the plane. Since we are free to choose the coordinate system we suppose that the field intensity equals

$$B(x, y) = B(x) = B + b(x)$$

with a fixed $B > 0$, where the function $b(\cdot)$ is bounded, piecewise continuous and such that $\operatorname{supp} b \subset [-t, t]$ for some $t > 0$. We choose

$$A(\vec{x}) = \big(0,\ Bx + a(x)\big), \quad a(x) := \int_0^x b(t)\,\mathrm{d}t\,,$$

as the corresponding vector potential which allows us to apply again partial Fourier transformation in the y-direction; we infer that H_A is unitarily equivalent to the direct integral $\int_{\mathbb{R}}^{\oplus} H(p)\,\mathrm{d}p$ with the fiber operator

$$H(p) = -\partial_x^2 + \big(p + Bx + a(x)\big)^2 \tag{7.16}$$

on $L^2(\mathbb{R})$. Since the function a is bounded, the potential is for a fixed $p \in \mathbb{R}$ dominated by the oscillator term, $\operatorname{Dom}(H(p)) = \operatorname{Dom}(-\partial_x^2) \cap \operatorname{Dom}(x^2)$, the spectrum of $H(p)$ is purely discrete and consists of a sequence of positive eigenvalues $\epsilon_n(p)$ accumulating only at infinity. For the asymptotic behavior of the band functions $\epsilon_n(\cdot)$ we have a result similar to that of *Proposition* 7.2.1d.

Lemma 7.2.2 *For any $n \in \mathbb{N}_0$ we have $\epsilon_n(p) \to (2n+1)B$ as $|p| \to \infty$.*

Proof Since $\operatorname{supp} b \subset [-t, t]$, there are constants $a^\pm$ such that $a(x) = a^\pm$ for $\pm x > t$. By the shift of variable, $x \mapsto x + \frac{p}{B}$, the operator $H(p)$ is unitarily equivalent to

$$\hat{H}(p) = -\partial_z^2 + \left(Bz + a\left(z - \tfrac{p}{B}\right)\right)^2$$

in $L^2(\mathbb{R})$. Consider for definiteness the limit $p \to -\infty$; we will show that $\hat{H}(p)$ converges in the strong resolvent sense to the operator $h_0^- = -\partial_z^2 + (Bz + a^-)^2$. Let $\psi \in \operatorname{Dom}(h_0^-)$ and choose a positive function $g \in C^\infty(\mathbb{R})$ such that $g(x) = 0$ for $x \le 0$ and $g(x) = 1$ for $x \ge 1$. Define $\psi_p(z) = \psi(z)\, g\left(\frac{p+z}{p}\right)$, then $\psi_p \to \psi$ holds in $L^2(\mathbb{R})$ as $p \to -\infty$ and

$$\begin{aligned}(\hat{H}(p)\,\psi_p)(z) = &-\partial_z^2\psi(z)\, g\left(\tfrac{p+z}{p}\right) - \tfrac{2}{p}\,\partial_z\psi(z)\, g'\left(\tfrac{p+z}{p}\right) - \tfrac{1}{p^2}\,\psi(z) g''\left(\tfrac{p+z}{p}\right) \\ &+ (Bz + a^-)^2\,\psi_p(z) + 2R_p(z)\, Bz\,\psi_p(z) + R_p^2(z)\,\psi_p(z)\,,\end{aligned}$$

where $R(z, p) \to 0$ as $p \to -\infty$ pointwise. Taking into account the fact that $g\left(\frac{p+z}{p}\right) \to 1$ and that ψ_p, $g'\left(\frac{p+z}{p}\right)$ and $g''\left(\frac{p+z}{p}\right)$ are uniformly bounded with respect to p, it is easily seen from the above formula that $\hat{H}(p)\,\psi_p \to h_0^-\,\psi$ in $L^2(\mathbb{R})$ as $p \to -\infty$. Hence the graph of $\hat{H}(p)$ converges strongly to the graph of h_0^-, which implies the strong resolvent convergence by [RS, Theorem 8.26]. Since the spectra of $\hat{H}(p)$ and h_0^- are discrete and simple, this yields the convergence of the eigenvalues. Moreover, the unitary transformation associated with the variable shift $z \mapsto \frac{z+a^-}{B}$ shows that the spectrum of h_0^- consists of the Landau levels, hence $\epsilon_n(p) \to (2n+1)B$ holds as $p \to -\infty$; the proof of the limit $p \to +\infty$ is analogous. ∎

To prove the absolute continuity of $\sigma(H_A)$ we adopt, in addition to the properties of b stated above, the following assumptions:

(i) $b(\cdot)$ is nonzero and does not change sign in $[-t, t]$,
(ii) let $a_l := \inf \operatorname{supp} b$ and $a_r := \sup \operatorname{supp} b$; there are $c_0, \delta > 0$ and $m \in \mathbb{N}$ such that one of the following conditions holds,

$$|b(x)| \ge c_0(x - a_j)^m \quad \text{or} \quad |b(x)| \ge c_0(a_r - x)^m$$

for $x \in [a_l, a_l + \delta)$ and $x \in (a_r - \delta, a_r]$, respectively.

Theorem 7.10 *Suppose that at least one of the assumptions (i), (ii) is satisfied. Then $|\partial_p\epsilon_n(p)| > 0$ holds for each $n \in \mathbb{N}_0$ and any $|p|$ large enough, and furthermore, the spectrum of H_A is purely absolutely continuous.*

To prove the claim we need a few more preliminary results.

Lemma 7.2.3 *The operator family $\{H(p),\ p \in \mathbb{R}\}$ is analytic of type (A) in the sense of Kato, in particular, each $\epsilon_n(\cdot)$ is a real analytic function on $\mathbb{R}$.*

The proof of this lemma is left to the reader (Problem 11); this result shows that it is sufficient to check the first claim of the theorem in one of the asymptotic directions because it would imply that $\epsilon_n(\cdot)$ is nowhere constant. Let further $\psi_n(\cdot, p)$ denote real-valued eigenfunctions of the fiber operator, i.e.

$$H(p)\psi_n(x, p) = \epsilon_n(p)\psi_n(x, p)\,.$$

Put $Q_{n,p}(x) = \big(p+Bx+a(x)\big)^2 - \epsilon_n(p)$ and $l_n(x, p) = \psi_n'(x, p)^2 - Q_{n,p}\psi_n(x, p)^2$. By standard results on decay of eigenfunctions of second-order elliptic operator [Ag], the function $\psi_n(\cdot, p)$ decays super-exponentially as $|x| \to \infty$ which implies that $Q_{n,p}(x)\,\psi_n(x, p) \to 0$ as $|x| \to \infty$ for each p. Since $\psi_n''(x, p) = Q_{n,p}(x)\psi_n(x, p)$ holds by assumption we conclude (Problem 12) that

$$\lim_{x\to\pm\infty} l_n(x, p) = 0. \tag{7.17}$$

Lemma 7.2.4 *Let $f_n(x, p) := (p + Bx + a(x))\psi_n(x, p)^2$. For any p large enough there exists a constant $c(p) > 0$ such that*

$$5c(p)\,\mathrm{e}^{-p(x-x_0)} \geq f_n(x, p) \geq \frac{c(p)}{7}\,\mathrm{e}^{-3p(x-x_0)}$$

holds for all $-t \leq x_0 \leq x \leq t$.

Proof Note that the spectrum of the unperturbed operator $-\partial_x^2 + (p + Bx)^2$ is given by the Landau levels $(2n + 1)B$; since the function a in (7.16) is bounded, it follows that for each n the eigenvalue $\epsilon_n(p)$ is uniformly bounded with respect to $p \in \mathbb{R}$. Hence there is a $p_0 > 0$ such that $Q_{n,p}(x) > 0$ holds for all $x \in [-t, t]$ and $|p| > p_0$, and moreover, $Q_{n,p}(x)$ grows for a fixed x as $|p| \to \infty$, which makes it possible to employ a semiclassical form for the tails of the eigenfunctions [Ol, Theorem 6.2.1]: for large enough p we have

$$\psi_n(x, p) = \frac{c_1(p)}{Q_{n,p}^{1/4}(x)} \exp\left\{-\int_{x_0}^{x} \sqrt{Q_{n,p}(\xi)}\,\mathrm{d}\xi\right\} \big(1 + q_{n,p}(x)\big) \tag{7.18}$$

and the error term satisfies $|q_{n,p}(x)| \leq \exp\left[\frac{1}{2}\int_{x_0}^{x} |F_{n,p}'(x')|\,\mathrm{d}x'\right] - 1$, where

$$F_{n,p}(x) := \int \left\{Q_{n,p}^{-1/4}(x)\frac{\mathrm{d}^2}{\mathrm{d}x^2}\left(Q_{n,p}^{-1/4}(x)\right)\right\} \mathrm{d}x\,.$$

Evaluating $F_{n,p}'(x)$ one can check that the integrand in the above estimate can be made arbitrarily small for large enough $|p|$. Consequently, to a fixed $\lambda > 1$ we can always find a p_λ such that

$$\exp\left[\frac{1}{2}\int_{x_0}^{x}|F'_{n,p}(x')|\mathrm{d}x'\right]\le\lambda \tag{7.19}$$

holds for all $p > p_\lambda$. The representation (7.18) is valid on the halfline $x \ge x_0$, hence the coefficient $c_1(p)$ is nonzero; without loss of generality we may suppose that it is positive. The behavior of $\psi_n(x,p)/\psi_n(x_0,p)$ is for $x, x_0 \in [-t,t]$ determined essentially by the exponential factor, because $\big(Q_{n,p}(x_0)/Q_{n,p}(x)\big)^{1/4}$ can be then included in the error term. Since $a(\cdot)$ is bounded and we consider x, x_0 from a bounded interval, one has $\frac{1}{2}\,p \le \sqrt{Q_{n,p}(\xi)} \le \frac{3}{2}\,p$ for all p larger than some $p_1 > 0$, and therefore

$$\frac{9c_1(p)^2}{2p}\,\mathrm{e}^{-p(x-x_0)} \ge \psi_n(x,p)^2 \ge \frac{c_1(p)^2}{6p}\,\mathrm{e}^{-3p(x-x_0)}$$

if $p > \max(p_1, p_{3/2})$, where $p_{3/2}$ refers to $\lambda = 3/2$ in (7.19). To conclude the proof, it is sufficient to notice that $\lim_{p\to\infty}(p+Bx+a(x))p^{-1} = 1$ holds for $x \in [-t,t]$ and to put $c(p) := c_1(p)^2$. ∎

Proof of Theorem 7.10: As indicated we have to check that $\epsilon_n(\cdot)$ is not constant for any $n \in \mathbb{N}_0$ and $|p|$ large enough. The Feynman-Hellmann formula implies

$$\partial_p\epsilon_n(p) = 2\int_{-\infty}^{\infty}\big(p+Bx+a(x)\big)\psi_n(x,p)^2\,\mathrm{d}x\,.$$

Let us check the contribution to the integral from the semi-infinite intervals $(-\infty,-t]$ and $[t,\infty)$. Since $l'_n(x,p) = -2(B+b(x))(p+Bx+a(x))\psi_n(x,p)^2$ and $b(x) = 0$ holds for $|x| > t$ by assumption, we can write

$$\begin{aligned}2\int_{(-\infty,-a]\cup[a,\infty)}&\big(p+Bx+a(x)\big)\psi_n(x,p)^2\,\mathrm{d}x\\ &= -\frac{1}{B}\int_{(-\infty,-a]\cup[a,\infty)} l'_n(x,p)\,\mathrm{d}x = \frac{1}{B}\big[l_n(a,p)-l_n(-a,p)\big]\,,\end{aligned}$$

where we have employed the property (7.17). Using the above expression of $l'_n(x,p)$ for the second time, we can rewrite the right-hand side of the last formula alternatively as $-\frac{2}{B}\int_{-a}^{a}(B+b(x))(p+Bx+a(x))\psi_n(x,p)^2\,\mathrm{d}x$ obtaining

$$\epsilon'_n(p) = -\frac{2}{B}\int_{-a}^{a} b(x)(p+Bx+a(x))\psi_n(x,p)^2\,\mathrm{d}x = -\frac{2}{B}\int_{-a}^{a} b(x)f_n(x,p)\,\mathrm{d}x\,.$$

Under the assumption *(i)* the first claim of the theorem follows immediately since $f_n(x,p)$ then has a definite sign in $[-t,t]$ for $|p|$ large enough.

Assume next that *(ii)* is valid and suppose that $|b(x)| \ge c_0(x-a_j)^m$ holds in $(a_l, a_l+\delta)$; the other cases can be treated in a similar way. In view of *Lemma* 7.2.4 the integral in the above expression of $\epsilon'_n(p)$ can be estimated as

$$\begin{aligned}&\int_{a_l}^{a_l+\delta} b(x) f_n(x,p)\,\mathrm{d}x + \int_{a_l+\delta}^{a_r} b(x) f_n(x,p)\,\mathrm{d}x \\ &\quad\geq \frac{c(p)}{7}\int_{a_l}^{a_l+\delta} b(x)\,\mathrm{e}^{-3p(x-a_l)}\,\mathrm{d}x - 5c(p)\int_{a_l+\delta}^{a_r} |b(x)|\,\mathrm{e}^{-p(x-a_l)}\,\mathrm{d}x \\ &\quad\geq \frac{1}{7}\,c_0 c(p)\int_0^{\delta} \xi^m\,\mathrm{e}^{-3p\xi}\,\mathrm{d}\xi - 10\,a c_0 c(p)\,\|b\|_\infty\,\mathrm{e}^{-p\delta}\,.\end{aligned}$$

The exponential function in the first integral on the right-hand side is bounded from below by max $\{0, 1-3p\xi\}$, hence we can conclude that

$$\epsilon_n'(p) < -\frac{2c_0 c(p)}{B}\left\{\frac{(3p)^{-m-1}}{7(m+1)(m+2)} - 10\,a\|b\|_\infty\,\mathrm{e}^{-p\delta}\right\} < 0$$

holds for all sufficiently large values of p. ∎

In view of the the direct-integral decomposition $\int_{\mathbb{R}}^{\oplus} H(p)\,\mathrm{d}p$ with the fiber operators (7.16) in combination with *Theorem* 7.10 the spectrum of H_A consists of absolutely continuous bands $I_n = \big[\inf_{p\in\mathbb{R}} \epsilon_n(p),\, \sup_{p\in\mathbb{R}} \epsilon_n(p)\big]$. They may or may not overlap; it is natural to ask how many of the gaps the perturbation leaves open. We are going to show that their number is finite provided

$$A[b] := \int_{-a}^{a} b(x)\,\mathrm{d}x \neq 0\,, \tag{7.20}$$

i.e. that the flux variation per unit length of the perturbation support is nonzero. Note that such a conclusion would not be surprising in case of *Theorem* 7.9 where the band functions have different asymptotics to the left and right, however, it is less self-evident for the localized perturbation we are discussing here.

We have to compare the counting functions $N(E,p)$ and $N_0(E)$ of the operators $H(p)$ and H_0, respectively, recall that they are defined as the numbers of eigenvalues of $H(p)$ and H_0 smaller than E. Let us remark that speaking of the latter one can have in mind the simple spectrum of the fiber operators $H_0(p)$ in the decomposition $H_0 = \int_{\mathbb{R}}^{\oplus} H_0(p)\,\mathrm{d}p$ which is, of course, p-independent.

Proposition 7.2.2 *Assume (7.20). Then for any number $m \in \mathbb{N}_0$ there are p_0 and $E(m,p_0)$ such that the inequality*

$$\big(N_0(E) - N(E,p_0)\big)\,\mathrm{sgn}\,A[b] > m$$

holds for all $E > E(m,p_0)$.

Proof The assumption $\int_{-a}^{a} b(x)\,\mathrm{d}x \neq 0$ is equivalent to the fact that the quantities $a^{\pm} := \int_0^{\pm t} b(x)\,\mathrm{d}x$ do not coincide; assume for definiteness that $A[b] < 0$, in other words, $a^- > a^+$. Since we are interested in the high-energy limit we may suppose that the field variation support lies in the classically allowed region, $E > \big(p_0 + Bx + a(x)\big)^2$ for any $x \in [-t, t]$, and to employ the Bohr-Sommerfeld quantization condition obtaining

$$\pi n(E, p_0) = \int_{x^l(E)}^{x^r(E)} \sqrt{E - \big(p_0 + Bx + a(x)\big)^2}\, \mathrm{d}x + \mathcal{O}(1)\,, \tag{7.21}$$

where the classical turning points satisfy by assumption the inequalities $x^l(E) = -B^{-1}(\sqrt{E} + p_0 - a^-) < -t$ and $x^r(E) = -B^{-1}(\sqrt{E} - p_0 - a^+) > t$.

We want to compare (7.21) with the analogous expression for H_0. Since the spectrum of $-\partial_z^2 + Bz^2$ is not affected by a shift of the potential, we make the replacement $z \mapsto z + \frac{a^-}{B}$ obtaining

$$\pi n_0(E) = \int_{x^l(E)}^{x_0^r(E)} \sqrt{E - \big(p_0 + Bx + a^-\big)^2}\, \mathrm{d}x + \mathcal{O}(1)\,;$$

we employed here the fact that the left turning point is the same for both potentials, whereas the right one is moved to $x_0^r(E) = -B^{-1}(\sqrt{E} - p_0 - a^-) > t$. Since $a^- > a^+$ we have $t < x_0^r(E) < x^r(E)$. Taking further into account that the two potentials coincide to the left of $-t$, we may write the sought difference $\pi\,[N(E, p_0) - N_0(E)]$ as

$$\begin{aligned}
&\int_{-a}^{a} \left\{\sqrt{E - (p_0 + Bx + a(x))^2} - \sqrt{E - (p_0 + Bx + a^-)^2}\right\} \mathrm{d}x \\
&+ \int_{a}^{x_0^r(E)} \left\{\sqrt{E - (p_0 + Bx + a^+)^2} - \sqrt{E - (p_0 + Bx + a^-)^2}\right\} \mathrm{d}x \\
&+ \int_{x_0^r(E)}^{x^r(E)} \sqrt{E - (p_0 + Bx + a^+)^2}\, \mathrm{d}x + \mathcal{O}(1)\,,
\end{aligned}$$

the last term being a positive number independent of E. In the first term we integrate over a fixed interval, hence it behaves as $\mathcal{O}(E^{-1/2})$ in the limit $E \to \infty$ and may be absorbed into the error term. Furthermore, choosing $p_0 \geq -a^+ - tB$ we achieve that the integrand in the second term is positive, which implies

$$\pi\,[n(E, p_0) - n_0(E)] \geq \int_{x_0^r(E)}^{x^r(E)} \sqrt{E - (p_0 + Bx + a^+)^2}\, \mathrm{d}x + \mathcal{O}(1)\,.$$

It remains to estimate the last integral. Since the function is non-negative, decreasing and vanishes only at $x = x^r(E)$, it is bounded from below by

$$\int_{x_0^r(E)}^{x^r(E)-\delta} \sqrt{E - (p_0 + Bx + a^+)^2}\, \mathrm{d}x \geq \sqrt{2B\delta\sqrt{E} - B^2\delta^2}\left(\frac{a^- - a^+}{B} - \delta\right)$$

for any $\delta \in \big(0, x^r(E) - x_0^r(E)\big)$; choosing the latter sufficiently small to make the last factor positive we get the sought result for E large enough. The inequality $N(E, p_0) < N_0(E) - c$ for $A[b] > 0$ is obtained in the same way. ∎

Corollary 7.2.2 *The number of open gaps in $\sigma(H_A)$ is finite provided $A[b] \neq 0$.*

Proof Suppose again $a^- > a^+$. Since $\epsilon_n(p) \to (2n+1)B$ holds as $p \to \infty$ for a fixed $n \in \mathbb{N}$ by *Lemma* 7.2.2, it is sufficient to find $\tilde{n}$ and $\tilde{p}$ such that the inequality $\epsilon_{n+1}(\tilde{p}) < (2n+1)B$ is valid for all $n \geq \tilde{n}$; this follows immediately from *Proposition* 7.2.2 with $m = 2$. In the opposite case, $a^- < a^+$, the inequality is replaced by $\epsilon_{n-1}(\tilde{p}) > (2n+1)B$ and the argument is analogous. ∎

7.3 Notes

Section 7.1 *Theorem* 7.1 is essentially due to [Ex95]; note that the assumption of a one-sided tilt excludes classical bound states in such a guide which in general may exist. The inequality $H_\Omega(F') \geq H_\Omega(F)$ for $F' \geq F$ implies by the minimax principle that the eigenvalues are nondecreasing functions of the field strength. It does not automatically imply the monotonicity of $N(F) := N(H_\Omega(F), \lambda_1(F))$ because the function $\lambda_1(\cdot)$ is increasing, however, one can prove it in particular situations. Having thus N noninteracting fermions bound in a locally curved tube, one is able to eject them sequentially by applying an increasing electric field—one can speak in this situation of a *quantum pipette*.

The Hardy-type inequality (7.3) as well as *Theorem* 7.3 come from [EkKo05], for the diamagnetic inequality see, e.g., [Ka73] or [RS, Theorem X.27]. As mentioned in Remark 7.1.1b, a local magnetic field also removes bound states induced by small local deformations of the waveguide, see Problem 3, or by the presence of a (sufficiently short) Neumann window [BEK05]. Note that similarly to the case of locally twisted three-dimensional tubes, the repulsive effect of the magnetic field in two-dimensional strips decays with the square of the distance from the support of the field.

Theorems 7.4 and 7.5 have been proven in [DEM01] where one can also find extensions of these results, in particular, a demonstration of analogous properties for potentially confined magnetic guides. Properties of the band functions $\nu_j(p)$ used in the proof of *Theorem* 7.4 are discussed in detail in [GS97], see also [BRS08].

Section 7.2 For the Feynman-Hellmann formula see, e.g., [Ka, Sect.VII.3.4]. *Theorem* 7.6 is due to [dBP99]; similar results hold in the situation when the hard wall described by Dirichlet boundary conditions is replaced by a smooth confining potential [MMP99]. The extension to more general planar domains contained in *Theorem* 7.7 was established in [FGW00]. The boundary regularity property can be weakened; it is enough to suppose that $\partial\Omega$ has the uniform C^3 property in the sense of [Ad]. The conjugate operator is in this case chosen as $\Pi = s(\vec{x}) - \nabla s \cdot \vec{\pi}^\perp$, where $\vec{\pi}^\perp$ is canonically conjugate momentum turned to the perpendicular direction.

As we have indicated the result may not hold if the wedge condition is violated, for instance, a small but non-decaying impurity potential in a magnetic strip may lead to Anderson localization due to interference between the involved edge states. Stability of edge states under various perturbations, including those coming from random potentials, was analyzed in [HS08a, HS08b], for the scattering theory of edge states see [BG06].

Theorem 7.8 was proved in [EJK99] where other properties of spectral bands of this system were also discussed. For properties of the confluent hypergeometric function used in the proof see [AS, Chap. 13], the argument showing that the Landau levels remain in the spectrum of the perturbed operator follows [DMP99].

The first treatment of translationally invariant magnetic fields including the proof of *Theorem* 7.9 comes from the paper [Iw85] where other situations are also treated including the case when $B(x)$ has the same limit at $\pm\infty$, see also [MP97] for extensions of Iwatsuka's results. The conjecture about universality of transport in the presence of non-constant, translationally invariant fields was put forth in [CFKS, Sect. 6.5]. Note that the magnetic field need not be sign-definite. If, for instance, the plane is divided into two halfplane supporting homogeneous fields of opposite orientation, the spectrum of H_A is absolutely continuous again; a classical charged particle then follows in the vicinity of the interface trajectories referred to as *snake orbits* [RP00]. For results concerning quantization of edge currents in the Iwatsuka model we refer to [DGR11].

Theorem 7.10 comes from [EKo00]. The Bohr-Sommerfeld quantization condition which yields Eq. (7.21) can be found, e.g., in [Ti, Theorem 7.5]. With respect to *Corollary* 7.2.2 we note that the case of a zero-mean field variation when (7.20) fails is more involved; a particular example of such a field is treated in [EKo00, Sect. 3].

7.4 Problems

1. Prove *Proposition* 7.1.1.
Hint: Use the diamagnetic inequality (7.2) to show that $\sigma_{\mathrm{ess}}(H_B^\Omega) \subset [1,\infty)$. To prove the opposite inclusion construct appropriate Weyl sequences.

2. Fill in the details of the proof of *Theorem* 7.3.

3. Let $H_B^{\Omega_\lambda}$ be the magnetic Hamiltonian in a weakly deformed strip, i.e. the operator associated with the form (7.6) in which the curved strip Ω is replaced by (6.17). Show that if the function $f \in C_0^\infty(\mathbb{R})$ and the magnetic field satisfies the assumptions of *Theorem* 7.3, $\sigma_{\mathrm{disc}}(H_B^{\Omega_\lambda}) = \emptyset$ holds for all λ small enough.
Hint: Cf. [EkKo05].

4. Modify the claim of *Theorem* 7.3 for the case of the *Aharonov-Bohm field* corresponding to the magnetic vector potential of the form

$$A(x,y) = \Phi\left(\frac{-(y-y_0)}{x^2+(y-y_0)^2}, \frac{x}{x^2+(y-y_0)^2}\right)$$

which gives rise to the magnetic flux $2\pi\Phi$ supported by the point $(0, y_0) \in \Omega_0$. Prove that for a non-integer Φ the inequality (7.3) holds with the right-hand side replaced by $c_{AB} \int_{\Omega_0} \frac{|u(x,y)|^2}{x^2+(y-y_0)^2}\, dx\, dy$ with a positive constant c_{AB}.
Hint: Cf. [EkKo05].
5. Fill in the details of the proof of *Theorem* 7.5. Derive the leading term in the resonance-width expansion in the case when both the weak potential and magnetic-field perturbations are present.
Hint: Cf. [DEM01].
6. The claim of *Theorems* 7.4 and 7.5 can be modified to the case of a guide with potential confinement, i.e. for the operator $H_\lambda(B) := (-i\partial_x - By)^2 - \partial_y^2 + V(x) + W(y) + \lambda U(\vec{x})$ on $L^2(\mathbb{R}^2)$, where U, V are as before and the confining potential W satisfies, e.g., the inequality $W(y) \geq cy^2$ for some $c > 0$.
7. Extend the transformation (1.7) to curved strips in a homogeneous magnetic field. Show that $(-i\nabla + A)^2$ on $L^2(\Omega)$ with Dirichlet conditions is unitarily equivalent to

$$H(B) = \Pi_s^* \Pi_s - \partial_u^2 + V(s,u)$$

on $L^2(\Omega_0)$ with Dirichlet condition at $|u| = a$, where $V(s,u)$ is the effective potential (1.8) and $\Pi_s := \frac{1}{1+u\gamma(s)}\{-i\partial_s^2 + Bu(1 + \frac{1}{2}u\gamma(s))\}$.
Hint: Use a suitable modification of the Landau gauge, cf. [Ex93].
8. Prove assertion (c) of *Proposition* 7.2.1.
Hint: Employ the Feynman-Hellmann formula.
9. The operator $h(p)$ defined in (7.8) is by an obvious change of variables unitarily equivalent to $\hat{h}(p) = -\partial_z^2 + B^2z^2$ in $L^2(-p/B, \infty)$ with Dirichlet boundary condition at $z = -\frac{p}{B}$. Check that $\hat{h}(p)$ converges to $-\partial_z^2 + B^2z^2$ in $L^2(\mathbb{R})$ in the strong resolvent sense as $p \to +\infty$.
Hint: Modify the argument used in the proof of *Lemma* 7.2.2.
10. Fill in the details of the proof of *Theorem* 7.6. In particular, show that the commutator $[iy, H_0] = [iy, H]$ is H_0-bounded and that $[iy, [iy, H_0]]$ is bounded.
11. Prove *Lemma* 7.2.3.
Hint: Mimic the proof of *Proposition* 7.2.1a; write $H(p) = H(0) + 2p(a(x) + Bx) + p^2$ and check that $a(x) + Bx$ is $H(0)$-bounded with relative bound zero.
12. Prove the limit (7.17).
Hint: Check first that if $f \in C^2(\mathbb{R})$ satisfies $\lim_{x\to\infty} f(x) = \lim_{x\to\infty} f''(x) = 0$, then $\lim_{x\to\infty} f'(x) = 0$, then employ the fact that $\psi_n''(x,p) = Q_{n,p}(x)\psi_n(x,p)$.

Chapter 8
Graph Limits of Thin Network Systems

... there is a risk of being lost in the maze of tangled structures and crevasses, sometimes reminiscent of jumbled colonnades, sometimes of petrified geysers.

Stanislaw Łem, Solaris

There are numerous situations when the dominating feature of a waveguide system is its essentially one-dimensional nature. We have seen examples in the previous chapters, for instance, in Sect. 1.6 where we have discussed the zero-width limit of a bent waveguide. The situation becomes more complicated when the system in question is not a single duct but rather a network of nontrivial topology the "skeleton" of which is a graph. The squeezing limit of such systems is the subject of this chapter; we will see that the limiting behavior depends crucially on the boundary conditions defining the Laplacian on the network.

Properties of quantum particles the motion of which is confined to a graph represents a rich topic which would deserve a separate monograph. Here we limit ourselves to presenting the basic notions and describing the the features we shall need for our purpose. They concern in the first place coupling of the wave functions at the graph vertices which makes the corresponding Schrödinger operator self-adjoint; we shall describe general coupling conditions with this property and various approximations of the couplings using regular and singular potentials. Armed with these tools we shall then investigate spectral convergence of Schrödinger operators on quasi-one-dimensional manifolds, which we can pictorially characterize as "fat graphs", shrinking to a given graph.

8.1 Quantum Graphs

By a ***quantum graph*** we mean a pair (Γ, H) where Γ is a metric graph consisting of a countable number of edges, i.e. one-dimensional line segments, connected by a countable number of vertices, and H is an operator on a Hilbert space associated with the graph Γ, typically $L^2(\Gamma)$, which plays here the role of the system Hamiltonian.

P. Exner and H. Kovařík, *Quantum Waveguides*,
Theoretical and Mathematical Physics, DOI 10.1007/978-3-319-18576-7_8

Let us look into these notions in more detail. A ***graph*** is a family of ***vertices*** $\mathcal{V} = \{v_i : i \in \mathcal{I}\}$, where $\mathcal{I}$ is at most countable, and ***edges*** $\mathcal{E} = \{e_{ij} : (i, j) \in \mathcal{I}_{\mathcal{E}} \subset \mathcal{I} \times \mathcal{I}\}$. The index set $\mathcal{I}_{\mathcal{E}}$ characterizes the adjacency of the graph telling us which vertices are connected by an edge. The ***degree of a vertex*** is defined as the number of edges connected to this vertex; we consider only graphs where all the vertex degrees are finite. One may suppose that each pair of vertices is connected by not more than one edge, otherwise we can always add a "dummy" vertex of degree two to any "superfluous" edge. The above definition is purely combinatorial, the second ingredient to add is a metric structure. We suppose that Γ is a ***metric graph***, which means that every edge e_n is isometric with a line segment $I_j := [0, \ell_j]$; the edge lengths ℓ_j may be finite or infinite. It often happens that the graphs we consider are subsets of a Euclidean space which induces the said local metric structure by embedding, however, it is not always the case and we shall not need such an assumption for most of this chapter.

The local metric makes it possible to introduce the Hilbert space $L^2(\Gamma) := \bigoplus_{j\in\mathcal{J}} L^2(I_j)$ with elements written as $\Psi = \{\psi_j\}_{j\in\mathcal{J}}$, or simply $\{\psi_j\}$. Since we are going to discuss here the dynamics of nonrelativistic and spinless particles, this will be the state space; in other situations it can be replaced by more complicated spaces such as $L^2(\Gamma; \mathcal{G})$ where $\mathcal{G}$ corresponds to the internal degrees of freedom, being $\mathbb{C}^2$ for electrons, etc. Assuming that the j-th edge supports a real-valued potential $V_j \in L^1_{\mathrm{loc}}(I_j)$ and putting $V := \{V_j\}$ we naturally define the corresponding Schrödinger operator on Γ by

$$H(V)\{\psi_j\} := \{-\psi_j'' + V_j\,\psi_j\}. \tag{8.1}$$

The definition needs to fix the domain, of course. It will consist of Ψ whose components ψ_j belong to the Sobolev space $H^2(I_j)$; the crucial point is to choose the coupling conditions for the wave functions at the graph vertices which will make $H(V)$ acting as (8.1) a self-adjoint operator.

The domain choice is not unique. In order to see how many self-adjoint operators can correspond to the symbol appearing in (8.1) one can employ the theory of self-adjoint extensions. The considerations are not affected by the potentials unless the latter are strongly singular at the edge endpoints; for the sake of simplicity we thus put $V = 0$. Suppose first that the graph is fully decoupled in the sense that the operator at each edge is determined by Dirichlet conditions at the edge endpoint, and the corresponding self-adjoint Laplacian on $L^2(\Gamma)$ is an orthogonal sum of its component operators.

Restricting this operator to functions with supports separated from the vertices we get a symmetric operator H_0 with deficiency indices (d, d) where d is the number of edge ends, and the sought Hamiltonians have to be chosen among its self-adjoint extensions. Not all of them are suitable, though, because we are not interested here in the dynamics which allows the particle to hop from one vertex to another. Consequently, we restrict ourselves to *local* couplings which relate boundary values of edges meeting in each vertex separately. The number of parameters characterizing such a self-adjoint extension is, of course, $\sum_{j\in I} d_j^2$ where d_j is the degree of the jth vertex.

Since the operators in question are differential of the second order, the best way to describe their self-adjoint extensions is through boundary conditions coupling the boundary values

$$\psi_j(0+) := \lim_{x\to 0+} \psi_j(x)\,, \quad \psi_j'(0+) := \lim_{x\to 0+} \psi_j'(x)\,, \tag{8.2}$$

where the variable x measures the distance of an edge point from the vertex; the limits make sense for functions $\psi_j \in H^2(I_j)$ which constitute the domain of the adjoint H_0^*. To simplify the discussion, let us consider a star-shaped Γ with a single vertex and n half-line edges joined there, $\mathcal{H} = \bigoplus_{j=1}^n L^2(\mathbb{R}_+)$. The self-adjoint extensions are maximal restrictions of H_0^* which annulate the boundary form $(\Phi, \Psi) \mapsto (\Phi, H_0^*\Psi) - (H_0^*\Phi, \Psi)$, or explicitly

$$(\Phi, \Psi) \mapsto \sum_{j=1}^{n} \Big(\overline{\phi}_j(0+)\psi_j'(0+) - \overline{\phi}_j'(0+)\psi_j(0+)\Big)\,. \tag{8.3}$$

The latter can alternatively be rewritten as a symplectic form $([\Phi], M[\Psi])$ on $\mathbb{C}^{2n}$, where $M = \begin{pmatrix} 0 & I \\ -I & 0 \end{pmatrix}$ and the symbol $[\Psi] := \begin{pmatrix} \Psi(0+) \\ \Psi'(0+) \end{pmatrix}$ denotes the $2n$-dimensional vector of boundary values. Thus one has to find the ***Lagrangean subspaces***, in other words, maximal subspaces in $\mathbb{C}^{2n}$ on which the form (8.3) vanishes.

Proposition 8.1.1 *(a) Any self-adjoint extension of H_0 is characterized by the conditions*

$$A\Psi(0+) + B\Psi'(0+) = 0\,, \tag{8.4}$$

where A, B are $n\times n$ matrices such that the $n\times 2n$ matrix $(A|B)$ has maximum rank and AB^* is Hermitean. Conversely, any pair of matrices with these properties determines through (8.4) a self-adjoint extension of H_0.
(b) The on-shell scattering matrix at momentum k for a star graph with the vertex coupling (8.4) equals $S(k) = -(A+ikB)^{-1}(A-ikB)$.
(c) Any self–adjoint extension of H_0 is uniquely characterized by (8.4) with $A = U - I$ and $B = i(U+I)$, where U is an $n\times n$ unitary matrix.

Proof is left to the reader (Problem 1).

The last claim removes the inherent non-uniqueness of the conditions (8.4). This is not the only way to characterize the couplings uniquely; another one will be mentioned in Sect. 8.2.3 below, see also Problem 2.

The higher the degree of the vertex, the larger the coupling condition family. Let us mention two subfamilies which will be important in the following.

Examples 8.1.1 (a) *δ-coupling*: choosing $U = \frac{2}{n+i\alpha}N - I$ with $\alpha \in \mathbb{R}\cup\{\infty\}$, where N is the $n \times n$ matrix whose entries are all equal to one, we get the coupling which can be cast into the form

$$\psi_1(0+) = \cdots = \psi_n(0+) =: \psi(0)\,, \qquad \sum_{j=1}^{n} \psi'_j(0+) = \alpha\,\psi(0)\,; \tag{8.5}$$

the name is inspired by the fact that for $n = 2$ the conditions describe the usual δ interaction on the line. Note that it is the only family of couplings with the wave functions continuous in the vertex. For $\alpha = \infty$ we have $U = -I$ and the requirement (8.5) is replaced by the Dirichlet condition, $\psi_j(0+) = 0\,,\ j = 1, \dots, n$. Another particular case of interest is $\alpha = 0$. The corresponding conditions (8.5) are called *free* or *Kirchhoff*.

(b) δ'_s*-coupling:* another possible choice of the matrix U is $I - \frac{2}{n-i\beta}N$ with $\beta \in \mathbb{R}\cup\{\infty\}$ which gives rise to the conditions

$$\psi'_1(0+) = \cdots = \psi'_n(0+) =: \psi'(0+)\,, \qquad \sum_{j=1}^{n} \psi_j(0+) = \beta\,\psi'(0+)\,. \tag{8.6}$$

They are a counterpart of (8.5) with the roles of the function and derivative boundary values interchanged. If $\beta = \infty$ we have $U = I$ which leads to the full Neumann decoupling. Note that both the δ and δ'_s-couplings have the property of being invariant with respect to edge permutations – cf. also Problem 4.

The appeal of the quantum graph concept is that it provides us with a wide family of models which are interesting both from the physical and mathematical point of view. We restrict ourselves here to the problem indicated in the opening of the chapter; in the notes we mention some other problems related to quantum graphs and give a guide to further reading.

8.2 Vertex Coupling Approximations

Before coming to the main topic of this chapter let us examine how different vertex couplings can be approximated on the graph itself, either using suitable families of potentials or by other means such as a local change of the graph topology. Since the coupling concerns a single vertex we again consider in this section a star-shaped Γ with a single vertex and n half-line edges joined there.

8.2.1 δ-Coupling

First we are going to analyze an approximation inspired by the way in which a δ-interaction on a line can be understood as a limit of Schrödinger operators with suitably scaled potentials. We shall employ the symbol $H_\alpha(V)$ for the operator (8.1) with the δ-coupling (8.5) in the vertex. Given an n-tuple of functions $W_j : \mathbb{R}_+ \to \mathbb{R}$, we construct a family of squeezed potentials

$$W_{\varepsilon,j} := \frac{1}{\varepsilon}\, W_j\left(\frac{x}{\varepsilon}\right), \quad j = 1, \dots, n\,. \tag{8.7}$$

Then we can make the following claim.

Theorem 8.1 *Suppose that the functions $V_j \in L^1_{\mathrm{loc}}(\mathbb{R}_+)$ are bounded below and $W_j \in L^1(\mathbb{R}_+)$ for $j = 1, \dots, N$. Let further $H(V + W_\varepsilon)$ have the limit-point property at infinity for any ε; then*

$$H_0(V + W_\varepsilon) \longrightarrow H_\alpha(V) \quad \text{as} \quad \varepsilon \to 0+ \tag{8.8}$$

holds in the norm resolvent sense, where $\alpha := \sum_{j=1}^N \int_0^\infty W_j(x)\,\mathrm{d}x$.

Proof Under the stated assumptions the involved operators are self-adjoint. To find their resolvent difference, we first find the integral kernel $G^{\alpha,V}_{j\ell}(x, y; k)$ of $(H_\alpha(V) - k^2)^{-1}$ which is a matrix valued operator on $L^2(\mathbb{R}_+; \mathbb{C}^n)$. Given k with $\mathrm{Im}\, k \geq 0$, $k^2 \in \rho(H_\infty(V))$, we denote by $u_j \equiv u_j(\cdot; k)$ and $v_j \equiv v_j(\cdot; k)$ the solutions of the equation

$$h_j(V_j)\psi_j := -\psi_j'' + V_j\psi_j = k^2\psi_j$$

such that $u_j(0; k) = 0$ and $v_j \in L^2$ at infinity. If $\alpha = \infty$, the edges are decoupled and the corresponding components of $H_\infty(V)$ are characterized by the resolvent kernels

$$g_j(x, y; k) := -\frac{u_j(x_<; k)v_j(x_>; k)}{W(u_j, v_j)}\,,$$

where as usual $x_< := \min\{x, y\}$ and $x_> := \max\{x, y\}$, while $W(u_j, v_j)$ is the Wronskian of the two solutions. The operators $H_\alpha(V)$ and $H_\infty(V)$ are self-adjoint extensions of the same symmetric operator with deficiency indices (n, n) and applying Krein's formula we easily find the relation

$$G^{\alpha,V}_{j\ell}(x, y; k) = \delta_{j\ell}\, g_j(x, y; k) + \frac{v_j(x; k)\, v_\ell(y; k)}{v_j(0; k)v_\ell(0; k)(\alpha - M(k))}\,, \tag{8.9}$$

where we have put $M(k) := \sum_{j=1}^N \frac{v_j'(0;k)}{v_j(0;k)}$ (Problem 5). Consider further the resolvent kernel $G^{0,V+W_\varepsilon}_{j\ell}(x, y; k)$ of the operator $H_0(V + W_\varepsilon)$. Using the resolvent formula we can rewrite it as

$$G^{0,V+W_\varepsilon}_{j\ell}(x,y;k) = G^{0,V}_{j\ell}(x,y;k) - \sum_{r,s} \int_0^\infty \int_0^\infty G^{0,V}_{j\ell}(x,y';k)\, W_{\varepsilon,r}(x')^{1/2}$$
$$\times \left(I + |W_\varepsilon|^{1/2}\left(H_0(V)-k^2\right)^{-1} W_\varepsilon^{1/2}\right)^{-1}_{rs}(x',x'')\,|W_{\varepsilon,r}(x'')|^{1/2}$$
$$\times\, G^{0,V}_{s\ell}(x'',y;k)\,\mathrm{d}x'\,\mathrm{d}x''.$$

Changing the integration variables to x'/ε and x''/ε we can rewrite the resolvent in question concisely as $-B_{k,\varepsilon}\left(I + C_{k,\varepsilon}\right)^{-1} \tilde{B}_{k,\varepsilon}$, where the involved operators are determined by their integral kernels,

$$\begin{aligned}
(B_{k,\varepsilon})_{j\ell}(x,y) &= G^{0,V}_{j\ell}(x,\varepsilon y;k)\, W_\ell(y)^{1/2},\\
(\tilde{B}_{k,\varepsilon})_{j\ell}(x,y) &= |W_j(x)|^{1/2}\, G^{0,V}_{j\ell}(\varepsilon x,y;k),\\
(C_{k,\varepsilon})_{j\ell}(x,y) &= |W_j(x)|^{1/2}\, G^{0,V}_{j\ell}(\varepsilon x,\varepsilon y;k)\, W_\ell(y)^{1/2},
\end{aligned}$$

which converge pointwise to

$$\begin{aligned}
(B_k)_{j\ell}(x,y) &= G^{0,V}_{j\ell}(x,0;k)\, W_\ell(y)^{1/2},\\
(\tilde{B}_k)_{j\ell}(x,y) &= |W_j(x)|^{1/2}\, G^{0,V}_{j\ell}(0,y;k),\\
(C_k)_{j\ell}(x,y) &= |W_j(x)|^{1/2}\, G^{0,V}_{j\ell}(0,0;k)\, W_\ell(y)^{1/2},
\end{aligned}$$

respectively, as $\varepsilon \to 0+$. The explicit form of the last operator makes it possible to find the kernel of its inverse,

$$(I + C_k)^{-1}_{j\ell}(x,y) = \delta(x-y)\delta_{j\ell} - \frac{|W_j(x)|^{1/2} W_\ell(y)^{1/2}}{\langle W\rangle - M(k)},$$

where $\langle W\rangle := \sum_{j=1}^N \int_0^\infty W_j(x)\,\mathrm{d}x$; we employ here the fact that $G^{0,V}_{j\ell}(0,0;k) = -M(k)^{-1}$ by (8.9). Thus, in the limit, the resolvent difference has the kernel

$$\begin{aligned}
&-\sum_r \int_0^\infty \mathrm{d}x'\, W_r(x')\, G^0_{jr}(x,0;k) G^0_{r\ell}(0,y;k)\\
&+\sum_{r,s} \int_0^\infty \int_0^\infty \mathrm{d}x'\,\mathrm{d}x''\, W_r(x') W_s(x'')\, \frac{G^0_{jr}(x,0;k) G^0_{s\ell}(0,y;k)}{\langle W\rangle - M(k)}\\
&= \frac{v_j(x;k) v_\ell(y;k)}{v_j(0;k) v_\ell(0;k)}\, \frac{\langle W\rangle}{M(k)(\langle W\rangle - M(k))},
\end{aligned}$$

which will coincide with the difference of $G^{\alpha,V}_{j\ell}(x,y;k)$ and $G^{0,V}_{j\ell}(x,y;k)$ given by (8.9) provided we set $\alpha = \langle W\rangle$. It is sufficient to check the norm-resolvent convergence for a particular value of k. Since all the operators $h_j(V_j)$ are bounded below by assumption, one may choose $k = i\kappa$ with κ large enough to get exponentially decaying solutions $v_j(\cdot; k)$ at the semi-infinite edges; the convergence is then easily established (Problem 5). ∎

8.2.2 δ_s'-Coupling

The approximation using scaled potentials discussed in the previous section yields a nontrivial result only if the wave functions are continuous at the vertex, and we need different ideas to deal with the other couplings. Since we already know how to approximate the δ-couplings, it is natural to try to use them as "building blocks" and to construct an approximation by amending the graph with additional vertices. We shall describe now how such an approximation can work for the coupling conditions (8.6).

We consider again a star-shaped graph Γ with n edges and suppose that no potential influences the particle. To approximate the Hamiltonian H_β on $L^2(\Gamma)$ with the coupling (8.6), we employ the scheme sketched in Fig. 8.1: we place a δ-interaction at a distance a at each edge, in other words an additional vertex of degree two, with the a-dependent coupling strength $c(a)$, and suppose that in the center we have the δ-coupling of strength $b(a)$. The corresponding Hamiltonian will be denoted by $H^{b,c}(a)$; the crucial point is the choice of the functions $b,\ c$.

Theorem 8.2 *Let* $b(a) = -\frac{\beta}{a^2}$ *and* $c(a) = -\frac{1}{a}$. *Then* $H^{b,c}(a) \to H_\beta$ *as* $a \to 0+$ *in the norm resolvent sense, the convergence rate being* $\mathcal{O}(a)$.

Proof We shall give a sketch only, leaving the details to the reader (Problem 6). We employ the fact that both the operators $H^{b,c}(a)$ and H_β exhibit a symmetry with respect to permutation of the edges, and thus they can be decomposed into a component acting on the subspace of functions which depend on the distance from the central vertex only, in other words, $\Psi = \{\psi_j\}$ such that

$$\psi_j(x) = \psi_k(x)\,, \quad j, k = 1, \ldots, n\,,$$

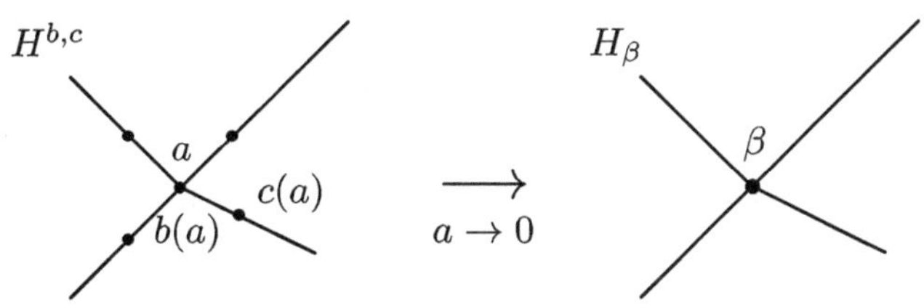

Fig. 8.1 Approximation scheme for the δ_s'-coupling

satisfying a mixed boundary condition at $x = 0$, and a part in the orthogonal complement which satisfies Dirichlet and Neumann conditions at the central vertex for the δ- and δ'_s-couplings, respectively. Since the same decomposition can be done for the corresponding resolvents, the task is reduced to the investigation of a pair of halfline problems. We shall describe the symmetric case; the other one is dealt with analogously (Problem 6). We have to write down Green's functions of the respective operators. For the halfline operator with the δ-coupling of strength b at the origin the resolvent kernel at energy $k^2 = -\kappa^2$ equals

$$G_\kappa^b(x,y) = \frac{e^{-\kappa x_>}}{\kappa\,(b+\kappa)}\big(b\sinh(\kappa x_<) + \kappa\cosh(\kappa x_<)\big)\,.$$

By Krein's formula we then obtain the part of the Green function of $H^{b,c}(a)$ acting on the above described symmetric subspace,

$$G_\kappa^{b,c}(x,y) = G_\kappa^b(x,y) + \frac{G_\kappa^b(x_<,a)\,G_\kappa^b(a,x_>)}{-c^{-1} - G_\kappa^b(a,a)}\,.$$

On the other hand, the symmetric part of the resolvent kernel of H_β equals

$$G_\kappa^\beta(x,y) = \frac{e^{-\kappa x_>}}{\kappa\,(N+\beta\kappa)}\,(N\sinh(\kappa x_<) + \beta\kappa\cosh(\kappa x_<)),\quad a \le x \le y\,.$$

Substituting $b = -\frac{\beta}{a^2}$ and $c = -\frac{1}{a}$ into the expression of $G_\kappa^{b,c}(x,y)$, we find by a straightforward computation that

$$\lim_{a\to 0+} G_\kappa^{b,c}(x,y) = G_\kappa^\beta(x,y)$$

holds for all $x, y > 0$. Since the two Green functions decay exponentially in both variables, one easily concludes that the resolvent difference converges to zero in the Hilbert-Schmidt norm, and therefore also in the operator norm. ∎

8.2.3 General Singular Vertex Coupling

The method we employed in the previous example can be extended to a wider class of vertex couplings, however, adding δ-type vertices alone is not sufficient to deal with the general singular conditions (see the notes). To approximate an arbitrary coupling (8.4), we need two more ingredients. First of all, we have to locally change the graph topology, adding not only vertices but also edges which would shrink to zero in the limit. In this way one can get (8.4) with real matrices A, B; to overcome this restriction one also has to introduce local magnetic fields, i.e. to place suitable vector potentials at the added edges.

First we will rewrite the coupling conditions (8.4) into another form which is again simple and unique but requires an appropriate edge numbering.

Proposition 8.2.1 *For a quantum graph vertex of degree n, the following is valid:* *(a) If* $S \in \mathbb{C}^{m,m}$ *with* $m \le n$ *is a Hermitean matrix and* $T \in \mathbb{C}^{m,n-m}$*, then the equation*

$$\begin{pmatrix} I^{(m)} & T \\ 0 & 0 \end{pmatrix} \Psi'(0+) = \begin{pmatrix} S & 0 \\ -T^* & I^{(n-m)} \end{pmatrix} \Psi(0+) \tag{8.10}$$

expresses admissible boundary conditions giving rise to a self-adjoint operator. *(b) Conversely, for any vertex coupling there is a number* $m \le n$ *and a numbering of edges such that the coupling is described by the conditions (8.10) with uniquely given matrices* $T \in \mathbb{C}^{m,n-m}$ *and* $S = S^* \in \mathbb{C}^{m,m}$*. If the edge numbering is given one can bring the coupling into the form (8.10) by a permutation* $(1, \dots, n) \mapsto (\Pi(1), \dots, \Pi(n))$ *of the edge indices with the matrices* S, T *uniquely determined by the permutation* Π*.*

Proof It is easy to check that the matrices

$$A = \begin{pmatrix} -S & 0 \\ T^* & -I^{(n-m)} \end{pmatrix}, \quad B = \begin{pmatrix} I^{(n)} & T \\ 0 & 0 \end{pmatrix},$$

satisfy the requirements of *Proposition* 8.1.1: the $n \times 2n$ matrix $(A|B)$ obviously has rank n and $AB^* = \mathrm{diag}(-S, 0)$ is Hermitean.

The task is to prove the first part of the claim (b), the second one will follow by a simultaneous permutation of the elements of Ψ and Ψ'. We shall start with a fixed coupling condition (8.4) and show that one can cast it into the form (8.10). This means that we have to find a number $m \le n$, a numbering of the edges and the corresponding matrices S and T, and moreover, we have to show that such a number m is the only possible and that S, T depend uniquely on the edge numbering. To this end we may use only manipulations that do not affect the meaning of the coupling, namely (i) simultaneous permutation of columns of the matrices A, B combined with corresponding simultaneous permutation of components in Ψ and Ψ', (ii) left multiplication by a regular matrix.

We see from (8.10) that m is the rank of the matrix applied to Ψ'. We note that the rank of this matrix and the other matrix is not influenced by any of the manipulations mentioned above, hence it is obvious that $m = \mathrm{rank}\, B$ is given uniquely. Then there is an m-tuple of linearly independent columns of the matrix B, let their indices be $j_1, \dots, j_m$. We permute simultaneously the columns of B and A so that those with indices $j_1, \dots, j_m$ are now at the positions $1, \dots, m$, and we do the same with the components of the vectors Ψ, Ψ'. Labeling the permuted matrices A, B and the vectors Ψ, Ψ' with tildes, we get $\tilde{A}\tilde{\Psi} + \tilde{B}\tilde{\Psi}' = 0$ where for simplicity we drop the argument. Furthermore, since $\mathrm{rank}(\tilde{B}) = \mathrm{rank}(B) = m$, there are m rows of $\tilde{B}$ that are linearly independent; let their indices be $i_1, \dots, i_m$. First we permute the rows in $\tilde{A}\tilde{\Psi} + \tilde{B}\tilde{\Psi}' = 0$ so those with the indices $i_1, \dots, i_m$ are moved to the positions $1, \dots, m$; note that this corresponds to a matrix multiplication of the whole by a permutation matrix from the left. In this way we pass from $\tilde{A}$, $\tilde{B}$ to matrices which we denote by $\check{A}$, $\check{B}$; it is obvious that this operation keeps the first m columns of the matrix $\check{B}$ linearly independent.

In the next step we add to each of the last $n-m$ rows of $\check{A}\tilde{\Psi}+\check{B}\tilde{\Psi}'=0$ a linear combination of the first m rows such that all of the last $n-m$ rows of $\check{B}$ vanish. This is possible, because the last $n-m$ rows of $\check{B}$ are linearly dependent on the first m rows. This is again an authorized operation, not changing the meaning of the boundary conditions; the resulting matrices on the left-hand side will be denoted by $\hat{B}$ and $\hat{A}$, and the condition will become $\hat{A}\tilde{\Psi}+\hat{B}\tilde{\Psi}'=0$. It is clear from the construction that the matrix $\hat{B}$ has the block form,

$$\hat{B}=\begin{pmatrix} \hat{\mathcal{B}}_{11} & \hat{\mathcal{B}}_{12} \\ 0 & 0 \end{pmatrix},$$

where $\hat{\mathcal{B}}_{12}\in\mathbb{C}^{m,n-m}$ and the square matrix $\hat{\mathcal{B}}_{11}\in\mathbb{C}^{m,m}$ is regular, because its columns are linearly independent. We proceed by multiplying the system $\hat{A}\tilde{\Psi}+\hat{B}\tilde{\Psi}'=0$ from the left by the matrix $\mathrm{diag}(\hat{\mathcal{B}}_{11}^{-1},I^{(n-m)})$ arriving at boundary conditions

$$\begin{pmatrix} \mathcal{A}_{11} & \mathcal{A}_{12} \\ \mathcal{A}_{21} & \mathcal{A}_{22} \end{pmatrix}\tilde{\Psi}+\begin{pmatrix} I^{(m)} & \mathcal{B}_{12} \\ 0 & 0 \end{pmatrix}\tilde{\Psi}'=0$$

with $\mathcal{B}_{12}:=\hat{\mathcal{B}}_{11}^{-1}\hat{\mathcal{B}}_{12}$. They are equivalent to (8.4), hence the involved matrices have to satisfy the requirements of *Proposition* 8.1.1. Let us begin with the second one: the corresponding matrix product is Hermitean if and only if $\mathcal{A}_{11}+\mathcal{A}_{12}\mathcal{B}_{12}^*$ is Hermitean and $\mathcal{A}_{21}+\mathcal{A}_{22}\mathcal{B}_{12}^*=0$. We infer that $\mathcal{A}_{21}=-\mathcal{A}_{22}\mathcal{B}_{12}^*$, hence the above condition acquires the form

$$\begin{pmatrix} \mathcal{A}_{11} & \mathcal{A}_{12} \\ -\mathcal{A}_{22}\mathcal{B}_{12}^* & \mathcal{A}_{22} \end{pmatrix}\tilde{\Psi}+\begin{pmatrix} I^{(m)} & \mathcal{B}_{12} \\ 0 & 0 \end{pmatrix}\tilde{\Psi}'=0\,.$$

Applying next the maximum rank requirement we see that rank $\left(-\mathcal{A}_{22}\mathcal{B}_{12}^*|\mathcal{A}_{22}\right)$ must equal $n-m$, and since $\left(-\mathcal{A}_{22}\mathcal{B}_{12}^*|\mathcal{A}_{22}\right)=-\mathcal{A}_{22}\cdot\left(\mathcal{B}_{12}^*|I^{(n-m)}\right)$ we obtain $\mathrm{rank}(\mathcal{A}_{22})=n-m$, i.e. $\mathcal{A}_{22}$ must be a regular matrix. This allows us to multiply the last condition from the left by the matrix $\begin{pmatrix} I^{(m)} & -\mathcal{A}_{12}\mathcal{A}_{22}^{-1} \\ 0 & -\mathcal{A}_{22}^{-1} \end{pmatrix}$ which is obviously well-defined and regular; this leads to the condition

$$\begin{pmatrix} \mathcal{A}_{11}+\mathcal{A}_{12}\mathcal{B}_{12}^* & 0 \\ \mathcal{B}_{12}^* & -I^{(n-m)} \end{pmatrix}\tilde{\Psi}+\begin{pmatrix} I^{(m)} & \mathcal{B}_{12} \\ 0 & 0 \end{pmatrix}\tilde{\Psi}'=0\,.$$

We have noted that the square matrix $\mathcal{A}_{11}+\mathcal{A}_{12}\mathcal{B}_{12}^*$ is Hermitean. We denote it by $-S$, rename the block $\mathcal{B}_{12}$ as T and transfer the term containing $\tilde{\Psi}'$ to the right-hand side arriving thus at the condition (8.10); the order of components in $\tilde{\Psi}$ and $\tilde{\Psi}'$ determines the appropriate numbering. ∎

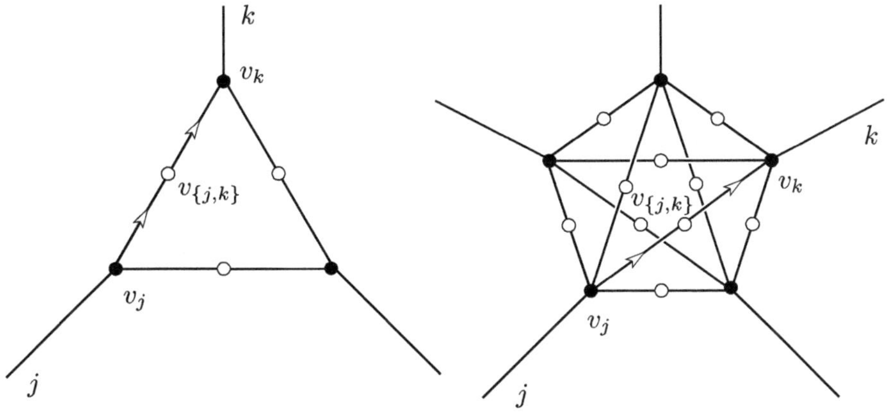

Fig. 8.2 The approximation scheme for a vertex of degree $n = 3$ and $n = 5$. The inner edges are of length $2d$, some may be missing depending on the choice of the matrices S and T. The *arrows* symbolize the vector potentials

After this preliminary we are able to construct an approximation of a general vertex coupling. We consider again a star graph of n edges; in view of the proposition we may suppose that the wave functions are coupled according to (8.10) renaming the edges if necessary. The construction is sketched in Fig. 8.2; we disconnect the edges and connect their endpoints by line segments carrying appropriate operators according to the following rules:
(i) As a convention, the rows of the matrix T are indexed from 1 to m, while the columns are indexed from $m+1$ to n. For brevity, we use the symbol $\hat{n} = \{1, \dots, n\}$ in the rest of this section.

(ii) The external semi-infinite edges of the approximating graph, each parametrized by $x \in \mathbb{R}_+$ are at their endpoints V_j connected to the inner edges by δ-coupling with the parameter $v_j(d)$ for each $j \in \hat{n}$.

(iii) Certain pairs V_j, V_k of external edge endpoints will be connected by segments of length $2d$. This will be the case if one of the following conditions is satisfied, taking into account the convention (i):

(1) $j \in \hat{m}, k \geq m+1$, and $T_{jk} \neq 0$ (or $j \geq m+1, k \in \hat{m}$, and $T_{kj} \neq 0$),
(2) $j, k \in \hat{m}$ and $(\exists l \geq m+1)(T_{jl} \neq 0 \wedge T_{kl} \neq 0)$,
(3) $j, k \in \hat{m}$, $S_{jk} \neq 0$, and the previous condition is not satisfied.

(iv) We denote the center of such a connecting segment by $W_{\{j,k\}}$ and place there a δ-interaction with a parameter $w_{\{j,k\}}(d)$. We adopt another convention: the connecting edges will be regarded as the union of two line segments of length d, with the variable running from zero at $W_{\{j,k\}}$ to d at V_j or V_k.

(v) Finally, we put a vector potential on each connecting segment. What matters is its component tangential to the edge; we suppose it is constant along the edge and denote its value between the points $W_{\{j,k\}}$ and V_j as $A_{(j,k)}(d)$, and between the points $W_{\{j,k\}}$ and V_k as $A_{(k,j)}(d)$; recall that the two half-segments have opposite orientation, thus $A_{(k,j)}(d) = -A_{(j,k)}(d)$ holds for any pair $\{j,k\}$.

As in *Theorem* 8.2, the choice of the dependence of $v_j(d)$, $w_{\{j,k\}}(d)$, and $A_{(j,k)}(d)$ on the length parameter d is crucial; we shall specify it below. We denote by $N_j \subset \hat{n}$ the set containing indices of all the external edges connected to the j-th one by an internal edge, i.e.

$$\begin{aligned} N_j &:= \{k \in \hat{m} : S_{jk} \neq 0\} \cup \{k \in \hat{m} : (\exists l \geq m+1)(T_{jl} \neq 0 \wedge T_{kl} \neq 0)\} \\ &\quad \cup \{k \geq m+1 : T_{jk} \neq 0\} \qquad\qquad \text{for } j \in \hat{m} \\ N_j &:= \{k \in \hat{m} : T_{kj} \neq 0\} \qquad\qquad\qquad \text{for } j \geq m+1 \end{aligned}$$

The definition of the set N_j has two simple consequences, namely

$$k \in N_j \Leftrightarrow j \in N_k \qquad \text{and} \qquad j \geq m+1 \Rightarrow N_j \subset \hat{m}\,.$$

We employ the following symbols for wave function components on the edges: those on the j-th external edge is denoted by ψ_j, while the wave function on the connecting segments is denoted by $\varphi_{(j,k)}$ on the interval between $W_{\{j,k\}}$ and V_j and $\varphi_{(k,j)}$ on the other half of the segment; the conventions concerning the parametrization of the intervals have been specified above.

Next we shall write the coupling conditions corresponding to the above described scheme, first without the vector potentials; for simplicity we shall refrain from indicating the dependence of the parameters v_j, $w_{\{j,k\}}$ on the distance d. The δ-interaction at the segment connecting the j-th and k-th outer edge (present for $j,k \in \hat{n}$ such that $k \in N_j$) is expressed through the conditions

$$\varphi_{(j,k)}(0) = \varphi_{(k,j)}(0) =: \varphi_{\{j,k\}}(0)\,, \quad \varphi'_{(j,k)}(0+) + \varphi'_{(k,j)}(0+) = w_{\{j,k\}}\varphi_{\{j,k\}}(0)\,,$$

while the δ-coupling at the endpoint of the j-th external edge, $j \in \hat{n}$, means

$$\psi_j(0) = \varphi_{(j,k)}(d) \quad \text{for all } k \in N_j\,, \quad \psi'_j(0) - \sum_{k \in N_j} \varphi'_{(j,k)}(d-) = v_j\psi_j(0)\,.$$

It is not difficult to modify these conditions to include the vector potentials: the continuity requirement is preserved, while the coupling parameter changes from v_j to $v_j + i\sum_{k\in N_j} A_{(j,k)}$ (cf. Problem 7). In other words, the impact of the added potentials results in the phase shifts $dA_{(j,k)}(d)$ and $dA_{(k,j)}(d)$, respectively, on the appropriate parts of the connecting segments.

To choose $v_j(d)$, $w_{\{j,k\}}(d)$, and $A_{(j,k)}(d)$ we insert the boundary values written as $\varphi_{(j,k)}(d) = e^{idA_{(j,k)}}\big(\varphi_{(j,k)}(0) + d\varphi'_{(j,k)}(0)\big) + \mathcal{O}(d^2)$ and $\varphi'_{(j,k)}(d) = e^{idA_{(j,k)}}\varphi'_{(j,k)}(0) + \mathcal{O}(d)$ for any $j,k \in \hat{n}$ into the above boundary conditions and fix the d-dependence in such a way that the limit $d \to 0$ yields (8.10). The procedure is rather lengthy and we just state the result referring to the notes for the original proof.

As for $A_{(j,k)}(d)$, we have the relations

$$A_{(j,l)}(d) = \begin{cases} \frac{1}{2d} \arg T_{jl} & \text{if } \mathrm{Re}\, T_{jl} \geq 0\,, \\ \frac{1}{2d} \left(\arg T_{jl} - \pi\right) & \text{if } \mathrm{Re}\, T_{jl} < 0 \end{cases} \tag{8.11}$$

for all $j \in \hat{m},\ l \in N_j \backslash \hat{m}$, while for $j \in \hat{m}$ and $k \in N_j \cap \hat{m}$ we put

$$A_{(j,l)}(d) = \begin{cases} \frac{1}{2d} \arg \left(d S_{jk} + \sum_{l=m+1}^{n} T_{jl} \overline{T_{kl}}\right) \\ \frac{1}{2d} \left[\arg \left(d S_{jk} + \sum_{l=m+1}^{n} T_{jl} \overline{T_{kl}}\right) - \pi \right] \end{cases} \tag{8.12}$$

depending similarly on whether Re $\left(d S_{jk} + \sum_{l=m+1}^{n} T_{jl} \overline{T_{kl}}\right)$ is non-negative or not. Concerning $w_{\{j,k\}}(d)$, we require that

$$w_{\{j,l\}}(d) = \frac{1}{d}\left(-2 + \frac{1}{\langle T_{jl} \rangle}\right) \quad \text{for all}\ \ j \in \hat{m},\ l \in N_j \backslash \hat{m} \tag{8.13}$$

and

$$\frac{1}{2 + d \cdot w_{\{j,k\}}} = -\left\langle d \cdot S_{jk} + \sum_{l=m+1}^{n} T_{jl} \overline{T_{kl}} \right\rangle \quad \text{for all}\ \ j \in \hat{m},\ k \in N_j \cap \hat{m}\,, \tag{8.14}$$

where we have employed the symbol $\langle c \rangle := \pm |c|$ for $\mathrm{Re}\, c \geq 0$ and $\mathrm{Re}\, c < 0$, respectively. Finally, the expressions for v_l are given by

$$v_l(d) = \frac{1 - \sharp N_l + \sum_{h=1}^{m} \langle T_{hl} \rangle}{d} \qquad \text{for all}\ \ l \geq m + 1 \tag{8.15}$$

and

$$v_j(d) = S_{jj} - \frac{\sharp N_j}{d} - \sum_{k=1}^{m} \left\langle S_{jk} + \frac{1}{d} \sum_{l=m+1}^{n} T_{jl} \overline{T_{kl}} \right\rangle + \frac{1}{d} \sum_{l=m+1}^{n} (1 + \langle T_{jl} \rangle) \langle T_{jl} \rangle \tag{8.16}$$

if $j \in \hat{m}$ and $k \in N_j \cap \hat{m}$.

The above choice of the parameters has been guided by the effort to obtain the "correct" coupling conditions (8.10), however, our real aim is analyze the convergence of the corresponding operators. Let us thus denote the free Hamiltonian on the star graph Γ with the coupling (8.10) in the vertex by H^{star}, while H_d^{approx} will stand for the operators of the described approximating family; the symbols $R^{\mathrm{star}}(k^2)$ and $R_d^{\mathrm{approx}}(k^2)$ will denote respectively the resolvents of those operators at the energy k^2. We have to keep in mind that they act on different spaces: $R^{\mathrm{star}}(k^2)$ maps $L^2(\Gamma)$ onto Dom H^{star}, while the domain of $R_d^{\mathrm{approx}}(k^2)$ is $L^2(\Gamma) \oplus L^2(\Gamma_d^{S,T})$, where $\Gamma_d^{S,T}$

is the family of connecting edges of length $2d$ described above. In order to compare the resolvents, we thus identify $R^{\text{star}}(k^2)$ with the orthogonal sum

$$R_d^{\text{star}}(k^2) := R^{\text{star}}(k^2) \oplus 0 \tag{8.17}$$

adding the zero operator acting on $L^2(\Gamma_d^{S,T})$. Then both operators act on the same space and one can estimate their difference; using explicit forms of the corresponding resolvent kernels one can check in a straightforward but rather tedious way (see the notes) the relation

$$\| R_d^{\text{star}}(k^2) - R_d^{\text{approx}}(k^2) \|_{\text{HS}} = \mathcal{O}(\sqrt{d}) \quad \text{as } d \to 0+$$

for the Hilbert-Schmidt norm. With the identification (8.17) in mind we can then state the sought approximation result.

Theorem 8.3 *Let v_j, $j \in \hat{n}$, $w_{\{j,k\}}$, $j \in \hat{n}$, $k \in \mathbb{N}_j$, and $A^{(j,k)}(d)$ depend on the length d according to (8.11)–(8.16). Then the family $H^{\text{approx}}(d)$ converges to H^{star} in the norm-resolvent sense as $d \to 0+$.*

8.3 An Abstract Convergence Result

As we have indicated our main task in this chapter is to show how one can approximate Laplace operators on quantum graphs by families of Schrödinger operators on certain graph-like manifolds dubbed "fat graphs" shrinking to a given graph. As another useful preliminary we will now describe an abstract convergence scheme which allows us to compare operators acting on different Hilbert spaces. In this section we consider a general pair of self-adjoint non-negative operators H and $\tilde{H}$ acting on Hilbert spaces $\mathcal{H}$ and $\tilde{\mathcal{H}}$, respectively.

8.3.1 Scale of Hilbert Spaces

Given a Hilbert space $\mathcal{H}$ with the norm $\|\cdot\|$ induced by the inner product $\langle\cdot,\cdot\rangle$ and a non-negative unbounded operator H we define the scale of Hilbert spaces

$$\mathcal{H}_k := \text{dom}(H+1)^{k/2}, \quad \|u\|_k := \|(H+1)^{k/2}u\|, \quad k \geq 0,$$

extending it to negative exponent values by duality, $\mathcal{H}_{-k} := \mathcal{H}_k^*$. Note that $\mathcal{H} = \mathcal{H}_0$ is naturally embedded into $\mathcal{H}_{-k}$ through $u \mapsto \langle u, \cdot\rangle$ since

$$\|\langle u, \cdot\rangle\|_{-k} = \|R^{k/2}u\|_0, \quad R := (H+1)^{-1},$$

where we have employed the standard identification $\mathcal{H} \simeq \mathcal{H}^*$ by $u \mapsto \langle u, \cdot\rangle$. Hence we have

$$\|u\|_{-k} = \sup_{v \in \mathcal{H}_k} \frac{|\langle u, v\rangle|}{\|v\|_k} \qquad \text{for any } k \in \mathbb{R}\,.$$

For the other Hilbert space $\tilde{\mathcal{H}}$ with inner product $\langle\cdot, \cdot\rangle$ and norm $\|\cdot\|$ together with a non-negative unbounded operator $\tilde{H}$ we define in the same way the scale of Hilbert spaces $\tilde{\mathcal{H}}_k$ with norms $\|\cdot\|_k$. These definitions obviously include the usual scale of Sobolev spaces as a particular case.

Suppose now we have two scales of Hilbert spaces, $\mathcal{H}_k$ and $\tilde{\mathcal{H}}_k$, associated to non-negative operators H and $\tilde{H}$ with the resolvents $R := (H+1)^{-1}$ and $\tilde{R} := (\tilde{H}+1)^{-1}$, respectively. The norm of an operator $A : \mathcal{H}_k \to \tilde{\mathcal{H}}_{-\tilde{k}}$ is given by

$$\|A\|_{k\to-\tilde{k}} := \sup_{u \in \mathcal{H}_k} \frac{\|A\,u\|_{-\tilde{k}}}{\|u\|_k} = \|\tilde{R}^{\tilde{k}/2}\, A\, R^{k/2}\|_{0\to 0}\,.$$

The norm of the adjoint $A^* : \tilde{\mathcal{H}}_{\tilde{k}} \to \mathcal{H}_{-k}$ then satisfies

$$\|A^*\|_{\tilde{k}\to-k} = \|A\|_{k\to-\tilde{k}}\,, \tag{8.18}$$

and moreover, we have

$$\|A\|_{k\to-\tilde{k}} \le \|A\|_{n\to-\tilde{n}} \quad \text{if} \quad k \ge n\,,\ \tilde{k} \ge \tilde{n}\,. \tag{8.19}$$

Next we will formulate the notion of *δ-closeness* of operators H and $\tilde{H}$ acting on $\mathcal{H}$ and $\tilde{\mathcal{H}}$, respectively, which will allow us to compare them by means of suitable identification maps. Suppose that we have linear operators

$$J : \mathcal{H} \to \tilde{\mathcal{H}}\,, \quad J_1 : \mathcal{H}_1 \to \tilde{\mathcal{H}}_1\,, \qquad J' : \tilde{\mathcal{H}} \to \mathcal{H}\,, \quad J_1' : \tilde{\mathcal{H}}_1 \to \mathcal{H}_1\,.$$

Let $\delta > 0$ and $k \ge 1$, then we say that $(H, \mathcal{H})$ and $(\tilde{H}, \tilde{\mathcal{H}})$ are δ-close with respect to the quasi-unitary maps (J, J_1) and (J', J_1') of order k if the following conditions are fulfilled,

$$\|J - J_1\|_{1\to 0} \le \delta\,, \qquad \|J' - J_1'\|_{1\to 0} \le \delta\,, \tag{8.20}$$

$$\|J - J'^*\|_{0\to 0} \le \delta\,, \tag{8.21}$$

$$\|\tilde{H} J_1 - J_1'^* H\|_{k\to -1} \le \delta\,, \tag{8.22}$$

$$\|\mathbb{1} - J'J\|_{1\to 0} \le \delta\,, \qquad \|\mathbb{1} - JJ'\|_{1\to 0} \le \delta\,, \tag{8.23}$$

$$\|J\|_{0\to 0} \le 2\,, \qquad \|J'\|_{0\to 0} \le 2\,, \tag{8.24}$$

where $\mathbb{1}$ is the identical map in the appropriate space. Note that if $\mathcal{H} = \tilde{\mathcal{H}}$ and the identifications are trivial, the above conditions provide us with a simple tool to prove norm-resolvent convergence (Problem 8).

8.3.2 Resolvent Convergence and Functional Calculus

Our primary aim is to use the notion of δ-closeness of order k in cases when $(H, \mathcal{H})$ and $(\tilde{H}, \tilde{\mathcal{H}})$ refer to different Hilbert spaces and to derive consequences for relations between the corresponding resolvents.

Theorem 8.4 *Put* $n := \max\{0, k-2\}$ *and assume (8.20)–(8.22). The resolvents* $R := (H+1)^{-1}$ *and* $\tilde{R} := (\tilde{H}+1)^{-1}$ *then satisfy the bounds*

$$\begin{aligned}\|\tilde{R}J - JR\|_{n\to 0} = \;& \|JH - \tilde{H}J\|_{n+2\to -2} \le 4\delta\,,\\ & \|\tilde{R}^j J - JR^j\|_{n\to 0} \le 4j\delta\,,\end{aligned}$$

the last one being valid for all $j \in \mathbb{N}$.

Proof To employ a suitable telescopic estimate, we start from the identity

$$JH - \tilde{H}J = (J - J'^*)H + (J' - J_1')^*H + (J_1'^*H - \tilde{H}J_1) + \tilde{H}(J_1 - J)\,,$$

Taking into account (8.18) and (8.19) we obtain from it the bound

$$\begin{aligned}&\|\tilde{R}J - JR\|_{n\to 0} = \|\tilde{R}(JH - \tilde{H}J)R\|_{n\to 0} = \|JH - \tilde{H}J\|_{2+n\to -2}\\ &\le \|J - J'^*\|_{n\to -2} + \|J' - J_1'\|_{2\to -n} + \|J_1'^*H - \tilde{H}J_1\|_{2+n\to -2} + \|J_1 - J\|_{2+n\to 0}\\ &\le \|J - J'^*\|_{0\to 0} + \|J' - J_1'\|_{1\to 0} + \|J_1'^*H - \tilde{H}J_1\|_{k\to -1} + \|J_1 - J\|_{1\to 0} \le 4\delta,\end{aligned}$$

which proves (8.25). To demonstrate the second claim, we use the identity $R^jJ - JR^j = \sum_{i=0}^{j-1} \tilde{R}^{j-1-i}(\tilde{R}J - JR)R^i$ which yields the estimate

$$\|\tilde{R}^j J - JR^j\|_{n\to 0} \le \sum_{i=0}^{j-1} \|\tilde{R}^{j-1-i}\|_{0\to 0}\|\tilde{R}J - JR\|_{n\to 0}\|R^i\|_{n\to n} \le 4j\delta$$

following from the previous result and the fact that $\|R\|_{n\to n} \le 1$ holds for any n, and similarly for $\tilde{R}$. ∎

The theorem has important consequences which we shall state without proofs referring to the notes for the source. To formulate the first one we introduce the space $C(\overline{\mathbb{R}}_+)$ of functions which are continuous on $[0, \infty]$.

Corollary 8.3.1 *Assume (8.20)–(8.22) and (8.24), then*

$$\|\varphi(\tilde{H})J - J\varphi(H)\|_{n\to 0} \le \eta_\varphi(\delta)$$

holds for any $\varphi \in C(\overline{\mathbb{R}}_+)$ where $\eta_\varphi(\delta) \to 0$ as $\delta \to 0$.

Corollary 8.3.2 *Given $U \subset [0,\infty]$, assume that $\psi : [0,\infty] \to \mathbb{C}$ is a bounded measurable function, continuous on U. Moreover, suppose that $\lim_{\lambda\to\infty}\psi(\lambda)$ exists. Then*

$$\|\psi(\tilde{H})J - J\psi(H)\|_{n\to 0} \le \eta_\psi(\delta)$$

holds for all pairs of non-negative operators and Hilbert spaces, $(H,\mathcal{H})$ and $(\tilde{H},\tilde{\mathcal{H}})$, which are δ close and such that $\sigma(H) \subset U$ or $\sigma(\tilde{H}) \subset U$, and in addition, $\eta_\psi(\delta) \to 0$ as $\delta \to 0$.

8.3.3 Spectral Convergence

Next we formulate some convergence results concerning the spectra of operators H and $\tilde{H}$. Let $E \subset \mathbb{R}$ and denote by $P = \chi_E(H)$, $\tilde{P} = \chi_E(\tilde{H})$ the corresponding spectral projectors of H and $\tilde{H}$, respectively.

Theorem 8.5 *Let E be a measurable and bounded subset of $\mathbb{R}$. Then there exists a $\delta_0 = \delta_0(E,k)$ such that for all $\delta \in (0,\delta_0)$ we have*

$$\dim P = \dim \tilde{P}$$

for all pairs of non-negative operators and Hilbert spaces $(H,\mathcal{H})$ and $(\tilde{H},\tilde{\mathcal{H}})$ which are δ close and satisfy the condition $\partial E \cap \sigma(H) = \emptyset$ or $\partial E \cap \sigma(\tilde{H}) = \emptyset$.

Proof Suppose that $f \in P\mathcal{H}$, then we have

$$\|f\|_n \le C_{E,n}\,\|f\|_0\,, \qquad C_{E,n} := \sup_{\lambda\in E}(1+\lambda)^{n/2} < \infty\,.$$

Moreover, from the triangle inequality and *Corollary* 8.3.2 we obtain

$$\|\tilde{P}Jf\|_0 \ge \|Jf\|_0 - \|\tilde{P}J - JP\|_{n\to 0}\|f\|_0 \ge (1 - \delta' C_{E,1} - \eta_{\chi_E}(\delta))\,\|f\|_0,$$

where $\delta' := \sqrt{3\delta}$; the last inequality follows from the estimate

$$\begin{aligned}\left|\|Jf\|_0^2 - \|f\|_0^2\right| &= |((f,(J^*J - \mathbb{1}))| \le |(f,(J^* - J')Jf)| + |(f,(J'J - \mathbb{1})f)|\\ &\le \|J^* - J'\|_{0\to 0}\|Jf\|_0\|f\|_0 + \|J'J - \mathbb{1}\|_{1\to 0}\|Jf\|_1\|f\|_0 \le 3\delta\|f\|_1^2.\end{aligned}$$

Hence there is a δ_0 such that $\|\tilde{P}Jf\|_0 \geq \frac{1}{2}\|f\|_0$ for $\delta \in (0, \delta_0)$ which means that $\tilde{P}J\restriction_{P\mathcal{H}}$ is injective. Consequently, if $f_1, \dots, f_n$ are linear independent in $P\mathcal{H}$, the same is true for $\tilde{P}Jf_1, \dots, \tilde{P}Jf_n$; this proves that $\dim P \leq \dim \tilde{P}$.

To prove the opposite inequality, take $u \in \tilde{P}\tilde{\mathcal{H}}$ and consider functions $\mu_i \in C(\overline{\mathbb{R}}_+)$ with values in $[0, 1]$ such that $\mu_1 + \mu_2 + \mu_3 = 1$. Suppose in addition that $\operatorname{supp}\mu_1$ and $\operatorname{supp}\mu_2$ are compact, $\operatorname{supp}\mu_1$ and $\operatorname{supp}\mu_3$ are disjoint, and $\operatorname{supp}\mu_2 \cap E = \emptyset$. Then we have $\mu_2(\tilde{H})\tilde{P} = 0$ and we infer from *Corollary* 8.3.2 that

$$\begin{aligned}
\|PJ^*u\|_{-n} &\geq \|J^*\tilde{P}u\|_{-n} - \|(\tilde{P}J^* - J^*\tilde{P})u\|_{-n} \\
&\geq \|\mu_1(H)J^*u\|_{-n} - \|\mu_2(H)J^*u\|_{-n} - \|\mu_3(H)J^*u\|_{-n} \\
&\quad -\|\tilde{P}J^* - J^*\tilde{P}\|_{n\to 0}\|u\|_0 \\
&\geq C'_{E,n}\|J^*\tilde{P}u\|_0 - \|(\mu_2(H)J^* - J^*\mu_2(\tilde{H}))\tilde{P}u\|_{-n} - \eta_{\chi_E}(\delta)\|u\|_0\,,
\end{aligned}$$

where $C'_{E,n} := \inf_{\{\lambda:\,\mu_1(\lambda)=1\}}(1+\lambda)^{-n/2} - \sup_{\lambda\in\operatorname{supp}\mu_3}(1+\lambda)^{-n/2}$. Since the supports are disjoint by assumption, we can choose the functions μ_j in such a way that $C'_{E,n} > 0$. The norm involving μ_2 can be estimated from above by $\eta_{\mu_2}(\delta)$ using *Corollary* 8.3.1. Moreover, by (8.21) and the triangle inequality we have

$$\|J^*u\|_0 \geq \|J'u\|_0 - \|(J^* - J')u\|_0 \geq (1 - C_{E,1}\,\delta' - \delta)\,\|u\|_0$$

with the same δ' as in the first part. In this way we have shown that

$$\|PJ^*u\|_{-n} \geq \Big(C'_{E,m}(1 - C_{E,1}\,\delta' - \delta) - \eta_{\mu_2}(\delta) - \eta_{\chi_E}(\delta)\Big)\,\|u\|_0\,,$$

which, by the same reasoning as above, implies $\dim P \geq \dim \tilde{P}$. ∎

8.4 The Squeezing Limit of Neumann Networks

Having derived the abstract convergence results we can apply them to the problem we are interested in, namely to find the convergence properties of operators on graph-like manifolds as they approach those of the corresponding quantum graphs with suitable vertex conditions. In what follows $\Gamma \equiv \Omega_0$ will be a fixed metric graph and $\{\Omega_\varepsilon\}_{0<\varepsilon\leq\varepsilon_0}$ a family of the corresponding "fat graphs", as sketched in Fig. 8.3, parametrized by ε being the diameter of the network "fibers". Before we analyze the network squeezing we have to describe the framework in which we shall work in more detail.

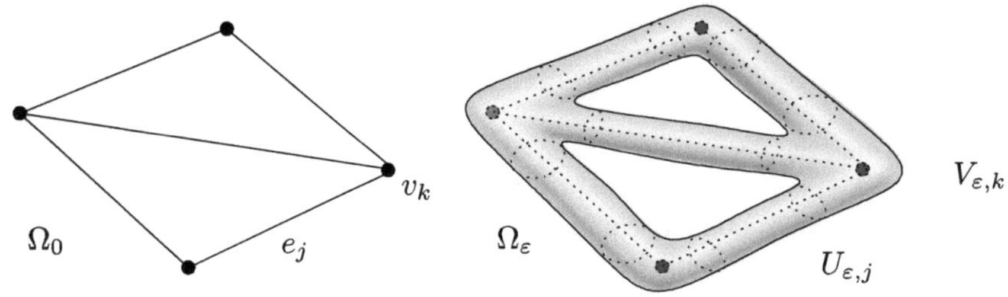

Fig. 8.3 A graph and the associated graph-like manifold

8.4.1 The Problem Setting

Let us start with the graph. Without loss of generality we may suppose that Ω_0 is connected with vertices $\{v_k\}_{k\in K}$ and edges $\{e_j\}_{j\in\mathcal{J}}$. We suppose that e_j has length $\ell_j > 0$, i.e. $e_j \simeq I_j := [0, \ell_j]$. While the basic operator on Ω_0 is the "plain" Laplacian as in Sect. 8.1, we can also think about more general Sturm-Liouville operators. With this aim in mind we make Ω_0 into a metric measure space with measure given by $p_j(x)\mathrm{d}x$ on the edge e_j, $j \in \mathcal{J}$, where $p_j : I_j \to (0, \infty)$ is a smooth density function; we suppose that the functions p_j are uniformly separated from both zero and infinity. Our Hilbert space will then be $\mathcal{H} = L^2(\Omega_0) = \bigoplus_{j\in\mathcal{J}} L^2(I_j, p_j(x)\,\mathrm{d}x)$ with the norm

$$\|u\|^2_{\Omega_0} = \sum_{j\in\mathcal{J}} \|u_j\|^2_{I_j} = \sum_{j\in\mathcal{J}} \int_{I_j} |u_j(x)|^2 p_j(x)\,\mathrm{d}x\,.$$

Let $H^1(\Omega_0)$ be the space of continuous functions on Ω_0 such that the Sobolev norm $\|u\|_{1,\Omega_0} := \sum_{j\in\mathcal{J}}(\|u_j\|^2_{I_j} + \|u'_j\|^2_{I_j})$, where u'_j is the weak derivative of u_j, is finite. Consider the quadratic form

$$u \mapsto \|u'\|^2_{\Omega_0} := \sum_{j\in\mathcal{J}} \|u'_j\|^2_{I_j}, \quad u \in H^1(\Omega_0)$$

which is closed and bounded below; the unique self-adjoint operator associated with it is the negative weighted Laplacian acting as

$$(-\Delta_{\Omega_0} u)(x) = -\frac{1}{p_j(x)}(p_j(x)u'_j)'(x) \tag{8.25}$$

on the domain consisting of functions which are locally H^2 and satisfy the (weighted) Kirchhoff boundary conditions at each vertex v_k, that is, they are continuous at v_k and

$$\sum_{\{j:\, e_j \text{ meets } v_k\}} p_j(v_k)u_j'(v_k) = 0\,, \tag{8.26}$$

where, as usual, the derivative is taken in the outward direction on each edge. The spectrum of Δ_{Ω_0} is purely discrete provided Ω_0 is finite, i.e. having a finite number of finite-length edges; otherwise it may have a continuous component or the discrete spectrum may even be void. If $\sigma_{\text{disc}}(-\Delta_{\Omega_0}) \neq \emptyset$, we denote the corresponding eigenvalues by $\lambda_k(\Omega_0)$, $k = 1, 2, \dots$, written in the ascending order and repeated according to their multiplicity.

Now we turn to the thickened-graph side of the problem. Given a Riemannian manifold X we denote by $L^2(X)$ the usual space of (equivalence classes of) square integrable functions on X with respect to the invariant volume measure $\mathrm{d}\sigma$. Similarly as in Sect. 4.1.1 the measure has density $(\det(g_{ij}))^{1/2}$ with respect to the local Lebesgue measure in a fixed chart; the norm of $L^2(X)$ will be denoted by $\|\cdot\|_X$. For compactly supported smooth functions u we set

$$\|\mathrm{d}u\|_X^2 = \int_X |\mathrm{d}u|^2 \mathrm{d}\sigma\,, \quad |\mathrm{d}u|^2 = \sum_{i,j} g^{ij}\, \partial_i \overline{u}\, \partial_j u\,,$$

where (g^{ij}) is the component representation of the inverse matrix $(g_{ij})^{-1}$. Suppose first that X has no boundary, then we define the negative Laplacian $-\Delta_X$ as the unique self-adjoint operator associated with the closure of the quadratic form $u \mapsto \|\mathrm{d}u\|_X^2$ defined above. On the other hand, if X has a piecewise smooth boundary $\partial X \neq \emptyset$ we introduce the Neumann Laplacian through the closure of the form $u \mapsto \|\mathrm{d}u\|_X^2$ defined on $C^\infty(X)$, the space of smooth functions with derivatives continuous up to the boundary of X. The spectrum of $-\Delta_X$, with any local boundary condition if $\partial X \neq \emptyset$, is purely discrete as long as X is compact, otherwise it can be partly or fully continuous. If $\sigma_{\text{disc}}(-\Delta_X) \neq \emptyset$, we denote the eigenvalues by $\lambda_k(X)$, $k \in \mathbb{N}$, again arranged in increasing order and repeated according to their multiplicity.

Let us now look specifically at the manifolds which can describe thickened graphs. We choose a positive, small enough ε_0 and for each $0 < \varepsilon \leq \varepsilon_0$ we associate with the graph Ω_0 a connected Riemannian manifold Ω_ε of dimension $d \geq 2$ equipped with a metric g_ε which we shall specify below. We suppose that Ω_ε is the union of subsets $U_{\varepsilon,j}$ and compact $V_{\varepsilon,k}$ such that the interiors of all of them are mutually disjoint for all possible combinations of $j \in \mathcal{J}$ and $k \in K$. We think of $U_{\varepsilon,j}$ as the thickened edge e_j and of $V_{\varepsilon,k}$ as the thickened vertex v_k. This is illustrated in Fig. 8.3; it is useful to keep in mind that while from the point of view of physical applications we think of Ω_ε as embedded in $\mathbb{R}^\nu$, $\nu \geq d$, mathematically one can analyze the approximation using intrinsic geometrical properties of Ω_ε only. In fact, we may assume that $U_{\varepsilon,j}$ and $V_{\varepsilon,k}$ are ε-independent as manifolds and implement the squeezing to the graph Ω_0 through a properly chosen family of metrics g_ε. For the edge regions we assume that $U_{\varepsilon,j}$ is diffeomorphic to $I_j \times F_j$ for all $0 < \varepsilon \leq \varepsilon_0$ where F_j is the tube cross section being a compact and connected manifold, with or without a boundary, of dimension $m := d - 1$. Similarly, for the vertex regions we assume that $V_{\varepsilon,k}$ is diffeomorphic to an ε-independent manifold V_k for $0 < \varepsilon \le \varepsilon_0$.

In analogy with the construction of Sect. 8.2.3 we employ a decomposition of $e_j \simeq I_j$ into two halves with reverted orientations. Now, however, we suppose that the variables increase away from the vertex. We collect all the halves $I_{j,k}$ sprouting from the vertex v_k, i.e. those with $j \in J_k := \{j \in \mathcal{J} : e_j \text{ meets } v_k\}$. We put $U_{j,k} := I_{j,k} \times F$, and furthermore, the midpoint of the edge $e_j \simeq I_j$ will be x_j^* and the endpoint of I_j corresponding to the edge v_k will be x_{jk}^0, so $I_{j,k} = [x_j^*, x_{jk}^0]$. For brevity, in the following, provided no confusion can occur, we shall omit the edge and vertex subscripts, similarly we write $U_{\varepsilon_0} = U$, etc.

Consider thus the thickened edge $U = I \times F$, assuming without loss of generality that $\operatorname{vol} F = 1$, with the metric which describes the shrinking of the cross section, being of the product form

$$\tilde{g}_\varepsilon := \operatorname{diag}\left(1, \varepsilon^2 r^2(x)\, h(y)\right), \quad (x, y) \in U = I \times F, \tag{8.27}$$

where y stands for suitable coordinates on F and $r(x) = r_j(x) := (p_j(x))^{1/m}$ is by assumption a smooth function on the edge e_j. It is important to notice that while we need not suppose that Ω_ε is embedded in the Euclidean space, we want such situations to be included. In that case we have to allow the length of the thickened edge to change with ε. We denote by G_ε and $\tilde{G}_\varepsilon$ the $d \times d$-matrices associated with the metrics g_ε and $\tilde{g}_\varepsilon$ in the coordinates (x, y). Obviously the two metrics coincide up to an error term as $\varepsilon \to 0$. More specifically

$$G_\varepsilon = \tilde{G}_\varepsilon + \begin{pmatrix} o(1) & o(\varepsilon) \\ o(\varepsilon) & o(\varepsilon^2) \end{pmatrix} = \begin{pmatrix} 1 + o(1) & o(\varepsilon) \\ o(\varepsilon) & \varepsilon^2 r_j\, h + o(\varepsilon^2) \end{pmatrix}, \tag{8.28}$$

so $g_{\varepsilon,xx} = 1 + o(1)$ and $g_{\varepsilon,y_\alpha y_\beta} = \varepsilon^2 r^2(x)\, h_{\alpha\beta}(y) + o(\varepsilon^2)$ while the off-diagonal terms are $g_{\varepsilon,xy_\alpha} = g_{\varepsilon,y_\alpha x} = o(\varepsilon)$. The fact that the metric g_ε is equal to the product metric $\tilde{g}_\varepsilon$ up to error terms is crucial for the construction.

The shrinking of a vertex region can again be realized through an ε-dependent metric on a fixed manifold $V_\varepsilon = V$. In general, it can occur with a rate different from that of the edge squeezing. Putting $g := g_{\varepsilon_0}$ we assume that

$$c_- \varepsilon^2 g \le g_\varepsilon \le c_+ \varepsilon^{2\alpha} g \tag{8.29}$$

for some constants $0 < c_- \le c_+$, more specifically such that an inequality holds for all diagonal components of the metric tensors in any local basis, where the power α satisfies the inequalities

$$\frac{d-1}{d} < \alpha \le 1\,. \tag{8.30}$$

Obviously, $\alpha \leq 1$ is needed for the above bound to make sense with $0 < \varepsilon \leq \varepsilon_0$. The vertex parts may shrink more slowly than the edge parts but not too slow; this assumption is again crucial for the result (see the notes).

8.4.2 Spectral Convergence: Kirchhoff Coupling

Now we can apply the abstract convergence result of Sect. 8.3 to prove the resolvent convergence of $-\Delta_{\Omega_\varepsilon}$ to $-\Delta_{\Omega_0}$ as $\varepsilon \to 0$ which will, in particular, mean that a part of the eigenvalues of the former operator will converge to those of the latter. For simplicity we suppose that the graph is finite, i.e. it has a finite number of finite edges. We set

$$\mathcal{H} = L^2(\Omega_0)\,, \quad \mathcal{H}_1 = H^1(\Omega_0)\,, \quad \tilde{\mathcal{H}} = L^2(\Omega_\varepsilon)\,, \quad \tilde{\mathcal{H}}_1 = H^1(\Omega_\varepsilon)\,.$$

The crucial thing is the choice of the identification operator J. We are interested in the behavior in the bottom part of the spectrum neglecting the effects associated with higher transverse modes. Consequently, we choose

$$(Ju)(z) := \begin{cases} 0 & \text{if } z \in V_k\,, \\ \varepsilon^{-m/2}\, u_j(x) & \text{if } z = (x,y) \in U_j \end{cases} \qquad u \in \mathcal{H} \tag{8.31}$$

recalling that $m = d - 1$, and

$$(J_1)u(z) := \begin{cases} \varepsilon^{-m/2}\, u(v_k) & \text{if } z \in V_k\,, \\ \varepsilon^{-m/2}\, u_j(x) & \text{if } z = (x,y) \in U_j \end{cases} \qquad u \in \mathcal{H}_1\,. \tag{8.32}$$

The mappings in the opposite direction will mean essentially projecting onto the lowest transverse mode. The corresponding eigenfunction is constant, be it the case of a manifold without boundary or with Neumann condition imposed at the boundary, hence we will employ the following averaging operators,

$$(Nu)(x) = (N_j u)(x) := \int_F u(x,y)\,\mathrm{d}y\,, \quad Cu = C_k u := \frac{1}{\operatorname{vol} V_k} \int_{V_k} u(y)\,\mathrm{d}y\,,$$

on $U_j = I_j \times F$ and V_k, respectively. Using them we define the operators

$$(J'u)_j(x) := \varepsilon^{m/2}\,(N_j u)(x)$$

and

$$(J_1'u)_j(x) := \varepsilon^{m/2}\left(N_j u(x) + \rho(x)(C_k u - N_j u(x^0)\right)\,,$$

where ρ is a smooth interpolation function with values in [0, 1] chosen in such a way that $\rho(x^0) = 1$ and $\rho(x) = 0$ for all $|x - x^0| \geq \frac{1}{2} \min_{j\in\mathcal{J}} |e_j|$, and $x^0 = x^0_{jk} \in \partial I_j$ denotes the edge point which can be identified with the vertex v_k; recall that $I_{j,k}$ denotes the half of the interval I_j adjacent with the vertex v_k and oriented in the direction away from v_k.

Using these notions we are able to state the main result of this section.

Theorem 8.6 *In the given setting the norm-resolvent convergence*

$$\lim_{\varepsilon\to 0} \|R_\varepsilon J - JR\| = 0\,,$$

is valid, where $R = (-\Delta_{\Omega_0} + 1)^{-1}$ *and* $R_\varepsilon = (-\Delta_{\Omega_\varepsilon} + 1)^{-1}$.

To prove the theorem we first collect some auxiliary estimates. In order to avoid the need to discuss geometric peculiarities of the vertex regions, we characterize them in a simple spectral way. To this end, we denote by $\lambda_2^N(X)$ the *second* eigenvalue of the Neumann Laplacian on a compact manifold X.

Lemma 8.4.1 *Let X be a connected and compact manifold with a smooth boundary. We associate with a given $u \in H^1(X)$ the constant function $u_0(x) := \frac{1}{\mathrm{vol} X}\int_X u(y)\,\mathrm{d}y$. Then we have $\|u_0\|_X \leq \|u\|_X$, and furthermore,*

$$\|u - u_0\|^2 \leq \frac{1}{\lambda_2^N(X)}\, \|\mathrm{d}u\|_X^2\,, \quad \|u\|_X^2 - \|u_0\|_X^2 \leq \frac{1}{\delta\,\lambda_2^N(X)} + \delta\,\|u\|_X^2$$

for any $\delta > 0$.

Proof The first estimate follows trivially from the Cauchy-Schwarz inequality. Concerning the second one, note that $u - u_0$ is orthogonal to the first eigenfunction of the Neumann Laplacian. By the minimax principle we have

$$\lambda_2^N(X)\|u - u_0\|_X^2 \leq \|\mathrm{d}(u - u_0)\|_X^2 = \|\mathrm{d}u\|_X^2\,,$$

and since X is connected by assumption, we have $\lambda_2^N(X) > 0$. Finally,

$$\Big|\|u\|_X^2 - \|u_0\|_X^2\Big| \leq 2\|u - u_0\|_X\,\|u\|_X \leq \frac{1}{\delta}\|u - u_0\|_X^2 + \delta\,\|u\|_X^2$$

holds for all $\delta > 0$ which concludes the proof. ∎

Next we compare the two metrics involved in our considerations, the "true" one, g_ε given in (8.27), with the product metric $\tilde{g}_\varepsilon$.

Lemma 8.4.2 *Suppose that g_ε, $\tilde{g}_\varepsilon$ are as in (8.27) and (8.28). Then*

$$\begin{aligned}(\det G_\varepsilon)^{1/2} &= (1 + o(1))\,(\det \tilde{G}_\varepsilon)^{1/2}\,,\\ g_\varepsilon^{xx} &:= (G_\varepsilon^{-1})_{xx} = 1 + o(1)\,,\end{aligned}$$

and furthermore, $|\mathrm{d}_x u|^2 = (1+o(1))\,|du|^2_{g_\varepsilon}$ *and* $|\mathrm{d}_F u|^2 = o(\varepsilon)\,|du|^2_{g_\varepsilon}$, *where* d_x *and* d_F *are the (exterior) derivative with respect to* $x \in I$ *and* $y \in F$, *respectively. All the estimates are uniform in* $(x, y) \in I \times F$ *as* $\varepsilon \to 0$.

Proof Using (8.28) we easily find $\det(G_\varepsilon \tilde{G}_\varepsilon^{-1}) = 1 + o(1)$ which yields the first relation. As for the second one, the upper left component of

$$G_\varepsilon^{-1} - \tilde{G}_\varepsilon^{-1} = -\tilde{G}_\varepsilon^{-1}(G_\varepsilon - \tilde{G}_\varepsilon)\tilde{G}_\varepsilon^{-1} + o(G_\varepsilon - \tilde{G}_\varepsilon)$$
$$= \begin{pmatrix} 1 & 0 \\ 0 & \mathcal{O}(\varepsilon^{-2}) \end{pmatrix} \begin{pmatrix} o(1) & o(\varepsilon) \\ o(\varepsilon) & o(\varepsilon^2) \end{pmatrix} \begin{pmatrix} 1 & 0 \\ 0 & \mathcal{O}(\varepsilon^{-2}) \end{pmatrix} + o(1)$$

is of order $o(1)$. To prove the next relation, we have to show that the inequality

$$\begin{pmatrix} 1 & 0 \\ 0 & \delta I \end{pmatrix} \leq (1 + o(1))\, G_\varepsilon^{-1}$$

holds in the sense of quadratic forms for some $\delta > 0$, where I is here the $m \times m$ unit matrix. It follows from (8.28) that the off-diagonal terms of G_ε are of order of $o(\varepsilon)$ while the diagonal ones 1 and $o(\varepsilon^2)$ from which the result is readily obtained; the remaining claim is checked in a similar way. ∎

Next we need a comparison of a function u and its derivative $\mathrm{d}u$ with the corresponding normal averages Nu and $\mathrm{d}_x Nu$, respectively.

Lemma 8.4.3 *For any function* $u \in H^1(U_\varepsilon)$ *we have*

$$\|u\|^2_{U_\varepsilon} - \|\varepsilon^{m/2} Nu\|^2_I \leq o(\varepsilon^{1/2}) \left(\|u\|^2_{U_\varepsilon} + \|\mathrm{d}u\|^2_{U_\varepsilon} \right).$$

Proof Applying *Lemma* 8.4.1 with $X = F$ we get

$$\|u(x,\cdot)\|^2_F - |(Nu)(x)|^2 \leq \frac{1}{\delta \lambda_2^N(F)} \|\mathrm{d}_F\, u(x,\cdot)\|^2_F + \delta\, \|u(x,\cdot)\|^2_F$$

for any $x \in I_j$, and integrating over I we obtain

$$\|u\|^2_{\tilde{U}_\varepsilon} - \varepsilon^m \|Nu\|^2_I \leq \frac{o(\varepsilon)}{\delta \lambda_2^N(F)} \int_{\tilde{U}_\varepsilon} |\mathrm{d}u|^2_{g_\varepsilon}(\omega)\, \mathrm{d}\omega + \delta \|u\|^2_{\tilde{U}_\varepsilon},$$

where we have used the last relation from *Lemma* 8.4.2. Since the eigenvalue $\lambda_2^N(F)$ is ε-independent we can put $\delta := \sqrt{o(\varepsilon)}$ and apply the first relation from the same lemma to obtain the result for the manifold U_ε. ∎

Lemma 8.4.4 *For any* $u \in H^1(U_\varepsilon)$ *we have*

$$\|\varepsilon^{m/2}(Nu)'\|^2_I - \|\mathrm{d}u\|^2_{U_\varepsilon} \leq o(1) \|\mathrm{d}u\|^2_{U_\varepsilon}.$$

Proof The claim follows from the estimate

$$\|\varepsilon^{m/2}(Nu)'\|_I^2 = \|\varepsilon^{m/2}N(\mathrm{d}_x u)\|_I^2 \le (1+o(1))\|\mathrm{d}_x u\|_{U_\varepsilon}^2 \le (1+o(1))\|\mathrm{d}u\|_{U_\varepsilon}^2,$$

where in the second step we have used the third relation from *Lemma* 8.4.2. ∎

Lemma 8.4.5 *For all* $u \in H^1(U_\varepsilon)$ *and* $x^0 \in \partial I$ *we have*

$$|Nu(x^0)|^2 \le O(\varepsilon^{-m})\left(\|u\|_{U_\varepsilon}^2 + \|\mathrm{d}u\|_{U_\varepsilon}^2\right).$$

Proof We have $\|u \upharpoonright_{\partial U_\varepsilon}\|_{\partial U}^2 \le c_1(\|u\|_U^2 + \|\mathrm{d}u\|_U^2)$ by standard Sobolev imbedding theorems, and using *Lemma* 8.4.2 we can estimate $|Nu(x^0)|^2$ from above by

$$\int_F |u(x^0,y)|^2\, dF(y) \le c_1(\|u\|_U^2 + \|\mathrm{d}u\|_U^2) \le \mathcal{O}(\varepsilon^{-m})(\|u\|_{U_\varepsilon}^2 + \|\mathrm{d}u\|_{U_\varepsilon}^2),$$

which gives the result. ∎

Proof of Theorem 8.6: We shall employ *Theorem* 8.4 with $J,\ J_1,\ J'$ and J_1' defined at the beginning of the section. Let $u \in \mathcal{H}_1$, then in view of the first relation from *Lemma* 8.4.2 we have

$$\begin{aligned}
\|u\|_{\Omega_0}^2 - \|J_1 u\|_{\Omega_\varepsilon}^2 &\le \sum_{j\in\mathcal{J}} (\|u\|_{I_j}^2 - \|J_1 u\|_{U_{\varepsilon,j}}^2) \\
&= \sum_{j\in\mathcal{J}} (\|u\|_{I_j}^2 - (1+o(1))\|J_1 u\|_{\tilde U_{\varepsilon,j}}^2) = o(1)\ \|u\|_{\Omega_0}^2,
\end{aligned}$$

and in the same way the second relation yields

$$\begin{aligned}
\|\mathrm{d}J_1 u\|_{\Omega_\varepsilon}^2 - \|u'\|_{\Omega_0}^2 &= \sum_{j\in\mathcal{J}} ((1+o(1))\|g_\varepsilon^{xx}\mathrm{d}_x J_1 u\|_{\tilde U_{\varepsilon,j}}^2 - \|u'\|_{I_j}^2) \\
&= \sum_{j\in\mathcal{J}} ((1+o(1))\|u'\|_{I_j}^2 - \|u'\|_{I_j}^2) = o(1)\ \|u'\|_{\Omega_0}^2.
\end{aligned}$$

To get the needed estimates for the identification operator J_1' we first note that using *Lemmata* 8.4.1 and 8.4.5 together with (8.29) and the estimate used in the proof of *Lemma* 8.4.5 one can check that for all $u \in \tilde{\mathcal{H}}_1$ we have

$$\begin{aligned}
|C_k\, u - N_j u(x^0)|^2 &\le \mathcal{O}(\varepsilon^{2\alpha-d})\ \|\mathrm{d}u\|_{V_{\varepsilon,k}}^2 \\
\|u - C\, u\|_{V_\varepsilon}^2 &\le \mathcal{O}(\varepsilon^{\beta})\ \|\mathrm{d}u\|_{V_\varepsilon}^2 \\
\|u\|_{V_\varepsilon}^2 &\le \mathcal{O}(\varepsilon^{d\alpha-m})\ (\|u\|_{U_\varepsilon\cup V_\varepsilon}^2 + \|\mathrm{d}u\|_{U_\varepsilon\cup V_\varepsilon}^2)
\end{aligned}$$

where $x^0 \in I_{j,k}$ corresponds to the vertex v_k and $\beta := (2+d)\alpha - d$. Applying the above inequalities in combination with *Lemma* 8.4.3 we arrive at the bound

$$\begin{aligned}\|u\|^2_{\Omega_\varepsilon} - \|J'_1 u\|^2_{\Omega_0} &\le \sum_{k\in K}\Big(\|u\|^2_{V_{\varepsilon,k}} + \sum_{j\in J_k}\big(\|u\|^2_{U_{\varepsilon,jk}} - \|\varepsilon^{m/2}Nu\|^2_{I_{j,k}}\big)\\ &\quad + \sum_{j\in J_k}\big(\delta\|\varepsilon^{m/2}Nu\|^2_{I_{j,k}} + \varepsilon^m\delta^{-1}\|\rho\|^2_{I_{j,k}}|Cu - Nu(x^0)|^2\big)\Big)\\ &\le o(1)(\|u\|^2_{\Omega_\varepsilon} + \|du\|^2_{\Omega_\varepsilon}),\end{aligned}$$

where $\delta := \varepsilon^{(2\alpha-1)/2}$. In the same way, using the above inequalities again, this time together with *Lemma* 8.4.4, we obtain

$$\|(J'_1 u)'\|_{\Omega_\varepsilon} - \|du\|_{\Omega_\varepsilon} \le o(1)(\|u\|^2_{\Omega_\varepsilon} + \|du\|^2_{\Omega_\varepsilon}).$$

Lemmata 8.4.1–8.4.5 thus allow us to conclude that the pairs $(\Delta_{\Omega_0}, \mathcal{H})$ and $(\Delta_{\Omega_\varepsilon}, \tilde{\mathcal{H}})$ are δ-close of order one in the sense of Sect. 8.3.1, and the sought norm-resolvent convergence follows from *Theorem* 8.4. ∎

The result can be extended to graphs which may not be finite (see the notes). It has a consequence for eigenvalue convergence; note that if the graph is finite the spectra of both the operators $-\Delta_{\Omega_0}$ and $-\Delta_{\Omega_\varepsilon}$ are discrete. We denote the eigenvalues by $\lambda_k(\Omega_\varepsilon)$ and $\lambda_k(\Omega_0)$, respectively, then *Theorem* 8.6 implies

Corollary 8.4.1 $\lambda_k(\Omega_\varepsilon) \to \lambda_k(\Omega_0)$ *holds as* $\varepsilon \to 0$ *for any* $k \in \mathbb{N}$.

Notice also that $-\Delta_{\Omega_\varepsilon}$ has "more" eigenvalues than the graph operator, those corresponding to higher transverse modes blow up to infinity in the limit.

8.4.3 Spectral Convergence: More General Couplings

We have seen that using the Neumann Laplacian on a squeezing family of networks Ω_ε one can obtain the Laplacian on the respective graph Ω_0 with Kirchhoff conditions at the vertices. If we want to get in this way different nontrivial couplings it is necessary to employ other operators; some inspiration can be found in the graph approximation results discussed in Sect. 8.2.

Let us begin with the δ-coupling. In view of *Theorem* 8.1, a natural idea is to add a suitably scaled potential to the Laplace operator on a graph-like manifold. For the sake of simplicity, let us consider the case of a star-shaped graph $\Omega_0 = \Gamma$ having one vertex v from which N edges e_j of length $\ell_j \in (0,\infty]$ sprout. The corresponding graph-like manifold Ω_ε is the union of one vertex neighborhood V_ε and N thickened edges,

$$\Omega_\varepsilon = V_\varepsilon \cup_{j=1}^N U_{\varepsilon,j}\,, \quad V_\varepsilon = \varepsilon V\,, \quad U_{\varepsilon,j} = \varepsilon\,(0,\ell_j)\times F_j\,,$$

where V is a fixed vertex neighborhood and F_j is the transverse manifold; we suppose now that the cross sections do not change along the edges. To simplify things further we suppose that all of them are the same; without loss of generality we may put $\mathrm{vol}_{d-1} F_j = 1$ for all $j = 1, \dots, N$. We denote by H_0 the Laplace operator on $L^2(\Omega_0)$ corresponding to δ-coupling (8.5) with the strength

$$q(v) := \int_V Q(x)\,\mathrm{d}x$$

at the vertex v, where we choose a fixed bounded and measurable function Q supported in the (unscaled) vertex neighborhood V. The operator H_0 is associated with the quadratic form

$$h_0[f] = \|f'\|^2_{\Omega_0} + q(v)\,|f(v)|^2, \qquad f \in H^1(\Omega_0)\,. \tag{8.33}$$

We begin again with a few auxiliary results.

Lemma 8.4.6 *For any* $f \in \mathrm{Dom}\,(H_0)$ *we have*

$$h_0[f] + \|f\|^2_{\Omega_0} \le 2\,\max\left\{C_{1/2}, \sqrt{2}\right\}\|(H_0 \pm i)f\|^2_{\Omega_0}\,,$$

where

$$C_\eta := 2\,\max\left\{\frac{|q(v)|^2}{\eta N^2}, \frac{|q(v)|}{\ell_0 N}\right\}, \qquad \ell_0 := \min_j\{\ell_j, 1\}\,.$$

Proof Since $H^1(\Omega_0)$ is continuously embedded into $L^\infty(\Omega_0)$, it follows that

$$|f(v)|^2 \le \frac{1}{N}\left(a\,\|f'\|^2_{\Omega_0} + \frac{2}{a}\,\|f\|^2_{\Omega_0}\right), \quad 0 < a \le \ell_0\,.$$

Hence

$$|h_0[f] - \|f'\|^2_{\Omega_0}| \le |q(v)|\,\frac{1}{N}\left(a\,\|f'\|^2_{\Omega_0} + \frac{2}{a}\,\|f\|^2_{\Omega_0}\right),$$

which in turn implies

$$\|f'\|^2_{\Omega_0} + \|f\|^2_{\Omega_0} \le 2\left|h_0[f] + \|f\|^2_{\Omega_0}\right| + 2C_{1/2}\|f\|^2_{\Omega_0}.$$

The first term on the right-hand side of the last inequality can be estimated by

$$\begin{aligned}|h_0[f] + \|f\|^2_{\Omega_0}|^2 &\le 2(h_0[f]^2 + \|f\|^4_{\Omega_0}) = 2\left|h_0[f] - i\|f\|^2_{\Omega_0}\right|\left|h_0[f] + i\|f\|^2_{\Omega_0}\right| \\ &= 2|(f, (H_0 \mp i)f)|^2 \le \|f\|^2_{\Omega_0}\,\|(H_0 \mp i)f\|^2.\end{aligned}$$

In view of the fact that $\|f\|_{\Omega_0} \le \|(H_0 \mp i)f\|$, this concludes the proof. ∎

Let us pass next to the description of the approximating family. On the manifold Ω_ε corresponding to the graph Γ we consider the operator

$$H_\varepsilon = -\Delta_{\Omega_\varepsilon} + Q_\varepsilon(x)\,, \quad Q_\varepsilon(x) = \frac{1}{\varepsilon} Q(x) \quad \text{on } L^2(\Omega_\varepsilon) \tag{8.34}$$

associated with the quadratic form

$$h_\varepsilon[u] = \|\mathrm{d}u\|^2_{\Omega_\varepsilon} + (u, Q_\varepsilon u)_{L^2(\Omega_\varepsilon)}\,, \qquad u \in H^1(\Omega_\varepsilon)\,.$$

In a way similar to the above estimates for the graph Hamiltonian H_0 we obtain the following results (see the notes).

Lemma 8.4.7 *To a given $\eta \in (0, 1)$ there exists an $\varepsilon_\eta = \mathcal{O}(\ell_0)$ such that*

$$|h_\varepsilon[u] - \|\mathrm{d}u\|^2_{\Omega_\varepsilon}| \le \eta\, \|\mathrm{d}u\|^2_{\Omega_\varepsilon} + \tilde{C}_\eta\, \|u\|^2_{\Omega_\varepsilon}$$

holds true for any $u \in H^1(\Omega_\varepsilon)$ and all $\varepsilon \in (0, \varepsilon_\eta]$, where

$$\tilde{C}_\eta \le c\, \|Q\|_\infty \max\left\{\frac{\|Q\|_\infty}{\eta}, \frac{1}{\ell_0}\right\}$$

and c is a constant independent of ε and ℓ_0.

Lemma 8.4.8 *For any $u \in \mathrm{Dom}\,(H_\varepsilon)$ and all $\varepsilon \in (0, \varepsilon_{1/2}]$ we have*

$$\|du\|^2_{\Omega_\varepsilon} + \|u\|^2_{\Omega_\varepsilon} \le 2 \max\left\{\tilde{C}_{1/2}, \sqrt{2}\right\} \|(H_\varepsilon \mp i)u\|^2_{\Omega_\varepsilon}\,.$$

With these preliminaries, we are ready to state and prove the result concerning approximations of δ-couplings.

Theorem 8.7 *The pairs $(H_0, \mathcal{H})$ and $(H_\varepsilon, \tilde{\mathcal{H}})$ referring to (8.33) and (8.34), respectively, are first-order δ-close in the sense of Sect. 8.3.1.*

Proof It suffices to check the inequality (8.22), since the remaining estimates follow in the same way as in the case of Kirchhoff boundary conditions. We start from the δ-closeness of the quadratic forms h_0 and h_ε, namely we will show that

$$h_0[J_1' u, f] - h_\varepsilon[u, J_1 f] \;\le\; \delta_\varepsilon\, \|u\|_{H^1(\Omega_\varepsilon)}\, \|f\|_{H^1(\Omega_0)}\,, \tag{8.35}$$

where $h_0[\cdot,\cdot]$ and $h_\varepsilon[\cdot,\cdot]$ are the sesquilinear forms generated by h_0 and h_ε, respectively, and $\delta_\varepsilon = \mathcal{O}(\varepsilon^{1/2})$ as $\varepsilon \to 0$. We have

$$\begin{aligned}\left|h_0[J_1' u, f] - h_\varepsilon[u, J_1 f]\right|^2 \le 2\,\varepsilon^{d-1}\Big(\Big|\sum_j p_j(C\bar{u} - N\bar{u}(v))(\chi_j', f')_{L^2(e_j)}\Big|^2 \\ + |q(v)C\bar{u} - (Qu, \mathbb{1}_v)_{L^2(V)}|^2 |f(v)|^2\Big),\end{aligned}$$

where $\mathbb{1}_v$ is the constant function on V such that $\|\mathbb{1}_v\|_{L^2(V)} = 1$ and χ_j is a piecewise affine linear function on e_j with $\chi_j(0) = 1$ and $\chi_j(\ell_j) = 0$; recall that C and N are the averaging maps defined in the previous section. The first term can be estimated as before using *Lemma* 8.4.1, for the second one we use the fact that

$$q(v)\, C\bar{u} = (u, C(Q\mathbb{1}_v))_{L^2(V)}$$

to estimate the quantity in question as

$$\begin{aligned} |q(v)C\bar{u} - (Qu, \mathbb{1}_v)_{L^2(V)}|^2 &= |(u, CQ - Q)_{L^2(V)}|^2 \\ &= |(u, P_V\, Q)_{L^2(V)}|^2 = |(P_v u, Q)_{L^2(V)}| \\ &\le \frac{1}{\lambda_2^N(V)}\, \|du\|^2_{L^2(V)}\, \|Q\|^2_{L^2(V)}\,, \end{aligned}$$

where $P_v u := u - Cu$; in doing that we have applied *Lemma* 8.4.1 with $X = V$. Collecting these estimates and taking into account the dependence of the constants on ℓ_0, we arrive at relation (8.35) with

$$\delta_\varepsilon \le c\sqrt{\varepsilon}\, \max\left\{\frac{\|Q\|_\infty}{\sqrt{\ell_0}},\, \frac{1}{\ell_0}\right\}. \tag{8.36}$$

Using the shorthands $R_0^\pm = (H_0 \mp i)^{-1}$ and $R_\varepsilon^\pm = (H_\varepsilon \mp i)^{-1}$ for the respective resolvents, and setting $f := R_0^\pm \tilde{f}$ and $u := R_\varepsilon^\pm \tilde{u}$, we get

$$\begin{aligned} (\tilde{u}, (JR_0^\pm - R_\varepsilon^\pm J)\tilde{f}) &= (\tilde{u}, Jf) - (u, J\tilde{f}) \\ &= (\tilde{u}, (J - J_1)f) + (h_\varepsilon[u, J_1 f] - h_0[J_1' u, f]) + ((J_1' - J^*)u, \tilde{f}) \\ &\quad -i((u, (J_1 - J)f) + ((J_1' - J^*)u, f))\,, \end{aligned}$$

where the scalar product is taken in $L^2(\Omega_\varepsilon)$. Hence from the first two and the last two of relations (8.20)–(8.24) in combination with *Lemmata* 8.4.6 and 8.4.8 we infer that

$$(\tilde{u}, (JR_0^\pm - R_\varepsilon^\pm J)\tilde{f}) \le \tilde{\delta}_\varepsilon\, \|\tilde{f}\|\, \|\tilde{u}\|\,, \quad \tilde{\delta}_\varepsilon := 2\,\delta_\varepsilon \max\{\tilde{C}_{1/2}\,, \sqrt{2}\},$$

where we have employed the fact that $C_{1/2} \le \tilde{C}_{1/2}$, and consequently

$$\|JR_0^\pm - R_\varepsilon^\pm J\| \le \tilde{\delta}_\varepsilon\,; \tag{8.37}$$

this concludes the proof. ∎

Corollary 8.4.2 *The spectrum of the "fat star" Hamiltonian H_ε converges to that of H_0 uniformly on any finite interval as $\varepsilon \to 0$.*

Corollary 8.4.3 *To any $\lambda \in \sigma_{\mathrm{disc}}(H)$ there exists a family $\{\lambda_\varepsilon\} \subset \sigma_{\mathrm{disc}}(H_\varepsilon)$ such that $\lambda_\varepsilon \to \lambda$ as $\varepsilon \to 0$ and the multiplicity is preserved. If λ is a simple eigen-*

value with normalized eigenfunction ϕ, then there is a family of simple normalized eigenfunctions $\{\phi_\varepsilon\}$ of H_ε such that $\|J\phi - \phi_\varepsilon\|_{\Omega_\varepsilon} \to 0$ as $\varepsilon \to 0$.

Remark 8.4.1 The claim of *Theorem* 8.7 can be generalized in various ways (see the notes for further reading). In particular, the cross section volumes $\mathrm{vol}_{d-1} F_j$ need not be the same. What is more important, the norm-resolvent convergence remains valid for any locally finite graph satisfying natural uniformity conditions, specifically, $\mathrm{vol}\, V/\mathrm{vol}\, \partial V$ and $\|Q \restriction_V \|_\infty$ must have an upper bound for all $v \in \mathcal{V}$, the edge lengths must be bounded uniformly from below, and the same has to be true for the second Neumann eignevalues $\lambda_2(V)$ and $\lambda_2(F)$ on all the vertices and edges.

From the introduction to this chapter we know that δ-couplings form an important but small subset in the family of all self-adjoint couplings, and we have to ask next how one can deal with those having wave functions discontinuous at the vertex. One can proceed using the same strategy as before, namely to employ the graph approximation results of Sect. 8.2 and to "lift" them to the manifolds. Let us see how it works in the case of the δ'_s−coupling.

To focus on the essential features of the approximation we again adopt several simplifying assumptions. We consider the star-shaped graph Γ with one vertex v_0 and N edges e_j, as sketched in the left picture of Fig. 8.4. Furthermore, we suppose that all the edges have the same length $\ell_j = 1$, and that all the transverse volumes of the corresponding "fat star" edges are the same and vol $F_j = 1$. The operator H^β on $L^2(\Gamma)$ is determined by the coupling conditions (8.6); the corresponding quadratic form is given by

$$h^\beta[f] = \|f'\|^2_{\Omega_0} + \frac{1}{\beta}\Big|\sum_j f_j(v_0)\Big|^2$$

with $\mathrm{Dom}\,(h^\beta) = H^1(\Omega_0)$ if $\beta \neq 0$, while for $\beta = 0$ we have $h^\beta[f] = \|f'\|^2_{\Omega_0}$ with $\mathrm{Dom}\,(h^\beta) = \{f \in H^1(\Omega_0) : \sum_j f_j(v_0) = 0\}$. For definiteness one has to fix boundary conditions at the loose ends of the edges; here we have chosen Neumann conditions, however, the choice is not important for the approximation.

The spectrum of H^β is easily found, in particular, its negative part is given by the following result, the proof of which is left to the reader (Problem 9).

Lemma 8.4.9 *H^β is positive if $\beta \geq 0$. On the other hand, if $\beta < 0$, then H^β has exactly one negative eigenvalue $\lambda = -\kappa^2$ determined by the condition*

$$\cosh\kappa + \frac{\beta\kappa}{N}\sinh\kappa = 0\,.$$

In order to approximate H^β on Ω_0 by Schrödinger operators on the "fat star" manifold, we employ the result of *Theorem* 8.2 and use the operator constructed there as an intermediate step denoting it now by $H^{\beta,a}$. We can regard it as a Hamiltonian on the graph Γ_a obtained from Γ by adding vertices v_j of degree two on each edge

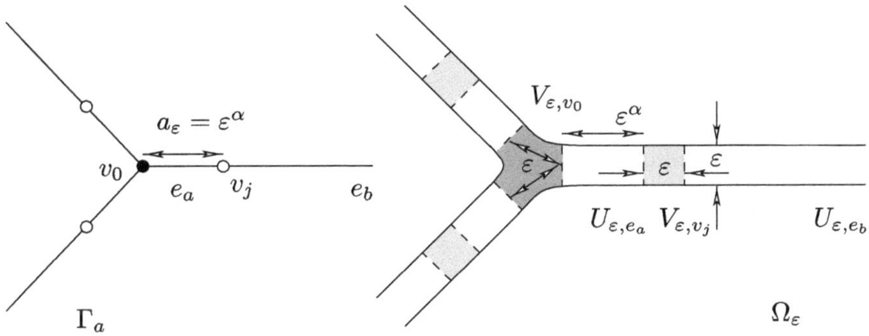

Fig. 8.4 The δ'_s approximation scheme, the *left picture* shows the graph approximation on Γ_a used as an intermediate step

$e = e_j$ at the distance $a \in (0, 1)$ from the central one, v_0, and imposing an attractive δ-coupling in each on them. Every edge is thus split into a pair of edges, $e_a = e_{a,j}$ and $e_b = e_{b,j}$, as indicated in the left part of Fig. 8.4.

The crucial point of the approximation is the proper, a-dependent choice of the involved δ-couplings parameters. If we assume that the graph-approximation operator $H^{\beta,a}$ is associated with the closed quadratic form

$$h^{\beta,a}[f] := \|f'\|^2_{\Gamma_a} - \frac{\beta}{a^2}\sum_j |f_j(v_j)|^2 - \frac{1}{a}\sum_j |f_j(v_0)|^2\,, \quad \mathrm{Dom}\,(h^{\beta,a}) = H^1(\Gamma_a)\,,$$

then by *Theorem* 8.2 we have

$$\|(H^{\beta,a} - z)^{-1} - (H^\beta - z)^{-1}\| = \mathcal{O}(a)$$

as $a \to 0$, where $\|\cdot\|$ denotes the operator norm in $L^2(\Gamma_a)$. We note in passing that $\inf \sigma(H^{\beta,a}) \to -\infty$ as $a \to \infty$ if $\beta \geq 0$ (Problem 10).

Next we pass to the manifold model approaching the intermediate Hamiltonian $H^{\beta,a}$ in the limit $\varepsilon \to 0$. We have two parameters now and we have to fix their relation. It is easy to figure out that the approximation requires the transverse squeezing to be faster than the vertex distance diminishing; we choose $a = a_\varepsilon := \varepsilon^\alpha$ with $\alpha \in (0, 1)$ to be specified later. The approximation of δ-coupling discussed above suggests how the manifold model Ω_ε of the graph Ω_0 has to be constructed; it is shown in the right part of Fig. 8.4. To the additional vertices of degree two cylinder parts of length ε and distance of order of $a_\varepsilon = \varepsilon^\alpha$ from v_0 will correspond; the edge e_{a_ε} thus has length a_ε. The potentials in the vertex regions are chosen to be constant and equal to

$$Q_{\varepsilon,v}(x) := \frac{1}{\varepsilon}\,\frac{q_\varepsilon(v)}{\mathrm{vol}\,V_v}\,, \quad x \in V_v \quad \text{for} \quad v = v_0 \quad \text{or} \quad v = v_e\,,$$

where in correspondence with the intermediate Hamiltonian $H^{\beta,a}$ we choose

$$q_\varepsilon(v_0) := -\beta\,\varepsilon^{-2\alpha} \quad \text{and} \quad q_\varepsilon(v_e) := -\varepsilon^{-\alpha}\,.$$

The resulting Hamiltonian H_ε^β in $L^2(\Omega_\varepsilon)$ is then defined as the operator associated with the quadratic form

$$h_\varepsilon^\beta[u] = \|\mathrm{d}u\|^2_{\Omega_\varepsilon} - \varepsilon^{-1-2\alpha}\,\frac{\beta}{\operatorname{vol} V_{v_0}}\,\|u\|^2_{V_{\varepsilon,v_0}} - \varepsilon^{-1-\alpha}\sum_e \|u\|^2_{V_{\varepsilon,v_e}}$$

defined on Dom $(h_\varepsilon^\beta) = H^1(\Omega_\varepsilon)$. Note that $\inf\sigma(H_\varepsilon^\beta)$ depends again on sgn β; for $\beta \geq 0$ this operator family becomes unbounded below as $\varepsilon \to 0$ (Problem 10).

Theorem 8.8 *Suppose that* $0 < \alpha < 1/13$, *then the relation*

$$\|J(H^\beta - z)^{-1} - (H_\varepsilon^\beta - z)^{-1}J\| \to 0 \quad \text{as} \quad \varepsilon \to 0$$

is valid for $z \notin \mathbb{R}$, *where* J *is the identification map (8.31), and the same is true for* $\|J(H^\beta - z)^{-1}J^* - (H_\varepsilon^\beta - z)^{-1}\|$. *It implies, in particular, the same spectral convergence as in Corollary 8.4.3 above.*

Proof Since negative numbers may belong to the spectrum of both the operators H^β and H_ε^β we have to take resolvents at non-real values. We employ the results of the previous section in combination with *Theorem* 8.2. Let $H^{\beta,\varepsilon} := H^{\beta,a_\varepsilon}$ be the ε-dependent intermediate Hamiltonian on the graph with the δ-potentials as defined above. The lower bounds on the tube lengths for the corresponding manifold model now depend explicitly on ε, namely $\ell_0 = a_\varepsilon = \varepsilon^\alpha$. From the definition of the constant $\tilde{C}_{1/2}$, the bound preceding (8.37), *Lemma* 8.4.8, and equation (8.36) we obtain

$$C_{1/2}(\varepsilon) = \mathcal{O}(\varepsilon^{-4\alpha})\,, \qquad \delta_\varepsilon = \mathcal{O}(\varepsilon^{(1-5\alpha)/2})\,.$$

Next we note that (8.37) implies

$$\|J(H^{\beta,\varepsilon} \pm i)^{-1} - (H_\varepsilon^\beta \pm i)^{-1}J\| \leq 10\,\delta_\varepsilon \max\{\tilde{C}_{1/2}(\varepsilon), \sqrt{2}\} = \mathcal{O}(\varepsilon^{(1-13\alpha)/2})\,.$$

Using a telescopic estimate for the norm of the resolvent difference in question in combination with *Theorem* 8.2 we infer that

$$\|J(H^{\beta,\varepsilon} \pm i)^{-1} - (H_\varepsilon^\beta \pm i)^{-1}J\| \leq \bar{\delta}_\varepsilon$$

with $\bar{\delta}_\varepsilon = \mathcal{O}(\varepsilon^{\max\{\alpha,(1-13\alpha)/2\}})$ which, in combination with the first resolvent formula, proves the sought convergence. Furthermore, we observe that (8.37) implies $\|(\mathbb{1} - J\,J^*)(H_\varepsilon^\beta \pm i)^{-1}\| \leq \bar{\delta}_\varepsilon$ so the two operators are $\bar{\delta}_\varepsilon$-close; this concludes the proof. ∎

In conclusion of this section we are going to describe a general result in the Neumann-type case; we state it without a proof referring to the notes for more information. For simplicity we again consider star graphs only. By *Theorem* 8.3 the Hamiltonian H on such a graph Γ with any vertex coupling, characterized by matrices S, T according to *Proposition* 8.2.1, can be approximated using a locally modified graph with added edges of lengths $2d$ as sketched in Fig. 8.2. We consider the corresponding "fat graph" associated with this approximating graph, consisting of tubes of diameter ε, and construct the appropriate magnetic Schrödinger operator. The procedure follows the same lines as in the δ_s' case above but it is slightly more complicated due to the presence of the inner edges. With any such edge e we associate the quadratic form

$$h_{\varepsilon,e}[u_e] := \int_{-d}^{d} \left(\|u_e'(s) + iA_e(d)u_\varepsilon(s)\|^2 + \|\mathrm{d}u_\varepsilon(s)\|^2_{L^2(\varepsilon F_e)}\right) \mathrm{d}s + \frac{w_e(d)}{2\varepsilon} \int_{-\varepsilon}^{\varepsilon} \|u_\varepsilon(s)\|^2 \, \mathrm{d}s$$

defined for H^1 functions on $e \times \varepsilon F_e$, where the potentials $A_e(d)$ and w_e are given by the relations (8.11)–(8.14). In a similar way one defines

$$h_{\varepsilon,e}[u_e] := \int_0^\infty \left(\|u_e'(s)\|^2 + \|\mathrm{d}u_\varepsilon(s)\|^2_{L^2(\varepsilon F_e)}\right) \mathrm{d}s$$

for an outer edge e, and

$$h_{\varepsilon,v}[u_v] := \|\mathrm{d}u_v(s)\|_{L^2(V_{\varepsilon,v})} + \frac{v_v(d)}{\varepsilon \operatorname{vol} V_{\varepsilon,v}} \|u_v\|_{L^2(V_{\varepsilon,v})},$$

where $v_v(d)$ is given by (8.15)–(8.16), in the manifold part related to a vertex v; the approximating operator H_ε is then associated with the quadratic form

$$h_\varepsilon[u] := \sum_e h_{\varepsilon,e}[u_e] + \sum_v h_{\varepsilon,v}[u_v],$$

where we sum over all the edges and vertices of the approximating graph.

To get a meaningful limit we have once again to relate the two parameters that control the squeezing process; we put $d = d_\varepsilon := \varepsilon^\alpha$ with a fixed $\alpha \in (0, 1)$. We denote the resolvent of H_ε by $R_\varepsilon(z)$ and recall the definition (8.17); then we are in position to make the following claim.

Theorem 8.9 *Assume that $R^{\mathrm{star}}(z)$ is the resolvent of a star-graph Hamiltonian with the vertex coupling characterized by matrices S and T and $R_\varepsilon(z)$ is the resolvent of the operator H_ε described above. If $0 < \alpha < 1/13$, the relation*

$$\|J R^{\mathrm{star}}_{d_\varepsilon}(z) J^* - R_\varepsilon(z)\| = \mathcal{O}\left(\varepsilon^{\min\{1-13\alpha,\alpha\}/2}\right)$$

holds for all $z \in \mathbb{C} \setminus \mathbb{R}$*, where* J *is the identification map (8.31). The spectral convergence implications are the same as in Corollary 8.4.3 above.*

Remark 8.4.2 Comparison of the last two theorems illustrates that the approximation of vertex couplings by squeezed networks is not unique. A δ'_s interaction is in view of (8.6) characterized by the condition $\frac{1}{\beta}Nf(0) - f'(0) = 0$, where N is the $n \times n$ matrix with all the entries equal to one, which corresponds to $S = \beta^{-1}N$ and $T = 0$ in the condition (8.10). According to the described scheme, all the inner edges are present, no magnetic field is needed, and we choose

$$w_{\{j,k\}}(d) = -\frac{\beta}{d^2} - \frac{2}{d} \quad \text{and} \quad v_j(d) = \frac{2-n}{\beta} - \frac{n-1}{d};$$

the corresponding Schrödinger operator then has a step-like potential of order $-\varepsilon^{-\alpha-1}$ near the vertex V_j and of order $-\beta\varepsilon^{-2\alpha-1}$ around the edge midpoint $W_{j,k}$. The approximation in *Theorem* 8.8 is different, using the original star graphs as the skeleton of the squeezed network, however, the behavior of the involved potential with respect to the length scale $d = \varepsilon^\alpha$ is the same in both cases up to multiplicative constants and higher order terms.

8.5 The Squeezing Limit of Dirichlet Networks

The fact that the network manifold had Neumann boundaries (or no boundaries at all) played an essential role in the considerations of the previous section; if we modify the boundary conditions the picture will change substantially. Let us now examine what the squeezing limit can produce if the network has Dirichlet boundary which, as we know, is the more important case from the viewpoint of applications to small semiconductor structures. First of all, one has to modify the limit using the energy renormalisation that we know from *Theorem* 1.6, since the threshold of the essential spectrum in an infinite tube blows up as its diameter tends to zero and we have to subtract this divergent quantity.

However, this is not the only difference. We have seen that the Laplacian on Neumann networks typically yields Kirchhoff coupling conditions in the limit. We may produce other couplings by adding properly scaled potentials but what is important is that the limit is generically nontrivial resulting in graph Hamiltonians with the edges coupled. This is not true in the Dirichlet case, where the generic limit in the vicinity of the moving threshold leads to a fully decoupled graph with Dirichlet conditions at the edge endpoints. However, it does not mean that the limit is always trivial. Specifically, if the network system possesses a threshold resonance, then the limiting procedure can result in a graph with nontrivial coupling conditions at the vertices.

To explain the meaning of the last claim let us discuss in detail the simplest nontrivial case when the squeezed network is represented by a bent waveguide which collapses onto the star graph Γ_0 consisting of two half-lines joined in a single vertex.

The starting point of our model is a planar strip $\Omega \subset \mathbb{R}^2$ of width $2a$ built over a smooth reference curve Γ of signed curvature γ in the way described in Sect. 1.1. Beginning with this, we shall construct a family of bent strips $\Omega_\varepsilon \subset \mathbb{R}^2$ changing simultaneously the strip width and its curvature as functions of the scaling parameter ε in such a way that

$$\gamma_\varepsilon(s) = \frac{\sqrt{\lambda(\varepsilon)}}{\varepsilon}\,\gamma\left(\frac{s}{\varepsilon}\right), \quad a_\varepsilon := \varepsilon^\alpha a\,, \tag{8.38}$$

with $\alpha > 1$ to be fixed later and $\lambda(\varepsilon)$ being a fixed positive function such that $\lambda(\varepsilon)$ is analytic near zero and

$$\lambda(\varepsilon) = 1 + \lambda'(0)\,\varepsilon + \mathcal{O}(\varepsilon^2)\,.$$

Our scaled family of bent planar waveguides can be characterized as follows,

$$\Omega_\varepsilon := \left\{(x,y) \in \mathbb{R}^2 : x = \xi_\varepsilon(s) - u\dot\eta_\varepsilon(s),\; y = \eta_\varepsilon(s) + u\dot\xi_\varepsilon(s),\; s \in \mathbb{R},\; |u| < a_\varepsilon\right\},$$

where $\xi_\varepsilon(\cdot)$ and $\eta_\varepsilon(\cdot)$ are determined by $\gamma_\varepsilon(\cdot)$ through Eqs. (1.4) and (1.5). As usual we suppose that $a_\varepsilon\|\gamma_\varepsilon\|_\infty < 1$; if γ is bounded it is surely true for all ε small enough. If γ is smooth and $\gamma(s) \to 0$ as $|s| \to \infty$, the family $\{\Omega_\varepsilon\}$ obviously shrinks to the graph Γ_0 mentioned above in the limit $\varepsilon \to 0$.

Our object of interest will be the family of Dirichlet Laplacians $-\Delta^D_{\Omega_\varepsilon}$ on the bent strips Ω_ε. Using the straightening transformation described in Sect. 1.1 we easily conclude that the Laplacian $-\Delta^D_{\Omega_\varepsilon}$ is unitarily equivalent to the operator H_ε on the (unscaled) straight strip $\mathbb{R} \times (-d, d)$ given by

$$H_\varepsilon = -\partial_s\,\frac{1}{(1+\varepsilon^\alpha u\,\gamma_\varepsilon(s))^2}\,\partial_s - \varepsilon^{-2\alpha}\,\partial_u^2 + \varepsilon^{-2}\,V_\varepsilon(s,u),$$

with Dirichlet conditions at $|u| = \varepsilon^\alpha a$ and the effective potential

$$V_\varepsilon(s,u) = -\frac{\lambda(\varepsilon)\,\gamma(s/\varepsilon)^2}{4(1+\varepsilon^\alpha u\,\gamma_\varepsilon(s))^2} + \frac{\varepsilon^{\alpha-1}\,u\,\sqrt{\lambda(\varepsilon)}\,\ddot\gamma(s/\varepsilon)}{2(1+\varepsilon^\alpha u\,\gamma_\varepsilon(s))^3} - \frac{5}{4}\,\frac{\varepsilon^{2\alpha-2}\,u^2\,\lambda(\varepsilon)\,\dot\gamma(s/\varepsilon)^2}{(1+\varepsilon^\alpha u\,\gamma_\varepsilon(s))^4}.$$

Let χ_n be elements of the orthonormal basis in $L^2(-a_\varepsilon, a_\varepsilon)$ analogous to (1.10), namely real-valued functions such that $\chi_n(\pm a_\varepsilon) = 0$ satisfying

$$-\chi_n''(u) = \kappa_{n,\varepsilon}^2\,\chi_n(u)\,, \qquad \kappa_{n,\varepsilon} = \frac{\pi n}{2a_\varepsilon}\,.$$

Consider now $(s,u,s',u') \mapsto (H_\varepsilon - k^2 - \kappa_{m,\varepsilon}^2)^{-1}(s,u,s',u')$, the integral kernel of the resolvent of H_ε taken at a point shifted by the running transverse threshold energy $\kappa_{m,\varepsilon}^2$, and define

$$R^{\varepsilon}_{nm}(k^2, s, s') := \int_{-a}^{a} \chi_n(u)\chi_m(u')(H_\varepsilon - k^2 - \kappa^2_{m,\varepsilon})^{-1}(s, u, s', u')\,\mathrm{d}u\,\mathrm{d}u' .$$

It is not difficult to verify that the integral operator $R^{\varepsilon}_{nm}(k^2) : L^2(\mathbb{R}) \to L^2(\mathbb{R})$ with this kernel is bounded and analytic in k^2 for $k^2 \in \mathbb{C} \setminus \mathbb{R}$ and $\operatorname{Im} k > 0$.

To proceed we need to recall some facts about one-dimensional Schrödinger operators on the line,

$$L = -\frac{\mathrm{d}^2}{\mathrm{d}s^2} + V(s) . \tag{8.39}$$

We say that L has a ***zero energy resonance*** if there is a function $\psi_r \in L^\infty(\mathbb{R})$, $\psi_r \notin L^2(\mathbb{R})$, such that $L\psi_r = 0$ holds in the sense of distributions. Moreover, if

$$\int_{\mathbb{R}} V(s)\,ds \neq 0 , \qquad \mathrm{e}^{\eta|\cdot|}\, V \in L^1(\mathbb{R}) , \tag{8.40}$$

holds for some $\eta > 0$, then exactly one of the following alternatives occurs:

(a) the operator L does not have a zero energy resonance, or
(b) the operator L does have a zero energy resonance; in this case ψ_r can be chosen real and the real constants c_1 and c_2 given by

$$c_1 := \left(\int_{\mathbb{R}} V(s)\,ds\right)^{-1} \int_{\mathbb{R}}\int_{\mathbb{R}} V(s)\,\frac{|s-s'|}{2}\, V(s')\,\psi_r(s')\,\mathrm{d}s\,\mathrm{d}s' ,$$
$$c_2 := -\frac{1}{2}\int_{\mathbb{R}} V(s)\, s\,\mathrm{d}s ,$$

cannot vanish at the same time.

The graph Γ_0 is isomorphic to the real line and the coupling is a generalized point interaction at the vertex. To describe the limit we introduce two Hamiltonians, H^{d} and H^{r}, on $\mathbb{R}$ acting as $f \mapsto -f''$ with different boundary conditions at $x = 0$. The first is the Dirichlet-decoupled operator H^{d} with the domain $\mathrm{Dom}\,(H^{\mathrm{d}}) := \{f \in H^2(\mathbb{R} \setminus \{0\}) \cap H^1(\mathbb{R}) :\ f(0) = 0\}$. The domain of H^{r} depends on c_1 and c_2 in the following way:

$$\mathrm{Dom}\,(H^{\mathrm{r}}) := \Big\{f \in H^2(\mathbb{R} \setminus \{0\}) : (c_1 + c_2)f(0+) = (c_1 - c_2)f(0-) ,$$
$$(c_1 - c_2)f'(0+) = (c_1 + c_2)f'(0-) + \frac{\hat\lambda}{c_1 + c_2}\, f(0-)\Big\}$$

for a fixed $\hat\lambda$ and c_1, c_2 such that $c_1 + c_2 \neq 0$, and

$$\mathrm{Dom}\,(H^{\mathrm{r}}) := \Big\{f \in H^2(\mathbb{R} \setminus 0) :\ f(0-) = 0,\ f'(0+) = \frac{\hat\lambda}{4c_1^2}\, f(0+)\Big\}$$

if $c_1+c_2 = 0$. Note that H^{r} couples the two halflines in a nontrivial way *iff* $|c_1| \neq |c_2|$ (see also Problem 11).

To match the introduced notions we put $V = -\frac{1}{4}\gamma^2$ in (8.39); if the operator has a zero energy resonance, we set

$$\hat{\lambda} = -\frac{1}{4}\lambda'(0) \int_{\mathbb{R}} \gamma^2(s)\, \psi_r(s)^2\, \mathrm{d}s\,.$$

Armed with these notions we can describe the outcome of the squeezing limit.

Theorem 8.10 *Assume that the curve Γ_ε has no self-intersections for all ε small enough. Let γ be piecewise C^2 with compact support and $\dot{\gamma}$, $\ddot{\gamma}$ bounded. In addition, suppose that $\alpha > 5/2$ and put $V = -\frac{1}{4}\gamma^2$ in (8.39).*

(a) If L does not have a zero energy resonance, then

$$\text{u-}\lim_{\varepsilon\to 0} R^{\varepsilon}_{nm}(k^2) = \delta_{nm}\,(H^{\mathrm{d}} - k^2)^{-1}\,, \qquad k^2 \in \mathbb{C}\setminus\mathbb{R}\,,\ \operatorname{Im} k > 0\,.$$

(b) If L has a zero energy resonance, then

$$\text{u-}\lim_{\varepsilon\to 0} R^{\varepsilon}_{nm}(k^2) = \delta_{nm}\,(H^{\mathrm{r}} - k^2)^{-1}\,, \qquad k^2 \in \mathbb{C}\setminus\mathbb{R}\,,\ \operatorname{Im} k > 0\,,$$

provided that, in addition, $k \neq k_0$ with k_0 given in Problem 11,

where u- lim *refers to convergence in the operator norm on $L^2(\mathbb{R})$.*

Proof of *Theorem* 8.10 is based on a pair of auxiliary results which concern the family of one-dimensional operators

$$L_\varepsilon = -\frac{d^2}{ds^2} + \frac{\lambda(\varepsilon)}{\varepsilon^2}\, V\left(\frac{s}{\varepsilon}\right).$$

Notice that we have already met similar one-dimensional Schrödinger operators – see, e.g., relation (8.7). There, however, the scaling was "natural", preserving the integral of the potential. Here, in contrast, we have something like a squared δ singularity, thus it is not surprising that the limit will be different.

Lemma 8.5.1 *Suppose that V satisfies condition (8.40) for some $\eta > 0$.*

(a) If L does not have a zero energy resonance, then

$$\text{u-}\lim_{\varepsilon\to 0} (L_\varepsilon - k^2)^{-1} = (H^{\mathrm{d}} - k^2)^{-1}\,, \qquad k^2 \in \mathbb{C}\setminus\mathbb{R}\,,\ \operatorname{Im} k > 0\,.$$

(b) If L has a zero energy resonance, then

$$\text{u-}\lim_{\varepsilon\to 0} (L_\varepsilon - k^2)^{-1} = (H^{\mathrm{r}} - k^2)^{-1}\,, \qquad k_0^2 \neq k^2 \in \mathbb{C}\setminus\mathbb{R}\,,\ \operatorname{Im} k > 0\,.$$

Proof We start by introducing the operator,

$$T(k) = (1 + u\, G_k v)^{-1}\,, \qquad G_k(s,s') = G_k(s-s') = \frac{i}{2k}\, \mathrm{e}^{ik|s-s'|}\,,$$

where $v(s) = |V(s)|^{1/2}$ and $w(s) = v(s)\,\mathrm{sgn}V(s)$ is the standard Birman-Schwinger factorization of the potential, cf. Remark 6.1.1a. We recall that the operator $T(k)$ admits a norm convergent expansion in k, namely

$$T(k) = \sum_{n=p}^{\infty} (ik)^n\, t_n\,, \tag{8.41}$$

where $p = 0$ if L has no zero energy resonance and $p = -1$ otherwise (see the notes for the references). We have $(L-k^2)^{-1} = G_k - G_k\, v\, T(k)\, w\, G_k$ and a simple scaling argument then shows that

$$(L_\varepsilon - k^2)^{-1} = G_k - \frac{1}{\varepsilon}\, A_\varepsilon(k) T(\varepsilon k)\, C_\varepsilon(k), \tag{8.42}$$

where $A_\varepsilon(k)$ and $C_\varepsilon(k)$ are operators defined through their respective integral kernels, $A_\varepsilon(k;s,s') := G_k(s-\varepsilon s')v(s')$ and $C_\varepsilon(k;s,s') := w(s)G_k(\varepsilon s - s')$.

Suppose first that L has a zero energy resonance and fix k with $\mathrm{Im}\, k > 0$. By (8.40) the operators $A_\varepsilon(k)$ and $C_\varepsilon(k)$ are Hilbert-Schmidt, and moreover

$$\begin{aligned} A_\varepsilon(k;s,s') &= \big(G_k(s) + ik\, G_k(s)\,(|s-\varepsilon s'| - |s|) + a_\varepsilon(k;s,s')\big)\, v(s')\,, \\ C_\varepsilon(k;s,s') &= w(s)\,\big(G_k(s') + ik\, G_k(s')\,(|s'-\varepsilon s| - |s'|) + c_\varepsilon(k;s,s')\big)\,, \end{aligned}$$

where

$$a_\varepsilon(k;s,s') = \left(-\frac{ik}{2}\, \mathrm{e}^{ik|s|} \int_0^{|s-\varepsilon s'|-|s|} \mathrm{e}^{ik\varsigma}\,(|s-\varepsilon s'| - |s| - \varsigma)\, \mathrm{d}\varsigma \right) v(s')$$

and a similar integral representation holds for $c_\varepsilon(k;s,s')$. It follows that the Hilbert-Schmidt norms of the operators $a_\varepsilon(k)\, v$ and $w\, c_\varepsilon(k)$ satisfy

$$\max\,\{\|a_\varepsilon(k)v\|_{\mathrm{HS}},\, \|w\, c_\varepsilon(k)\|_{\mathrm{HS}}\} \le \frac{|k|}{4}\, \frac{1}{\sqrt{\mathrm{Im}\, k}}\, \|s^2 w\|_2\, \varepsilon^2.$$

Scaling the power-series expansion (8.41) we get

$$T(\varepsilon k) = \frac{1}{ik\varepsilon}\, t_{-1} + t_0 + ik\,\varepsilon\, t_1 + b_\varepsilon(k)$$

with the remainder satisfying $\|b_\varepsilon(k)\| \le c\,\varepsilon^2$; this allows us to write

$$
\begin{aligned}
(A_\varepsilon(k)T(\varepsilon k)C_\varepsilon(k))(s,s') = \int_{\mathbb{R}^2} \Big[& \big(G_k(s) + ik\,G_k(s)\,(|s-\varepsilon\tau| - |s|\big)v(\tau) \\
& \times\big(\frac{1}{ik\varepsilon} t_{-1}(\tau,\tau') + t_0(\tau,\tau') + ik\varepsilon\, t_1(\tau,\tau')\big) w(\tau') \\
& \times\big(G_k(s') + ik\,G_k(s')\,(|s'-\varepsilon\tau'| - |s'|\big)\Big]\,\mathrm{d}\tau\,\mathrm{d}\tau' \\
& + r_\varepsilon^1(s,s')
\end{aligned}
$$

with $\|r_\varepsilon^1\| \le c\,\varepsilon^2$. Next we need some properties of t_{-1}, t_0 and t_1, namely

$$
t_{-1}w = 0\,, \quad t_{-1}^* v = 0\,, \quad (v, t_0\, w) = 0\,, \quad (\cdot\, v, t_{-1}\, w\,\cdot) = \frac{c_2^2}{c_1^2 + c_1^2 + i\hat{\lambda}/2k}\,,
$$

$$
(\cdot\, v, t_0\, w) = (v, t_0\, w\,\cdot) = \frac{c_1 c_2}{c_1^2 + c_1^2 + i\hat{\lambda}/2k}\,, \qquad (v, t_1\, w) = \frac{c_2^2 + i\hat{\lambda}/2k}{c_1^2 + c_1^2 + i\hat{\lambda}/2k}\,,
$$

where $(\cdot\, v, t_0\, w) = \int s\bar{v}(s)\,(t_0 w)(s)\,ds$, etc. (see Problem 12 and the references given in the notes). With the help of them we can express the integral kernel $(A_\varepsilon(k)T(\varepsilon k)C_\varepsilon(k))(s,s')$ as

$$
\begin{aligned}
\varepsilon\Bigg(& \frac{2k\,c_2^2}{i(c_1^2 + c_1^2 + i\hat{\lambda}/2k)}\, G_k(s)\,G_k(s') - \frac{2i(c_2^2 + i\hat{\lambda}/2k)}{k(c_1^2 + c_1^2 + i\hat{\lambda}/2k)}\, G_k'(s)G_k'(s') \\
& - \frac{2c_1c_2}{c_1^2 + c_1^2 + i\hat{\lambda}/2k}\,\big(G_k(s)G_k'(s') + G_k'(s)G_k(s')\big)\Bigg) + r_\varepsilon^2(s,s')
\end{aligned}
$$

with $\|r_\varepsilon^2\| \le c\,\varepsilon^{3/2}$, where $G_k'(t)$ denotes the derivative of $G_k(t)$ with respect to t. The last equation in combination with (8.42) shows that $(L_\varepsilon - k^2)^{-1}$ tends as $\varepsilon \to 0$ in the operator norm to $R^r(k^2)$ with the integral kernel equal to

$$
\begin{aligned}
G_k(s-s') + & \frac{2ik\,c_2^2}{c_1^2 + c_1^2 + i\hat{\lambda}/2k}\, G_k(s)\,G_k(s') + \frac{2i}{k}\,\frac{c_2^2 + i\hat{\lambda}/2k}{c_1^2 + c_1^2 + i\hat{\lambda}/2k}\, G_k'(s)\,G_k'(s') \\
& + \frac{2c_1c_2}{c_1^2 + c_1^2 + i\hat{\lambda}/2k}\,\big(G_k(s)\,G_k'(s') + G_k'(s)\,G_k(s')\big)\,.
\end{aligned}
$$

The second claim of the lemma will follow if we verify that $R^r(k^2)$ is the resolvent of H^{r}. Let $f \in L^2(\mathbb{R})$ and examine the vector $g_f = R^r(k^2)f$ which can be expressed from the last relation using the fact that $\int_{\mathbb{R}} G_k'(t)f(t)\,dt = -(G_k' f)(0)$. As the resolvent maps $L^2(\mathbb{R})$ into the domain of the operator, we have to check that g_f satisfies the boundary conditions from the definition of the operator H^{r} which can be done by a direct calculation (Problem 12).

On the other hand, if the operator L has no zero energy resonance, the expansion of $T(\varepsilon k)$ starts from the zero order in ε and the above properties of its lowest terms have to be replaced by

$$(v, t_0\, w) = 0\,, \quad (\cdot\, v, t_0\, w) = (v, t_0\, w\, \cdot) = 0\,, \quad (v, t_1\, w) = -2\,.$$

In the same way as above we then obtain that $(L_\varepsilon - k^2)^{-1}$ converges as $\varepsilon \to 0$ in the operator norm to $R^d(k^2)$ with the kernel

$$R^d(k^2; s, s') = G_k(s - s') + 2ik\, G_k(s)\, G_k(s')$$

which is nothing but the integral kernel of the resolvent of H^{d}. ∎

To proceed with the demonstration of *Theorem* 8.10 we also need the following technical lemma for the proof of which we refer to the notes.

Lemma 8.5.2 *Let V fulfil condition (8.40) for some $\eta > 0$ and $\int_{\mathbb{R}} V(s)\,\mathrm{d}s \neq 0$. Then for any k with $k^2 \in \mathbb{C} \setminus \mathbb{R}$, $\operatorname{Im} k > 0$, there is a constant C_k such that*

$$\max\left\{\limsup_{\varepsilon\to 0} \varepsilon^{1/2}\, \|\partial_s\, (L_\varepsilon - k^2)^{-1}\|_{L^2\to L^2},\ \limsup_{\varepsilon\to 0}\, \|(L_\varepsilon - k^2)^{-1}\|_{L^2\to L^\infty}\right\} \le C_k\,.$$

In order to apply the results of the above one-dimensional analysis to the family $\{H_\varepsilon\}$ we will introduce an intermediate operator with separated variables,

$$H_\varepsilon^0 = -\partial_s^2 - \varepsilon^{-2\alpha}\, \partial_u^2 + \frac{\lambda(\varepsilon)}{\varepsilon^2}\, V\left(\frac{s}{\varepsilon}\right)$$

on $L^2(\mathbb{R} \times (-a, a))$ subject to the Dirichlet condition at $|u| = a$. We note that its matrix resolvent kernel with energy shifted by transverse threshold values,

$$R_{nm}^{\varepsilon,0}(k^2; s, s') := \int_{-d}^{d} \chi_n(\tau)\, (H_\varepsilon^0 - k^2 - \kappa_{m,\varepsilon}^2)^{-1}(k^2, s, \tau, s', \tau')\, \chi_m(\tau')\, \mathrm{d}\tau\, \mathrm{d}\tau'\,,$$

is nothing but $\delta_{nm}\, (L_\varepsilon - k^2)^{-1}(s, s')$. We have the following approximation result.

Lemma 8.5.3 *Under the conditions of Theorem 8.10 we have*

$$\text{u-}\lim_{\varepsilon\to 0} \left(R_{nm}^{\varepsilon,0}(k^2) - R_{nm}^{\varepsilon}(k^2)\right) = 0\,, \qquad k_0^2 \neq k^2 \in \mathbb{C} \setminus \mathbb{R},\ \operatorname{Im} k > 0\,.$$

Proof The result will follow if we show that for any $f, g \in C_0^\infty(\mathbb{R})$ the inequality

$$\left|\left(g, \left(R_{nm}^{\varepsilon,0}(k^2) - R_{nm}^{\varepsilon}(k^2)\right) f\right)\right| \le c\, \varepsilon^{\alpha - 5/2}\, \|g\|_2\, \|f\|_2 \tag{8.43}$$

holds with a constant c independent of f, g. From the resolvent identity we can express the difference $(H_\varepsilon - k^2 - \kappa_{m,\varepsilon}^2)^{-1} - (H_\varepsilon^0 - k^2 - \kappa_{m,\varepsilon}^2)^{-1}$ as

$$(H_\varepsilon - k^2 - \kappa_{m,\varepsilon}^2)^{-1} \left[\varepsilon^{\alpha-2} b\left(\frac{s}{\varepsilon}, u\right) \partial_s + \varepsilon^{-2} W_\varepsilon(s,u)\right] (H_\varepsilon^0 - k^2 - \kappa_{m,\varepsilon}^2)^{-1},$$

where we have introduced the shorthands $W_\varepsilon(s,u) := V_\varepsilon(s,u) - \frac{1}{4}\gamma^2\left(\frac{s}{\varepsilon}\right)$ and $b(s,u) = -2u\dot{\gamma}(s)(1+\varepsilon^{\alpha-1}u\gamma(s))^{-3}$. Hence it suffices to estimate the quantities

$$I_1 := \left(g \otimes \chi_n, (H_\varepsilon - k^2 - \kappa_{m,\varepsilon}^2)^{-1} \varepsilon^{\alpha-2} b\left(\frac{\cdot}{\varepsilon}, \cdot\right) \partial_s((L_\varepsilon - k^2)^{-1} f) \otimes \chi_m\right),$$
$$I_2 := \left(g \otimes \chi_n, (H_\varepsilon - k^2 - \kappa_{m,\varepsilon}^2)^{-1} \frac{1}{\varepsilon^2} W_\varepsilon(s,u)((L_\varepsilon - k^2)^{-1} f) \otimes \chi_m\right).$$

By the Cauchy-Schwarz inequality and *Lemma* 8.5.2 we infer that

$$\begin{aligned}|I_1| &\le \varepsilon^{\alpha-2}\,|\mathrm{Im}\,k^2|^{-1}\|g\|_2\|b\|_\infty \left\|\partial_s\,(L_\varepsilon - k^2)^{-1} f\right\|_2 \\ &\le c\,\varepsilon^{\alpha-5/2}\,|\mathrm{Im}\,k^2|^{-1}\|g\|_2\,\|f\|_2,\end{aligned}$$

where we have used the assumptions imposed on γ. As for I_2, from the explicit expression for V_ε, *Lemma* 8.5.2 and the Cauchy-Schwarz inequality we conclude that

$$\begin{aligned}|I_2| &\le |\mathrm{Im}\,k^2|^{-1}\|g\|_2\,\varepsilon^{-2}\left\|W_\varepsilon(\cdot,\cdot)((L_\varepsilon - k^2)^{-1} f) \otimes \chi_m\right\|_2 \\ &\le |\mathrm{Im}\,k^2|^{-1}\|g\|_2\,\varepsilon^{-2}\left\|W_\varepsilon\right\|_2 \|((L_\varepsilon - k^2)^{-1} f) \otimes \chi_m\|_\infty \\ &\le c\,|\mathrm{Im}\,k^2|^{-1}\|g\|_2\,\|f\|_2\,\varepsilon^{\alpha-5/2};\end{aligned}$$

this implies (8.43) and completes the proof. ∎

Theorem 8.10 is now a direct consequence of *Lemmata* 8.5.1 and 8.5.3. ∎

Remark 8.5.1 (a) If $\lambda(s) = 1$ in (8.38) the coupling resulting from the squeezing limit is a scale-invariant point interaction, see Problem 11. On the other hand, for $c_1 = 1, c_2 = 0$ the limit yields the usual δ-interaction of strength $\hat{\lambda}$.
(b) Notice that in *Theorem* 8.10 we have actually proved more than we wanted covering not only the physically important case $n = m = 1$ but also an analogous convergence in the vicinity of higher transverse thresholds. It is not surprising because we have repeatedly seen that these modes become asymptotically decoupled as $a \to 0$.

8.6 Notes

Section 8.1 The origins of the quantum graph model can be traced back to the foundation period of quantum mechanics. Already in the 1930s Linus Pauling suggested that the hexagon pictures everybody knows from organic chemistry textbooks can in fact provide a realistic description of aromatic hydrocarbons, with some of the electrons building the molecule "frame" and the others "living" on it. The idea was later worked out by Ruedenberg and Scherr [RSch53] but then the concept fell into oblivion, and if it appeared in the following three decades it was regarded rather as an obscure textbook example. The situation changed in the late 1980 s with the progress in experimental solid-state physics. The diminishing size of artificially fabricated structures reached the state when the electron transport in them became dominantly ballistic and quantum graphs suddenly reemerged as a useful model. It also became clear that quantum graphs are attractive from the theoretical point of view, as a laboratory to study various properties of quantum systems: on the one hand they are mathematically accessible because they typically involve ordinary differential operators, on the other hand they make it possible to consider systems with a nontrivial geometry and topology. The bibliography concerning quantum graphs is nowadays indeed extensive. The most complete discussion of the subject can be found in the recent monograph [BK]; a good introduction with numerous references can also be obtained from the review paper [Ku04] and proceedings volumes [BCFK, EKKST].

Quantum graphs are used to model systems of a different physical nature. At the beginning it included microstructures fabricated from semiconductor or metallic materials, later carbon nanotubes were added; the latter became particularly interesting when their branching became experimentally possible [LPX00]. It is worth mentioning, however, that the model is also suitable for studying systems of a non-quantum nature, at least as long as a stationary situation is considered and the problem reduces, for instance, to a spectral analysis of the Laplacian on the appropriate graph—one can recall the investigation of microwave phenomena in optical cable networks [HBPSŻ04].

Even within the realm of quantum physics various equations of motion are considered on graph-shaped configuration spaces. Most often it is Schrödinger's equation, either free or with the addition of potentials corresponding to external electric or magnetic fields. One can also add internal degrees of freedom such as spin, etc. On the other hand, one can also study Dirac operators on graphs. Such a model, too, was for a long time regarded as a theoretician's exercise and only attracted limited attention [BH03, BT90]. The situation changed recently with the discovery of *graphene*, a single-layer sheet of carbon atoms, in which electrons behave effectively as massless relativistic particles, and numerous papers devoted to this subject have appeared.

Metric graphs are not the only type of graphs on which Laplace and similar operators are studied. A lot is known about this problem on *combinatorial* graphs, for a survey and bibliography we refer to Sunada's review in [EKKST]. The two subjects are connected, there is a *duality* relation between equations on metric and combinatorial graphs. In physics this was noticed in de Gennes-Alexander theory of superconductivity; for a more general and rigorous formulation see [Ex97a, Ca97]. In particular, a unitary equivalence between the corresponding Hamiltonians can be demonstrated provided the metric graphs are equilateral [Pa12].

A lot of attention has been paid to spectral properties of quantum graphs and their relations to the underlying graph geometry. Weakly coupled bound states of Schrödinger operators on graphs were for the first time considered in [Ex96b] and later in a more general context in [Ko07, EEK10]. Other properties of the discrete spectrum, in particular, Lieb-Thirring and Cwikel-Lieb-Rosenbljum-type inequalities, were studied in [So09, DH10, EFK11]. Spectral properties of quantum graphs differ in some aspects from those of usual Schrödinger operators. For instance, the unique continuation property may not hold here, so Hamiltonians on infinite graphs can have compactly supported eigenfunctions [Ku05]; if they correspond to eigenvalues embedded in the continuous spectrum, a perturbation can turn them into resonances [EŠe94, DET08, EL10b]. Another spectral property concerns infinite periodic graphs where one can open spectral gaps by "decoration", i.e. attaching a fixed compact graph to a periodic subfamily of graph vertices [AS00, Ku05]. A different mechanism of gap opening was observed on radial *tree graphs* with equal edge lengths by Solomyak and collaborators [NS00, SoS02]. On the other hand, the spectrum of a "sparse" radial tree graph, with growing edge lengths, is generically singular [BF09, EL10a].

Attention has also been paid to various *inverse problems*. Gutkin and Smilansky modified the famous question of Mark Kac asking whether one can hear the shape of a graph [GS01]. In the work that followed some conditions under which it is possible to reconstruct the quantum graph topology and vertex coupling from its spectral and/or scattering data were given. It was noted that the symmetries may spoil the reconstruction [BK05] and examples of isospectral graphs were constructed [BPB09].

Another often asked question concerns spectral behavior over large intervals of energy, in particular, the asymptotics. For compact graphs various *trace formulæ* expressing the spectrum in terms of periodic orbits on the graph were derived starting from the work of Roth [Ro83]. Kottos and Smilansky proposed such a formula as a tool to study quantum chaos on graphs [KSm99]; the formula for graphs with a general vertex coupling has been derived in [BE09]. On infinite graphs high-energy behavior of resonances is of interest; it appears that for particular graph topologies and coupling conditions their distribution may not follow the usual Weyl law [DP11, DEL10, EL11].

Quantum graphs are also used to study the effect of *randomness*, in particular, the existence of Anderson localization at the bottom of the spectrum when the geometry, coupling, or potential on the graph edges is random, see for instance [CMV06, EHS07, HP09, KP09]. On the other hand, under some circumstances the absolutely continuous spectrum on graphs can be stable under weak disorder, see e.g. [Kl94, ASW06].

Returning to the subject of the present section, the construction of self-adjoint graph Hamiltonians using self-adjoint extensions was proposed in [EŠ89b]. The standard form of the coupling conditions from *Proposition* 8.1.1 comes from [KoS99], the unique form of the claim (c) from [Ha00, KoS00], see also [FT00], however, it was known before in the general theory of boundary forms [GG]. For the notions of δ- and δ'_s-couplings see [EŠ89b, Ex96a]. The "projection form" of the coupling conditions described in Problem 2 comes from [FKW07], see also [Ku04].

Section 8.2 The approximation of a δ-coupling by a family of squeezing potentials, *Theorem* 8.1, is taken from [Ex96b]; the proof follows the scheme used in [AGHH, Sect. I.3.2] for one-dimensional point interactions. The key idea concerning the approximation of more general couplings with discontinuous wave functions at the vertex was proposed formally by Cheon and Shigehara [CS98] in the case $n = 2$, and proved to yield a norm-resolvent convergence in [AN00, ENZ01]. Note that the choice of the approximating operator is rather subtle: as shown in [AN00] the resolvents involved are highly singular as $a \to 0$ but in their difference the first four orders cancel giving a convergent result. An extension to vertices of higher degrees was done for the δ'_s-coupling in [CE04]; *Theorem* 8.2 is taken from this paper. The method based on adding δ vertices also works in cases without an edge-permutation symmetry giving at most a $2n$ parameter family of vertex couplings [ET07].

The general approximation described in Section 8.2.3 including the alternative form of the coupling conditions from *Proposition* 8.2.1 was formulated in [CET10]; we refer to this paper for the detailed proof of *Theorem* 8.3. The described construction is not unique, however, the choice of the parameters is highly non-generic.

Section 8.3 The abstract convergence results of quasi-one-dimensional spaces are due to [Po06], we also refer to this paper for the proofs of *Corollaries* 8.3.1 and 8.3.2. A more general discussion of this problem as well as of the approximations discussed in the following section and various related results can be found in the book [Po11].

Section 8.4 A formal argument showing that the Neumann Laplacian on a squeezing network should converge to the corresponding graph Laplacian with Kirchhoff coupling was already presented in [RSch53]. The first rigorous treatment of the problem appeared much later in the paper [FW93] followed by [Sa00], [KZ01], and [RuS01]. While the later works use mostly PDE techniques, the original proof in [FW93] was based on an analysis of diffusion processes in shrinking neighborhoods of a graph. Our exposition is taken essentially from the paper [EP05] where the convergence was proved in terms of the intrinsic geometry of the manifold only. Note also that the compact graph assumption in *Theorem* 8.6 can be abandoned, see [Po06]. In the same setting one can also demonstrate convergence of resonances, see [EP07].

While the main interest concerns uniformly shrinking networks, *Theorem* 8.6 is stated in a way which admits the tube cross sections varying along graph edges, this feature being expressed through the metric tensor (8.27) and leading to the weighted graph Laplacian (8.25) in the limit, and moreover, the vertex regions of the graph-like manifold Ω_ε may shrink at a rate different from that of the edges.

The difference, however, must not be too large as the condition (8.30) shows. It is demonstrated in [EP05] that a slow vertex region shrinking, $0 < \alpha < (d-1)/d$, leads to fully decoupled edges with Dirichlet boundary conditions. In the borderline case, $\alpha = (d-1)/d$, one gets a nontrivial coupling at the vertices, see [EP05] for details.

The approximation of δ-coupling at the vertices by Neumann-type Laplacians by means of suitably scaled potentials on thin branched manifolds, *Theorem* 8.7, is proved in [EP09] in a more general form allowing the tube cross sections to have different volumes. Furthermore, one can find there the generalization of the δ-coupling approximation to locally finite graphs mentioned in Remark 8.4.1. The same paper treats the case of δ_s'-coupling approximation discussed here; we also refer to [EP09] for proofs of *Lemmata* 8.4.7 and 8.4.8. An attentive reader might have noticed that the argument of the approximation potential (8.34) is not the same as in (8.7). It is nevertheless the same thing, the variable scaling being hidden in the used metric. *Theorem* 8.9 which provides a general solution of the problem in the Neumann-type situation comes from [EP13]; we refer to this paper for the proof using again the technique introduced in Sect. 8.3 and a discussion of other aspects of this approximation.

Section 8.5 Subtraction of the first transverse eigenvalue is the most natural renormalisation but not the only one. For an alternative approach see [MV07] where the reference value is chosen in a different way, specifically between the first and the second transverse threshold. In such a case the squeezing limit gives generically a nontrivial result and the coupling conditions in the vertex are determined by the scattering matrix of the associated "fat star".

Theorem 8.10 was proved in [ACF07] in the situation when the curvature is scaled naturally, i.e. $\lambda(s) = 1$ and the bending of the strip is in view of (1.4) preserved at the scaling; in that paper the reader can also find the proof of *Lemma* 8.5.2. The extension to the case when the bending angle may "wiggle" was done in [CaE07]. Properties of zero energy resonances needed to state and prove the theorem come from [BGW85]; their application to scaling with $\lambda(s) = 1$ in (8.38) was done in [ACF07], the extension to the more general situation is worked out in [CaE07].

Squeezing limits of more complicated, branched Dirichlet networks, with a renormalisation referring to the first transverse eigenvalue, have also been considered. If the vertex regions are more narrow than the edge regions, the limit is trivial [Po05]. The mechanism which can produce a nontrivial result is the same as in the example discussed here; one has to start from an operator having a threshold resonance. This was discussed in [Gr08], a different but related approach using graphs with finite edges was suggested in [DAC10]; the limit leads to weighted Kirchhoff conditions. The analysis of these approximations has still to be completed and a procedure which would allow the approximation a general coupling condition by squeezed Dirichlet networks, analogous to *Theorem* 8.9 in the Neumann case, is not known in the present.

8.7 Problems

1. Prove *Proposition* 8.1.1.
Hint: (a) A pair $A,\ B$ such that the $n \times 2n$ matrix $(A|B)$ has maximum rank defines a Lagrangean subspace through (8.4) *iff* AB^* is Hermitean. (c) The claim can be checked directly. Alternatively, note that the squared norms $\|\Psi(0+) \pm i\ell\Psi'(0+)\|^2$ with a fixed $\ell > 0$ coincide *iff* the boundary form (8.3) with $\Phi = \Psi$ vanishes.

2. For any coupling in a graph vertex of degree n there are orthogonal and mutually orthogonal projections $P,\ Q$ in $\mathbb{C}^n$ and an invertible Hermitean Λ acting on the range of $C := I - P - Q$ such that the coupling conditions can be written in the form $P\Psi(0+) = 0$, $Q\Psi'(0+) = 0$, and $C\Psi'(0+) = \Lambda C\Psi(0+)$.

3. The Laplacians acting on $\bigoplus_{j=1}^n L^2(\mathbb{R}_+)$ with the coupling conditions described in Examples 8.1.1 have $\mathbb{R}_+$ as their essential spectrum and a simple eigenvalue equal to $-(\alpha/n)^2$ and $-(n/\beta)^2$ if α and β, respectively, are negative. Find the spectral properties in case the star graph has finite edge lengths.

4. Vertex couplings of *Proposition* 8.1.1*c* which are invariant w.r.t.permutations of the edges form a two-parameter family characterized by matrices $U = aI + bJ$ with complex coefficients a, b satisfying the relations $|a| = 1$ and $|a + nb| = 1$.

5. Fill in the details of the proof of *Theorem* 8.1.

6. Fill in the details of the proof of *Theorem* 8.2.

7. (a) Let $H = \left(-i\frac{\mathrm{d}}{\mathrm{d}x} - A\right)^2 + V(x)$ be a magnetic Schrödinger operator on $L^2(I)$, where $I = (0, L)$ is a finite or semiinfinite interval, with a potential V, vector potential A, and boundary conditions at the endpoints of I which make H_A self-adjoint. Let $\psi_A^{s,t}$ denote the solution of the equation $H_A\psi = k^2\psi$ with the boundary values $\psi_A^{s,t}(0) = s$ and $(\psi_A^{s,t})'(0) = t$. Check that it is related to the analogous solution in the nonmagnetic case, $A = 0$, by $\psi_A^{s,t}(x) = \mathrm{e}^{i\int_0^x A(z)\mathrm{d}z}\,\psi_0^{s,t}(x)$ for all $x \in (0, L)$.
(b) Consider a quantum graph vertex with n outgoing edges indexed by $j = 1, \dots, n$ and parametrized by $x \in (0, L_j)$. Suppose that the j-th edge carries a constant vector potential A_j. Using claim (a) find how the δ-coupling with a parameter α modifies in the magnetic case, namely that it includes continuity, $\psi_j(0) = \psi_k(0) =: \psi(0)$ for all $j, k = 1, \dots, n$, together with the condition $\sum_{j=1}^n \psi_j'(0) = \left(\alpha + i\sum_{j=1}^n A_j\right)\psi(0)$.

8. Consider a non-negative operator H on a Hilbert space $\mathcal{H}$ and a sequence of non-negative operators $\tilde{H} = H_n$ on $\mathcal{H}$ which all have the same form domain as H. Suppose that H and H_n are δ_n-close of order one with respect to trivial identification maps, where $\delta_n \to 0$ as $n \to \infty$, then $H_n \to H$ holds in the norm-resolvent sense. On the other hand, such a convergence does not imply the appropriate closeness.
Hint: The assumed δ_n-closeness implies $\|H - H_n\|_{1\to-1} \to 0$ while the norm-resolvent convergence requires $\|H - H_n\|_{2\to-2} \to 0$.

9. Prove *Lemma* 8.4.9.

10. Consider the Hamiltonian $H^{\beta,a}$ defined in Sect. 8.4.3. Check that these operators are uniformly bounded from below as $a \to 0$ if $\beta < 0$, while for $\beta \geq 0$ we have $\inf \sigma(H^{\beta,a}) \to -\infty$ as $a \to 0$. Furthermore, show that the operators H^{β}_{ε} used in the δ'_s-approximation have the same property.

11. The coupling in the definition of the operator H^{r} appearing in *Theorem* 8.10 is described by the boundary conditions of *Proposition* 8.1.1*c* with the matrix

$$U = \frac{1}{2(c_1^2+c_2^2)+i\hat{\lambda}} \begin{pmatrix} -4c_1c_2 - i\hat{\lambda} & 2(c_1^2-c_2^2) \\ 2(c_1^2-c_2^2) & 4c_1c_2 - i\hat{\lambda} \end{pmatrix}.$$

$\sigma_{\mathrm{ess}}(H^{\mathrm{r}})$ is absolutely continuous for any $\hat{\lambda} \in \mathbb{R}$ and coincides with $[0,\infty)$. For $\hat{\lambda} > 0$ there are no eigenvalues, while for $\hat{\lambda} < 0$ there is just one negative eigenvalue equal to $k_0^2 = -\frac{1}{4}\hat{\lambda}^2(c_1^2+c_2^2)^{-2}$ and the corresponding normalized eigenfunction is

$$\psi_0(s) = \sqrt{\frac{|\hat{\lambda}|}{2}} \frac{1}{c_1^2+c_2^2} \left\{ \begin{matrix} (c_1-c_2)e^{ik_0 s} & s>0 \\ (c_1+c_2)e^{-ik_0 s} & s<0 \end{matrix} \right\}, \quad k_0 = \frac{i|\hat{\lambda}|}{2(c_1^2+c_2^2)}.$$

Finally, for $\hat{\lambda} = 0$ the operator H^{r} has a zero energy resonance. The on-shell scattering matrix at energy k^2, $k \geq 0$, is given by $S(k) = \begin{pmatrix} t^l(k) & r^r(k) \\ r^l(k) & t^r(k) \end{pmatrix}$ with the amplitudes

$$t^{\{l,r\}}(k) = \frac{2k(c_1^2-c_2^2)}{2k(c_1^2+c_2^2)+i\hat{\lambda}}, \qquad r^{\{l,r\}}(k) = \pm\frac{4kc_1c_2 \mp i\hat{\lambda}}{2k(c_1^2+c_2^2)+i\hat{\lambda}}.$$

12. Fill in the details of the proof of *Lemma* 8.5.1.

Chapter 9
Periodic and Random Systems

Most spectral and transport properties discussed in the previous chapters concerned perturbations which have been in some sense localized, although there were exceptions, for instance, the effects connected with motion in a homogeneous magnetic field treated in Chap. 7. Now we are going to discuss two important situations in which the geometric variations are infinitely extended, namely periodic and random quantum waveguides and layers.

Periodic systems are ubiquitous in nature, in particular, due to the existence of (natural or artificial) crystal structures and their spectra have typical properties such as the absolute continuity and band-and-gap character. The question to ask is whether the same can be said about periodically curved tubes; we will analyze it in the simplest case of periodic planar waveguides. Apart from the geometry, a periodic structure may also come from a potential. We have encountered an example of such a system speaking about non-classical edge states in Sect. 7.2.2, here we are going to analyze spectral properties caused by periodic impurities modeled by point interactions in straight strips or layers.

On the other hand, a periodic structure is in fact a very particular arrangement and in most real physical systems a disorder coming from randomly distributed impurities must be taken into account. Its presence has important spectral consequences. By a deep insight of Philip Anderson which subsequently had a profound influence on both physics and mathematics, the interference caused by interaction with randomly positioned scatterers leads to a localization of the particle. The latter can be manifested in various ways, for example, by the occurrence of a dense pure point spectrum, at least in the vicinity of the spectral threshold, or by the so-called Lifschitz tails of the integrated density of states. We are going to illustrate these effects in the model of a randomly deformed two-dimensional waveguide.

P. Exner and H. Kovařík, *Quantum Waveguides*,
Theoretical and Mathematical Physics, DOI 10.1007/978-3-319-18576-7_9

9.1 Periodic Waveguides

For simplicity we limit ourselves to the two-dimensional situation and discuss the spectrum of periodic planar waveguides. As in the previous chapters it is convenient to regard them as deformations of a straight waveguide. Since the properties we are going to discuss do not change under scaling, we may take, for instance, $\Omega_0^+ = \mathbb{R} \times (0, \pi)$ and consider a function $h : \Omega_0^+ \to \mathbb{R}^2$ such that

$$h(x_1 + 2\pi, x_2) = \big(h_1(x_1, x_2) + 2\pi,\ h_2(x_1, x_2)\big) \quad \text{for all } \vec{x} \in \Omega_0^+\,, \tag{9.1}$$

where $\vec{x} = (x_1, x_2)$ as usual. The periodic waveguide Ω to be considered is then the image of the straight strip, $\Omega = h(\Omega_0^+)$. Note that this may describe various geometries such as strips which are periodically curved, have a periodically changing width, and other possibilities.

9.1.1 Absolute Continuity

Our aim is to prove the absolute continuity of the spectrum. First we need to recall a result which guarantees this property for a certain class of second-order operators in $\mathbb{R}^2$ with periodic coefficients, both as a motivation and as a tool.

Theorem 9.1 *Consider the lattice $\Gamma = (2\pi\mathbb{Z}) \times (2\pi\mathbb{Z})$ in $\mathbb{R}^2$ and the operator*

$$H = \sum_{j,k=1}^{2} \frac{1}{\mu}\, (i\partial_j - A_j)\, G_{jk}\, (i\partial_k - A_k) + \frac{1}{\mu}\, V$$

in $L^2(\mathbb{R}^2, \mu\, \mathrm{d}\vec{x})$. Let the functions $G_{jk}, A_j, \mu, V,\ j, k = 1, 2$, be Γ−periodic and assume that $G_{jk}, A_j \in W^{1,\infty}(\mathbb{R}^2)$ and $\mu, V \in L^\infty(\mathbb{R}^2)$. Suppose, moreover, that there are positive constants c and C such that

$$c\, |\xi|^2 \le \langle \xi, G(\vec{x})\xi \rangle_{\mathbb{R}^2} \le C\, |\xi|^2 \quad \text{and} \quad c \le \mu(\vec{x}) \le C$$

holds for all $\xi \in \mathbb{R}^2$ and $\vec{x} \in \mathbb{R}^2$. Then $\sigma(H)$ is absolutely continuous.

We refer the reader to the notes for sources concerning the history and the proof of this theorem. Our main goal here is to demonstrate the validity of a similar claim for periodic waveguides:

Theorem 9.2 *Let $h \in W^{3,\infty}(\Omega_0^+)$ be bijective and such that the Jacobian $J_h(\vec{x})$ does not vanish for any $\vec{x} \in \Omega_0^+$. Under the periodicity condition (9.1) the spectrum of the Dirichlet-Laplacian $-\Delta_D^\Omega$ in $L^2(\Omega)$ corresponding to the periodically deformed strip $\Omega = h(\Omega_0^+)$ is absolutely continuous.*

The strategy behind the proof of *Theorem* 9.2 will be to reduce the problem to the analysis of an operator on Ω_0^+ with periodic boundary conditions and $A_1 = A_2 = V = 0$, and to apply *Theorem* 9.1. The reduction will be done in several steps which we describe below in a series of auxiliary results.

As we did repeatedly before, in the first step we translate the geometry into the coefficients of the operator by transforming the problem into an equivalent one on the straight strip Ω_0^+. To this end, we employ the inverse mapping $f = h^{-1} : \Omega \to \Omega_0^+$ and introduce the function

$$\tilde{G}(\vec{x}) = (J_f^{-1} D_f\, (D_f)^T \circ f^{-1})(\vec{x}), \quad \tilde{\mu}(\vec{x}) = J_f(f^{-1}(\vec{x}))^{-1}, \tag{9.2}$$

where D_f is the Jacobi matrix of f and J_f denotes its determinant. Next we define the unitary operator $\tilde{U}$ from $L^2(\Omega)$ onto $L^2(\Omega_0^+, \tilde{\mu}\, d\vec{x})$ by

$$(\tilde{U}\psi)(\vec{x}) = \psi(f^{-1}(\vec{x})) \quad \text{for } \psi \in L^2(\Omega)\,.$$

Finally, let $H_D(\tilde{G}, \tilde{\mu})$ be the operator on $L^2(\Omega_0^+,\, \tilde{\mu}\, \mathrm{d}\vec{x})$ associated with the sesquilinear form defined on $H_0^1(\Omega_0^+)$ by

$$Q_D(\psi, \phi) = \int_{\Omega_0^+} \langle \nabla\psi, \tilde{G}\, \nabla\phi\rangle_{\mathbb{R}^2}\, \mathrm{d}\vec{x}\,.$$

Lemma 9.1.1 *$\tilde{U}\,(-\Delta_D^{\Omega})\,\tilde{U}^{-1} = H_D(\tilde{G}, \tilde{\mu})$ holds in $L^2(\Omega_0^+,\, \tilde{\mu}\, \mathrm{d}\vec{x})$.*

Proof We denote by $\vec{\boldsymbol{x}} = f^{-1}(\vec{x})$ the variables in Ω. Let $B(\vec{\boldsymbol{x}}) = J_f^{-1} D_f\, (D_f)^T(\vec{\boldsymbol{x}})$ so that $\tilde{G}(\vec{x}) = B(f^{-1}(\vec{x}))$. For any $\psi, \phi \in H_0^1(\Omega_0^+)$ we then have

$$\int_\Omega \nabla_{\vec{\boldsymbol{x}}}\, (\tilde{U}^{-1}\psi)(\vec{\boldsymbol{x}}) \cdot \nabla_{\vec{\boldsymbol{x}}}\, (\tilde{U}^{-1}\phi)(\vec{\boldsymbol{x}})\, \mathrm{d}\vec{\boldsymbol{x}} = \int_\Omega J_f(\vec{\boldsymbol{x}})\, \langle \nabla_{\vec{x}}\, \psi(\vec{x}),\, B(\vec{\boldsymbol{x}})\, \nabla_{\vec{x}}\, \phi(\vec{x})\rangle_{\mathbb{R}^2}\, \mathrm{d}\vec{\boldsymbol{x}}$$
$$= \int_{\Omega_0^+} \langle \nabla_{\vec{x}}\, \psi(\vec{x}),\, \tilde{G}(\vec{x})\, \nabla_{\vec{x}}\, \phi(\vec{x})\rangle_{\mathbb{R}^2}\, \mathrm{d}\vec{x} = Q_D(\psi, \phi)\,,$$

which is nothing else than the stated unitary equivalence. ∎

In the next step we are going to replace the matrix $\tilde{G}$ with a new one which is diagonal on the boundary.

Lemma 9.1.2 *There exists a bijective mapping $w \in W^{2,\infty}(\Omega_0^+)$ from Ω_0^+ onto itself with J_w uniformly positive, satisfying the periodicity requirement (9.1), and such that if $\tilde{G}$ is given by (9.2), then the matrix-valued function*

$$\hat{G}(\vec{x}) = (J_w^{-1} D_w\, \tilde{G}\, (D_w)^T \circ w^{-1})(\vec{x})$$

is 2π-periodic with respect to x_1, belongs to $W^{1,\infty}(\Omega_0^+)$, and satisfies

$$\hat{G}_{12}(x_1, 0) = \hat{G}_{12}(x_1, \pi) = 0 \quad \text{for all} \quad x_1 \in \mathbb{R}.$$

Proof Note that $\det \tilde{G} = 1$ holds in view of (9.2), hence we can factorize $\tilde{G}$ as $G = M^T G_0 M$ using the matrices

$$G_0 = \begin{pmatrix} \tilde{G}_{22}^{-1} & 0 \\ 0 & \tilde{G}_{22} \end{pmatrix}, \quad M = \begin{pmatrix} 1 & 0 \\ m & 1 \end{pmatrix}, \quad m := \tilde{G}_{12}\, \tilde{G}_{22}^{-1}.$$

Fix a function $\eta \in C_0^\infty(\mathbb{R})$ satisfying $\eta(s) = 1$ for $|s| \le 1$ and a number $\delta > 0$; then we define the map $w : \Omega_0^+ \to \Omega_0^+$ by $w(\vec{x}) = \vec{y} = (x_1 - b(\vec{x}), x_2)$, where

$$b(\vec{x}) := m(x_1, 0)\, x_2\, \eta\Big(\frac{x_2}{\delta}\Big) + m(x_1, \pi)(x_2 - \pi)\, \eta\Big(\frac{x_2 - \pi}{\delta}\Big).$$

It is easy to check that for δ small enough w is bijective, belongs to $W^{2,\infty}(\Omega_0^+)$, and that $J_w = 1 + \mathcal{O}(\delta)$, hence J_w is uniformly positive for such δ. It is also obvious that w satisfies the requirement (9.1), which in turn implies that $\hat{G}$ is 2π-periodic with respect to x_1. Moreover, from the fact that $D_w \in W^{1,\infty}(\Omega_0^+)$ we deduce that $\hat{G} \in W^{1,\infty}(\Omega_0^+)$. Finally, a direct calculation using the construction of w shows that $M(D_w)^T$ is the unit matrix for $x_2 = 0$ and $x_2 = \pi$, hence we get $\hat{G}(\vec{x}) = (G_0 \circ w^{-1})(\vec{x})$ for these values of x_2 thus concluding the proof. ∎

In view of *Lemma* 9.1.1, to prove *Theorem* 9.2 it suffices to show that the spectrum of $H_D(\tilde{G}, \tilde{\mu})$ in $L^2(\Omega_0^+, \tilde{\mu}\, d\vec{x})$ is absolutely continuous. To do so we extend the operator $H_D(\tilde{G}, \tilde{\mu})$ from Ω_0^+ into

$$S = \overline{\Omega_0^+ \cup \Omega_0^-}, \qquad \Omega_0^- = \mathbb{R} \times (-\pi, 0).$$

For this purpose we have to introduce appropriate extension operators. Let us denote by $L^p_+(S)$ and $L^p_-(S)$ the subspaces of symmetric and antisymmetric functions (with respect to the axis $x_2 = 0$) from $L^p(S)$. Here $p \in [2, \infty]$. The respective projections $P_\pm$ on these subspaces are given by

$$P_\pm \psi(\vec{x}) = \frac{1}{2}\, (\psi(x_1, x_2) \pm \psi(x_1, -x_2)).$$

The extension operators $T_\pm : L^p(\Omega_0^+) \to L^p(S)$ are then given by

$$(T_\pm \psi)(\vec{x}) = \begin{cases} 2^{-1/p}\, \psi(x_1, x_2) & \text{if } \vec{x} \in \Omega_0^+, \\ -2^{-1/p}\, \psi(x_1, -x_2) & \text{if } \vec{x} \in \Omega_0^-. \end{cases}$$

To relate these extensions to 2π-periodic functions with respect the variable x_2, we define $\mathcal{C} = \mathbb{R}^2/\gamma_2$, where $\gamma_2 = \{2\pi m\,(0,1) \,:\, m \in \mathbb{Z}\}$. It follows that

$$W^{1,p}_{\pm}(\mathcal{C}) := P_{\pm} W^{1,p}(\mathcal{C}) = W^{1,p}(\mathcal{C}) \cap L^p_{\pm}(\mathcal{C})\,. \tag{9.3}$$

The proof of the following lemma is left to the reader (Problem 1).

Lemma 9.1.3 *We have*

$$T_+\, W^{1,p}(\Omega_0^+) = W^{1,p}_+(\mathcal{C}) \quad \text{and} \quad T_-\, W^{1,p}(\Omega_0^-) = W^{1,p}_-(\mathcal{C})\,.$$

In addition, for any $\psi_+ \in W^{1,p}(\Omega_0^+)$ and $\psi_- \in W^{1,p}_0(\Omega_0^+)$ we also have

$$\partial_1(T_\pm \psi_\pm) = T_\pm(\partial_1 \psi_\pm) \quad \text{and} \quad \partial_2(T_\pm \psi_\pm) = T_\mp(\partial_2 \psi_\pm)\,. \tag{9.4}$$

Now we are able to transform the problem to the investigation of a *periodic* operator on the doubled strip S. Let $\tilde{\mu}$ be given by (9.2). With *Lemma* 9.1.2 in mind we assume that

$$\tilde{G}_{12}(x_1, 0) = \tilde{G}_{12}(x_1, \pi) \quad \text{for all } x_1 \in \mathbb{R}\,. \tag{9.5}$$

We employ the operator $H_p(G,\mu)$ in $L^2(S, \mu\, d\vec{x})$ associated with the following sesquilinear form,

$$Q_p[\psi, \phi] = \int_S \langle \nabla\psi, G\, \nabla\phi \rangle_{\mathbb{R}^2}\, d\vec{x}\,, \quad \psi, \phi \in D(Q_p) = W^{1,2}(\mathcal{C} \restriction S)\,,$$

where G and μ are given by

$$G_{jj} = T_+\, \tilde{G}_{jj}\,, \quad G_{12} = T_-\, \tilde{G}_{12}\,, \quad \mu = T_+\, \tilde{\mu}\,. \tag{9.6}$$

In view of (9.3) we have $P_\pm D(Q_p) = D(Q_p) \cap L^2_\pm(\mathcal{C})$. Next we show that

$$Q_p[P_+\psi, P_-\phi] = 0 \qquad \forall\, \psi,\, \phi \in D(Q_p)\,. \tag{9.7}$$

Indeed, using the notation $\psi_+ = P_+\psi$ and $\phi_- = P_-\phi$, we obtain

$$Q_p[\psi_+, \phi_-] = \sum_{j,k=1}^{2} \int_S \langle \partial_j \psi_+, G_{jk}\, \partial_k \phi_- \rangle_{\mathbb{R}^2}\, d\vec{x}\,.$$

It follows from (9.6) that G_{jj} and μ are even functions in the variable x_2. This in combination with *Lemma* 9.1.3, see Eq. (9.4), shows that all the integrands on the right-hand side of the last equation are odd with respect to x_2. Hence (9.7) holds true which implies that $L^2_\pm(S)$ are invariant subspaces of the operator $H_p = H_p(G, \mu)$. This means that

$$H_p = H_p^+ \oplus H_p^- \,, \tag{9.8}$$

where $H_p^\pm$ are the self-adjoint operators associated with the sesquilinear forms obtained from $Q_p[\cdot\,,\cdot]$ by restricting the domain $D(Q_p)$ to $W^{1,2}_\pm(\mathcal{C} \upharpoonright S)$.

It is now clear how to proceed further. We shall use *Theorem* 9.1 to show that the spectrum of H_p is absolutely continuous, and from (9.8) we can then conclude that the same is true for the operator H_p^-. Note that the latter is related to $H_D(\tilde{G}, \tilde{\mu})$ in the following way.

Lemma 9.1.4 *Under the condition (9.5) we have*

$$T_-\, H_D(\tilde{G}, \tilde{\mu})\, T_-^* = H_p^- \,.$$

Proof In view of *Lemma* 9.1.3 it suffices to check that the corresponding sesquilinear forms coincide for any $\psi, \phi \in W^{1,2}_-(\mathcal{C} \upharpoonright S)$. Using the definition of T_- we easily see that

$$(T_-^* u)(\vec{x}) = \sqrt{2}\, u(\vec{x}) \quad \text{for all } \vec{x} \in \Omega_0^+ \text{ and } u \in L^2_-(\mathcal{C} \upharpoonright S)\,;$$

if we now take into account the symmetry properties of $G_{j,k}$ we find that the sesquilinear form of the operator $T_-\, H_D(\tilde{G}, \tilde{\mu})\, T_-^*$ satisfies

$$\int_S \langle \nabla(T_-^*\psi), \tilde{G}\, \nabla(T_-^*\phi)\rangle_{\mathbb{R}^2}\, d\vec{x} = 2 \sum_{j,k=1}^{2} \int_{\Omega_0^+} \langle \partial_j \psi, \tilde{G}_{jk}\, \partial_k \phi\rangle_{\mathbb{R}^2}\, d\vec{x}$$
$$= \sum_{j,k=1}^{2} \int_S \langle \partial_j \psi, G_{jk}\, \partial_k \phi\rangle_{\mathbb{R}^2}\, d\vec{x} \,.$$

Since the last expression coincides with the sesquilinear form of the operator H_p^-, we obtain the result. ∎

Proof of Theorem 9.2: In view of *Lemma* 9.1.2 it suffices to prove the claim for $\tilde{G}$ which verifies (9.5). Since the coefficients G, μ defined by (9.6) satisfy the assumptions of *Theorem* 9.1, it follows that the spectrum of H_p is absolutely continuous. Hence the same is true for its orthogonal parts H_p^+ and H_p^-, see (9.8). From *Lemma* 9.1.4 we then conclude that $H_D(\tilde{G}, \tilde{\mu})$ is absolutely continuous, which in turn by *Lemma* 9.1.1 implies the claim of *Theorem* 9.2. ∎

9.1.2 Periodically Curved Waveguides

Let us now investigate in more detail a particular case of a periodic waveguide discussed above, namely a two-dimensional periodically curved strip Ω of width d. It is convenient to choose one the boundaries as the reference curve, in other words, we suppose that Ω is the image of $\Omega_0 = \mathbb{R} \times (0, d)$ by the map (1.3),

$$h(s,u) = (\xi(s) - u\dot\eta(s),\, \eta(s) + u\dot\xi(s))\,, \quad (s,u) \in \Omega_0\,,$$

expressed in terms of the curvilinear coordinated (s, u) introduced in Sect. 1.1. As we have demonstrated there, the Dirichlet Laplacian $-\Delta_D^\Omega$ on $L^2(\Omega)$ is unitarily equivalent to the operator

$$H_d = -\partial_s(1+u\gamma)^{-2}\partial_s - \partial_u^2 + V(s,u) \quad \text{in} \quad L^2(\Omega_0)$$

with the effective potential $V(s, u)$ given by (1.8); we label the symbol with the strip width d which will be important in the following. In contrast to Sect. 1.1, the signed curvature $\gamma(s) = \dot\eta(s)\ddot\xi(s) - \dot\xi(s)\ddot\eta(s)$ is now supposed to be a *periodic* function; for the sake of simplicity we suppose that its period is 2π. In such a case the function $V(\cdot, u)$ is 2π-periodic for a fixed u as well, and by the standard Bloch-Floquet decomposition H_d is unitarily equivalent to a direct integral,

$$H_d = U^{-1}\Big(\int_{[0,1]}^{\oplus} H_{d,\theta}\,\mathrm{d}\theta\Big)\,U\,, \tag{9.9}$$

where the fiber operators on $L^2(\Lambda_d)$ with the cell $\Lambda_d := (0, 2\pi) \times (0, d)$ act as

$$H_{d,\theta} = -\partial_s(1+u\gamma)^{-2}\partial_s - \partial_u^2 + V(s,u)$$

on Dom $(H_{d,\theta})$ consisting of all functions $f \in H^2(\Lambda_d)$ which satisfy the conditions $f(\cdot, 0) = f(\cdot, d) = 0$ and $\partial_s^j f(2\pi, \cdot) = \mathrm{e}^{2\pi i\theta}\,\partial_s^j f(0, \cdot)$ for $j = 0, 1$. The quasi-momentum θ runs through the Brillouin zone $[0, 1]$, and the unitary equivalence (9.9) is mediated by

$$(Uf)(s,u,\theta) = \sum_{j\in 2\pi\mathbb{Z}} \mathrm{e}^{i\theta}\, f(s-j,u)\,, \quad (s,u) \in \Lambda_d\,.$$

Since the operator $H_{d,\theta}$ has compact resolvent, its spectrum is purely discrete; we denote its eigenvalues by $E_n(\theta, d)$. In view of decomposition (9.9) we have

$$\sigma(-\Delta_D^\Omega) = \sigma(H_d) = \overline{\bigcup_{n=1}^{\infty} I_n(d)}\,, \quad I_n(d) := \{E_n(\theta,d) :\, \theta \in [0,1]\}\,.$$

In the following we shall assume that the strip bending over a period is zero,

$$\int_0^{2\pi} \gamma(s)\,\mathrm{d}s = 0\,, \tag{9.10}$$

and that γ is sufficiently regular, $\gamma \in C^2(\mathbb{R})$. With the help of parametrization (1.5) it is easy to check that the mapping h then satisfies the assumptions of *Theorem* 9.2, which means that the spectrum of $-\Delta_D^\Omega$ is absolutely continuous, and the same holds, of course, for the operator H_d. This means that none of the intervals $I_n(d)$ defined above degenerate into a single point, and consequently, the spectrum of H_d is the union of spectral bands,

$$I_n(d) = \left[\min_{\theta\in[0,1]} E_n(\theta,d),\ \max_{\theta\in[0,1]} E_n(\theta,d)\right], \quad n = 1, 2, \dots .$$

It is natural to ask whether there are open gaps between these bands, and if so, where they are located. We are going to show that the spectrum contains at least one open gap provided the strip width d is sufficiently small.

It is obvious that the existence of open gaps depends on properties of the band functions $E_n(\theta, d)$. In analogy with the considerations of Sect. 1.6 we shall use the fact that in a thin strip the spectral properties are determined in the leading order by a one-dimensional Schrödinger operator of the type (1.44). Hence we compare the behavior of $E_n(\cdot, d)$ in the limit $d \to 0$ with that of the eigenvalues of the reference operator

$$T_\theta = -\frac{\mathrm{d}^2}{\mathrm{d}s^2} - \frac{1}{4}\gamma^2(s)$$

on $L^2(0, 2\pi)$. The domain of T_θ consists of all $f \in H^2(0, 2\pi)$ satisfying the conditions $f(2\pi) = \mathrm{e}^{2\pi i\theta} f(0)$ and $f'(2\pi) = \mathrm{e}^{2\pi i\theta} f'(0)$. Let $\{t_j(\theta)\}_{j\in\mathbb{N}}$ be the eigenvalues of T_θ counted with their multiplicities.

Proposition 9.1.1 *Suppose that $\gamma \in C^2(\mathbb{R})$ is 2π-periodic and such that Ω is not self-intersecting for all d small enough. Then the expansion*

$$E_j(\theta, d) = \kappa_1^2 + t_j(\theta) + \mathcal{O}(d) \tag{9.11}$$

holds as $d \to 0$ with the error term uniform with respect to $\theta \in [0, 1]$, where as usual $\kappa_1 = \pi/d$.

Proof We write conventionally $\gamma = \gamma_+ - \gamma_-$ with $\gamma_\pm = \frac{1}{2}(|\gamma| \pm \gamma)$ and put $c_\pm := \|\gamma_\pm\|_\infty$. Using furthermore $c_1 := \|\dot\gamma\|_\infty$ and $c_2 := \|\ddot\gamma\|_\infty$ we define

$$V_+(s) := \frac{1}{2}\,(1 - dc_-)^{-3} dc_2 - \frac{1}{4}\,(1 + dc_+)^{-2}\,\gamma(s)^2$$

and

$$V_-(s) := -\frac{1}{2}(1-dc_-)^{-3}dc_2 + -\frac{5}{4}(1-dc_-)^{-4}d^2c_1^2 - \frac{1}{4}(1-dc_-)^{-2}\gamma(s)^2 .$$

Using these notations, we introduce the operators

$$H^{\pm}_{\theta,d} = -(1 \mp dc_{\mp})^{-2}\,\partial_s^2 - \partial_u^2 + V_{\pm}(s)$$

on $L^2(\Lambda_d)$ acting on Dom $(H^{\pm}_{\theta,d}) =$ Dom $(H_{\theta,d})$ which estimate $H_{\theta,d}$ satisfying the inequalities

$$H^{-}_{\theta,d} \le H_{\theta,d} \le H^{+}_{\theta,d} \tag{9.12}$$

in the sense of quadratic forms on $H^1(\Lambda_d)$. We also introduce the operators

$$T^{\pm}_{\theta,d} = -(1 + d\gamma_{\mp})^{-2}\,\partial_s^2 + V_{\pm}(s)$$

on $L^2(0, 2\pi)$ with Dom $(T^{\pm}_{\theta,d}) =$ Dom (T_θ) and denote by $\epsilon_j^{\pm}(\theta, d)$ their eigenvalues counted with multiplicities. A straightforward calculation shows that

$$(1 \mp dc_{\mp})^{-2}\,T_\theta + \mathcal{O}(d) \le T^{\pm}_{\theta,d} \le (1 \mp dc_{\mp})^{-2}\,T_\theta + \mathcal{O}(d)$$

for $d \to 0$, where the error terms are independent of s and θ, and consequently, the minimax principle implies that

$$\epsilon_j^{\pm}(\theta, d) = k_j(\theta) + \mathcal{O}_j(d) \quad \text{as } d \to 0 \tag{9.13}$$

with the error terms uniform with respect to $\theta \in [0, 1]$. On the other hand, by separation of variables one can check easily that the eigenvalues $\lambda^{\pm}_{j,k}(\theta, d)$ of the operators $H^{\pm}_{\theta,d}$ are given by

$$\lambda^{\pm}_{j,k}(\theta, d) = \kappa_k^2 + \epsilon_j^{\pm}(\theta, d)\,, \tag{9.14}$$

where as usual $\kappa_k = k\kappa_1$. Our goal is to show that for every $j_0 \in \mathbb{N}$ there is a $\tilde{d}(j_0)$ such that for any $d \le \tilde{d}(j_0)$ we have

$$\lambda^{\pm}_{j,k}(\theta, d) \ge \lambda^{\pm}_{j_0,1}(\theta, d)$$

for all $k \ge 2,\ j \ge 1$ and $\theta \in [0, 1]$. Indeed, from (9.14) and (9.13) we get

$$\begin{aligned}\lambda^{\pm}_{j,k}(\theta, d) - \lambda^{\pm}_{j_0,1}(\theta, d) &\ge 3\kappa_1^2 + \epsilon_1^{\pm}(\theta, d) - \epsilon_{j_0}^{\pm}(\theta, d)\\ &\ge 3\kappa_1^2 + \min_{\theta\in[0,1]} t_1(\theta) - \max_{\theta\in[0,1]} t_{j_0}(\theta) + \mathcal{O}_{j_0}(d)\end{aligned}$$

for all $k \geq 2$, $j \geq 1$ and $\theta \in [0, 1]$ which implies the sought inequality. The latter, in turn, shows that for any $j_0 \in \mathbb{N}$ there is a $\tilde{d} = \tilde{d}(j_0)$ such that for any $d < \tilde{d}$ and any $n \leq j_0$ the n-th eigenvalue of $H^{\pm}_{\theta,d}$ is equal to $\lambda^{\pm}_{n,1}(\theta, d)$. The minimax principle in combination with (9.12) then shows that

$$\lambda^{-}(n, 1, \theta, d) \leq E_n(\theta, d) \leq \lambda^{+}(n, 1, \theta, d) \qquad \forall\, n \leq j_0,\ \forall\, \theta \in [0, 1]$$

holds for any $d < \tilde{d}$. In view of (9.14) and (9.13) we thus obtain the asymptotic expansion (9.11) with the error term independent of $\theta \in [0, 1]$. ∎

In order to prove the existence of open gaps we also need the following classical result from the theory of inverse Sturm-Liouville problems.

Theorem 9.3 *Let* W *be a real-valued piecewise continuous function on* $[0, 2\pi]$. *Denote by* $\{\lambda^{\pm}_j\}_{j\in\mathbb{N}}$ *the non-decreasing sequences of eigenvalues of the operators*

$$A^{\pm} = -\frac{d^2}{ds^2} + W(s) \quad in \quad L^2(0, 2\pi)$$

with $\mathrm{Dom}\,(A^{\pm}) = \{f \in H^2(0, 2\pi) :\ f(2\pi) = \pm f(0),\ f'(2\pi) = \pm f'(0)\}$. *If*

$$\lambda^{+}_j = \lambda^{+}_{j+1} \quad \text{for all even } j \quad \text{and} \quad \lambda^{-}_j = \lambda^{-}_{j+1} \quad \text{for all odd } j\,,$$

then the potential W *is constant on* $[0, 2\pi]$.

Now we are able to prove the existence of an open spectral gap for the operator H_d, and hence also for the periodically curved strip Hamiltonian $-\Delta^{\Omega}_{D}$.

Theorem 9.4 *Suppose that* γ *satisfies the assumptions of Proposition 9.1.1. Assume, moreover, that condition (9.10) is valid and that the strip is not straight,* $\gamma \neq 0$. *Then there exists an* $m \in \mathbb{N}$ *and a constant* $C_m > 0$ *such that*

$$\min_{\theta\in[0,1]} E_{m+1}(\theta, d) - \max_{\theta\in[0,1]} E_m(\theta, d) = C_m + \mathcal{O}(d)$$

holds true as $d \to 0$.

Proof Consider the eigenvalues $t_j(\theta)$, $\theta \in [0, 1]$, of the operator T_θ. It is well known that each function $t_j(\cdot)$ is continuous and even with respect to the center of the Brillouin zone, $k_j(1-\theta) = k_j(\theta)$ (see the notes). Moreover, for every j odd (even) $k_j(\cdot)$ is strictly increasing (decreasing) as θ increases from 0 to 1/2. In particular, we have

$$t_{2j-1}(0) < t_{2j-1}(1/2) \leq t_{2j}(1/2) < t_{2j}(0) \quad \text{for all } j \in \mathbb{N}\,. \tag{9.15}$$

We denote by

$$\mathcal{G}_j = \min_{\theta\in[0,1]} t_{j+1}(\theta) - \max_{\theta\in[0,1]} t_j(\theta)$$

the jth spectral gap of $T = \int_{[0,1]}^{\oplus} T_\theta \, \mathrm{d}\theta$. In view of *Proposition* 9.1.1 we have

$$\min_{\theta\in[0,1]} E_{n+1}(\theta, d) - \max_{\theta\in[0,1]} E_n(\theta, d) = \mathcal{G}_n + \mathcal{O}(d) \tag{9.16}$$

for each $n \in \mathbb{N}$ as $d \to 0$. Moreover, from (9.15) it follows that

$$\mathcal{G}_j = k_{j+1}(1/2) - k_j(1/2) \quad \text{for } j \text{ odd}, \quad \mathcal{G}_j = k_{j+1}(0) - k_j(0) \quad \text{for } j \text{ even}.$$

Hence we can apply *Theorem* 9.3 with $A^+ = T_0$ and $A^- = T_{1/2}$. Assuming that $\mathcal{G}_j = 0$ for all $j \in \mathbb{N}$, we come to the conclusion that γ is constant on $[0, 2\pi]$. Condition (9.10) then implies $\gamma = 0$ which contradicts, however, our hypotheses on γ. Consequently, there is an $m \in \mathbb{N}$ such that $\mathcal{G}_m > 0$, and the stated asymptotic relation follows from (9.16). ∎

Knowing that the spectrum of $-\Delta_D^\Omega$ contains at least one open gap for d small enough, one naturally asks where such open gaps are located, in other words, what is the value of m in *Theorem* 9.4. In order to give at least a partial answer to this question we consider gently curved waveguides passing to the strip $\Omega(\varepsilon)$ which results from replacing the reference curve (boundary) by the one with the scaled curvature, $\gamma \mapsto \varepsilon\gamma$, where $\varepsilon > 0$ is a small parameter. As long as γ is 2π-periodic the Dirichlet Laplacian $-\Delta_D^{\Omega(\varepsilon)}$ admits the Bloch-Floquet decomposition; we denote the corresponding band functions by $E_n(\theta, d, \varepsilon)$. Let now α_n be the Fourier coefficients of the 2π-periodic function γ^2,

$$\gamma(s)^2 = \frac{1}{\sqrt{2\pi}} \sum_{n\in\mathbb{Z}} \alpha_n \, \mathrm{e}^{ins} ;$$

it turns out that their values are linked to the spectral-gap localization of the operator $-\Delta_D^{\Omega(\varepsilon)}$ (see the notes).

Theorem 9.5 *Suppose that γ satisfies the assumptions of the previous theorem, and moreover, assume that $\Omega(\varepsilon)$ is not self-intersecting for all d and ε small enough. If $\alpha_m \neq 0$ holds for some $m \in \mathbb{N}$, then there is an $\varepsilon_0 > 0$ such that for any $0 < \varepsilon \le \varepsilon_0$ the asymptotic expansion*

$$\min_{\theta\in[0,1]} E_{m+1}(\theta, d, \varepsilon) - \max_{\theta\in[0,1]} E_m(\theta, d, \varepsilon) = C_\varepsilon + \mathcal{O}(d)$$

holds true as $d \to 0$ with some $C_\varepsilon > 0$.

9.2 Periodic Point Perturbations

A waveguide can also have a periodic structure coming from a potential rather than from the geometry. The spectral analysis of such systems can be made more explicit if the perturbation consists of point interactions which we have treated in Chap. 5. Let us see what happens if they are periodically arranged.

9.2.1 Point Perturbations in a Strip

Consider first the situation of Sect. 5.1 when Ω is a straight strip. As before we assume that its width is $d = \pi$, and since this fixes the scale, we keep the longitudinal period as a parameter to be fixed. In other words, we suppose that the set of point perturbations $\{\alpha, \vec{a}\} \equiv \{[\alpha_j, \vec{a}_j] : \; j = 1, 2, \dots\}$ in $\Omega = \mathbb{R} \times (0, \pi)$ is countably infinite and has a periodic pattern with a period $\ell > 0$; the number of perturbations in a period cell is assumed to be finite.

The point perturbations determine the operator $H(\alpha, \vec{a})$ through boundary conditions (5.2) as in Sect. 5.1 which is again self-adjoint (see the notes). The periodicity makes it possible to reduce the problem to the spectral analysis of systems with a finite number of perturbations by means of the appropriate Bloch-Floquet decomposition: there is a unitary operator from $L^2(\Omega)$ to the direct integral of the spaces $L^2(\Lambda_\ell)$, where Λ_ℓ is the period cell such that

$$UH(\alpha, \vec{a})U^{-1} = \frac{\ell}{2\pi} \int_{|\theta\ell| \le \pi}^{\oplus} H(\alpha, \vec{a}; \theta)\, \mathrm{d}\theta\,,$$

where $H(\alpha, \vec{a}; \theta)$ is the corresponding point-interaction Hamiltonian on $L^2(\Lambda_\ell)$, i.e. the Laplacian defined on functions from $H^2(\Lambda_\ell)$ satisfying the boundary conditions (5.2) at $\vec{x} = \vec{a}_j$ together with $\psi(x, 0) = \psi(x, d) = 0$ and

$$\psi(\ell, y) = \mathrm{e}^{i\theta\ell}\psi(0, y)\,, \quad \partial_x\psi(\ell, y) = \mathrm{e}^{i\theta\ell}\, \partial_x\psi(0, y)\,.$$

With a slight abuse of notation we shall use the symbol $(\alpha, \vec{a})$ for the subset of the point interactions in Λ_ℓ and suppose that their number is N.

In the absence of point perturbations, $\alpha_j = \infty$, the spectrum of $H(\alpha, \vec{a}; \theta)$ is easily found by separation of variables: the eigenvalues

$$\epsilon_{mn}(\theta) = \left(\frac{2\pi m}{\ell} + \theta\right)^2 + n^2\,, \quad m \in \mathbb{Z}\,,\; n = 1, 2, \dots\,, \tag{9.17}$$

correspond to the eigenfunctions $\eta_m^\theta \otimes \chi_n$, where χ_n are the transverse modes (1.10), and

$$\eta_m^\theta(x) = \frac{1}{\sqrt{\ell}}\, e^{i(2\pi m+\theta\ell)x/\ell}, \quad m \in \mathbb{Z}.$$

From here one obtains by a direct computation the free resolvent kernel,

$$G_0(\vec{x}_1, \vec{x}_2; \theta; z) = \sum_{n=1}^{\infty} \frac{\sinh((\ell-|x_1-x_2|)\sqrt{n^2-z}) + e^{2i\eta\theta\ell}\sinh(|x_1-x_2|\sqrt{n^2-z})}{\cosh(\ell\sqrt{n^2-z}) - \cos(\theta\ell)} \times \frac{\sin(ny_1)\sin(ny_2)}{\pi\sqrt{n^2-z}},$$

where $\eta := \mathrm{sgn}\,(x_1 - x_2)$. Then we have the following result (Problem 3).

Proposition 9.2.1 *The resolvent kernel of $H(\alpha, \vec{a}; \theta)$ is expressed by the formula analogous to that of Proposition 5.1.4 with*

$$\Lambda_{jm}(\alpha, \vec{a}, \theta; z) = \big(\alpha_j - \xi(\vec{a}_j, \theta; z)\big)\,\delta_{jm} - G_0(\vec{a}_j, \vec{a}_m; \theta; z)(1-\delta_{jm}),$$

where the regularized diagonal part is given by

$$\xi(\vec{a}_j, \theta; z) = \frac{1}{\pi}\sum_{n=1}^{n[z]} \left(\frac{\sin(\ell\sqrt{z-n^2})}{\cos(\ell\sqrt{z-n^2}) - \cos\theta\ell}\, \frac{\sin^2 nb_j}{\sqrt{z-n^2}} - \frac{1}{2n} \right) + \frac{1}{\pi}\sum_{n=n[z]+1}^{\infty} \left(\frac{\sinh(\ell\sqrt{n^2-z})}{\cosh(\ell\sqrt{n^2-z}) - \cos\theta\ell}\, \frac{\sin^2 nb_j}{\sqrt{n^2-z}} - \frac{1}{2n} \right) \tag{9.18}$$

with $n[z] := \max\{0, [\sqrt{z}]\}$. The above quantity is defined everywhere except at the points where $z = \epsilon_{mn}(\theta)$ and $\sin nb_j \neq 0$; as a function of real z it is increasing between neighboring singularities.

To find the eigenvalues of $H(\alpha, \vec{a}; \theta)$ which determine the band spectrum of the original operator $H(\alpha, \vec{a})$ one has to find the resolvent singularities which are determined by the condition $\det \Lambda(\alpha, \vec{a}, \theta; z) = 0$. Let us see how the solution looks like in the simplest case when we have a *single array of perturbations*, $N = 1$, and the spectral condition simplifies to

$$\xi(\vec{a}, \theta; z) = \alpha\,.$$

According to *Proposition* 9.2.1 the function on the left-hand side is for real z monotonically increasing between its singularities. This means that for fixed α, θ there is a sequence $\{E_r(\alpha, \vec{a}, \theta)\}_{r\in\mathbb{N}}$ of eigenvalues arranged in the ascending order; each of them depends, in fact, only on the y-component b of the vector $\vec{a}$. The lowest one satisfies $E_1(\alpha, \vec{a}, \theta) < \epsilon_{01}(\theta) = 1 + \theta^2$ and between any two neighboring singularities there is just one of the other eigenvalues. It is also clear that each of the $E_r(\alpha, \vec{a}, \theta)$ is continuous with respect to the parameters and $E_r(\cdot, \vec{a}, \theta)$ is increasing for fixed b and θ. The eigenvalues need not be monotonous with respect to θ, however, the implicit-function theorem tells us that

$$\frac{\partial E_r(\alpha, \vec{a}, \theta)}{\partial \theta} = - \frac{\partial \xi(\vec{a}, \theta; z)}{\partial \theta} \left(\frac{\partial \xi(\vec{a}, \theta; z)}{\partial z} \right)^{-1} \Bigg|_{(\theta, E_r)}$$

whenever the denominator is nonzero. Leaving aside the thresholds, $z = n^2$, and the eigenvalues (9.17), a straightforward differentiation shows that $\xi(\vec{a}, \cdot; \cdot)$ is analytic in both variables. Since the function ξ is not identically zero, the derivative $\frac{\partial E_r(\alpha,\vec{a},\theta)}{\partial \theta}$ does not vanish in an open set, which means that the spectrum of $H(\alpha, \vec{a})$ is absolutely continuous.

Let us ask next about the number of open gaps. The spectrum below $z = 1$ may be estimated by means of extrema of the function ξ: we have

$$\xi_+(\vec{a}, z) := \max_{|\theta\ell| \le \pi} \xi(\vec{a}, \theta; z) = \frac{1}{\pi} \sum_{n=1}^{\infty} \left(\frac{\sin^2 nb}{\kappa_n(z)} \coth\left(\frac{1}{2} \kappa_n(z) \ell \right) - \frac{1}{2n} \right),$$

where again $\kappa_n(z) = \sqrt{n^2 - z}$, and a similar formula for the minimum, $\xi_-(\vec{a}, z)$, with coth replaced by tanh. Both functions are increasing and have the same logarithmic asymptotics as $\xi(\vec{a}, z)$ of Sect. 5.1 for $z \to -\infty$ (cf. Problem 5.5). On the other hand, $\xi_+(\vec{a}, \cdot)$ diverges as $z \to 1-$, while $\xi_-(\vec{a}, \cdot)$ has a finite limit. This shows that an open gap exists provided

$$\alpha < \xi_-(\vec{a}, 1-) = \frac{\ell}{2\pi} \sin^2 b - \frac{1}{2\pi} + \frac{1}{\pi} \sum_{n=2}^{\infty} \left(\frac{\sin^2 nb}{\kappa_n(1)} \tanh\left(\frac{1}{2} \kappa_n(1) \ell \right) - \frac{1}{2n} \right).$$

The condition is satisfied for a strong enough coupling, or alternatively, for any fixed α and the point-interaction spacing ℓ large enough.

Let us further check how many gaps can be open in the spectrum of such systems. Recall that we have already encountered periodic arrays of point interactions when we discussed the edge states having no classical analogue in Sect. 7.2.2. The corresponding Hamiltonian has an infinite number of gaps, however, their existence comes from the magnetic field responsible for the occurrence of Landau levels in the spectrum of the unperturbed system. In the non-magnetic case an array of point interactions in the plane, sometimes called a ***straight polymer model***, has one gap provided the coupling is strong enough (see the notes). Now we are going to show that for a "coated polymer" we are discussing here a stronger result is valid, namely that for a suitable choice of the parameters it can have any finite number of open gaps.

To this end, we consider the energy between the first and the second threshold, $z \in (1,2)$. We suppose that $\ell \gg 1$ and rewrite the right-hand side of the relation (9.18) as $\xi(\vec{a},\theta;z) = \xi_0(\vec{a},\theta;z) + \eta(\vec{a},\theta;z)$, where

$$\xi_0(\vec{a},\theta;z) := \frac{\sin(\ell\sqrt{z-1})}{\cos(\ell\sqrt{z-1}) - \cos\theta\ell}\,\frac{\sin^2 b}{\pi\sqrt{z-1}}\,.$$

Similarly as above one can check that $\eta(\vec{a},\theta;\cdot)$ is monotonically increasing with a bounded derivative everywhere in the chosen interval of energies. Moreover,

$$\eta(\vec{a},\theta;z) \le \eta_+(z) := -\frac{1}{2\pi} + \frac{1}{\pi}\sum_{n=2}^{\infty}\left(\frac{\sin^2 nb}{\sqrt{n^2-z}}\coth\left(\frac{\ell}{2}\sqrt{n^2-z}\right) - \frac{1}{2n}\right),$$

and the minimum, $\eta_-(z)$, is obtained when coth is replaced by tanh. The upper and lower estimate become close to each other when ℓ is large; using the inequality $\coth u - \tanh u < 5\,\mathrm{e}^{-2u}$ for $2u \ge 1$, we find

$$\eta_+(z) - \eta_-(z) < \frac{5}{\pi}\sum_{n=2}^{\infty}\frac{\sin^2 nb}{\sqrt{n^2-z}}\,\mathrm{e}^{-\ell\sqrt{n^2-z}} < \frac{5}{\pi}\sum_{n=2}^{\infty}\mathrm{e}^{-\ell(n-1)}$$

for $z \in (1,2)$, so $\eta_+(z) - \eta_-(z) < \frac{5}{\pi}\,\mathrm{e}^{-\ell}(1-\mathrm{e}^{-\ell})^{-1}$ and the allowed corridor shrinks exponentially with increasing ℓ. On the other hand, the function

$$g_\theta(u) := \frac{\sin u}{\cos u - \cos\theta\ell}$$

is increasing between any two zeros of its denominator. In the intervals, where it is positive, it is estimated by the appropriate branch of $\tan\left(\frac{u}{2} + \pi m\right)$ from below; if it is negative, we have a similar estimate from above with tan replaced by $-\cot$. Hence independently of θ we have

$$\pm\xi_0(\vec{a},\theta;z) \ge \frac{\sin^2 b}{\pi\sqrt{z-1}}\begin{pmatrix}\tan\\ \cot\end{pmatrix}\left(\frac{\pi}{2}\left\{\frac{\ell}{\pi}\sqrt{z-1}\right\}\right),$$

where $\{\cdot\}$ denotes the fractional part. Putting these estimates together, we see that the oscillating part dominates, hence for sufficiently large $|\alpha|$ there are gaps having $1 + \left(\frac{\pi m}{\ell}\right)^2$ as one endpoint provided the latter lies in the chosen interval. In addition, we have $\tan u + \cot u \ge 2$, which means that the gap between the lower and the upper bound to $\xi(\vec{a},\theta;z)$ never closes for $z \in (1,2)$ provided $5\,\mathrm{e}^{-\ell}(1-\mathrm{e}^{-\ell})^{-1} < \sqrt{2}\sin^2 b$. Summarizing the above considerations, we come to the following conclusion.

Proposition 9.2.2 *The spectrum of $H(\alpha, \vec{a})$ describing a single array of point perturbations can have for a fixed $\alpha \in \mathbb{R}$ an arbitrary finite number of gaps provided the distance ℓ is large enough.*

In contrast to the magnetic case one expects that from some energy on the spectral bands start overlapping. However, as for the geometric perturbations considered in the previous section, the validity of the Bethe-Sommerfeld conjecture for waveguides with periodic potential or point-interaction remains an open problem (see the notes).

9.2.2 Magnetic Layers with Periodic Point Perturbations

In Sect. 5.3 we treated motion in a planar layer $\Omega = \mathbb{R}^2 \times (0, d)$ with a finite number of point perturbations. In the same way as above we can extend this analysis to Hamiltonians describing periodic lattices of point interactions; we leave this task to the reader (Problem 4). Now we are going to discuss what will happen if such a system is under the influence of a homogeneous magnetic field. For the sake of simplicity we assume that the field is perpendicular to the layer and each elementary cell of the lattice contains only one point interaction.

We adopt the notation of Sect. 5.3 and denote by $\vec{x} = (x, y) \in \Omega$ the coordinates in the layer, where $x = (x_1, x_2) \in \mathbb{R}^2$ and $y \in (0, d)$; as before we put $d = \pi$. The magnetic field is supposed to be $\vec{B} = (0, 0, B)$ with $B \neq 0$. We start with the unperturbed operator,

$$H_0 = \big(- i\partial_1 + A_1\big)^2 + \big(- i\partial_2 + A_2\big)^2 - \partial_y^2$$

on $L^2(\Omega)$ with the domain

$$\mathrm{Dom}\,(H_0) = \big\{ f \in H^2(\Omega) : \; H_0 f \in L^2(\Omega)\,,\; f(x,0) = f(x,d) = 0\,,\; x \in \mathbb{R}^2 \big\}.$$

We use the circular gauge, $\vec{A}(x) = (A_1(x), A_2(x)) = \frac{1}{2}(-Bx_2, Bx_1)$. Due to the presence of the magnetic field the decomposition (5.13) now becomes

$$H_0 \simeq \bigoplus_{n=1}^{\infty} h_n \otimes I_n\,, \quad h_n := \big(- i\partial_1 - \frac{B}{2}x_2\big)^2 + \big(- i\partial_2 + \frac{B}{2}x_1\big)^2 + n^2\,, \quad (9.19)$$

where I_n is the unit operator on $L^2(0, \pi)$ and $\simeq$ denotes unitary equivalence; the operator on the right-hand side acts on $\bigoplus_{n=1}^{\infty} L^2(\mathbb{R}^2) \otimes \{\chi_n\}$ with $\chi_n(x_3) = \sqrt{\frac{2}{\pi}}\,\sin(nx_3)$ (Problem 5). It follows that the spectrum of H_0 is given by

$$\sigma(H_0) = \sigma_{\mathrm{ess}}(H_0) = \big\{ |B|(2l - 1) + n^2 \;:\; l, n \in \mathbb{N} \big\},$$

in other words, it consists of eigenvalues of infinite multiplicity being sums of the Landau levels $|B|(2l+1)$ and the transverse-mode energies. The eigenvalue corresponding to the l-th Landau level and the n-th transverse mode will be denoted by $\epsilon(l,n) = |B|(2l-1) + n^2$.

Let Λ be the infinite lattice with the basis consisting of vectors $\vec{a} = (a_1, 0, 0)$ and $\vec{b} = (b_1, b_2, 0)$ with $b_2 \neq 0$. By assumption the corresponding elementary cell $\{sa + tb : (s,t) \in [0,1)\} \times (0,\pi)$ of Ω contains a single point potential; without loss of generality we may suppose that it is placed at the point $\vec{\rho} = (0,0,\rho_3)$ and define $\Gamma = \Lambda + \{\vec{\rho}\}$. In order to demonstrate how the spectrum of H_0 changes under the influence of periodically distributed point interactions, we need to check how the system behaves with respect to translations. In particular, we have to determine the two-dimensional translation group with respect to which the perturbed operator is invariant. This will allow us to write a formula that will replace the Bloch-Floquet decomposition used in the non-magnetic case.

We consider the group of discrete magnetic translations over the lattice Λ,

$$W(\xi,\Lambda) = \{(\vec{\lambda},\zeta) : \vec{\lambda} \in \Lambda,\ \zeta = \mathrm{e}^{i\pi\eta m},\ m \in \mathbb{Z}\},$$

where $\eta := \frac{1}{2\pi} a_1 b_2 B$ is the flux of the field $\vec{B}$ through the elementary cell. The group law in the coordinates referring to the basis $\{\vec{a}, \vec{b}\}$ of Λ has the form

$$(\vec{\lambda},\zeta)(\vec{\lambda}',\zeta') = \left(\vec{\lambda}+\vec{\lambda}',\ \zeta\zeta' \exp\left(\pi i\eta(\lambda_a\lambda'_b - \lambda_b\lambda'_a)\right)\right);$$

notice that groups $W(\xi,\Lambda)$ corresponding to different values of ξ and different two-dimensional lattices Λ_i having nevertheless the same value of η are isomorphic, hence we will denote the group simply by W_η. It has a representation in the state space $L^2(\Omega)$ of our system given by

$$(T(\vec{\lambda},\zeta)f)(\vec{x}) = \zeta \exp\left(\frac{i}{2}(\vec{B}\times\vec{\lambda})\cdot\vec{x}\right) f(\vec{x}-\vec{\lambda}), \quad f \in L^2(\Omega),$$

i.e. acting as translation with an additional exponential factor. If the flux through the elementary cell is rational, $\eta = N/M$, then any unitary representation of the group W_η decomposes uniquely into an orthogonal sum of irreducible representations (see the notes) by means of the ***Landau-Zak transformation***

$$\mathcal{L}_n : L^2(\mathbb{R}^2) \otimes \ell^2(\mathbb{N}) \to L^2(T^2_\eta) \otimes \mathbb{C}^M \otimes \mathbb{C}^N \otimes \ell^2(\mathbb{N}) \otimes \ell^2(\mathbb{N})$$

with the torus $T^2_\eta = [0, M^{-1}) \times [0,1)$, where $\mathcal{L}_n$ acts in the following way,

$$(\mathcal{L}_n f)(p,j,k,l,n) = N^{-1/2} \sum_{m\in\mathbb{Z}} \exp\left(2\pi i m\,\frac{p_2+k}{N}\right) \times \int_{\mathbb{R}^2} f(x,n)\,\overline{\psi}_0(x, p_1+\eta j+m, l)\,\mathrm{d}x\,. \tag{9.20}$$

Here $p = (p_1, p_2) \in T_\eta^2$ and $\psi_0(x, q, l)$ with $q \in \mathbb{R}$ and $l \in \mathbb{N}_0$ are generalized eigenfunctions of the operator $h_n - n^2$ associated with the lattice Λ,

$$\psi_0(x, q, l) = \Big(\frac{b_2}{\eta}\frac{\pi^{3/2}}{|B|^{3/2}} 2^{l+1} l!\Big)^{-1/2} \exp\Big(\frac{i\pi b_1}{a_1 \eta} q^2\Big) \exp\Big(\frac{2i\pi x_1}{a_1}\big(\frac{\eta x_2}{2b_2} + q\big)\Big)$$
$$\times \exp\Big(- |B|\big(x_2 + \frac{b_2}{\eta} q\big)^2\Big) H_l\Big(|B|^{1/2}\big(x_2 + \frac{b_2}{\eta} q\big)\Big),$$

where H_l denotes the Hermite polynomial of order l.

Consider now the perturbation of H_0 by point interactions placed at the points of the lattice Γ. The construction is the same as in Chap. 5: one starts from the symmetric operator obtained by restriction of $\mathrm{Dom}\,(H_\Gamma^{(0)}) = \{f \in \mathrm{Dom}\,(H_0) : f(\vec{w}) = 0,\ \vec{w} \in \Gamma\}$ which makes sense in view of the Sobolev embedding, and identify the perturbed operator $H_{\mathcal{A},\Gamma}$ as a suitable self-adjoint extension of $H_\Gamma^{(0)}$. We introduce the generalized boundary values from the behavior

$$f(\vec{x}) = L_0(f, \vec{w}) \frac{1 + i\vec{A}(\vec{w}) \cdot (\vec{x} - \vec{w})}{4\pi|\vec{x} - \vec{w}|} + L_1(f, \vec{w}) + \mathcal{O}(|\vec{x} - \vec{y}|)$$

of the functions $f \in D\big(H_\Gamma^{(0)*}\big)$ in the vicinity of the lattice points modified to take the presence of the magnetic field into account. To get a self-adjoint extension, the vectors $L_j(f) := \{L_j(f, \vec{w}) : \vec{w} \in \Gamma\}$, $j = 0, 1$ have to satisfy the condition

$$L_1(f) - \mathcal{A} L_0(f) = 0\,, \tag{9.21}$$

where $\mathcal{A}$ is a self-adjoint operator on $\ell^2(\Gamma)$. In analogy with (5.2) we are interested primarily in the physically relevant situation where the point perturbations are local, i.e. the matrix representation of $\mathcal{A}$ is diagonal. However, we may consider a more general class of $\mathcal{A}$ satisfying a "short-range" condition

$$|\mathcal{A}(\vec{w}, \vec{w}')| \le c_1\, \mathrm{e}^{-c_2|\vec{w} - \vec{w}'|} \tag{9.22}$$

for some positive c_1 and c_2 and all $\vec{w}, \vec{w}' \in \Gamma$. The Green function of $H_{\mathcal{A},\Gamma}$ can be again expressed by means of Krein's formula (Problems 6 and 7). To this end we introduce the matrix-valued function

$$Q(\vec{w}, \vec{w}', z) = \begin{cases} G_0(\vec{w}, \vec{w}', z) & \text{if } \vec{w} \ne \vec{w}' \\ Q_0(w_3, z) & \text{if } \vec{w} = \vec{w}' \end{cases}$$

where the regularized Green function of $H_\Gamma^{(0)}$,

$$Q_0(\vec{w}, z) := \lim_{\vec{x} \to \vec{w}} \Big(G_0(\vec{x}, \vec{w}, z) - \frac{1}{4\pi\, |\vec{x} - \vec{w}|} \Big)$$

depends in fact on the third component of $\vec{w}$ only, $Q_0(\vec{w}, z) = Q_0(w_3, z)$. We then have the following result (Problem 6).

Proposition 9.2.3 *Assume that $\mathcal{A}$ is a self-adjoint operator on $\ell^2(\Gamma)$ which satisfies (9.22) and which is invariant with respect to the group W_η. Then there is a unique self-adjoint extension $H_{\mathcal{A},\Gamma}$ of the operator $H_\Gamma^{(0)}$ with the resolvent kernel $G(\vec{x}, \vec{y}, z) = (H_\Gamma - z)^{-1}(\vec{x}, \vec{y})$ given by*

$$G(\vec{x}, \vec{y}; z) = G_0(\vec{x}, \vec{y}; z) + \sum_{\vec{w}, \vec{w}' \in \Gamma} [\mathcal{A} - Q(z)]^{-1}(\vec{w}, \vec{w}')\, G_0(\vec{x}, \vec{w}; z)\, G_0(\vec{w}', \vec{y}; z)\,.$$

The operator-valued function $Q(\cdot)$ is meromorphic in $z \in \mathbb{C}$ and all its poles coincide with the spectrum of H_0. Moreover, the operator H_Γ is invariant with respect to W_η and the functions from its domain satisfy the condition (9.21).

The operator $\mathcal{A} - Q(z)$ plays the same role as the matrix $\Lambda(z)$ in the Krein formula expressions of Chap. 5. In the physically interesting case when $\mathcal{A}$ is diagonal the invariance with respect to W_η means that all the point-interaction coupling constants are the same, $\mathcal{A}(\vec{w}, \vec{w}') = \delta_{\vec{w},\vec{w}'}\alpha$ with a fixed $\alpha \in \mathbb{R}$.

Using the above result we are able to describe the spectral properties of the full operator $H_{\mathcal{A},\Gamma}$. It turns out that the structure of the spectrum depends on the value of the magnetic flux through the elementary cell. Suppose first that the flux is an integer, $\eta = N \in \mathbb{N}$ and $M = 1$, in which case the index j in the Landau-Zak transformation (9.20) can be omitted. Given a vector $\vec{\lambda} \in \Lambda$ we denote by (λ_a, λ_b) its coordinates relative to the basis $\{\vec{a}, \vec{b}\}$ and introduce the Fourier transformation $\mathcal{F}_\eta : \ell^2(\Gamma) \mapsto L^2(T_\eta^2)$ by

$$(\mathcal{F}_\eta\, \phi)(p) = \sum_{\vec{\lambda} \in \Lambda} \phi(\vec{\lambda} + \vec{\rho})\, e_\lambda(p),$$

with $e_\lambda(p) := \exp\big(-2\pi i(\lambda_a p_1 + \lambda_b p_2 + \frac{1}{2} N \lambda_a \lambda_b)\big)$. Since the unperturbed Green function G_0 is W_η invariant, the operator $Q(z)$ transforms as follows,

$$\tilde{Q}(p, z) := (\mathcal{F}_\eta\, Q(z)\, \mathcal{F}_\eta^{-1})(p) = \sum_{\vec{\lambda} \in \Lambda} Q(\vec{\lambda}, \vec{\rho}, z)\, e_\lambda(p)\,, \qquad (9.23)$$

which is well defined since $|Q(\vec{w}, \vec{w}', z)| \le c_3\, \mathrm{e}^{-c_4|\vec{w} - \vec{w}'|}$ holds for all $\vec{w}, \vec{w}' \in \Gamma$ and some $c_3, c_4 > 0$; this follows from the definition of $Q(z)$ and the properties of the free Green function G_0 (Problem 7).

The periodicity can be used to simplify the spectral analysis in a way analogous to the Bloch-Floquet expansion. Since H_Γ is invariant with respect to W_η, it is unitarily equivalent to the direct integral

$$\tilde{H}_\Gamma = \int_{T_\eta^2}^{\oplus} \tilde{H}_\Gamma(p)\, \mathrm{d}p\,,$$

where the fiber operator $\tilde{H}_\Gamma(p)$ acts on the space $\mathbb{C}^N \otimes \ell^2(\mathbb{N}) \otimes \ell^2(\mathbb{N})$ as

$$(\tilde{H}_\Gamma(p)f)(k,l,n) = \big(\mathcal{L}_n\, h_n(\mathcal{L}_n^{-1} f)\big)(p,1,k,l,n)$$

with the indices $k = 1, \dots N$ and $n, l \in \mathbb{N}$, cf. (9.19). The direct-integral decomposition means that the spectrum of H_Γ coincides with the closure of $\bigcup_{p\in T_\eta^2} \sigma(\tilde{H}_\Gamma(p))$, hence we have to analyze the spectrum of $\tilde{H}_\Gamma(p)$ for a fixed value of the quasi-momentum $p \in T_\eta^2$. This can be done by investigating its Green function $\tilde{G}(p)$. Using *Proposition* 9.2.3 in combination with a straightforward calculation we express its matrix elements as

$$\begin{aligned}\tilde{G}(p,k,k',l,l',n,n';z) &= \delta_{kk'}\,\delta_{ll'}\,\delta_{nn'}\,\frac{1}{\epsilon(l,n)-z} \\ &+ [\tilde{\mathcal{A}}(p) - \tilde{Q}(p,z)]^{-1}\,\frac{\tilde{\delta}_0(p,k,l)\,\tilde{\delta}_0(p,k',l')}{(\epsilon(l,n)-z)(\epsilon(l',n')-z)}\,\chi_n(\rho_3)\,\chi_{n'}(\rho_3)\,,\end{aligned} \tag{9.24}$$

where $\tilde{\mathcal{A}}(p) := \sum_{\vec{\lambda}\in\Lambda} \mathcal{A}(\vec{\lambda}+\vec{p})\, e_\lambda(p)$ and

$$\tilde{\delta}_0(p,k,l) := N^{-1/2} \sum_{m\in\mathbb{Z}} \exp\Big(2\pi i\, m\, \frac{p_2+k}{N}\Big)\, \bar{\psi}_0(0, p_1+m, l)$$

is the $\mathcal{L}_n$-transformed delta function in $\mathbb{R}^2$. In this way the problem is reduced to an analysis of the function $\tilde{Q}(p,z)$ given by (9.23). To this end we denote the eigenvalues of H_0 arranged in ascending order by ϵ_i, $i = 0, 1\dots$; if there is more than one pair of numbers (l,n) such that $\epsilon(l,n) = \epsilon_i$ then we denote this set of pairs by $J(\epsilon_i)$ and their number by $|J(\epsilon_i)|$.

By examining the residua of the Green function (9.24) at $z = \epsilon_i \in \sigma(H_0)$ it is straightforward to determine whether a given eigenvalue of H_0 stays in the spectrum of $\tilde{H}_\Gamma(p)$ and if so, what is its multiplicity:

Lemma 9.2.1 *Assume that the flux $\eta = N \in \mathbb{N}$ and fix $\epsilon_i \in \sigma(H_0)$ and $p \in T_\eta^2$. Then one of the following alternatives occurs:*
(a) There is at least one pair $(l,n) \in J(\epsilon_i)$ with $\tilde{\delta}_0(p,\cdot,l) \neq 0$ and $\chi_n(\rho_3) \neq 0$, then ϵ_i is an eigenvalue of $\tilde{H}_\Gamma(p)$ with multiplicity $N\,|J(\epsilon_i)| - 1$.
(b) $\tilde{\delta}_0(p,\cdot,l)\chi_n(\rho_3) = 0$ holds for all indices $(l,n) \in J(\epsilon_i)$ and $\tilde{Q}(p,\epsilon_i) \neq \tilde{\mathcal{A}}(p)$, then ϵ_i is an eigenvalue of $\tilde{H}_\Gamma(p)$ with multiplicity $N\,|J(\epsilon_i)|$.
(c) $\tilde{\delta}_0(p,\cdot,l)\chi_n(\rho_3) = 0$ holds for all indices $(l,n) \in J(\epsilon_i)$ and $\tilde{Q}(p,\epsilon_i) = \tilde{\mathcal{A}}(p)$, then ϵ_i is an eigenvalue of $\tilde{H}_\Gamma(p)$ with multiplicity $N\,|J(\epsilon_i)| + 1$.

In particular, if $\eta = N \geq 2$, then any eigenvalue of H_0 remains in the spectrum of H_Γ, while for the single flux quantum through the lattice cell, $\eta = N = 1$, the eigenvalues $z_0 \in \sigma(H_0)$ for which there is just one pair of indices $(l,n) \in J(z_0)$ and $\chi_n(\rho_3) \neq 0$ have to be removed.

On the other hand, the perturbation not only removes eigenvalues; the spectrum of H_Γ can also contain other points than eigenvalues of H_0. In order to find them we consider the implicit equation

$$\tilde{Q}(p,E)=\tilde{A}(p)\,. \tag{9.25}$$

Its solutions, for which the $\tilde{\delta}_0$ in the numerator does not vanish, are poles of the Green function (9.24), and therefore determine points of the spectrum of H_Γ.

To find the solutions of (9.25) as functions of p let us first exclude the "orphan" eigenvalues of H_0, i.e. those $\epsilon_{i'}$ for which there is no pair of indices $(l,n)\in J(\epsilon_{i'})$ satisfying $\chi_n(\rho_3)\neq 0$. It is obvious from (9.24) that they cannot be poles of $\tilde{Q}(p,\cdot)$, and therefore equation (9.25) may have no solution in the intervals $(\epsilon_{i'-1},\epsilon_{i'})$ and $(\epsilon_{i'},\epsilon_{i'+1})$. Eliminating these "orphan" points we obtain a subsequence of eigenvalues which, for simplicity, we again denote by $\{\epsilon_j\}$; note that this subsequence is still infinite. For notational convenience we set $\epsilon_{-1}=-\infty$. For a given $l\in\mathbb{N}$ we also put $U_l=\{p\in T^2_\eta:\ \tilde{\delta}_0(p,\cdot,l)\neq 0\}$. Fix now a $p\in T^2_\eta$, then for any interval $(\epsilon_{j-1},\epsilon_j)$ there are two pairs of indices $(l_1,n_1)\in J(\epsilon_{j-1})$ and $(l_2,n_2)\in J(\epsilon_j)$ with $\chi_{n_1}(\rho_3)\neq 0$ and $\chi_{n_2}(\rho_3)\neq 0$. The function $\tilde{Q}(p,\cdot)$ then diverges at the points of the subsequence $\{\epsilon_j\}$, and since

$$\partial_z\tilde{Q}(p,z)=\sum_{l=1}^{\infty}\sum_{n=l}^{\infty}\frac{\chi_n^2(\rho_3)}{|\epsilon(l,n)-z|^2}\sum_{k=0}^{N-1}|\tilde{\delta}_0(p,k,l)|^2\,,$$

it is strictly increasing in each of the intervals $(\epsilon_{j-1},\epsilon_j)$. Hence there exists a unique solution $E_j(p)$ to the implicit equation (9.25) in the interval $(\epsilon_{j-1},\epsilon_j)$. Moreover, $\tilde{Q}(p,\cdot)$ and $\tilde{A}(p)$ are real-analytic which means that the function $E_j(\cdot)$ is also real-analytic on the set $U_{l_1}\cap U_{l_2}$. Since $U_{l_1}\cap U_{l_2}$ has a full measure in T^2_η, the function $E_j(\cdot)$ extends by continuity to the entire torus T^2_η.

Lemma 9.2.2 *The function $E_j(\cdot)$ defined above has the following properties:*

(a) If $\epsilon_{j-1}<E_j(p)<\epsilon_j$, then $E_j(p)$ is the unique solution of equation (9.25).
(b) If ϵ_j is a pole of the function $\tilde{Q}(p,\cdot)$, then $E_j(p)<\epsilon_j<E_{j+1}(p)$.
(c) If ϵ_j is not a pole of the function $\tilde{Q}(p,\cdot)$, then

$$\begin{aligned}
\tilde{Q}(p,\epsilon_j)<\tilde{A}(p)\ &\Rightarrow\quad E_j(p)=\epsilon_j<E_{j+1}(p)\\
\tilde{Q}(p,\epsilon_j)>\tilde{A}(p)\ &\Rightarrow\quad E_j(p)<\epsilon_j=E_{j+1}(p)\\
\tilde{Q}(p,\epsilon_j)=\tilde{A}(p)\ &\Rightarrow\quad E_j(p)=\epsilon_j=E_{j+1}(p)
\end{aligned}$$

Proof Let $(l_1,n_1)\in J(\epsilon_{j-1})$ and $(l_2,n_2)\in J(\epsilon_j)$ be pairs of indices such that $\chi_{n_1}(\rho_3)\neq 0$ and $\chi_{n_2}(\rho_3)\neq 0$; recall that such pairs exist by construction of the sequence $\{\epsilon_j\}$. Define $\mu(p,z):=\tilde{Q}(p,z)-\tilde{A}(p)$.

Assume first that $E_j(p)$ does not coincide with either of the endpoints ϵ_{j-1} and ϵ_j. From the the equation $\mu(p, E_j(p)) = 0$ and the joint continuity of $\mu(\cdot, \cdot)$ in the vicinity of $p, E_j(p))$ we infer that $\mu(p, E_j(p)) = 0$ holds for all $p \in T^2_\eta$; the claim (a) then follows from the fact that $\mu(p, \cdot)$ is strictly increasing. The point $E_j(p)$ defined above is not a pole of $\mu(p, \cdot)$. Indeed, consider the function $z \mapsto \beta(p, z) := \mu(p, z)(z - \epsilon_{j-1})(z - \epsilon_j)$, which is analytic in an interval $(\epsilon_{j-1} - \delta, \epsilon_j + \delta)$ for $\delta > 0$ small enough. By continuity of β we then conclude that $\beta(p, E_j(p)) = 0$ holds for all $p \in T^2_\eta$. Hence $E_j(p)$ cannot be a pole of $\mu(p, \cdot)$, which proves (b). To check (c) we will need the following implications:

$$
\begin{aligned}
E_j(p) = \epsilon_j &\Rightarrow \lim_{z \to E_j(p)} \mu(p, z) \le 0\,, \\
E_j(p) = \epsilon_{j-1} &\Rightarrow \lim_{z \to E_j(p)} \mu(p, z) \ge 0\,.
\end{aligned}
\tag{9.26}
$$

Recall that the limits exist in view of the monotonicity of $\mu(p, \cdot)$. To prove the first implication assume that $E_j(p) = \epsilon_j$, but $\lim_{z \to E_j(p)} \mu(p, z) > 0$. Then there are $E_0 \in (\epsilon_{j-1}, \epsilon_j)$ with $\mu(p, E_0) > 0$ and p_0 such that $\mu(p_0, E_0) > 0$ and $E_0 < E_j(p_0)$. From the monotonicity of $\mu(p, \cdot)$ we then get a contradiction, $0 < \mu(p_0, E_0) < \mu(p_0, E_j(p_0)) = 0$. The second implication in (9.26) can be demonstrated in a similar way.

Assume now that ϵ_j is not a pole of $\tilde{Q}(p, \cdot)$ and $\mu(p, \epsilon_j) < 0$. Then by (9.26) it follows that $\epsilon_j < E_{j+1}(p)$. To proceed we consider a sequence $\{p_n\}_{n \in \mathbb{N}} \subset U_{l_1} \cap U_{l_2}$ such that $p_n \to p$ as $n \to \infty$. Then there is an n_0 such that $\mu(p_n, \epsilon_j) < 0$ for all $n > n_0$, which in view of $\mu(p_n, E_j(p_n)) = 0$ and the monotonicity of $\mu(p, \cdot)$ implies that $\epsilon_j < E_j(p_n)$ for all $n > n_0$. Since $E_j(p_n) \to E_j(p)$, it follows that $\epsilon_j \le E_j(p)$ and hence $\epsilon_j = E_j(p)$. The second implication in (c) is treated in the same way. Finally, consider the last case in (c) when $\mu(p, \epsilon_j) = 0$. Assume, for instance, that $E_j(p) < \epsilon_j$. Then from the claim (a) and (9.26) it follows that $\lim_{z \to E_j(p)} \mu(p, z) \ge 0$; hence there exists an $E_1 \in (E_j(p), \epsilon_j)$ such that $0 \le \mu(p, E_j(p)) < \mu(p, E_1) < \mu(p, \epsilon_j) = 0$, which is impossible. It follows that $E_j(p) = \epsilon_j$. The same argument shows that $\epsilon_j = E_{j+1}(p)$. ∎

Now we are able to describe the spectrum of the periodically perturbed magnetic layer with a single perturbation in each lattice cell.

Theorem 9.6 *Suppose that the flux through lattice cell $\eta = n \in \mathbb{N}$. Then the spectrum of H_Γ consists of two parts:*
(a) The first part is the union of spectral bands I_j, $j = 0, 1, \dots,$ where I_j is the range of the function $E_j(\cdot)$, defined by the implicit equation (9.25), over the torus T^2_η.

Two adjacent bands I_j and I_{j+1} have a common endpoint ϵ_j if and only if there are $p, p' \in T^2_\eta$ such that $\tilde{Q}(p,\cdot)$ and $\tilde{Q}(p',\cdot)$ do not have a pole at ϵ_j and

$$\tilde{Q}(p,\epsilon_j) \geq \tilde{\mathcal{A}}(p) \quad \wedge \quad \tilde{Q}(p',\epsilon_j) \leq \tilde{\mathcal{A}}(p') .$$

Moreover, there is at most one degenerate band corresponding to a constant $E_j(\cdot)$, in particular, such a degeneracy is excluded in the physically interesting case where the operator $\mathcal{A}$ is diagonal.
(b) The point spectrum consists, apart from the possible degenerate band mentioned above, of the eigenvalues of H_0 which persist under the perturbation. This concerns all the eigenvalues of H_0 if $\eta = N \geq 2$. On the other hand, in the case $\eta = 1$ the eigenvalues z_0 of H_0 for which there is just one pair of indices $(l,n) \in J(z_0)$ and $\chi_n(\rho_3) \neq 0$ do not belong to the spectrum of H_Γ.

Proof With *Lemmata* 9.2.1 and 9.2.2 in mind it only remains to prove the statement about the degenerate band in part (a). Suppose that there are two different degenerate bands $E, E' \in \mathbb{R}$ separated from the rest of the spectrum of H_0. Then we have $\tilde{Q}(p,E) - \tilde{\mathcal{A}}(p) = \tilde{Q}(p,E') - \tilde{\mathcal{A}}(p) = 0$ for all $p \in T^2_\eta$, which implies

$$Q(\vec{w},\vec{\rho},E) = \mathcal{A}(\vec{w},\vec{\rho}) = Q(\vec{w},\vec{\rho},E')$$

for all $\vec{w} \in \Lambda$. However, since the asymptotic expansion

$$Q(\vec{w},\vec{\rho},E) = C(E)\, \exp\Big(-\frac{|B|\,|\vec{w}|^2}{4}\Big)\, |\vec{w}|^{\frac{E-|B|}{|B|}} \left(1 + \mathcal{O}(|\vec{w}|^{-\nu})\right)$$

holds as $|\vec{w}| \to \infty$ with $\nu = \min\left\{2, \frac{3}{|B|}\right\}$, we arrive for $E \neq E'$ at a contradiction with the above identity for $|\vec{w}|$ large enough. ∎

Let us finally mention briefly the case of a general rational flux, $\eta = N/M$, when the index j in (9.20) runs from 1 to M and $T^2_\eta = [0, M^{-1}) \times [0,1)$. Consequently, the implicit equation (9.25) must be replaced by

$$\det \tilde{Q}(p,j,E) = \tilde{\mathcal{A}}(p,j) .$$

This equation defines in each interval $(\epsilon_{i-1}, \epsilon_i)$ the dispersion functions $E_i^{(r,j)}(\cdot)$ on T^2_η indexed by $r = 1, \dots \min\{M, N\}$. In addition to the point spectrum, which is given again by the persisting eigenvalues of H_0, the spectrum of H_Γ contains spectral bands given by the ranges of the dispersion functions $E_i^{(r,j)}(\cdot)$.

It turns out, however, that the matrices $\tilde{Q}(p,j,E) - \tilde{\mathcal{A}}(p,j)$ for different indices $j = 1, \dots, M$ are unitarily equivalent, which leads to the M-fold degeneracy of the spectral bands. This is the main difference to the case of an integer flux; we refer the reader to the notes for more details.

9.3 Random Waveguides

Let us now turn our attention to waveguides with a random geometry. Properties of such systems are far from being fully understood and we limit ourselves here to a discussion of random deformations of the straight strip $\Omega_0 = \mathbb{R} \times (0, d_{\max})$. More specifically, we are going to discuss a waveguide in $\mathbb{R}^2$ with one straight boundary whose width changes in a piecewise linear way as we move in the longitudinal direction; without loss of generality we may suppose that the deformations have the same sign, i.e. one can regard them as random one-sided dents in the strip of the maximum width $d_{\max}$.

To be more precise, we fix a $d \in (0, d_{\max})$ and characterize the strip boundary by an element of an infinite dimensional cube, $\omega \in \Sigma := [0, d]^{\mathbb{Z}}$, putting

$$d_i(\omega) := d_{\max} - \omega(i)\,,$$

where $\omega(i)$ is the ith coordinate of ω. Obviously, $d_i(\omega)$ lies between $d_{\min} := d_{\max} - d$ and $d_{\max}$. Let $k(\omega) : \mathbb{R} \to [d_{\min}, d_{\max}]$ be the polygonal curve in $\mathbb{R}^2$ joining the points $\{(i, d_i(\omega)) : i \in \mathbb{Z}\}$; using it, we define the waveguide $\Omega(\omega)$ as

$$\Omega(\omega) = \{(x_1, x_2) \in \mathbb{R}^2 : 0 < x_2 < k(\omega)(x_1)\}\,.$$

It becomes random if the variable ω characterizing its upper boundary is a random variable. We fix a probability measure μ on $[0, d]$. We suppose that $\operatorname{supp}\mu$ contains the point 0 and that the measure is nontrivial, $\operatorname{supp}\mu \neq \{0\}$, and denote by $\mathbb{P} = \mu^{\mathbb{Z}}$ the associated probability measure on Σ. In other words, the coordinates $\omega(i)$ are independent random variables, each with the probability distribution given by μ. The object of our interest is the Dirichlet Laplacian

$$H(\omega) = -\Delta_D^{\Omega(\omega)}$$

in $L^2(\Omega(\omega))$ as the Hamiltonian of the waveguide. Since the probability distribution of each $\omega(i)$ is the same, the measure $\mathbb{P}$ is ***ergodic***, i.e. it does not change under shifts, $\omega(i) \mapsto \omega(i + j)$ with $j \in \mathbb{Z}$. As a consequence, the spectrum of the family $\{H(\omega)\}_{\omega\in\Sigma}$ is deterministic, in other words, there is a set $J \subset \mathbb{R}$ such that $\sigma(H(\omega)) = J$ holds for $\mathbb{P}$-almost every $\omega \in \Sigma$ (see the notes). Moreover, since $0 \in \operatorname{supp}\mu$ by assumption, we have

$$\inf \sigma(H(\omega)) = \kappa_1^2 := \frac{\pi^2}{d_{\max}^2} \quad \text{for } \mathbb{P}\text{−almost every } \omega \in \Sigma\,. \tag{9.27}$$

A quantity of particular interest is the integrated density of states of the operator $H(\omega)$. To define it we need to fix some notation. We consider a cut part $\Omega_l(\omega) = \Omega(\omega) \cap \big((-\frac12 l, \frac12 l) \times (0, d_{\max})\big)$ of the strip and denote by $H_l^D(\omega)$ the

Dirichlet Laplacian on it. Similarly, $H_l^N(\omega)$ will be the Laplacian on the same domain with Neumann boundary conditions at the cuts, $\Omega(\omega) \cap \big(\{\pm\frac{1}{2}l\}\} \times (0, d_{\max})\big)$ and Dirichlet boundary conditions on the remaining part of $\partial\Omega_l(\omega)$. We adopt the notation introduced in Sect. 3.1 and denote by $N(H_l^b(\omega), \lambda)$ the number of eigenvalues (counted with multiplicities) of $H_l^b(\omega)$ with $b = D, N$ below λ. By the ergodic theorem the limits

$$\lim_{l\to\infty} \frac{1}{l}\, N(H_l^N(\omega), \lambda) = \lim_{l\to\infty} \frac{1}{l}\, N(H_l^D(\omega), \lambda) =: n(H(\omega), \lambda) \tag{9.28}$$

exist $\mathbb{P}$–almost surely. The quantity $n(H(\omega), \lambda)$ is the ***integrated density of states*** of $H(\omega)$ relative to the reference point λ which can be understood as the number of levels with energy less than λ per unit strip length.

The randomness influences the spectrum in various ways. In particular, it is manifested in the asymptotic behavior of $n(H(\omega), \lambda)$ as λ approaches κ_1^2. In the absence of the random deformations the spectral properties of $H(\omega)$ near the vicinity of the threshold are easily found: for the strip $\Omega_0 = \mathbb{R} \times (0, d_{\max})$ of a fixed width $d_{\max}$ we have

$$n(-\Delta_D^{\Omega_0}, \kappa_1^2 + \lambda) = \frac{1}{\pi}\sqrt{\lambda} + o(\sqrt{\lambda}) \quad \text{as} \quad \lambda \to 0+ \tag{9.29}$$

(Problem 8). In the random model the integrated density of states is expected to decay much faster. A heuristic reasoning behind this conjecture is that the existence of spectral points close to the threshold requires that the dent depths $\omega(i)$ vanish over a long stretch of the strip, which is a highly improbable event once the measure μ is nontrivial. It turns out that it is indeed the case.

Theorem 9.7 *Let $H(\omega)$ be as above, then there is a constant $C > 0$ such that*

$$\limsup_{\lambda\to 0+} \lambda^{1/2} \ln n(H(\omega), \kappa_1^2 + \lambda) \leq -C\,. \tag{9.30}$$

To prove the theorem we need some preliminary results. The first one is related to the way the eigenvalues behave under domain perturbations. Consider a twice continuously differentiable function $p : (-\frac{1}{2}l, \frac{1}{2}l) \to [0, d]$, where $d = d_{\max} - d_{\min}$. For a fixed $t \in [0, 1]$ we then define the bounded domain

$$\Omega_t^l = \big\{(x_1, x_2) \in \mathbb{R}^2 : |x_1| < \tfrac{1}{2}l,\ 0 < x_2 < d_{\max} - tp(x_1)\big\}.$$

Let H_t^l be the Laplace operator on $L^2(\Omega_t^l)$ with Neumann boundary conditions on the vertical parts of the boundary, $x_1 = \pm\frac{1}{2}l$, and Dirichlet boundary conditions elsewhere. Obviously, the spectrum of H_t^l is purely discrete. Let $\epsilon_1^l(t)$ be the lowest eigenvalue of H_t^l. By the analytic perturbation theory it follows that $\epsilon_1^l(\cdot)$ is continuously differentiable. Since $\epsilon_1^l(0) = \kappa_1^2$, we have

$$\epsilon_1^l(t) = \kappa_1^2 + t\,\partial_t \epsilon_1^l(0) + o(t) \quad \text{as} \quad t \to 0\,,$$

where the error term can be estimated explicitly.

Lemma 9.3.1 *There exists a constant* $\tau = \tau(d_{\max}, \|p\|_\infty, \|p'\|_\infty)$ *such that for all* $l \geq \frac{1}{\sqrt{3}}\, d_{\max}$ *and* $0 \leq t \leq \tau l^{-2}$ *we have*

$$\left|\epsilon_1^l(t) - \kappa_1^2 - t\,\partial_t \epsilon_1^l(0)\right| \leq \frac{\pi^2 l^2}{4\tau^2}\, t^2\,.$$

Proof Similarly to the case of deformed waveguides we transform the problem to the rectangle $\Omega_0^l = (-\frac{1}{2}l, \frac{1}{2}l) \times (0, d_{\max})$ using the transformation

$$\psi \mapsto (U_t\,\psi)(x_1, x_2) = \left(\frac{d_{\max} - tp(x_1)}{d_{\max}}\right)^{1/2} \psi\left(x_1,\, \frac{d_{\max} - tp(x_1)}{d_{\max}}\, x_2\right)$$

which maps $L^2(\Omega_t^l)$ unitarily onto $L^2(\Omega_0^l)$, hence H_t^l is unitarily equivalent to the operator $\widehat{H}_t^l$ on $L^2(\Omega_0^l)$ associated with the quadratic form

$$Q_t[\psi] = \int_{\Omega_t^l} |\nabla (U_t^{-1}\psi)|^2\, \mathrm{d}\vec{x}\,, \quad \mathrm{Dom}(Q_t) = H^1(-\tfrac{1}{2}l, \tfrac{1}{2}l) \otimes H_0^1(0, d_{\max})\,.$$

Since ψ can be supposed, without loss of generality, to be real-valued, a direct calculation yields the expression

$$\begin{aligned} Q_t[\psi] = \int_{\Omega_0^l} \Big[& (\partial_1\psi)^2 + \frac{tp'(x_1)}{d_{\max} - tp(x_1)}\, \psi\partial_1\psi \\ & + \left(\frac{tp'(x_1)}{d_{\max} - tp(x_1)}\right)^2 \psi^2 + \frac{2tp'(x_1)\,x_2}{d_{\max} - tp(x_1)}\, \partial_1\psi\,\partial_2\psi \\ & + \frac{t^2\,(p'(x_1))^2\,x_2}{(d_{\max} - tp(x_1))^2}\, \psi\,\partial_2\psi + \left(\frac{d_{\max}}{d_{\max} - tp(x_1)}\right)^2 (\partial_2\psi)^2 \Big]\, \mathrm{d}\vec{x}\,; \end{aligned} \tag{9.31}$$

the sought estimate of the Taylor expansion remainder can then be obtained by the standard analytic perturbation theory. We note that

$$Q_t[\psi] = Q_0[\psi] + t\,Q_0^{(1)}[\psi] + t^2 Q_0^{(2)}[\psi] + \dots\,,$$

where

$$
\begin{aligned}
Q_0^{(k)}[\psi] = \int_{\Omega_0^l} \Big[& \frac{p'(x_1)\, p(x_1)^{k-1}(x_1)}{d_{\max}^k}\, \psi \partial_1 \psi + \frac{p'^2(x_1)\, p^{k-2}(x_1)\,(k-1)}{4\, d_{\max}^k}\, \psi^2 \\
& + \frac{2x_2\, p'(x_1)\, p(x_1)^{k-1}(x_1)}{d_{\max}^k}\, \partial_1\psi\, \partial_2\psi + \frac{p^k(x_1)\,(k+1)}{d_{\max}}\, (\partial_2\psi)^2 \\
& + \frac{x_2 p'(x_1)\, p(x_1)^{k-2}(x_1)\,(k-1)}{d_{\max}^k}\, \psi \partial_2 \psi \Big]\, \mathrm{d}\vec{x}
\end{aligned}
$$

holds for all $k \geq 1$. It is now easy to see that there is a constant C which depends on $d_{\max}$, $\|p\|_\infty$, and $\|p'\|_\infty$ only, such that

$$
Q_0^{(k)}[\psi] \leq \left(\frac{2\|p\|_\infty}{d_{\max}} \right)^{k-1} C \left(\|\psi\|^2_{L^2(\Omega_0^l)} + \|\nabla\psi\|^2_{L^2(\Omega_0^l)} \right).
$$

This means that the operators $\widehat{H}_t^l$ form a type (B) holomorphic family for $t < t_0$, where $t_0 = t_0(d_{\max}, \|p\|_\infty, \|p'\|_\infty)$, and as a consequence, the perturbation series for the eigenvalues of $\widehat{H}_t^l$ has the convergence radius bounded from below by $c\pi^2/l^2$, where c depends on $d_{\max}$, $\|p\|_\infty$ and $\|p'\|_\infty$ only (Problem 9). If we now put $\tau = \frac{1}{2}c\pi^2$, a straightforward application of the second-order perturbation theory yields the sought result. ∎

Remark 9.3.1 It is not difficult to deduce from the expression (9.31) that

$$
\partial_t \epsilon_1^l(0) = \frac{2\pi^2}{d_{\max}^3\, l} \int_{-l/2}^{l/2} p(x_1)\, \mathrm{d}x_1\,, \tag{9.32}
$$

which can be regarded as a version of the well-known Hadamard perturbation formula (Problem 10).

The second main ingredient of the proof is related to the heuristic argument mentioned above and expresses the fact that the probability of finding an eigenvalue close to the threshold in long slices of the waveguide is rather small.

Lemma 9.3.2 *Let $m = \int_0^d s\, d\mu(s)$. Then there is a constant $a = a(d_{\max}, d)$ such that for all $l = 2i+1$ with $i \in \mathbb{N}$ and every $b \leq \min\left\{\frac{\pi^2}{4}, \frac{m^2}{a^2}\right\}$ we have*

$$
\mathbb{P}\{\epsilon_1(H_l^N(\omega)) \leq \kappa_1^2 + b\, l^{-2}\} \leq 4\, \exp\left(-l\, \frac{(m - a\sqrt{b})^2}{4\, d^2} \right), \tag{9.33}
$$

where $\epsilon_1(H_l^N(\omega))$ denotes the lowest eigenvalue of $H_l^N(\omega)$.

Proof Let $\varphi \in C_0^2(-\frac{1}{2}l, \frac{1}{2}l)$ be a function satisfying $0 \leq \varphi(s) \leq 1 - |s|$ for all $s \in (-\frac{1}{2}l, \frac{1}{2}l)$. We extend φ by zero to the whole real axis and define

$$\Omega_l(\omega, t) = \left\{(x_1, x_2) : |x_1| < \tfrac{1}{2}l \,,\ 0 < x_2 < d_{\max} - tp(\omega, x_1)\right\},$$

where

$$p(\omega, t) := \sum_{|i|<l/2} \omega(i)\varphi(x_1 - i)\,.$$

Let $H_l(\omega, t)$ be the Laplace operator on $L^2(\Omega_l(\omega, t))$ with Neumann boundary conditions on the vertical parts of the boundary and Dirichlet conditions elsewhere. Denote by $\epsilon_1^l(\omega, t)$ its lowest eigenvalue; by construction we have

$$\epsilon_1(H_l^N(\omega)) \geq \epsilon_1^l(\omega, t)$$

for all $t \in (0, 1)$. From *Lemma* 9.3.1 and equation (9.32) it follows that

$$|\epsilon_1^l(\omega, t) - \kappa_1^2 - t\,\partial_t\epsilon_1^l(\omega, 0)| \leq \frac{\pi^2 l^2}{4\tau^2}\, t^2\,,$$

and furthermore,

$$\partial_t\epsilon_1^l(\omega, 0) = \frac{c}{l} \sum_{|i|<l/2} \omega(i)$$

with the constants c and τ depending on $d_{\max}$ only. Assume now that the eigenvalue satisfies $\epsilon_1(H_l^N(\omega)) \leq \kappa_1^2 + bl^{-2}$ for $b \leq \frac{1}{4}\pi^2$. Inserting $t = s\tau l^{-2}$ with $s \in (0, 1)$ into the above inequalities we get

$$\partial_t\epsilon_1^l(\omega, 0) \leq \frac{\pi^2 s}{4\tau} + \frac{b}{\tau s}\,,$$

and optimizing with respect to s we arrive at $\partial_t\epsilon_1^l(\omega, 0) \leq \frac{\pi}{\tau}\sqrt{b}$, which yields

$$\frac{1}{l} \sum_{|i|<l/2} \omega(i) \leq \frac{\pi}{c\tau}\sqrt{b}\,.$$

Now we put $a := \pi/(c\tau)$. Assuming that $0 \leq b \leq m^2/a^2$, we get the estimate

$$\begin{aligned}\mathbb{P}\{\epsilon_1(H_l^N(\omega)) \leq \kappa_1^2 + b\,l^{-2}\} &\leq \mathbb{P}\Big\{\frac{1}{l} \sum_{|i|<l/2} \omega(i) \leq a\sqrt{b}\Big\} \\ &\leq \mathbb{P}\Big\{\Big|\frac{1}{l} \sum_{|i|<l/2} \omega(i) - m\Big| \geq m - a\sqrt{b}\Big\}\end{aligned}$$

and the last probability is by [T95, eq. (13.11)] bounded from above by the right-hand side of (9.33) which concludes the proof. ∎

Proof of Theorem 9.7: Fix $\lambda > 0$. By Neumann bracketing we infer that

$$\begin{aligned} n(H(\omega), \kappa_1^2 + \lambda) &\le \frac{1}{l}\,\mathbb{E}\big\{N(H_l^N(\omega), \kappa_1^2 + \lambda)\big\} \\ &\le \frac{1}{l}\int_{\{\omega:\epsilon_1(H_l^N(\omega))\le\kappa_1^2+\lambda\}} N(H_l^N(\omega), \kappa_1^2 + \lambda)\,\mathrm{d}\mathbb{P}(\omega) \\ &\le c'\,\mathbb{P}\big\{\epsilon_1(H_l^N(\omega)) \le \kappa_1^2 + \lambda\big\} \end{aligned}$$

holds for any $l > 0$, where $\mathbb{E}(\cdot)$ denotes expectation with respect to the probability $\mathbb{P}$ and $c' > 0$ is a suitable constant. If we put $\lambda = b\,l^{-2}$, then in view of the above estimate and *Lemma* 9.3.2 we get

$$n(H(\omega), \kappa_1^2 + b\,l^{-2}) \le 4C' \exp\Big(-l\,\frac{(m - a\sqrt{b})^2}{4\,d^2}\Big),$$

where a and m have been defined in the said lemma. This in turn implies that

$$\limsup_{\lambda\to 0+} \lambda^{1/2} \ln n(H(\omega), \kappa_1^2 + \lambda) \;\le\; -\,\frac{(m - a\sqrt{b})^2\,\sqrt{b}}{4d^2}$$

and the claim follows since the right-hand side is bounded for $b \in \big[0, \frac{m^2}{a^2}\big]$. ∎

Remark 9.3.2 The estimate (9.30) that we have proved implies

$$\limsup_{\lambda\to 0+} \frac{\ln(-\ln n(H(\omega), \kappa_1^2 + \lambda))}{\ln\lambda} \le -\frac{1}{2}\,.$$

In fact, under additional assumptions one can get a stronger result (see the notes): if there are positive c, r such that $\mu([0, \delta]) \le c\delta^r$, then the opposite inequality also holds, i.e.

$$\lim_{\lambda\to 0+} \frac{\ln(-\ln n(H(\omega), \kappa_1^2 + \lambda))}{\ln\lambda} = -\frac{1}{2}\,. \tag{9.34}$$

The exponential decay of the integrated density of states towards the edge of the spectrum is usually referred to as a ***Lifshitz tail***.

Another typical feature of random systems concerns the spectral character. It appears that randomness can partially or fully prevent unrestricted motion of a particle which is manifested by a ***localization***, meaning the appearance of a pure point spectrum, most often in the vicinity of a spectral gap endpoint usually referred to as a ***fluctuation boundary***. In the model of a random waveguide considered here the fluctuation boundary coincides with the bottom of the spectrum and localization is established by the following theorem; we refer to the notes for references to the proof and more details.

Theorem 9.8 *Assume that there exist $c, r > 0$ such that $\mu(I) \le c|I|^r$ holds for every interval $I \subset [0, d]$. Then there is a $\delta > 0$ such that the spectrum of $H(\omega)$ in the interval $[\kappa_1^2, \kappa_1^2 + \delta]$ is $\mathbb{P}$-almost surely pure point and the corresponding eigenfunctions are exponentially decaying in the x_1-direction.*

9.4 Notes

Section 9.1 *Theorem* 9.1 about the absolute continuity of periodic Schrödinger operators is of great importance in the theory of scattering in crystals and has a long history. It was first proven by Thomas in [Th73] for $G = I$ and $A = 0$. Later on, the Thomas approach was further generalized; in [BiSu97] a magnetic vector potential A was included, still for the trivial weight G, and in [Mor98] the claim was proved for a smooth, generally non-constant G. The version with the weaker regularity conditions stated here is due to Sobolev and Walthoe [SW02].

The absolute continuity of periodic waveguides expressed by *Theorem* 9.2 comes from [SW02]. In fact, a more general result is demonstrated in this paper: it is shown there that the claim also remains valid if one includes a periodic electromagnetic field and, under certain additional assumptions, even for the Neumann boundary conditions. Note, however, that the approach used in the proof of *Theorem* 9.2 relies heavily on the two-dimensional character of the waveguide, and indeed, in higher dimensions the problem remains open; there is only a partial result applicable to periodically curved waveguides that are sufficiently thin [BDE03]. The absolute continuity of Schrödinger operators describing motion in layers and cylinders under the influence of periodic potentials was studied in [FK11].

Theorem 9.4 is due to [Yo98]; we also refer to this paper for the proof of *Theorem* 9.5 and of the decomposition (9.9) (see also Problem 2). We stress that the result holds for thin enough periodic strips. In the general case the question remains open and one can conjecture that there may be nontrivial periodically curved waveguides without open gaps. The spectral properties of the operator T_θ, which imply equation (9.15) used in the proof of *Theorem* 9.4, can be found, for instance, in [RS, Theorem XIII.89]. The problem solved in *Theorem* 9.3 attracted a lot of attention and there are various ways to demonstrate the claim. The first proof was given by Borg [Bo46]; alternative proofs of his result were found later in [U61] and [Ho65]. A band-gap structure of the spectrum also appears in waveguides coupled by a periodic system of small Neumann windows. For such a model it has been shown in [BP13] that by varying the widths of the waveguides and the distance between the windows one can control the number of the open gaps.

One of the main questions in the analysis of the band-gap structure of periodic operators is the famous *Bethe-Sommerfeld conjecture* which states that the number of gaps in the spectrum of a Schrödinger operator $-\Delta + V(\vec{x})$ on $L^2(\mathbb{R}^d)$ with a periodic potential V is finite whenever $d \ge 2$. This conjecture was formulated in the early days of quantum mechanics, see [SB33], however, it turned out that for general potentials this problem is rather difficult, and the first rigorous results only appeared in the eighties. We refer to [Pa08] for a proof of the Bethe-Sommerfeld conjecture for smooth potentials in arbitrary dimension, a description of earlier work and references to papers covering the case with less regular potentials. Quantum waveguides as systems of a mixed dimension are not covered by these general results and no proof of the Bethe-Sommerfeld conjecture for periodic waveguides is known at present.

Section 9.2 The analysis of periodic point interactions in a strip comes from [EGŠT96]. The self-adjointness of the operator $H(\alpha, \vec{a})$ with periodic families of point perturbations can be established as in [AGHH, Sect. III.1.1]. An explicit solution of the spectral problem for the planar straight-polymer model is given in [AGHH, Sect. III.4].

For the notion of a lattice see [RS, Sect. XIII.16]. The magnetic translation group was introduced in [Za64], a discussion of its representations can be found in [OT69]. *Theorem* 9.6 is taken from [EN03a] where one can also find a precise description of the spectral bands in the case of a rational flux. On the other hand, the behavior for an irrational flux through the lattice cell is not known, but it is natural to conjecture that it will be substantially different from the one discussed here; the analogy with the *almost Mathieu operator* [AJ09] seems to show that convincingly.

A planar version of the model treated in this section, namely a two-dimensional Schrödinger operator with a periodic point potential and a perpendicular homogeneous magnetic field, was studied in [G92]. The main difference between the two situations is the possible existence of a spectral gap containing the whole interval $(\epsilon_{i-1}, \epsilon_i)$ between adjacent unperturbed eigenvalues for some integer i. While this cannot happen in the planar model, it occurs in the magnetic layer if the positions of the point potentials coincide with a node of each transverse mode corresponding to ϵ_i, in other words if $\chi_n(\kappa_3) = 0$ holds for all $(l, n) \in J(\epsilon_i)$ and at the same time $\tilde{Q}(p, \epsilon_i) \leq \tilde{\mathcal{A}}(p)$ for all $p \in T^2_\eta$. Note also that Krein's formula for Hamiltonians with an infinite number of point potentials arranged in a general way can be found in [Pos01].

Section 9.3 A thorough discussion of random-operator spectra can be found in the monographs [PF], [CL], [St], in particular, we refer to Sect. 2B of the former for a demonstration of the deterministic character of the spectra in case of an ergodic (metrically transitive) probability measure. To check this property in the present case requires the use of unitarily equivalent operators on $L^2(\mathbb{R} \times (0, d_{\max}))$ obtained by a coordinate transformation similar to that employed in Sect. 5.2.3, see [KS00]. *Theorem* 9.7 is taken from the indicated paper, where the reader can also find the proof of the lower bound in formula (9.34) and of *Theorem* 9.8.

It should be stressed that randomly deformed waveguides feature spectral properties very different from those coming from the various "deterministic" perturbations we have considered earlier. If the effective interaction was of a local and attractive character as in the case of bends, bulges, etc., we typically had isolated eigenvalues below the essential spectrum, while for periodic perturbations the spectrum was typically only essential and absolutely continuous. It turns out that in the random waveguide model discussed here $\sigma(H(\omega)) = [\kappa_1^2, \infty)$ holds $\mathbb{P}$-almost surely provided $\operatorname{supp} \mu = [0, d]$, see [KS00]. Hence the point spectrum the existence of which is established by *Theorem* 9.8 must be *generically* dense in the vicinity of the threshold κ_1^2; the cases where this is not true, such as periodic waveguides, occur with probability zero.

Apart from localization, the difference between periodic and random systems is manifested in the asymptotic behavior of the integrated density of states at the bottom of the spectrum. This again holds for a much wider class of operators: for Schrödinger operators $-\Delta + V_{\rm per}$ on $L^2(\mathbb{R}^d)$ with periodic operators it is known that

$$n(-\Delta + V_{\rm per},\, \kappa_1^2 + \lambda) \sim c\,\lambda^{d/2} \quad \text{as} \quad \lambda \to 0$$

with some $c > 0$, see [KS87]. On the other hand, for disordered systems with random potentials $V_{\rm rand}$ in $L^2(\mathbb{R}^d)$ one typically observes the Lifschitz-tail asymptotics,

$$n(-\Delta + V_{\rm rand},\, \kappa_1^2 + \lambda) \sim c_1 \exp(-c_2\lambda^{-d/2}) \quad \text{as} \quad \lambda \to 0$$

with $c_1,\, c_2 > 0$. Hence the randomness of the potential forces the integrated density of states to decay generically much faster to zero when the threshold is approached. Note that the asymptotics (9.34) is similar to the above formula with $d = 1$, which corresponds well with the fact that at low energies the waveguide character is dominantly one-dimensional as we observed earlier, for instance, in Chap. 6.

9.5 Problems

1. Prove *Lemma* 9.1.3.

2. Check that the map $U\,:\, L^2(\Omega_0) \to \int_{[0,1]}^{\oplus} L^2(\Lambda_d)\,\mathrm{d}\theta$ appearing in (9.9) is an isometry on $C_0^\infty(\Omega_0)$ and that it extends uniquely to a unitary operator.
Hint: Cf. [Yo98, Sect. 2].

3. (a) Prove *Proposition* 9.2.1.
(b) Analyze the spectral properties of a charged particle confined to a cylinder of a finite height supporting a finite number of point interactions and threaded by a magnetic field parallel to the cylinder axis.
Hint: (a) Employ Krein's formula. (b) Note that the problem is equivalent to the fiber operator analysis for a strip with periodic point interactions with the quasimomentum replaced by the magnetic flux through the cylinder.

4. Analyze the operator $H(\alpha, \vec{a})$ describing the Dirichlet Laplacian in a straight layer perturbed by a periodic family of point interactions.

5. Prove the decomposition (9.19). More precisely, show that

$$H_0 = \bigoplus_{n=1}^{\infty} (h_n \otimes I_n)\Pi_n\,,$$

where $\Pi_n\,:\, L^2(\Omega) \to L^2(\mathbb{R}^2)\otimes\{\chi_n\}$ acts as $(\Pi_n\, u)(\vec{x}) = \chi_n(x_3) \int_0^\pi \chi_n(s)\, u(x,s)\,\mathrm{d}s$.

6. Prove *Proposition* 9.2.3.

7. Use the result of Problem 5 to derive the explicit form of the free Green function of the magnetic layer Hamiltonian, namely

$$G_0(\vec{x},\vec{x}',z) = \frac{1}{2\pi^2}\exp\Big(-\frac{iB}{2}\,x\wedge x' - \frac{|B|}{4}\,|x-x'|^2\Big)\sum_{n=1}^{\infty}\Gamma\Big(\frac{|B|+n^2-z}{2\,|B|}\Big)$$
$$\times\, U\Big(\frac{|B|+n^2-z}{2\,|B|}\,,\,1\,;\,\frac{|B|}{2}\,|x-x'|^2\Big)\,\sin(nx_3)\,\sin(nx_3')\,,$$

where $x\wedge x' = x_1x_2' - x_1'x_2$ and U is the irregular confluent hypergeometric function. This formula in combination with the properties of $U(\cdot)$ yields the exponential decay of the matrix elements $Q(\vec{w},\vec{w}',z)$ of the operator appearing in (9.23).

8. Prove the asymptotic behavior (9.29).

9. Fill in the details of the proof of *Lemma* 9.3.1.

Hint: Use [Ka, eq. (VII.4.47)] to show that the convergence radius of the Taylor series of $\epsilon_1^l(t)$ is bounded from below by a nonzero multiple of ρ, the distance between κ_1^2 and the rest of the spectrum of H_0^l, and check that under the assumptions of the lemma we have $\rho = \pi^2/l^2$.

10. Prove relation (9.32).

Hint: Use the fact that $\partial_t\epsilon_1^l(0) = Q_0^{(1)}[u_0]$, where u_0 is the normalized eigenfunction of H_0^l corresponding to the eigenvalue $\epsilon_1^l(0)$.

Chapter 10
Leaky Waveguides

Most of the material in the previous chapters dealt with particles strictly confined to areas understood as guides or networks. While useful in many respects, such models are not fully realistic. Consider a microscopic semiconductor wire the boundary of which consists of an interface of two materials, and as such corresponds to a finite potential jump. The latter is often high enough so that one can justify the approximation in which it is replaced with Dirichlet boundary conditions, however, in this approach one neglects effects such as quantum tunneling between two wires placed close enough to each other.

This chapter is devoted to investigation of a class of models which overcome this difficulty. The configuration space here will not be restricted and the confinement of the particle to a guide, or a system of guides will be realized instead by an attractive interaction supported by such a geometric object. On the other hand, to make the problem simpler we will suppose that those guides have zero thickness, in other words, we are going to consider Hamiltonians which can be formally written as

$$-\Delta - \alpha(x)\,\delta(x-\Gamma)\,, \tag{10.1}$$

where Γ is a curve or a graph, with $\alpha(x) > 0$ to make the interaction attractive; we shall refer to such systems as ***leaky graphs*** (***curves, surfaces***, etc.). Apart from the opposite sign, we have encountered similar operators in Sects. 1.5.2 and 6.3.2. In contrast to those, we consider no outer hard-wall boundaries, on the other hand, we suppose that the attraction is position independent, $\alpha(x) = \alpha > 0$, and focus on relations between the spectral properties of such operators and the geometry of the interaction support.

P. Exner and H. Kovařík, *Quantum Waveguides*,
Theoretical and Mathematical Physics, DOI 10.1007/978-3-319-18576-7_10

10.1 Leaky Graph Hamiltonians

First of all we have to define the operators on $L^2(\mathbb{R}^d)$ corresponding to the heuristic expression (10.1). For simplicity we consider first the case $d = 2$; we shall comment on extensions to higher dimensions later. We suppose that the singular potential is supported by a graph $\Gamma \subset \mathbb{R}^2$ with the following properties:

(*i*) Edge smoothness: each edge $e_j \in \Gamma$ is a graph of a C^1 function $\Gamma_j : I_j \to \mathbb{R}^2$ where I_j is a closed interval (finite, semi-infinite, or the whole $\mathbb{R}$). Moreover, without loss of generality we may suppose that edges are parametrized by the arc length, $|\dot{\Gamma}_j(s)| = 1$.
(*ii*) Cusp absence: at the vertices of Γ the edges meet at nonzero angles.
(*iii*) Local finiteness: each compact subset of $\mathbb{R}^2$ contains at most a finite number of edges and vertices of Γ.

One way to define the singular Schrödinger operators we are interested in is to employ the results of Sect. 6.3.1, in particular, the quadratic form (6.20). To this end we define the measure

$$m_\Gamma : m_\Gamma(M) = \ell_1(M \cap \Gamma) \tag{10.2}$$

for any Borel set $M \subset \mathbb{R}^2$, where ℓ_1 is the one-dimensional Hausdorff measure given in our case by the edge-arc length. One can check that m_Γ belongs to the generalized Kato class (Problem 1) and consequently, the inequality (6.19) holds for a constant function α. The quadratic form (6.20) is thus closed and bounded from below, corresponding to a unique self-adjoint operator $H_{-\alpha m_\Gamma}$ for which we will employ in this chapter the symbol $H_{\alpha,\Gamma}$.

Since the edges of Γ are smooth by assumption we can use an alternative way which involves boundary conditions. We consider the operator acting as (negative) Laplacian, $(\dot{H}_{\alpha,\Gamma}\psi)(x) = -(\Delta\psi)(x)$ for any ψ which belongs to $H^2(\mathbb{R}^2 \setminus \Gamma)$, is continuous at each edge $e_j \in \Gamma$ with the normal derivatives having there a jump, namely

$$\frac{\partial\psi}{\partial n_+}(x) - \frac{\partial\psi}{\partial n_-}(x) = -\alpha\psi(x)\,, \quad x \in \operatorname{int} e_j\,; \tag{10.3}$$

the normal vector exists obviously at each inner point of an edge. It is easy to check that $\dot{H}_{\alpha,\Gamma}$ is e.s.a. and its closure is identical with the operator $H_{\alpha,\Gamma}$ introduced above by means of a quadratic form (Problem 2).

The fact that $H_{\alpha,\Gamma}$ can be defined by means of the quadratic form of the type (6.20) allows us to use other results derived in Sect. 6.3.1, in particular, the generalized Birman-Schwinger formula for its resolvent given by *Theorem* 6.7c which reduces the spectral analysis of the original operator to the investigation of the integral operator $R^k_{m,m}$. One can determine in this way not only the eigenvalues of $H_{\alpha,\Gamma}$ but also its eigenfunctions (Problem 3).

Before proceeding further with our investigation of the operators $H_{\alpha,\Gamma}$ we want to show that they can be regarded as a weak-coupling (or equivalently, low-energy) approximation of a class of regular Schrödinger operators; we shall do this assuming that the interaction support consists of a single infinite edge, i.e. that $\Gamma : \mathbb{R} \to \mathbb{R}^2$, $\Gamma = (\Gamma_x, \Gamma_y)$, is a C^2 function; with the usual abuse of notation we employ the same symbol for the function and its graph.

We suppose that the signed curvature $\gamma := \dot{\Gamma}_x \ddot{\Gamma}_y - \dot{\Gamma}_y \ddot{\Gamma}_x$ is bounded along the curve, $|\gamma(s)| < c_+$ for some $c_+ > 0$ and all $s \in \mathbb{R}$. Moreover, we assume that Γ has neither self-intersections nor "near-intersections", i.e. that there is a $c_- > 0$ such that $|\Gamma(s) - \Gamma(s')| \geq c_-$ holds for any s, s' with $|s - s'| \geq c_-$. This makes it possible to define in the vicinity of Γ the standard locally orthogonal system of coordinates (s, u) described in Sect. 1.1 which is unique in the strip neighborhood $\Sigma_\varepsilon := \{x(s, u) = \Gamma(s) + n(s)u : (s, u) \in \Sigma_\varepsilon^0\}$ with $\Sigma_\varepsilon^0 := \{(s, u) : s \in \mathbb{R}, |u| < \varepsilon\}$ as long as the condition $2\varepsilon < c_-$ is valid. With these prerequisites we are able to construct the approximating operator family. Given $W \in L^\infty(-1, 1)$, we define for all $\varepsilon < \frac{1}{2}c_-$ the transversally scaled potential,

$$V_\varepsilon(x) := \begin{cases} \frac{1}{\varepsilon} W\left(\frac{u}{\varepsilon}\right) & \text{if } x \in \Sigma_\varepsilon \\ 0 & \text{otherwise} \end{cases}$$

and put

$$H_\varepsilon(W, \Gamma) := -\Delta + V_\varepsilon \,. \tag{10.4}$$

The operators $H_\varepsilon(W, \Gamma)$ are obviously self-adjoint on $D(-\Delta) = H^2(\mathbb{R}^2)$ for any $\varepsilon \in (0, \frac{1}{2}c_-)$ and we have the following approximation result:

Theorem 10.1 *Under the stated assumptions, $H_\varepsilon(W, \Gamma) \to H_{\alpha,\Gamma}$ holds as $\varepsilon \to 0$ in the norm-resolvent sense, where $\alpha := \int_{-1}^{1} W(t)\,\mathrm{d}t$.*

Proof Since $H_\varepsilon(W, \Gamma)$ is bounded from below uniformly in ε, by taking $k = i\kappa$ with κ positive and large enough we can ensure that $k^2 \in \rho(H_\varepsilon(W, \gamma)) \cap \rho(-\Delta)$ for all $\varepsilon \leq \frac{1}{2} c_-$. The resolvents of both involved operators then can be expressed explicitly: for $H_\varepsilon(W, \Gamma)$ we have the Birman-Schwinger-type formula,

$$\begin{aligned} \left(H_\varepsilon(W, \Gamma) - k^2\right)^{-1} &= \left(-\Delta - k^2\right)^{-1} \\ &- \left(-\Delta - k^2\right)^{-1} V_\varepsilon^{1/2} \left[I + |V_\varepsilon|^{1/2} \left(-\Delta - k^2\right)^{-1} V_\varepsilon^{1/2}\right]^{-1} |V_\varepsilon|^{1/2} \left(-\Delta - k^2\right)^{-1}, \end{aligned}$$

where we use the standard convention mentioned in Remark 6.1.1a, namely $V_\varepsilon^{1/2} := |V_\varepsilon|^{1/2} \mathrm{sgn}\,(V_\varepsilon)$, while for $H_{\alpha,\Gamma}$ it is given by *Theorem* 6.7c. The first terms on the right-hand sides of these relations are ε-independent and subtract mutually when the difference is taken. The second term in the resolvent of $H_\varepsilon(W, \Gamma)$ acts on a vector $\psi \in L^2(\mathbb{R}^2)$ as

$$
\begin{aligned}
&-\iiint_{\mathbb{R}^2} G_k(x-x')V_\varepsilon^{1/2}(x')\left[I+|V_\varepsilon|^{1/2}R_0^k\,V_\varepsilon^{1/2}\right]^{-1}(x',x'')|V_\varepsilon|^{1/2}(x'')\\
&\qquad\times G_k(x''-x''')\,\psi(x''')\,\mathrm{d}x'\,\mathrm{d}x''\,\mathrm{d}x'''\\
&=\iint_{\Sigma_\varepsilon^0}\int_{\mathbb{R}^2} G_k\left(x-x(s',u')\right)\,\frac{1}{\varepsilon}\,W^{1/2}\left(\frac{u'}{\varepsilon}\right)\\
&\qquad\times\varepsilon\left[I+|V_\varepsilon|^{1/2}R_0^k V_\varepsilon^{1/2}\right]^{-1}(s',u';s'',u'')\,\frac{1}{\varepsilon}\left|W\left(\frac{u''}{\varepsilon}\right)\right|^{1/2}\\
&\qquad\times G_k\left(x'''-x(s'',u'')\right)\,(1+u'\gamma(s'))(1+u''\gamma(s''))\\
&\qquad\times\psi(x''')\,\mathrm{d}s'\,\mathrm{d}u'\,\mathrm{d}s''\,\mathrm{d}u''\,\mathrm{d}x''',
\end{aligned}
$$

where $(1+u\gamma(s))$ is the Jacobian of the transformation between the Cartesian and curvilinear coordinates similarly as in Sect. 1.1. Changing the integration variables in the last expression to $t' := u'/\varepsilon$ and $t'' := u''/\varepsilon$ we can rewrite the operator in question as the product $B_\varepsilon(I-C_\varepsilon)^{-1}\tilde{B}_\varepsilon$, where $\tilde{B}_\varepsilon : L^2(\mathbb{R}^2)\to L^2(\Sigma_1^0)$, $C_\varepsilon : L^2(\Sigma_1^0)\to L^2(\Sigma_1^0)$ and $B_\varepsilon : L^2(\Sigma_1^0)\to L^2(\mathbb{R}^2)$, are integral operators with kernels

$$
\begin{aligned}
B_\varepsilon(x;s',t') &:= G_k\left(x-x(s',\varepsilon t')\right)\,(1+\varepsilon t'\gamma(s'))W\left(t'\right)^{1/2},\\
\tilde{B}_\varepsilon(s,t;x') &:= |W(t)|^{1/2}(1+\varepsilon t\gamma(s))\,G_k\left(x'-x(s,\varepsilon t)\right),\\
C_\varepsilon(s,t;s',t') &:= |W(t)|^{1/2}G_k(x(s,\varepsilon t)-x(s',\varepsilon t'))W\left(t'\right)^{1/2},
\end{aligned}
$$

see Problem 4. We have $\|C_\varepsilon\|\le\|W\|_\infty\|P_1R_0^kP_1\|\le\|W\|_\infty|k|^{-2}$, where P_1 is the projection onto $L^2(\Sigma_1^0)\subset L^2(\mathbb{R}^2)$, hence $\|C_\varepsilon\|<1$ holds for κ large enough uniformly with respect to ε, and the operator can written as a sandwiched geometric series,

$$
B_\varepsilon(I-C_\varepsilon)^{-1}\tilde{B}_\varepsilon=\sum_{j=0}^{\infty}B_\varepsilon C_\varepsilon^j\,\tilde{B}_\varepsilon\,.
$$

Consider next the resolvent of $H_{\alpha,\Gamma}$. Since the operator $I-\alpha R_{m,m}^{i\kappa}$ is by *Theorem* 6.7b boundedly invertible for κ large enough, we can again use a geometric-series inverse. Since $\alpha=(W^{1/2},|W|^{1/2})$ holds by assumption, we have

$$
\begin{aligned}
&\alpha R_{\mathrm{dx},m}^k\sum_{j=0}^{\infty}\left(\alpha R_{m,m}^k\right)^j R_{m,\mathrm{dx}}^k=R_{\mathrm{dx},m}^k(W^{1/2},|W|^{1/2})R_{m,\mathrm{dx}}^k\\
&\quad+R_{\mathrm{dx},m}^k(W^{1/2},|W|^{1/2})R_{m,m}^k(W^{1/2},|W|^{1/2})R_{m,\mathrm{dx}}^k+\cdots=\sum_{j=0}^{\infty}BC^j\tilde{B}\,,
\end{aligned}
$$

where $B, C, \tilde{B}$ are operators mapping between the same spaces as their indexed counterparts above, determined by their integral kernels:

$$\begin{aligned} B(x;s',t') &:= G_k\left(x-\gamma(s')\right) W\left(t'\right)^{1/2}, \\ \tilde{B}(s,t;x') &:= |W(t)|^{1/2} G_k\left(x'-\gamma(s)\right), \\ C(s,t;s',t') &:= |W(t)|^{1/2} G_k(\gamma(s)-\gamma(s')) W\left(t'\right)^{1/2}. \end{aligned}$$

Note that while these operators depend on W, the expression $\sum_j BC^j\tilde{B}$ contains just the integral of the function, thus it does not depend on a particular shape of the approximating potential. In the same way as above one can check that $\max\{\|B\|, \|B_\varepsilon\|, \|C\|, \|C_\varepsilon\|, \|\tilde{B}\|, \|\tilde{B}_\varepsilon\|\} \le c_3$ holds for any $\varepsilon \in (0,1)$ with some positive $c_3 < 1$ provided $-k^2$ is large enough. Combining it with the estimate

$$\begin{aligned} &\left\| B_\varepsilon(I-C_\varepsilon)^{-1}\tilde{B}_\varepsilon - B(I-C)^{-1}\tilde{B} \right\| \\ &\quad \le \left\{ \|B_\varepsilon - B\| + \|\tilde{B}_\varepsilon - \tilde{B}\| \right\} \sum_{n=0}^{\infty} c_3^{n+1} + \|C_\varepsilon - C\| \sum_{n=0}^{\infty} n\, c_3^{n+1}, \end{aligned}$$

which can be obtained by the telescopic trick (Problem 4), we see that it is sufficient to assess the three norms involved here. The first one satisfies

$$\|B_\varepsilon - B\| \le \|W\|_\infty^{1/2} \left\{ (1+\|\gamma\|_\infty) \left\| R^k_{\Sigma,\varepsilon} - R^k_{\Sigma,0} \right\| + \varepsilon\|\gamma\|_\infty \left\| R^k_{\Sigma,0} \right\| \right\},$$

where $R^k_{\Sigma,\varepsilon}$, $R^k_{\Sigma,0}$ with $\Sigma = \Sigma_0^1$ are shorthands for the resolvent factors in this expression, in other words, integral operators from $L^2(\Sigma_1^0)$ to $L^2(\mathbb{R}^2)$ with the kernels

$$R^k_{\Sigma,\varepsilon}(x, x(s',\varepsilon t')) = G_k\left(x - x(s',\varepsilon t')\right) = \frac{1}{2\pi} K_0\left(\kappa|x - x(s',\varepsilon t')|\right)$$

and

$$R^k_{\Sigma,0}(x, \Gamma(s')) = G_k\left(x - \Gamma(s')\right) = \frac{1}{2\pi} K_0\left(\kappa|x - \Gamma(s')|\right),$$

respectively, where K_0 is the Macdonald function. To show that $R^k_{\Sigma,\varepsilon} \to R^k_{\Sigma,0}$ holds in the operator-norm topology, let us write the kernel of the difference as

$$\begin{aligned} &G_k\left(x - x(s',\varepsilon t')\right) - G_k\left(x - \Gamma(s')\right) = \frac{1}{2\pi}\left[K_0\left(\kappa|x - x(s',\varepsilon t')|\right) - K_0\left(\kappa|x - \Gamma(s')|\right) \right] \\ &= -\frac{\varepsilon t'}{2\pi} \int_0^1 K_1\left(\kappa|x - \Gamma(s') - n(s')\varepsilon t'\vartheta|\right) \kappa \left(\frac{\mathrm{d}}{\mathrm{d}\vartheta} \operatorname{dist}(x, \Gamma(s') + n(s')\varepsilon t'\vartheta) \right) \mathrm{d}\vartheta, \end{aligned}$$

where we have used the relation $K_0'(z) = -K_1(z)$. The last factor in the above integral does not exceed one in modulus, and therefore

$$\left| R^k_{\Sigma,\varepsilon}(x, x(s',\varepsilon t') - R^k_{\Sigma,0}(x, \Gamma(s')) \right| \le \frac{\varepsilon\kappa|t'|}{2\pi} \int_0^1 K_1\left(\kappa|x - \gamma(s') - n(s')\varepsilon t'\vartheta|\right) \mathrm{d}\vartheta.$$

This shows that

$$h_\infty := \sup_{x\in\mathbb{R}^2} \int_{\mathbb{R}} \mathrm{d}s' \int_{-1}^{1} \mathrm{d}t' \left| \left(R^k_{\Sigma,\varepsilon} - R^k_{\Sigma,0} \right) (x, x(s', \varepsilon t')) \right|$$
$$\le \frac{\varepsilon\kappa}{2\pi} \sup_{x\in\mathbb{R}^2} \int_{\Sigma_1^0} K_1 \left(\kappa |x - x(\sigma')| \right) \mathrm{d}\sigma' \le \frac{\varepsilon\kappa}{2\pi} \, \| K_1(\kappa|\cdot|) \|_{L^1(\mathbb{R}^2)} \,,$$

where the right-hand side is finite, because the function $K_1(\kappa|\cdot|)$ decays exponentially at large distances and has the integrable singularity $|\cdot|^{-1}$ at the origin. In the same way we find

$$h_1 := \sup_{x'\in\Sigma_1} \int_{\mathbb{R}^2} \left| \left(R^k_{\Sigma,\varepsilon} - R^k_{\Sigma,0} \right) (x, x') \right| \mathrm{d}x \le \frac{\varepsilon\kappa}{2\pi} \, \| K_1(\kappa|\cdot|) \|_{L^1(\mathbb{R}^2)} \,,$$

and consequently, the Schur-Holmgren bound (Problem 6.14) implies

$$\left\| R^k_{\Sigma,\varepsilon} - R^k_{\Sigma,0} \right\| \le (h_1 h_\infty)^{1/2} \le \frac{\varepsilon\kappa}{2\pi} \, \| K_1(\kappa|\cdot|) \|_{L^1(\mathbb{R}^2)} \,,$$

where the right-hand side tends to zero as $\varepsilon \to 0$. Similarly one verifies the convergence of $\|\tilde{B}_\varepsilon - \tilde{B}\|$ and $\|C_\varepsilon - C\|$ which concludes the proof. ∎

Let us briefly mention extensions of the concept to higher dimensions. This is straightforward for any $d > 2$ provided Γ is of codimension one and sufficiently regular to allow us to use the results of Sect. 6.3.1 (Problem 2c). Singular Schrödinger operators corresponding to the formal expression (10.1) can also be constructed if $\operatorname{codim}\Gamma = 2, 3$. Note that the above definition using boundary conditions (10.3) can be rephrased in the following way: we first restrict the Laplacian to a symmetric operator defined on functions which vanish in the vicinity of Γ, and afterwards we choose a particular self-adjoint extension specified by appropriate boundary conditions. A similar construction can be used for $\operatorname{codim}\Gamma = 2, 3$ provided we replace the boundary values entering the definition by generalized ones analogous to (5.1) and (5.11), respectively.

For the sake of simplicity, we restrict our attention to the situation when Γ is a curve in three-dimensional space being the graph of a C^2 function $\Gamma(s) : \mathbb{R} \to \mathbb{R}^3$; as in the case $\operatorname{codim}\Gamma = 1$ we suppose that the curve has neither self-intersections nor "near-intersections", and that $|\dot{\Gamma}(s)| = 1$. We also assume that the curve possesses a piecewise global Frenet frame (t, n, b) in the sense of Problem 1.12. For a fixed nonzero $\rho \in \mathbb{R}^2$ we then define the curve Γ_ρ as the graph of the function

$$\Gamma_\rho(s) := \Gamma(s) + \rho_1 b(s) + \rho_2 n(s) \,;$$

the distance between the two curves is by construction $r := |\rho|$ for small enough $|\rho|$, in particular, Γ and Γ_ρ do not intersect. Since any function $\psi \in H^2_{\mathrm{loc}}(\mathbb{R}^3 \setminus \Gamma)$ is continuous in $\mathbb{R}^3 \setminus \Gamma$ its restriction $\psi|_{\Gamma_\rho}$ is well defined as a distribution from $D'(\mathbb{R})$. We denote by $\mathcal{D}$ the set of functions $f \in H^2_{\text{loc}}(\mathbb{R}^3 \setminus \Gamma) \cap L^2(\mathbb{R}^3)$ such that the limits

$$L_0(f)(s) := -\lim_{r\to 0} \frac{2\pi}{\ln r} \psi|_{\Gamma_\rho}(s)\,, \quad L_1(f)(s) := \lim_{r\to 0} \left[\psi|_{\Gamma_\rho}(s) + L_0(f)(s)\,\frac{\ln r}{2\pi}\right]$$

exist a.e. in $\mathbb{R}$, are independent of the direction $\frac{1}{r}\rho$, and define functions from $L^2(\mathbb{R})$; the limits are understood here in the sense of the $D'(\mathbb{R})$ topology. Now we are able to define the singular Schrödinger operator: it acts as

$$H_{\alpha,\Gamma}\psi = -\Delta\psi \quad \text{for} \quad x \in \mathbb{R}^3 \setminus \Gamma \tag{10.5}$$

on the domain $D(H_{\alpha,\Gamma}) := \{\psi \in \mathcal{D} :\; L_1(\psi)(s) = \alpha L_0(\psi)(s)\}$ and it is indeed a well-defined Hamiltonian which we seek.

Proposition 10.1.1 *Under the said assumptions $H_{\alpha,\Gamma}$ is self-adjoint for any $\alpha \in \mathbb{R}$.*

We refer to the notes for references to the proof and further properties of the operators (10.5). A comparison with Sect. 5.1 shows the natural meaning of the definition as describing a point interaction in the normal plane to Γ.

10.2 Geometrically Induced Properties

It is natural to expect that a particle whose dynamics is governed by the Hamiltonian (10.1) will be likely to be found in the vicinity of the interaction support Γ, at least as long as its energy is sufficiently low. In contrast to the graph model discussed in Sect. 8.1, however, the spectral properties depend now not only on the edge lengths but in general on the whole geometry of Γ.

10.2.1 Effects of Curvature and Local Deformations

To demonstrate this claim we are first going to show that the curvature effects which we discussed *in extenso* in Chaps. 1 and 3 are robust and occur even if the confinement does not come from hard walls. Consider the operator $H_{\alpha,\Gamma}$ in $L^2(\mathbb{R}^2)$ corresponding to an infinite planar curve Γ. If $\Gamma = \Gamma_0$ is a straight line, $\Gamma_0(s) = as + b$ for some $a, b \in \mathbb{R}^2$ with $|a| = 1$, we can separate variables and prove easily that the spectrum is purely absolutely continuous and that

$$\sigma(H_{\alpha,\Gamma_0}) = \left[-\tfrac{1}{4}\alpha^2, \infty\right). \tag{10.6}$$

We are going to show that bends give rise to a non-void discrete spectrum. To be specific, we assume that the function $\Gamma : \mathbb{R} \to \mathbb{R}^2$ is continuous and piecewise C^1 satisfying the following assumptions:

(*i*) Non-triviality, i.e. $\Gamma \neq \Gamma_0$. Recall that the curve is parametrized by its arc length which means that $|\Gamma(s) - \Gamma(s')| \leq |s - s'|$ holds true; our first assumption is equivalent to the requirement that the inequality is sharp at least for some $s, s' \in \mathbb{R}$.

(*ii*) There is a number $c \in (0, 1)$ such that $|\Gamma(s) - \Gamma(s')| \geq c|s - s'|$, in particular, Γ has no cusps and self-intersections, and its asymptotes, if they exist, are not parallel to each other.

(*iii*) Asymptotic straightness: there are numbers $r > 0$, $\tau > \frac{1}{2}$, and $\omega \in (0, 1)$ such that the inequality

$$1 - \frac{|\Gamma(s) - \Gamma(s')|}{|s - s'|} \leq r\left[1 + |s + s'|^{2\tau}\right]^{-1/2}$$

holds true in the sector $S_\omega := \left\{(s, s') : \omega < \frac{s}{s'} < \omega^{-1}\right\}$.

The quantity used in assumption *(iii)* obviously marks the difference between Γ and the straight line; under a stronger regularity hypothesis one can use an alternative sufficient condition (Problem 5b).

Theorem 10.2 *Let $\alpha > 0$ and suppose that $\Gamma : \mathbb{R} \to \mathbb{R}^2$ satisfies the above assumptions. Then the essential spectrum is preserved, $\sigma_{\mathrm{ess}}(H_{\alpha,\Gamma}) = \left[-\frac{1}{4}\alpha^2, \infty\right)$, but $H_{\alpha,\Gamma}$ has at least one isolated eigenvalue below $-\frac{1}{4}\alpha^2$.*

Proof By the generalized Birman-Schwinger principle the spectral information is encoded in the operator $\mathcal{R}^{\kappa}_{\alpha,\Gamma} := \alpha R^{i\kappa}_{m,m}$ on $L^2(\mathbb{R})$ which is an integral operator with the kernel

$$\mathcal{R}^{\kappa}_{\alpha,\Gamma}(s, s') = \frac{\alpha}{2\pi} K_0\left(\kappa\,|\Gamma(s) - \Gamma(s')|\right), \quad \kappa > 0. \tag{10.7}$$

In particular, the eigenvalues of $H_{\alpha,\Gamma}$ are by *Theorem* 6.7d associated with solutions of the equation $\mathcal{R}^{\kappa}_{\alpha,\Gamma}\psi = \psi$.

Let us start with the essential spectrum. The Fourier transformation maps $K_0(\kappa x)$ to $(\pi/2)^{1/2}(p^2+\kappa^2)^{-1/2}$. The relation $f(-i\nabla)\psi = (2\pi)^{-1/2}(\mathcal{F}^{-1}f)*\psi$ then shows that $\mathcal{R}^{\kappa}_{\alpha,\Gamma_0}$ is unitarily equivalent to the operator of multiplication by $\frac{1}{2}\alpha(p^2+\kappa^2)^{-1/2}$ on $L^2(\mathbb{R})$, thus it is absolutely continuous and its spectrum is $[0, \frac{\alpha}{2\kappa}]$ as expected from $\sigma(H_{\alpha,\Gamma_0}) = \left[-\frac{1}{4}\,\alpha^2, \infty\right)$. By a compactness argument presented below we also have

$$\sigma_{\mathrm{ess}}(\mathcal{R}^{\kappa}_{\alpha,\Gamma}) = \left[0, \frac{a}{2\kappa}\right]$$

for any curve Γ satisfying the assumptions *(ii)* and *(iii)*, and consequently, $\sigma_{\mathrm{ess}}(H_{\alpha,\Gamma}) \supset \left[-\frac{1}{4}\alpha^2, 0\right]$. By the same compactness argument we find that apart from a possible discrete set corresponding to eigenvalues of a finite multiplicity, the points $-\kappa^2$ with $\kappa > \frac{1}{2}\alpha$ belong to $\rho\left(H_{\alpha,\gamma}\right)$, hence the interval $(-\infty, -\frac{1}{4}\alpha^2)$ is not contained in the essential spectrum. To prove the first claim of the theorem, it remains to check that $\sigma_{\rm ess}(H_{\alpha,\Gamma}) \supset [0,\infty)$ which can done by constructing a suitable Weyl sequence for any $k^2 \geq 0$ (Problem 5a).

Next we will regard the difference between Γ and Γ_0 as a perturbation. The key observation is that the kernel of $\mathcal{D}_\kappa := \mathcal{R}^\kappa_{\alpha,\Gamma} - \mathcal{R}^\kappa_{\alpha,\Gamma_0}$ is sign-definite, namely

$$\mathcal{D}_\kappa(s,s') := \frac{\alpha}{2\pi}\Big(K_0\left(\kappa|\Gamma(s)-\Gamma(s')|\right) - K_0\left(\kappa|s-s'|\right)\Big) \geq 0$$

in view of $|\Gamma(s) - \Gamma(s')| \leq |s-s'|$ and the monotonicity of K_0.

Step 1: We check that $\sup\sigma\left(\mathcal{R}^\kappa_{\alpha,\Gamma}\right) > \frac{\alpha}{2\kappa}$ holds if Γ is not straight. To this end it is enough to find a real-valued $\psi \in \mathcal{S}(\mathbb{R})$ such that $\left(\psi, \mathcal{R}^\kappa_{\alpha,\Gamma}\psi\right) - \frac{\alpha}{2\kappa}\|\psi\|^2 > 0$ holds, which is equivalent to the inequality

$$\frac{2\kappa}{\alpha}\int_{\mathbb{R}^2}\mathcal{D}_\kappa(s,s')\,\psi(s)\psi(s')\,\mathrm{d}s\,\mathrm{d}s' + \int_{\mathbb{R}}\frac{\kappa\,|\hat\psi(p)|^2}{\sqrt{p^2+\kappa^2}}\,\mathrm{d}p - \int_{\mathbb{R}}|\hat\psi(p)|^2\,\mathrm{d}p > 0\,.$$

Since $\mathcal{D}_\kappa(s,s') > 0$ for at least some values of s and s' by assumption (i), and since Γ is supposed to be piecewise C^1, it follows that $\mathcal{D}_\kappa(s,s') > 0$ holds on a subset of $\mathbb{R}^2$ of positive Lebesgue measure. Hence using trial functions $\psi(s) = \sqrt[4]{\frac{2\lambda^2}{\pi}}\,\mathrm{e}^{-\lambda^2 s^2}$ with $\lambda > 0$ one can check by a direct computation that the above inequality holds for λ small enough (Problem 5a).

Step 2: Next we check that $\mathcal{D}_\kappa$ is Hilbert-Schmidt as a map from $L^2(\mathbb{R})$ to $L^2(\mathbb{R})$. For the sake of brevity we define $\varrho := \kappa|\Gamma(s)-\Gamma(s')|$ and $\sigma := \kappa|s-s'|$ and estimate $K_0(\varrho) - K_0(\sigma)$ using the convexity of K_0 together with the relation $K_0'(z) = -K_1(z)$ as follows,

$$K_1(\sigma)(\sigma-\varrho) \leq K_0(\varrho) - K_0(\sigma) \leq \varrho\, K_1(\varrho)\frac{\sigma-\varrho}{\varrho}\,.$$

Using *(ii)* it is easy to conclude from here that the kernel of $\mathcal{D}_\kappa$ is bounded. Moreover, there is a constant $c_1 > 0$ such that $\varrho K_1(\varrho) \leq c_1\,\mathrm{e}^{-\varrho/2} \leq c_1\,\mathrm{e}^{-c\sigma/2}$, and at the same time assumption *(iii)* yields the inequality

$$\frac{\sigma-\varrho}{\varrho} \leq \frac{\sigma-\varrho}{c\,\sigma} \leq \frac{r}{c}\left[1+|s+s'|^{2\tau}\right]^{-1/2}$$

valid in the sector S_ω. Combining these estimates one can check that the integral $\int_{\mathbb{R}^2}\mathcal{D}_\kappa(s,s')^2\,\mathrm{d}s\,\mathrm{d}s'$ is finite for $\tau > \frac{1}{2}$ (Problem 5a). It follows that the operator $\mathcal{D}_\kappa$ is compact. This means, in particular, that the spectrum of $\mathcal{R}^\kappa_{\alpha,\Gamma}$ above $\frac{\alpha}{2\kappa}$ is discrete; at the same time the compactness justifies *a posteriori* the above reasoning concerning the essential spectrum.

Step 3: Finally we have to check that the function $\kappa \mapsto \mathcal{R}^{\kappa}_{\alpha,\Gamma}$ is operator-norm continuous and satisfies $\mathcal{R}^{\kappa}_{\alpha,\Gamma} \to 0$ as $\kappa \to \infty$. This is easily done for $\Gamma = \Gamma_0$ using the unitary equivalence with the multiplication operator, and using the explicit form of the kernel we find the same for the perturbation, since $\|\mathcal{D}_\kappa - \mathcal{D}_{\kappa'}\|_{\mathrm{HS}} \to 0$ holds as $\kappa' \to \kappa$ and $\|\mathcal{D}_\kappa\|_{\mathrm{HS}} \to 0$ as $\kappa \to \infty$.

The first two steps allow us to conclude that $\mathcal{R}^{\kappa}_{\alpha,\Gamma}$ has at least one isolated eigenvalue $\lambda(\kappa)$ above $\sup\sigma\left(\mathcal{R}^{\kappa}_{\alpha,\Gamma}\right) = \frac{\alpha}{2\kappa}$ which is a continuous function of κ and tends to zero as $\kappa \to \infty$. This implies the existence of a $\kappa_0 > \frac{\alpha}{2\kappa}$ such that $\lambda(\kappa_0) = 1$, and the eigenvalue of $H_{\alpha,\Gamma}$ associated with $\lambda(\kappa_0)$ by *Theorem* 6.7d thus lies below the essential-spectrum threshold $-\frac{1}{4}\alpha^2$. ∎

As a consequence of this result it is often possible to establish the existence of the discrete spectrum in a way reminiscent to the result of Problem 1.6.

Corollary 10.2.1 *Suppose that Γ has a subgraph in the form of an infinite curve satisfying the assumptions of the previous theorem, and $\sigma_{\mathrm{ess}}(H_{\alpha,\Gamma}) = \left[-\frac{1}{4}\alpha^2, \infty\right)$, then the discrete spectrum of $H_{\alpha,\Gamma}$ is non-empty.*

The claim follows easily from the minimax principle; the hypothesis can be in some cases verified directly (Problem 5c).

Example 10.2.1 (leaky star graphs) Consider a star graph $\Gamma \subset \mathbb{R}^2$ with $N \geq 2$ edges. One can characterize it by an $(N-1)$-tuple $\beta = \{\beta_1, \dots, \beta_{N-1}\}$ of angles amended by $\beta_N := 2\pi - \sum_{j=1}^{N-1} \beta_j$ which is assumed to be positive. We set $\vartheta_j := \sum_{i=1}^{j} \beta_j$ and $\vartheta_0 := 0$, and denote by L_j be the radial halfline with the endpoint at the origin, $L_j := \{x \in \mathbb{R}^2 :\ \arg x = \vartheta_j\}$, naturally parametrized by its arc length $s = |x|$; the star graph can then be identified with $\Gamma_\beta := \bigcup_{j=0}^{N-1} L_j$.

By Problem 5c the essential spectrum of the corresponding operator $H_N(\beta) := H_{\alpha,\Gamma_\beta}$ coincides with the interval $[-\frac{1}{4}\alpha^2, \infty)$. The existence and properties of the discrete spectrum depend on the angle family; in view of *Corollary* 10.2.1 we have $\sigma_{\mathrm{disc}}(H_N(\beta)) \neq \emptyset$ unless $N = 2$ and $\beta = \pi$. The spectrum can be found explicitly only in a particular case: for a cross-shaped Γ corresponding to $H_4(\beta_c)$ with $\beta_c = \left\{\frac{\pi}{2}, \frac{\pi}{2}, \frac{\pi}{2}\right\}$ the variables separate and there is a single isolated eigenvalue $-\frac{1}{2}\alpha^2$ corresponding to the eigenfunction $(2\alpha)^{-1}\mathrm{e}^{-\alpha(|x|+|y|)/2}$.

On the other hand, it is possible to derive some general properties of the discrete spectrum. We mention two of them (see also Problem 6):

(a) Fix N and a positive integer n. One can achieve that $\sharp\sigma_{\mathrm{disc}}(H_N(\beta)) \geq n$ by choosing one of the angles small enough. In particular, the number of bound states can exceed independently of α any fixed integer for N large enough.

It is obviously sufficient to check the claim for $H_2(\beta)$, where Γ is a broken line. We choose the coordinates in such a way that the two half-lines correspond to $\arg\theta = \pm\frac{1}{2}\beta$ and employ trial functions of the form $\Phi(x, y) = f(x)g(y)$ supported in the strip $L \leq x \leq 2L$, with $f \in C^2$ satisfying $f(L) = f(2L) = 0$, and

$g(y) = \chi_{|y|\le 2d_0}(y) + \chi_{|y|\ge 2d_0}(y)\,\mathrm{e}^{-\alpha(|y|-2d_0)}$, with $d_0 := L\tan\frac{\beta}{2}$. There will be at least n eigenvalues if we manage to make the value of the form

$$q[\Phi] := \|\nabla\Phi\|^2 - \frac{2\alpha}{\cos\frac{\beta}{2}}\|f\|^2 + \frac{1}{4}\alpha^2\|\Phi\|^2$$

negative for n linearly independent functions. Since we have $\|g\|^2 = 4d_0 + \alpha^{-1}$ and $\|g'\|^2 = \alpha$, this requirement can be reformulated as

$$\left(\frac{\pi n}{d_0}\tan\frac{\beta}{2}\right)^2 = \inf_{M_n^\perp}\sup_{M_n}\frac{\|f'\|^2}{\|f\|^2} < \alpha^2\,\frac{2\sec\frac{\beta}{2} - \alpha d_0 - \frac{5}{4}}{1+4\alpha d_0},$$

where M_n is an n-dimensional subspace on $L^2(L,2L)$, which is certainly satisfied for β small enough. In fact, one can see from here that the number of bound states for a sharply broken line is roughly proportional to the inverse angle, $n \gtrsim \beta^{-1}$. Note that the tunnel effect is at the root of this result: in the region where the two half-lines are close to each other the depth of the transverse potential well is effectively doubled.

(b) Let us next consider the Birman-Schwinger formulation of the spectral problem for these graphs. We use the distances

$$d_{ij}(s,s') \equiv d_{ij}^{\beta}(s,s') = \sqrt{s^2 + s'^2 - 2ss'\cos|\vartheta_j - \vartheta_i|}$$

with $\vartheta_j - \vartheta_i = \sum_{l=i+1}^{j}\beta_l$, in particular, $d_{ii}(s,s') = |s-s'|$, to define the corresponding operators on $L^2(\mathbb{R}^+)$ with the integral kernels $\mathcal{R}_{ij}^{\kappa}(s,s';\beta) := \frac{\alpha}{2\kappa}K_0\big(\kappa d_{ij}(s,s')\big)$; then the (discrete part of the) spectral problem for the operator $H_N(\beta)$ is by *Theorem* 6.7d equivalent to the matrix integral-operator equation

$$\sum_{j=1}^{N}\Big(\mathcal{R}_{ij}^{\kappa}(\beta) - \delta_{ij}I\Big)\phi_j = 0\,, \quad i = 1,\dots,N\,,$$

on $\bigoplus_{j=1}^{N} L^2(\mathbb{R}^+)$ and the associated eigenfunctions are expressed by the formula given in Problem 3. Notice that the "entries" of the above kernel have the monotonicity property, $\mathcal{R}_{ij}^{\kappa}(\beta) > \mathcal{R}_{ij}^{\kappa}(\beta')$ if $|\vartheta_j - \vartheta_i| < |\vartheta'_j - \vartheta'_i|$. This has the following consequence: each isolated eigenvalue $\lambda_n(\beta)$ of $H_2(\beta)$ is an increasing function of the angle β between the two half-lines in $(0,\pi)$.

The existence of curvature-induced bound states is not limited to planar curves. As a higher-dimensional analogue, consider operator $H_{\alpha,\Gamma}$ from *Proposition* 10.1.1 associated with an infinite piecewise C^1 curve Γ in $\mathbb{R}^3$. If the latter is a straight line, $\Gamma = \Gamma_0$, the spectrum is found by separation of variables,

$$\sigma(H_{\alpha,\Gamma}) = \sigma_{\mathrm{ac}}(H_{\alpha,\Gamma}) = [\zeta_\alpha,\infty)\,,$$

where $\zeta_\alpha = -4e^{-4\pi\alpha - 2\gamma_E}$ is the eigenvalue of the two-dimensional point interaction with $\gamma_E = -\psi(1) \approx 0.577$ being as usual Euler's constant. On the other hand, if the curve Γ is not straight but remains asymptotically straight we have a result similar to *Theorem* 10.2 above.

Theorem 10.3 *Fix $\alpha \in \mathbb{R}$ and suppose that $\Gamma : \mathbb{R} \to \mathbb{R}^3$ has the properties (i), (ii) stated above. Assume, moreover, that there exist numbers $r > 0$, $\tau > \frac{1}{2}$, and $\omega \in (0, 1)$ such that*

$$1 - \frac{|\Gamma(s) - \Gamma(s')|}{|s - s'|} \leq r \, \frac{|s - s'|}{(1 + |s - s'|)(1 + (s^2 + s'^2)^\tau)^{1/2}}$$

holds true in the sector S_ω. Then $\sigma_{\mathrm{ess}}(H_{\alpha,\Gamma}) = [\zeta_\alpha, \infty)$ and the operator $H_{\alpha,\Gamma}$ has at least one isolated eigenvalue in the interval $(-\infty, \zeta_\alpha)$.

The proof follows the same scheme as in the previous theorem, but the argument is slightly more involved due to the more singular character of the perturbation; we leave it to the reader (Problem 7). Another higher dimensional extension of *Theorem* 10.2 concerning the situation when the interaction support is a curved surface in $\mathbb{R}^3$ will be mentioned in Sect. 10.3 below. Note also that the results described in the last two theorems have a discrete analogue (Problem 8).

The geometry of Γ does not only influence the discrete spectrum. Before leaving the topic, let us briefly describe how it manifests itself in scattering. We limit our attention to the simple situation where $\Gamma \subset \mathbb{R}^2$ is a local modification of a straight line, in other words, we suppose that $\Gamma \setminus \Gamma_0$ is a finite graph satisfying assumptions *(i)* and *(ii)* of Sect. 10.1. Furthermore, we shall be interested primarily in the negative part of the spectrum where the states are kept by the interaction in the vicinity of Γ being guided along the graph edges.

For definiteness we choose coordinates in the plane in such a way that $\Gamma_0 = \{(x_1, 0) : x_1 \in \mathbb{R}\}$. The spectrum of H_{α,Γ_0} is absolutely continuous, being given by (10.6) and the generalized eigenfunctions corresponding to $\lambda \in (-\frac{1}{4}\alpha^2, 0)$ are

$$\omega_\lambda(x_1, x_2) = e^{ik_\alpha(\lambda)x_1}\, e^{-\alpha|x_2|/2} \tag{10.8}$$

and its complex conjugate $\bar{\omega}_\lambda$, both corresponding to waves with the effective momentum $k_\alpha(\lambda) := \left(\lambda + \frac{1}{4}\alpha^2\right)^{1/2}$ moving in opposite directions. The perturbation can be regarded as a singular interaction supported by the set

$$\Lambda = \Lambda_0 \cup \Lambda_1 \quad \text{with} \quad \Lambda_0 := \Gamma_0 \setminus \Gamma\,, \;\; \Lambda_1 := \Gamma \setminus \Gamma_0 = \bigcup_{i=1}^{N} \Gamma_i\,,$$

where Λ_0 is the removed part of the line and Γ_i, $i = 1, \dots, N$, are added edges. We use the fact that the resolvent of a singularly perturbed Laplacian can be expressed by means of *Theorem* 6.7c in order to quantify the difference between the resolvents of $H_{\alpha,\Gamma}$ and H_{α,Γ_0} using the latter as the comparison operator. We shall employ the symbol $\mu := m_{\Gamma_0}$ for the measure associated with the line, and similarly we introduce the Dirac measures on Λ corresponding to the perturbation, $\nu = \nu_0 + \sum_{i=1}^N \nu_i$, where ν_0 refers to the removed edge Λ_0 and ν_i to Γ_i; we associate with those edges an auxiliary Hilbert space $\mathcal{G} := L^2(\nu)$, which decomposes into $\mathcal{G} = \mathcal{G}_0 \oplus \mathcal{G}_1$ with $\mathcal{G}_0 := L^2(\nu_0)$ and $\mathcal{G}_1 := \bigoplus_{i=1}^N L^2(\nu_i)$. We will then reformulate the problem again in terms of an integral operator acting on the space $\mathcal{G}$.

By *Theorem* 6.7c we have $R^k_{\Gamma_0} = R^k_0 + \alpha R^k_{\mathrm{d}x,\mu}(I - \alpha R^k_{\mu,\mu})^{-1} R^k_{\mu,\mathrm{d}x}$ for any k with $\operatorname{Im} k > 0$ such that $k^2 \in \mathbb{C} \setminus [-\frac{1}{4}\alpha^2, \infty)$. The action of $R^k_{\Gamma_0}$ can be written explicitly: by a direct computation we find that its integral kernel equals

$$R^k_{\Gamma_0}(x,y) = G_k(x-y) + \frac{\alpha}{4\pi^3} \int_{\mathbb{R}^3} \frac{e^{i(p\cdot x - p'\cdot y)}}{(p^2-k^2)(p'^2-k^2)} \frac{\tau_k(p_1)}{2\tau_k(p_1)-\alpha}\, \mathrm{d}p\, \mathrm{d}p'_2$$

with $G_k(x-y) := \frac{1}{2\pi} K_0(-ik|x-y|)$ and $\tau_k(p_1) := (p_1^2 - k^2)^{1/2}$. Using this result we can introduce another trace map, namely

$$R^k_{\Gamma_0,\nu} : \mathcal{G} \to L^2(\mathbb{R}^2)\,, \quad R^k_{\Gamma_0,\nu} f = R^k_{\Gamma_0} * f\nu \quad \text{for } f \in \mathcal{G}\,, \tag{10.9}$$

together with its adjoint $(R^k_{\Gamma_0,\nu})^* : L^2(\mathbb{R}^2) \to \mathcal{G}$ and the map $R^k_{\Gamma_0,\nu\nu}$ which is the operator-valued matrix in $\mathcal{G}$ with the "block elements" $R^k_{\Gamma_0,ij} : L^2(\nu_j) \to L^2(\nu_i)$ defined as the appropriated embeddings of $R^k_{\Gamma_0}(\cdot,\cdot)$. To express the resolvent difference we introduce an operator-valued matrix in $\mathcal{G} = \mathcal{G}_0 \oplus \mathcal{G}_1$ given by

$$\Theta^k_\alpha := -(\alpha^{-1}\mathbb{I} + R^k_{\Gamma_0,\nu\nu}) \quad \text{with} \quad \mathbb{I} := \begin{pmatrix} I_0 & 0 \\ 0 & -I_1 \end{pmatrix},$$

where I_i are the unit operators in $\mathcal{G}_i$.

Proposition 10.2.1 *(a) The operator $R^{i\kappa}_{\Gamma_0,\nu}$ is bounded for any $\kappa \in (\frac{1}{2}\alpha, \infty)$ and to any $\sigma > 0$ there is a $\kappa_\sigma > 0$ such that $\|R^{i\kappa}_{\Gamma_0,\nu\nu}\| < \sigma$ holds for all $\kappa > \kappa_\sigma$.*
(b) Suppose that Θ^k_α is invertible for a given $k \in \mathbb{C}^+$ and the operator

$$R^k_\Gamma = R^k_{\Gamma_0} + R^k_{\Gamma_0,\nu}(\Theta^k_\alpha)^{-1}(R^k_{\Gamma_0,\nu})^*$$

is defined everywhere in $L^2(\mathbb{R}^2)$. Then k^2 belongs to $\rho(H_{\alpha,\Gamma})$ and the resolvent $(H_{\alpha,\Gamma} - k^2)^{-1}$ coincides with R^k_Γ.

Proof of these claims follows the same lines as the demonstration of *Theorem* 6.7 and we leave the details to the reader (Problem 9).

The result of *Proposition* 10.2.1 makes it possible to treat scattering in this system. The existence and completeness of wave operators can be checked by the trace-class method (Problem 10). In addition, one has to find the on-shell S-matrix relating the incoming and outgoing asymptotic solutions. We are especially interested in the negative part of the spectrum where these solutions are combinations of ω_λ and $\bar{\omega}_\lambda$ given by (10.8). These generalized eigenfunctions and their analogues ω_z for complex values of the spectral parameter, $z \in \rho(H_{\alpha,\Gamma_0})$, are square-integrable only locally, but as usual we can approximate them by regularized functions, for instance, $\omega_z^\delta(x) = \mathrm{e}^{-\delta x_1^2}\omega_z(x)$ with $\delta > 0$, which naturally belong to the domain of H_{α,Γ_0}.

We are looking for a function ψ_z^δ such that $(-\Delta_\Gamma - z)\psi_z^\delta = (-\Delta_{\Gamma_0} - z)\omega_z^\delta$. Computing the right-hand side and taking the limit $\lim_{\varepsilon\to 0}\psi_{\lambda+i\varepsilon}^\delta =: \psi_\lambda^\delta$ in the L^2 topology we find that ψ_λ^δ still belongs to Dom $(H_{\alpha,\Gamma})$, and moreover

$$\psi_\lambda^\delta = \omega_\lambda^\delta + R_{\Gamma_0,\nu}^{k_\alpha(\lambda)}(\Theta^{k_\alpha(\lambda)})^{-1} I_\Gamma \omega_\lambda^\delta\,,$$

where I_Γ is the standard embedding $H^1(\Gamma) \hookrightarrow \mathcal{G} = L^2(\nu)$ and $R_{\Gamma_0,\nu}^{k_\alpha(\lambda)}$ is the integral operator acting on $\mathcal{G}$ according to (10.9), with the kernel

$$R_{\Gamma_0}^{k_\alpha(\lambda)}(x,y) := \lim_{\varepsilon\to 0+} R_{\Gamma_0}^{k_\alpha(\lambda+i\varepsilon)}(x,y)\,;$$

similarly $\Theta^{k_\alpha(\lambda)} := -\alpha^{-1}\mathbb{I} - R_{\Gamma_0,\nu\nu}^{k_\alpha(\lambda)}$ are the maps $\mathcal{G} \to \mathcal{G}$ with $\mathrm{R}_{\Gamma_0,\nu\nu}^{k_\alpha(\lambda)}$. When we remove the regularization, the pointwise limit $\psi_\lambda := \lim_{\delta\to 0}\psi_\lambda^\delta$ ceases to be square integrable, however, it still belongs locally to L^2 and yields the generalized eigenfunction of $H_{\alpha,\Gamma}$, namely

$$\psi_\lambda = \omega_\lambda + R_{\Gamma_0,\nu}^{k_\alpha(\lambda)}(\Theta^{k_\alpha(\lambda)})^{-1} J_\Lambda \omega_\lambda\,,$$

where $J_\Lambda\omega_\lambda$ denotes the trace of ω_λ in $L^2(\nu)$. The sought on-shell S-matrix can be then found by inspecting the asymptotic behavior of the function ψ_λ as $|x_1| \to \infty$. Using the explicit form of the kernel $R_{\Gamma_0}^{k_\alpha(\lambda)}(\cdot,\cdot)$ we arrive at the following conclusion (Problem 10).

Theorem 10.4 *(a) Under the stated assumptions, wave operators for the pair $(H_{\alpha,\Gamma}, H_{\alpha,\Gamma_0})$ exist and are complete.*
(b) For a fixed $\lambda \in \big(-\frac14\alpha^2, 0\big)$ the generalized eigenfunctions of $H_{\alpha,\Gamma}$ behave as

$$\psi_\lambda(x) = \begin{cases} t(\lambda)\, e^{ik_\alpha(\lambda)x_1 - \alpha|x_2|/2}\,(1+o(1)) & \text{as } x_1 \to +\infty \\ \big(e^{ik_\alpha(\lambda)x_1} + r(\lambda)\, e^{-ik_\alpha(\lambda)x_1}\big)\, \mathrm{e}^{-\alpha|x_2|/2}\,(1+o(1)) & \text{as } x_1 \to -\infty \end{cases}$$

where $k_\alpha(\lambda) := \big(\lambda + \frac14\alpha^2\big)^{1/2}$ is the effective momentum and $t(\lambda)\,, r(\lambda)$ are the transmission and reflection amplitudes, respectively, given by

$$r(\lambda) = 1 - t(\lambda) = \frac{i\alpha}{8k_\alpha(\lambda)}\Big(J_\Lambda\bar{\omega}_\lambda, (\Theta^{k_\alpha(\lambda)})^{-1} J_\Lambda\omega_\lambda\Big)_{\mathcal{G}}\,.$$

10.2.2 Hiatus Perturbations

Let us return to the discrete spectrum and consider another type of perturbation coming now from the removal of a small part of the interaction support. In case of a planar graph this means making a cut in a graph edge, but one can consider the general codimension-one situation, i.e. Schrödinger operator on $L^2(\mathbb{R}^d)$ with the interaction supported by a $(d-1)$-dimensional surface having a "puncture".

Suppose thus that $\Gamma \subset \mathbb{R}^d$ is a compact C^m-smooth surface with $m \geq \frac{1}{2}d$; without loss of generality we may suppose that $0 \in \Gamma$. Let further $\{\mathcal{P}_\varepsilon\}_{\varepsilon\geq 0}$ be a family of subsets of Γ which obeys the following requirements:

(i) Measurability: $\mathcal{P}_\varepsilon$ is measurable with respect to the $(d-1)$-dimensional Lebesgue measure on Γ for any ε small enough.
(ii) Shrinking: $\sup_{x\in\mathcal{P}_\varepsilon} |x| = \mathcal{O}(\varepsilon)$ holds as $\varepsilon \to 0$.

Consider the operators $H_{\alpha,\Gamma}$ and $H_{\alpha,\Gamma_\epsilon}$ corresponding to $\Gamma_\varepsilon := \Gamma \setminus \mathcal{P}_\varepsilon$. Since Γ_ϵ is compact, one easily finds that $\sigma_{\text{ess}}(H_{\alpha,\Gamma_\varepsilon}) = [0,\infty)$ holds for all ε and $\sharp\sigma_{\text{disc}}(H_{\alpha,\Gamma}) < \infty$. Furthermore, the discrete spectrum is nonempty for any $\alpha > 0$ if $d = 2$, while in higher dimensions it is true only for $\alpha > \alpha_0$ with some critical $\alpha_0 > 0$ depending on Γ (see the notes).

Suppose thus that $H_{\alpha,\Gamma}$ has N isolated eigenvalues. It is not difficult to see that the perturbed operator $H_{\alpha,\Gamma_\varepsilon}$ has for all ε small enough the same number of negative eigenvalues, $\lambda_1(\varepsilon) < \lambda_2(\varepsilon) \leq \cdots \leq \lambda_N(\varepsilon)$, which satisfy

$$\lambda_j(\varepsilon) \to \lambda_j(0) \quad \text{as } \varepsilon \to 0\,, \quad 1 \leq j \leq N\,.$$

Let $\{\phi_j^N\}_{j=1}$ be an orthonormal system of eigenfunctions of $H_{\alpha,\Gamma}$ corresponding to these eigenvalues; without loss of generality we may suppose that $\phi_1(x) > 0$ in $\mathbb{R}^d$. By Sobolev's trace theorem, each function ϕ_j is continuous on Γ in the vicinity of the origin. In general, the spectrum may not be simple; given $\zeta \in \sigma_{\text{disc}}(H_{\alpha,\Gamma})$ we denote by $m(\zeta)$ and $n(\zeta)$ the smallest and largest index value j, respectively, for which $\zeta = \lambda_j(0)$, and introduce the positive matrix

$$C(\zeta) := \left(\overline{\phi_i(0)}\,\phi_j(0)\right)_{m(\zeta)\leq i,j\leq n(\zeta)}$$

denoting by $s_{m(\zeta)} \leq s_{m(\zeta)+1} \leq \cdots \leq s_{n(\zeta)}$ its eigenvalues. If $\zeta = \lambda_j(0)$ is a simple eigenvalue of $H_{\alpha,\Gamma}$, in particular, we have $m(\zeta) = n(\zeta) = j$ and $s_j = |\phi_j(0)|^2$. Then we have the following result:

Theorem 10.5 *Assume (i), (ii), and suppose that the coupling constant $\alpha > \alpha_0$. For a given $\zeta \in \sigma_{disc}(H_{\alpha,\Gamma})$ and $m(\zeta) \leq j \leq n(\zeta)$ the asymptotic formula*

$$\lambda_j(\varepsilon) = \zeta + \alpha\, m_\Gamma(\mathcal{P}_\varepsilon) s_j + o(\varepsilon^{d-1})$$

holds as $\varepsilon \to 0$, where $m_\Gamma(\cdot)$ is the $(d-1)$-dimensional Lebesgue measure on Γ.

We refer the reader to the notes for the proof sketch. We also remark that the compactness of Γ plays essentially no role in the argument which also allows us to treat other situations. As an example, consider, a smooth curve $\Gamma \subset \mathbb{R}^2$ and a family of curves Γ_ε with a hiatus described by the same function Γ where, however, the argument runs over $\mathbb{R} \setminus (-\varepsilon, \varepsilon)$ only.

Corollary 10.2.2 *Suppose that Γ satisfies the assumptions of Theorem 10.2, then the eigenvalues of $H_{\alpha,\Gamma_\varepsilon}$ with $m(\zeta) \le j \le n(\zeta)$ obey the asymptotic formula*

$$\lambda_j(\varepsilon) = \lambda_j(0) + 2\alpha\varepsilon\, s_j + o(\varepsilon) \quad \textit{as} \;\; \varepsilon \to 0\,.$$

A way to visualize these results is to realize that, up to an error term, the eigenvalue shift resulting from removing an ε–neighborhood of a hypersurface point is the same as that of adding a repulsive δ-interaction at this point, with the coupling constant proportional to the puncture "area". Let us add, however, that this is true only in the case when $\operatorname{codim}\Gamma = 1$. If we have, for instance, a simple eigenvalue of $H_{\alpha,\Gamma}$ with the eigenfunction ϕ, where Γ is a curve in $\mathbb{R}^3$, and perturb the latter by making a 2ε–hiatus in it, the leading term in the perturbation expansion is again proportional to $|\phi(0)|^2$, but this time it comes multiplied not by ε but rather by $\varepsilon \ln \varepsilon$ (see the notes).

10.2.3 Isoperimetric Problem

The ways in which the interaction support can influence the spectral properties of leaky-graph Hamiltonians are not exhausted by the above described results. Let us mention one more which is analogous to the problem discussed in Sect. 3.2.3; for simplicity we limit ourselves again to the two-dimensional situation and suppose that $\Gamma \subset \mathbb{R}^2$ is a loop of a fixed length. We know that the discrete spectrum of $H_{\alpha,\Gamma}$ is then nonempty for any fixed $\alpha > 0$, in particular, $\lambda_1(\alpha, \Gamma) := \inf \sigma(H_{\alpha,\Gamma}) < 0$ is a simple isolated eigenvalue. We ask about the shape of Γ which makes this ground-state eigenvalue maximal.

To state the problem, suppose that $\Gamma : [0, L] \to \mathbb{R}^2$ is a closed C^1, piecewise C^2 smooth curve, $\Gamma(0) = \Gamma(L)$; we allow self-intersections provided the curve meets itself at a non-zero angle. Furthermore, we divide such loops into equivalence classes: Γ and Γ' are equivalent if one can be obtained from the other by a Euclidean transformation of the plane; the spectral properties of the corresponding operators $H_{\alpha,\Gamma}$ and $H_{\alpha,\Gamma'}$ are obviously the same. The above assumptions are satisfied, in particular, by the circle $\mathcal{C} := \{\, (\frac{L}{2\pi} \cos \frac{2\pi s}{L}, \frac{L}{2\pi} \sin \frac{2\pi s}{L}) : s \in [0, L]\,\}$, and its equivalence class. Then we have the following result:

Theorem 10.6 *Within the above described class of loops, the ground-state eigenvalue $\lambda_1(\alpha, \Gamma)$ is for any fixed $\alpha > 0$ and $L > 0$ sharply maximized by the circle.*

To prove the theorem we need an auxiliary geometric result.

Lemma 10.2.1 *Let Γ have the properties described above, then we have*

$$\int_0^L |\Gamma(s+u)-\Gamma(s)|^p\,\mathrm{d}s \le \frac{L^{1+p}}{\pi^p}\sin^p\frac{\pi u}{L} \quad \textit{for } p\in(0,2]\,.$$

Proof is left to the reader (Problem 11); note that the right-hand side of the inequality is nothing else than the value of the integral when Γ is a circle.

Proof of Theorem 10.6 We shall employ again the generalized Birman-Schwinger principle in order to reformulate the question into the eigenvalue problem, $\mathcal{R}^{\kappa}_{\alpha,\Gamma}\phi = \phi$ on $L^2(0,L)$ with $\mathcal{R}^{\kappa}_{\alpha,\Gamma}$ defined by equation (10.7). We note that the operator-valued function $\kappa\mapsto\mathcal{R}^{\kappa}_{\alpha,\Gamma}$ is decreasing in $(0,\infty)$ and $\|\mathcal{R}^{\kappa}_{\alpha,\Gamma}\|\to 0$ holds as $\kappa\to\infty$. By the positivity improving property, the maximum eigenvalue of $\mathcal{R}^{\kappa}_{\alpha,\Gamma}$ is simple, and the same is true by *Theorem* 6.7d for the ground state of $H_{\alpha,\Gamma}$. If Γ is a circle, the latter eigenfunction exhibits rotational symmetry, and using Problem 3 we see that the respective eigenfunction of $\mathcal{R}^{\tilde\kappa_1}_{\alpha,\mathcal{C}}$ corresponding to the unit eigenvalue is constant, $\tilde\phi_1(s) = L^{-1/2}$. Then we can write

$$\max\sigma(\mathcal{R}^{\tilde\kappa_1}_{\alpha,\mathcal{C}}) = (\tilde\phi_1,\mathcal{R}^{\tilde\kappa_1}_{\alpha,\mathcal{C}}\tilde\phi_1) = \frac{1}{L}\int_0^L\int_0^L \mathcal{R}^{\tilde\kappa_1}_{\alpha,\mathcal{C}}(s,s')\,\mathrm{d}s\,\mathrm{d}s'\,,$$

while for a general loop Γ of length L a simple variational estimate gives

$$\max\sigma(\mathcal{R}^{\tilde\kappa_1}_{\alpha,\Gamma}) \ge (\tilde\phi_1,\mathcal{R}^{\tilde\kappa_1}_{\alpha,\Gamma}\tilde\phi_1) = \frac{1}{L}\int_0^L\int_0^L \mathcal{R}^{\tilde\kappa_1}_{\alpha,\Gamma}(s,s')\,\mathrm{d}s\,\mathrm{d}s'\,;$$

hence to check that the circle is a maximizer it is sufficient to show that

$$\int_0^L\int_0^L K_0\big(\kappa|\Gamma(s)-\Gamma(s')|\big)\,\mathrm{d}s\,\mathrm{d}s' \ge \int_0^L\int_0^L K_0\big(\kappa|\mathcal{C}(s)-\mathcal{C}(s')|\big)\,\mathrm{d}s\,\mathrm{d}s'$$

holds for *any* $\kappa>0$ and Γ of the considered class. By a simple change of variables we find that this is equivalent to the positivity of the functional

$$F_\kappa(\Gamma) := \int_0^{L/2}\mathrm{d}u\int_0^L\mathrm{d}s\Big[K_0\big(\kappa|\Gamma(s+u)-\Gamma(s)|\big) - K_0\big(\kappa|\mathcal{C}(s+u)-\mathcal{C}(s)|\big)\Big]\,,$$

where the second term is equal to $K_0\big(\frac{\kappa L}{\pi}\sin\frac{\pi u}{L}\big)$. Now we employ the (strict) convexity of K_0 which yields by means of the Jensen's inequality the estimate

$$\frac{1}{L}F_\kappa(\Gamma) \ge \int_0^{L/2}\left[K_0\left(\frac{\kappa}{L}\int_0^L|\Gamma(s+u)-\Gamma(s)|\mathrm{d}s\right) - K_0\left(\frac{\kappa L}{\pi}\sin\frac{\pi u}{L}\right)\right]\mathrm{d}u\,,$$

where the inequality is sharp unless $|\Gamma(s+u) - \Gamma(s)|$ is independent of s. Finally, we note that K_0 is decreasing in $(0, \infty)$, hence it is sufficient to apply *Lemma* 10.2.1 with $p = 1$ to the argument of the first term on the right-hand side. ∎

10.3 Strong Coupling Asymptotics

We have encountered repeatedly, in Sect. 1.6 and elsewhere, situations when the particle was strongly localized transversally and the motion became effectively lower dimensional. In the present context it is the coupling constant α which determines how much is the particle attracted to the graph determining, in particular, the "spread" of possible eigenfunctions in the direction transverse to the edges; it is thus natural to ask what happens if the attraction is strong. We have seen in Chap. 8, however, that the problem is significantly more complicated if the particle is localized in regions which are non-smooth and/or branched; this is the reason why we limit ourselves here to the situations where the interaction support is sufficiently smooth.

10.3.1 Interactions Supported by Curves

While for the definition of $H_{\alpha,\Gamma}$ the codimension of the interaction support was important, for the asymptotic behavior considered here it is the dimension which matters. As usual, we begin with the case of planar curves, at first finite ones.

Theorem 10.7 *Suppose that $\Gamma : [0, L] \to \mathbb{R}^2$ is a C^4 smooth function, $|\dot\Gamma| = 1$, which defines a curve Γ without self-intersections; then the relation*

$$\sharp\, \sigma_{disc}(H_{\alpha,\Gamma}) = \frac{\alpha L}{2\pi} + \mathcal{O}(\ln \alpha)$$

holds as $\alpha \to \infty$. In addition, if Γ is a closed curve, C^4 smooth at $\Gamma(0) = \Gamma(L)$, then the jth eigenvalue of the operator $H_{\alpha,\Gamma}$ behaves asymptotically as

$$\lambda_j(\alpha) = -\frac{1}{4}\alpha^2 + \mu_j + \mathcal{O}(\alpha^{-1} \ln \alpha)\,,$$

where μ_j is the jth eigenvalue of the operator $S_\Gamma := -\frac{\mathrm{d}^2}{\mathrm{d}s^2} - \frac{1}{4}\gamma(s)^2$ on $L^2(0, L)$ with periodic boundary conditions, counted with multiplicity, and $\gamma(s)$ is the signed curvature of Γ. If, on the other hand, Γ is a finite arc with regular ends, the formula holds again with μ_j referring instead to the analogous operator with Dirichlet conditions at the endpoints of the interval $[0, L]$.

Proof Suppose first that Γ is a closed curve without self-intersections, and consider its strip neighborhood analogous to that used in proof of *Theorem* 10.1 with the appropriate curvilinear coordinates, in other words, the set Σ_a onto which the function $\Phi_a : [0, L) \times (-a, a) \to \mathbb{R}^2$ defined by

$$(s, u) \mapsto (\Gamma_1(s) - u\Gamma_2'(s), \Gamma_2(s) + u\Gamma_1'(s)) \tag{10.10}$$

is a diffeomorphism for all $a > 0$ small enough. We are going to estimate $H_{\alpha,\Gamma}$ using Dirichlet-Neumann bracketing, imposing additional conditions at the boundary of Σ_a, which yields the inequalities

$$(-\Delta_{\Lambda_a}^{\mathrm{N}}) \oplus L_{a,\alpha}^- \le H_{\alpha,\Gamma} \le (-\Delta_{\Lambda_a}^{\mathrm{D}}) \oplus L_{a,\alpha}^+ , \tag{10.11}$$

where $\Lambda_a = \mathbb{R}^2 \setminus \overline{\Sigma}_a =: \Lambda_a^{\mathrm{in}} \cup \Lambda_a^{\mathrm{out}}$ is the exterior domain, and $L_{a,\alpha}^\pm$ are the self-adjoint operators associated with the forms

$$q_{a,\alpha}^\pm[f] = \|\nabla f\|_{L^2(\Sigma_a)}^2 - \alpha \int_\Gamma |f(\Gamma(s))|^2 \,\mathrm{d}s ,$$

where f belongs to $H_0^1(\Sigma_a)$ or to $H^1(\Sigma_a)$ for the $\pm$ sign, respectively. Importantly, the exterior domain Λ_a does not contribute to the negative part of the spectrum, thus for our purpose we may consider $L_{a,\alpha}^\pm$ only. Using the curvilinear coordinates (s, u), we pass from $L_{a,\alpha}^\pm$ to unitarily equivalent operators associated with the quadratic forms

$$\begin{aligned} b_{a,\alpha}^+[f] &= \int_0^L \int_{-a}^a (1 + u\gamma(s))^{-2} \left|\frac{\partial f}{\partial s}\right|^2 (s, u) \,\mathrm{d}u \,\mathrm{d}s + \int_0^L \int_{-a}^a \left|\frac{\partial f}{\partial u}\right|^2 (s, u) \,\mathrm{d}u \,\mathrm{d}s \\ &\quad + \int_0^L \int_{-a}^a V(s, u)|f(s, u)|^2 \,\mathrm{d}u \,\mathrm{d}s - \alpha \int_0^L |f(s, 0)|^2 \,\mathrm{d}s \end{aligned} \tag{10.12}$$

defined for $f \in H^1((0, L) \times (-a, a))$ satisfying periodic boundary conditions in the variable s and Dirichlet conditions at $u = \pm a$, and

$$b_{a,\alpha}^-[f] = b_{a,\alpha}^+[f] - \sum_{j=0}^{1} \frac{1}{2}(-1)^j \int_0^L \frac{k(s)}{1 + (-1)^j ak(s)} |f(s, (-1)^j a)|^2 \,\mathrm{d}s \tag{10.13}$$

without the Dirichlet condition at $u = \pm a$; the symbol V in (10.12) denotes the curvature-induced potential (1.8). In the next step we replace inequality (10.11) by rougher bounds squeezing $H_{\alpha,\Gamma}$ between operators with separated variables. To do so we introduce self-adjoint operators $U_a^\pm$ in $L^2(0, L)$ given by

$$U_a^\pm = -(1 \mp a\gamma_+)^{-2} \frac{\mathrm{d}^2}{\mathrm{d}s^2} + V_\pm(s) \tag{10.14}$$

with periodic boundary conditions, where

$$V_\pm(s) = \pm\frac{1}{2}(1 - a\gamma_+)^{-3}a\ddot{\gamma}_+ - \frac{5}{4}(1 \pm a\gamma_+)^{-4}a^2\dot{\gamma}_+^2 - \frac{1}{4}(1 \pm a\gamma_+)^{-2}\gamma(s)^2$$

with $\gamma_+ := \|\gamma\|_\infty$, $\dot{\gamma}_+ := \|\dot{\gamma}\|_\infty$, and $\ddot{\gamma}_+ := \|\ddot{\gamma}\|_\infty$. Furthermore, let $T_{\alpha,a}^\pm$ be the transverse operators associated with the quadratic forms

$$t_{a,\alpha}^+[f] = \int_{-a}^a |f'(u)|^2\,du - \alpha|f(0)|^2\,, \qquad \mathrm{Dom}\,(t_{a,\alpha}^+) = H_0^1(-a,a)\,,$$
$$t_{a,\alpha}^-[f] = t_{a,\alpha}^-[f] - \gamma_+\,(|f(a)|^2 + |f(-a)|^2)\,,\ \mathrm{Dom}\,(t_{a,\alpha}^-) = H^1(-a,a)\,.$$

The negative spectra of the operators $T_{\alpha,a}^\pm$ can be localized with an exponential precision (Problem 12): there is a number $c > 0$ such that $T_{\alpha,a}^\pm$ has a single negative eigenvalue $\xi_{\alpha,a}^\pm$ satisfying

$$-\frac{\alpha^2}{4}\left(1 + c\,\mathrm{e}^{-\alpha a/2}\right) < \xi_{\alpha,a}^- < -\frac{\alpha^2}{4} < \xi_{\alpha,a}^+ < -\frac{\alpha^2}{4}\left(1 - 8\,\mathrm{e}^{-\alpha a/2}\right) \tag{10.15}$$

provided α is large enough. If we now define

$$\tilde{H}_{\alpha,a}^\pm = U_a^\pm \otimes I + I \otimes T_{\alpha,a}^\pm\,,$$

then $\lambda_j(\alpha)$ is by the minimax principle squeezed between the corresponding eigenvalues of $\tilde{H}_{\alpha,a}^-$ and $\tilde{H}_{\alpha,a}^+$, hence for α large enough we have

$$\mu_j^-(a) + \xi_{\alpha,a}^- \le \lambda_j(\alpha) \le \mu_j^+(a) + \xi_{\alpha,a}^+\,, \tag{10.16}$$

where $\mu_j^\pm(a)$ is the jth eigenvalue of the operator $U_a^\pm$. On the other hand, from the definition of $V_\pm$ we infer that $\|V_\pm(\cdot) - V(\cdot\,,\cdot)\|_\infty = \mathcal{O}(a)$ as $a \to 0$, hence by a simple perturbation argument there is a C_j such that

$$|\mu_j^\pm(a) - \mu_j| \le C_j a \tag{10.17}$$

holds for a small enough. To conclude the argument, we choose $a = 6\alpha^{-1}\ln\alpha$ as the strip neighborhood halfwidth; the stated asymptotic formula for $\lambda_j(\alpha)$ then follows from (10.15) and (10.16).

If Γ is not closed, the same can be done with the comparison operators $S_\Gamma^{\mathrm{D,N}}$ having the appropriate boundary conditions, Dirichlet or Neumann, respectively, at the endpoints of Γ. This gives the estimate on $\sharp\,\sigma_{\mathrm{disc}}(H_{\alpha,\Gamma})$ and an upper bound on $\lambda_j(\alpha)$, however, it fails to provide a precise enough lower bound. If Γ has regular ends, i.e. it can be regarded as the restriction of a C^4 smooth curve $\Gamma : [-a, L+a] \to \mathbb{R}^2$ for some $a > 0$, one is able to perform the Neumann estimate on such an extended curve. This bound no longer has separated variables, but one can express the eigenfunctions of $H_{\alpha,\Gamma}$ using Problem 3 and employ decay properties of the Green function. The estimate is somewhat involved and we provide a reference to the complete argument in the notes. ∎

The case of a finite curve in $\mathbb{R}^3$ is similar. We consider a C^4 smooth function $\Gamma : [0, L] \to \mathbb{R}^3$ with $|\dot\Gamma(s)| = 1$. We assume that it has a piecewise global Frenet frame (t, n, b) in the sense of Problem 1.12 and define the straightening transformation using the map $\Phi_a : [0, L] \times \mathcal{B}_a \to \mathbb{R}^3$,

$$\Phi_a(s, r, \theta) = \Gamma(s) - r\,[n(s)\cos(\theta-\beta(s)) + b(s)\sin(\theta-\beta(s))]\,,$$

where $\mathcal{B}_a$ is the disc of radius a centered at the origin; for small enough a it is a diffeomorphism of a tubular neighborhood Σ_a of Γ that does not intersect itself. If the function β is chosen to satisfy the Tang condition (1.18), $\dot\beta = \tau$, then the longitudinal and transverse variable decouple and we can proceed as in the two-dimensional case (Problem 13) arriving at the following result.

Theorem 10.8 *For curves Γ without self-intersections described above, we have*

$$\sharp\,\sigma_{disc}(H_{\alpha,\Gamma}) = \frac{L}{\pi}(\zeta_\alpha)^{1/2}(1 + \mathcal{O}(e^{\pi\alpha}))$$

as $\alpha \to \infty$, where $\zeta_\alpha := 4\,\mathrm{e}^{2(-2\pi\alpha+\psi(1))}$. In addition, if Γ is a closed curve, the jth eigenvalue of the operator $H_{\alpha,\Gamma}$ behaves asymptotically as

$$\lambda_j(\alpha) = \zeta_\alpha + \mu_j + \mathcal{O}(\mathrm{e}^{\pi\alpha})\,,$$

where μ_j is the jth eigenvalue of the operator S_Γ on $L^2(0, L)$ with periodic boundary conditions described in the previous theorem.

The technique used to derive these results can be applied in other situations as well. If Γ is an infinite curve, the threshold of the essential changes in general and the estimates on $\sharp\,\sigma_{\rm disc}(H_{\alpha,\Gamma})$ are no longer relevant. On the other hand, the eigenvalue asymptotic formulaæ remain valid under stronger assumptions on the curvature and torsion.

Corollary 10.3.1 *Let $\Gamma : \mathbb{R} \to \mathbb{R}^d$, $d = 2, 3$, satisfy the hypotheses of Theorems 10.2 and 10.3, respectively. In addition, assume that $\dot\gamma(s)$ and $\ddot\gamma(s)^{1/2}$ are $\mathcal{O}(s^{-1-\varepsilon})$ as $|s| \to \infty$, and $\tau, \dot\tau \in L^\infty(\mathbb{R})$ for $d = 3$. Then the asymptotic expansions from the said theorems hold for all the isolated eigenvalues $\lambda_j(\alpha)$ of $H_{\alpha,\Gamma}$, where $S_\Gamma := -\frac{\mathrm{d}}{\mathrm{d}s^2} - \frac14\gamma(s)^2$ is now the operator on $L^2(\mathbb{R})$ with domain $H^2(\mathbb{R})$.*

Note that in this case we need not care about the multiplicity when counting the eigenvalues because the spectrum of S_Γ in $L^2(\mathbb{R})$ is simple. On the other hand, the smoothness requirement is essential for the above results.

Example 10.3.1 Consider again the operator $H_2(\beta)$ of Example 10.2.1. In this case Γ is a broken line, and as such it is self-similar. Using a scaling transformation we find easily that μ_j in the asymptotic formula of *Theorem* 10.7 should be replaced by $\left(\lambda_j + \frac{1}{4}\right)\alpha^2$, where λ_j is the jth eigenvalue of $H_2(\beta)$ corresponding to $\alpha = 1$.

10.3.2 Interactions Supported by Surfaces

The method used for curves also works for interactions supported by surfaces, but the geometric part is naturally different. Consider first a C^4 smooth compact and closed Riemann surface $\Sigma \subset \mathbb{R}^3$ of a finite genus g. As in Sect. 4.1.1, its geometry is encoded in the metric tensor $g_{\mu\nu}$ and the Weingarten tensor $h_\mu^{\ \nu}$; the eigenvalues $k_\pm$ of the latter are the principal curvatures which determine the Gauss curvature $K = \det(h_\mu^{\ \nu})$ and mean curvature $M = \frac{1}{2}\operatorname{Tr}(h_\mu^{\ \nu})$. For a compact Σ the essential spectrum is $[0, \infty)$ and we ask about the asymptotic behavior of the negative eigenvalues as $\alpha \to \infty$. This time it will be expressed in terms of a comparison operator of the form

$$S_\Sigma := -\Delta_\Sigma + K - M^2$$

on $L^2(\Sigma, d\sigma)$, where $\Delta_\Sigma = -g^{-1/2}\partial_\mu g^{1/2} g^{\mu\nu}\partial_\nu$ is the Laplace-Beltrami operator on Σ. The jth eigenvalue μ_j of S_Σ is bounded from above by that of Δ_Σ because $K - M^2 = -\frac{1}{4}(k_+ - k_-)^2 \le 0$, in particular, the two coincide if Σ is a sphere.

Theorem 10.9 *Under the stated assumptions,* $\sharp\,\sigma_{disc}(H_{\alpha,\Sigma}) \ge j$ *holds for any fixed integer* j *provided* α *is sufficiently large and the counting function behaves asymptotically as*

$$\sharp\,\sigma_{disc}(H_{\alpha,\Sigma}) = \frac{|\Sigma|}{16\pi}\alpha^2 + \mathcal{O}(\alpha)\,,$$

where $|\Sigma|$ *is the Riemann area of the surface* Σ. *Moreover, the* j*th eigenvalue* $\lambda_j(\alpha)$ *of* $H_{\alpha,\Sigma}$ *has the expansion*

$$\lambda_j(\alpha) = -\frac{1}{4}\alpha^2 + \mu_j + \mathcal{O}(\alpha^{-1}\ln\alpha)$$

as $\alpha \to \infty$, *where* μ_j *is the* j*th eigenvalue of* S_Σ.

Proof We employ a bracketing argument again. To construct the needed neighborhoods of Σ we use the field $\{n(x) \,:\, x \in \Sigma\}$ of unit vectors normal to the manifold, which exists globally because Σ is orientable, and define a map $\mathcal{L}_a \,:\, \Sigma \times (-a, a) \to \mathbb{R}^3$ by $\mathcal{L}_a(x, u) = x + un(x)$. Since Σ is smooth it is a diffeomorphism for all a small enough mapping onto the layer neighborhood $\Omega_a = \{x \in \mathbb{R}^3 \,:\, \operatorname{dist}(x, \Gamma) < a\}$. By bracketing we get a two-sided estimate for the negative spectrum of $H_{\alpha,\Sigma}$ in terms of Dirichlet and Neumann operators referring to the Ω_a component. They can be analyzed by means of the curvilinear coordinates as in Sect. 4.1.1; one arrives at estimates through operators with decoupled variables, $S_a^{\pm} \otimes I + I \otimes T_{\alpha,a}^{\pm}$ with

$$S_a^{\pm} := -C_{\pm}(a)\Delta_\Gamma + C_{\mp}^{-2}(a)(K - M^2) \pm va$$

and the transverse part which is the same as in the proof of *Theorem* 10.7. Here $C_{\pm}(a) := (1 \pm a\varrho^{-1})^2$ with $\varrho := \max(\{\|k_+\|_\infty, \|k_-\|_\infty\})^{-1}$ and v is a suitable constant. The rest of the argument is the same as in the two-dimensional case; to get the counting function one has to employ the Weyl formula (Problem 14) for the comparison operator S_Σ. ∎

Note that Σ need not be simply connected; the claim remains valid if it is a finite disjoint union of C^4 smooth compact Riemann surfaces of finite genera. Moreover, the asymptotic formula for the counting function is preserved for surfaces Σ which have a nonempty and smooth boundary.

The technique used to derive the above asymptotic expansions can also be applied to singular interactions supported by infinite surfaces provided we adopt additional assumptions, for instance

(i) Injectivity: the map $\mathcal{L}_a : \Sigma \times (-a, a) \to \Sigma_a \subset \mathbb{R}^3$ defined above is injective for all a small enough.
(ii) Uniform ellipticity: $c_-\delta_{\mu\nu} \le g_{\mu\nu} \le c_+\delta_{\mu\nu}$ holds for some $c_\pm > 0$.
(iii) Asymptotic planarity: $K, M \to 0$ as the geodesic radius $r \to \infty$.

The assumptions allow for modifications, in particular, the first and the third one can be replaced by another hypothesis (Problem 15). Then we have the following result, the proof of which is left to the reader.

Theorem 10.10 *Under assumptions (i) and (iii) we have* $\inf \sigma_{\rm ess}(H_{\alpha,\Sigma}) = \epsilon(\alpha)$, *where* $\epsilon(\alpha) + \frac{1}{4}\alpha^2 = \mathcal{O}(\alpha^2 e^{-\alpha a/2})$ *as* $\alpha \to \infty$. *In addition, if (ii) is valid and* Σ *is not a plane, then* $\sigma_{disc}(H_{\alpha,\Sigma}) \neq \emptyset$ *holds for all* α *large enough and the* j*th eigenvalue has the asymptotic expansion,*

$$\lambda_j(\alpha) = -\frac{1}{4}\alpha^2 + \mu_j + \mathcal{O}(\alpha^{-1}\ln\alpha)$$

as $\alpha \to \infty$, *where* μ_j *is the* j*th eigenvalue of the comparison operator* S_Σ *on* $L^2(\Sigma, d\sigma)$ *introduced above, counted with multiplicity.*

Note that this result establishes the existence of curvature-induced bound states for surfaces Σ which are asymptotically planar provided the coupling constant α is sufficiently large. The eigenfunctions are in such a case strongly localized around the surface, which allows us to regard this claim as an analogue of *Theorem* 4.2b. The question of under which conditions this is still valid beyond the strong-coupling regime remains open.

10.3.3 Periodic and Magnetic Systems

So far we have dealt with situations where the interaction support was either compact or at least its geometrically nontrivial part was localized. Let us now see what happens if Γ is periodic. We start again with a planar curve assumed to be C^4 smooth, hence its signed curvature γ is a C^2 function. We shall assume:

(i) Curvature periodicity: there is an $L > 0$ such that $\gamma(s+L) = \gamma(s)$.
(ii) Curve periodicity: in analogy with (9.10) we assume $\int_0^L \gamma(s)\,\mathrm{d}s = 0$. Without loss of generality we may suppose that the normal at $s = 0$ is $(1, 0)$, then $\Gamma(\cdot + L) - \Gamma(\cdot) = (l_1, l_2)$ where the period-shift components are $l_j := \int_0^L \sin\left(\frac{\pi}{2}(2-j) - \int_0^t \gamma(u)\,\mathrm{d}u\right) dt$ and we may assume that $l_1 > 0$.
(iii) Period cell match: the map (10.10) is injective for all a small enough and $\Phi_a((0, L) \times (-a, a)) \subset \Lambda := (0, l_1) \times \mathbb{R}$.

We proceed as in Sect. 9.1.2 and perform the Bloch-Floquet decomposition. The operator $H_{\alpha,\Gamma}(\theta)$ on $L^2(\Lambda)$ is for a θ from the Brillouin zone $\mathcal{B} := \left[-\frac{\pi}{l_1}, \frac{\pi}{l_1}\right]$ defined through the quadratic form (6.20) with $\Omega = \mathbb{R}^2$; its domain consists of functions $u \in H^1(\Lambda)$ satisfying the boundary conditions $u(l_1, l_2 + \cdot) = \mathrm{e}^{i\theta} u(0, \cdot)$. Identifying the Hilbert space $L^2(\mathbb{R}^2)$ with $\int_{\mathcal{B}}^{\oplus} L^2(\Lambda)\,\mathrm{d}\theta$ we can write with a slight abuse of notation the direct-integral decomposition

$$H_{\alpha,\Gamma} = \frac{l_1}{2\pi} \int_{\mathcal{B}}^{\oplus} H_{\alpha,\Gamma}(\theta)\,\mathrm{d}\theta \quad \text{implying} \quad \sigma(H_{\alpha,\Gamma}) = \bigcup_{\theta\in\mathcal{B}} \sigma(H_{\alpha,\Gamma}(\theta)) \,.$$

Note that the decomposition is similar to that of Sect. 7.2.2, with the magnetic field absent and the single point interaction replaced by the one supported by the curve segment $\Gamma((0, L))$. Since the latter is finite, it is easy to check that $\sigma_{\mathrm{ess}}(H_{\alpha,\Gamma}(\theta)) = [0, \infty)$; for the discrete spectrum we have the following result:

Theorem 10.11 *To any $j \in \mathbb{N}$ there is an $\alpha_j > 0$ such that $\sharp\,\sigma_{disc}(H_{\alpha,\Gamma}(\theta)) \geq j$ holds for $\alpha > \alpha_j$ and any $\theta \in \mathcal{B}$. The jth eigenvalue of $H_{\alpha,\Gamma}(\theta)$ counted with multiplicity has the asymptotic expansion*

$$\lambda_j(\alpha, \theta) = -\frac{1}{4}\alpha^2 + \mu_j(\theta) + \mathcal{O}(\alpha^{-1} \ln \alpha)$$

as $\alpha \to \infty$, where $\mu_j(\theta)$ is the jth eigenvalue of $S_\Gamma(\theta) = -\frac{\mathrm{d}^2}{\mathrm{d}s^2} - \frac{1}{4}\gamma(s)^2$ with the domain consisting of functions $u \in H^2(0, L)$ that satisfy $u(L) = \mathrm{e}^{i\theta} u(0)$ and $u'(L) = \mathrm{e}^{i\theta} u'(0)$, and the error term is uniform with respect to $\theta \in \mathcal{B}$.

Proof The proof follows closely the line of argument used to demonstrate *Theorem* 10.7, based on the bracketing technique. Let

$$\begin{aligned} R^{+}_{a,\theta} &= \{u \in H^1(\Sigma_a) : u|_{\partial\Sigma_a \cap \Lambda} = 0,\ u(l_1, \cdot) = \mathrm{e}^{i\theta} u(0, \cdot) \text{ on } (-a, a)\}, \\ R^{-}_{a,\theta} &= \{u \in H^1(\Sigma_a) : u(l_1, \cdot) = \mathrm{e}^{i\theta} u(0, \cdot) \text{ on } (-a, a)\}, \end{aligned}$$

and define the quadratic forms

$$q^{\pm}_{a,\alpha,\theta}[f] = \|\nabla f\|^2_{L^2(\Sigma_a)} - \alpha \int_{\Gamma((0,L))} |f(x)|^2 \, \mathrm{d}s\,, \quad \mathrm{Dom}\,(q^{\pm}_{a,\alpha,\theta}) = R^{\pm}_{a,\theta}\,.$$

Let $L^{\pm}_{a,\alpha,\theta}$ be the self-adjoint operators associated with the forms $q^{\pm}_{a,\alpha,\theta}$, respectively, and put again $a(\alpha) = 6\alpha^{-1} \ln \alpha$. By imposing additional Dirichlet respectively Neumann boundary conditions at the boundary of Σ_a and using the bracketing argument we find that

$$\kappa^{-}_j(\alpha, \theta) \le \lambda_j(\alpha, \theta) \le \kappa^{+}_j(\alpha, \theta)\,, \tag{10.18}$$

where $\kappa^{\pm}_j(\alpha, \theta)$ is the jth eigenvalue of the operator $L^{\pm}_{a(\alpha),\alpha,\theta}$. Following the proof of *Theorem* 10.7, we estimate $\kappa^{\pm}_j(\alpha, \theta)$ by the eigenvalues of suitably chosen operators with separated variables. Specifically, let $U^{\pm}_{a(\alpha),\theta}$ be given by the right-hand side of (10.14) acting on the domain

$$P_\theta = \{u \in H^2((0, L)); \quad u(L) = e^{i\theta} u(0)\,,\ u'(L) = \mathrm{e}^{i\theta} u'(0)\}\,,$$

and let $\mu^{\pm}_j(\alpha, \theta)$ denote their eigenvalues. We use the curvilinear coordinates (s, u) and pass from $L^{\pm}_{a,\alpha}$ to unitarily equivalent operators associated with the quadratic forms (10.12) and (10.13) with the respective form domains

$$Q^{-}_{a,\theta} = \{\varphi \in H^1((0, L) \times (-a, a)) : \varphi(l_1, \cdot) = \mathrm{e}^{i\theta} \varphi(0, \cdot) \text{ on } (-a, a)\}$$

and $Q^{+}_{a,\theta} = \{\varphi \in Q^{-}_{a,\theta} : \varphi(\cdot, \pm a) = 0 \text{ on } (0, L)\}$. The minimax principle then implies that $\kappa^{\pm}_j(\alpha, \theta)$ are sandwiched between the corresponding eigenvalues of the operators $U^{\pm}_{a(\alpha),\theta} \otimes I + I \otimes T^{\pm}_{\alpha,a(\alpha)}$, where $T^{\pm}_{\alpha,a}$ have been introduced in the proof of *Theorem* 10.7. More precisely, we have

$$\xi^{-}_{\alpha,a(\alpha)} + \mu^{-}_j(\alpha, \theta) \le \kappa^{-}_j(\alpha, \theta) \le \kappa^{+}_j(\alpha, \theta) \le \xi^{+}_{\alpha,a(\alpha)} + \mu^{+}_j(\alpha, \theta)\,. \tag{10.19}$$

On the other hand, equation (10.17) implies in view of our choice of a that

$$|\mu^{\pm}_j(\alpha, \theta) - \mu_j| = \mathcal{O}(\alpha^{-1} \ln \alpha) \quad \text{as } \alpha \to \infty\,;$$

this asymptotic behavior in combination with relations (10.15), (10.18), and (10.19) completes the proof. ∎

Combining the last result with Borg's theorem on the inverse problem for Hill's equation, we can make a claim about gaps of $\sigma(H_{\alpha,\Gamma})$ similar to *Theorem* 9.4.

Corollary 10.3.2 *Suppose that* $\Gamma : \mathbb{R} \to \mathbb{R}^d$, $d = 2, 3$, *is not a straight line,* $\gamma \neq 0$, *and satisfies the assumptions of Theorem 10.11 or Problem 16b, respectively, then the spectrum of* $H_{\alpha,\Gamma}$ *contains open gaps for all* α *large enough.*

Remark 10.3.1 The interaction support need not consist of a single curve, in a similar way one can treat finite or infinite families of curves periodic in $n \leq d$ directions, the only restriction is that their components have to satisfy individually the listed assumptions and the distances between them must have a uniform positive lower bound. There is a difference between the cases $n = d$ and $n < d$, however, because in the former the basic cell of the system is pre-compact, and therefore the spectrum of each $H_{\alpha,\Gamma}(\theta)$ is purely discrete. A particular situation occurs when a periodic Γ consists of disjoint compact components. The described asymptotic expansions are valid again but now the fiber comparison operator $S_\Gamma(\theta)$ is independent of the quasi-momentum θ, in other words

$$\lambda_j(\alpha, \theta) = -\frac{1}{4}\alpha^2 + \mu_j + \mathcal{O}(\alpha^{-1} \ln \alpha)$$

holds for $d = 2$ and the respective expansion for $d = 3$. Note that in the latter case the topology may again be nontrivial, e.g., for chains of interlocked rings.

Corollary 10.3.3 *Suppose that the curve* $\Gamma \subset \mathbb{R}^d$ *satisfies the assumptions of Theorem 10.11 or Problem 16b, respectively. In the case* $d = 2$ *to any* $\lambda > 0$ *there is an* $\alpha_\lambda > 0$ *such that the spectrum of the operator* $H_{\alpha,\Gamma}$ *is absolutely continuous in* $(-\infty, -\frac{1}{4}\alpha^2 + \lambda]$ *as long as* $\alpha > \alpha_\lambda$. *The same is true for* $d = 3$ *with* $-\frac{1}{4}\alpha^2$ *replaced by* ζ_α *provided* $-\alpha > \alpha_\lambda$.

Proof It is straightforward to check that $\{H_{\alpha,\Gamma}(\theta) : \theta \in \mathcal{B}\}$ is a type (A) analytic family. The spectral interval in question contains a finite number of eigenvalue branches, each being a real analytic function which has the above described asymptotic expansion. The functions $\mu_j(\cdot)$ are nonconstant, hence the same is true for $\lambda_j(\alpha, \cdot)$ provided $(-1)^d\alpha$ is large enough. ∎

In a similar way one can treat strong-coupling asymptotics for operators $H_{\alpha,\Sigma}$ describing singular interaction supported by a periodic surface Σ. We consider discrete translations of $\mathbb{R}^3$ generated by an r-tuple $\{l_\mu\}$, where $r = 1, 2, 3$, and decompose both Σ, assumed to be a C^4 smooth Riemann surface, not necessarily connected, and the ambient space $\mathbb{R}^3$ into period cells Σ_{p} and Λ, respectively, assuming again that they match mutually. The Bloch-Floquet decomposition is as above: we define the fiber operators $H_{\alpha,\Sigma}(\theta)$ on $L^2(\Lambda)$ through quadratic forms defined on functions satisfying the appropriate boundary condition and express $H_{\alpha,\Sigma}$ and its spectrum through them as

$$H_{\alpha,\Sigma} = \frac{1}{(2\pi)^r} \prod_{\mu=1}^{r} l_\mu \int_{\mathcal{B}}^{\oplus} H_{\alpha,\Sigma}(\theta)\,\mathrm{d}\theta\,, \quad \sigma(H_{\alpha,\Sigma}) = \bigcup_{\theta\in\mathcal{B}} \sigma(H_{\alpha,\Sigma}(\theta))\,,$$

where $\mathcal{B} = \mathsf{X}_{\mu=1}^{r}\big[-\frac{\pi}{l_\mu}, \frac{\pi}{l_\mu}\big]$ is the Brillouin zone. The spectrum of $H_{\alpha,\Gamma}(\theta)$ depends on r; it is purely discrete if $r = 3$ and Λ is compact, while for $r = 1, 2$ we have $\sigma_{\mathrm{ess}}(H_{\alpha,\Gamma}(\theta)) = [0, \infty)$. The eigenvalues are continuous functions of the quasimomentum components θ_μ.

As before we need a family of comparison operators on $L^2(\Sigma_{\mathrm{p}}, d\sigma)$. One way to describe them is to write their action as

$$S_\Sigma(\theta) := g^{-1/2}(-i\partial_\mu + \theta_\mu)g^{1/2}g^{\mu\nu}(-i\partial_\nu + \theta_\nu) + K - M^2$$

with the domain consisting of $\phi \in H^1(\Sigma_{\mathrm{p}})$ such that $\Delta_\Sigma\phi$ in the sense of distributions belongs to $L^2(\Sigma_{\mathrm{p}}, d\sigma)$ and satisfies periodic boundary conditions on $\mathcal{B}$. If Σ_{p} is precompact and the curvatures K, M are bounded, the spectrum of $S_\Sigma(\theta)$ is purely discrete for each $\theta \in \mathcal{B}$; we denote the jth eigenvalue, with the multiplicity taken into account, by $\mu_j(\theta)$.

Theorem 10.12 *Under the stated assumptions on the surface* Σ*, the following claims are valid:*
(a) Fix λ *as an arbitrary number if* $r = 3$ *and a non-positive one for* $r = 1, 2$*. To any* $j \in \mathbb{N}$ *there is an* $\alpha_j > 0$ *such that* $H_{\alpha,\Sigma}(\theta)$ *has at least* j *eigenvalues below* λ *for any* $\alpha > \alpha_j$ *and* $\theta \in \mathcal{B}$ *and the* j*th eigenvalue* $\lambda_j(\alpha, \theta)$ *has the expansion*

$$\lambda_j(\alpha, \theta) = -\frac{1}{4}\alpha^2 + \mu_j(\theta) + \mathcal{O}(\alpha^{-1}\ln\alpha)$$

as $\alpha \to \infty$*, where the error term is uniform with respect to* θ*.*
(b) If the set $\sigma(S) := \bigcup_{\theta\in\mathcal{B}} \sigma(S_\Gamma(\theta))$ *has a gap separating a pair of bands, then the same is true for* $\sigma(H_{\alpha,\Gamma})$ *provided* α *is large enough.*

Proof is left to the reader (Problem 16b).

Let us finally mention the strong-coupling asymptotics for planar loops threaded by a magnetic flux which shows a formal similarity with the case of a periodic curve discussed above. The physical interest in such systems is related, in particular, to the existence of ***persistent currents*** in mesoscopic rings. For a charged particle (typically an electron) confined to a loop Γ this effect is manifested by the dependence of the corresponding eigenvalues λ_n on the flux ϕ through the loop, conventionally measured in the units of flux quanta, $2\pi\hbar c|e|^{-1}$; the derivative $\frac{\partial\lambda_n}{\partial\phi}$ then equals $-\frac{1}{c}I_n$, where I_n is the persistent current in the nth state. If the particle is strictly confined to the loop, the eigenvalues in the absence of other than the magnetic potential are easily seen to be proportional to $(n + \phi)^2$ so the currents depend linearly on the applied field. The question is what can be said when the loop is leaky, i.e. the confinement comes from an attractive singular interaction supported by the curve.

Assuming for simplicity that the magnetic field is homogeneous and choosing the circular gauge, $A = \frac{1}{2}B(-x_2, x_1)$, we replace the formal expression (10.1) by

$$H_{\alpha,\Gamma}(B) := (-i\nabla + A)^2 - \alpha\delta(x - \Gamma)$$

in $L^2(\mathbb{R}^2)$; to define it properly we use quadratic form analogous to (6.20),

$$\psi \mapsto \left\| \left(-i\partial_1 - \frac{1}{2}Bx_2\right)\psi \right\|^2 + \left\| \left(-i\partial_2 + \frac{1}{2}Bx_1\right)\psi \right\|^2 - \alpha \int_{\mathbb{R}^2} |(I_m\psi)(x)|^2 \,\mathrm{d}x$$

with the domain $H^1(\mathbb{R}^2)$. It is straightforward to check that the form is closed and below bounded; we identify the (unique) self-adjoint operator associated to it with $H_{\alpha,\Gamma}(B)$. One can employ the same technique as above, bracketing and estimating the operator in the strip neighborhood of Γ using curvilinear coordinates. The thing to be changed is the comparison operator, S_Γ of *Theorem* 10.7 now being replaced by

$$S_\Gamma(B) = -\frac{\mathrm{d}}{\mathrm{d}s^2} - \frac{1}{4}\gamma(s)^2$$

on $L^2(0, L)$ with the domain consisting of $H^2(0, L)$ functions satisfying the boundary condition $\psi(L-) = \mathrm{e}^{iB|\Omega|}\psi(0+)$ and $\psi'(L-) = \mathrm{e}^{iB|\Omega|}\psi'(0+)$, where Ω is the area encircled by the curve Γ. The argument is now analogous to the proof of *Theorem* 10.11 with the quasi-momentum replaced by the magnetic flux. In this way we obtain the following result which establishes, in particular, the existence of persistent currents on a leaky loop for α large enough.

Theorem 10.13 *Let Γ be a C^4-smooth curve without self-intersections. For a fixed $j \in \mathbb{N}$ and a compact interval I we have $\sharp\,\sigma(H_{\alpha,\Gamma}(B)) \geq j$ for $B \in I$ if α is large enough, and the jth eigenvalue behaves in the limit $\alpha \to \infty$ as*

$$\lambda_j(\alpha, B) = -\frac{1}{4}\alpha^2 + \mu_j(B) + \mathcal{O}(\alpha^{-1}\ln\alpha)\,,$$

where $\mu_j(B)$ is the jth eigenvalue of $S_\Gamma(B)$ and the error term is uniform in B. This implies, in particular, that the function $\lambda_j(\alpha, \cdot)$ with a fixed j and α large enough cannot be constant.

10.4 Notes

Section 10.1 Most of the material in this chapter is taken from the review paper [Ex08]. Schrödinger operators with interactions supported by manifolds of a lower dimension were studied first in examples with a particular symmetry [AGS87, Sha88], a more systematical investigation began with the papers [BT92, BEKŠ94].

Note that the operators discussed here may also have other applications. A prominent example comes from studies of high contrast optical systems used to model photonic crystals. The physical interpretation is different in this case, the roles of the coupling and spectral parameters being switched, see [FK96] and the review [Ku01]. Operators of the type (10.1) are also encountered when one deals with contact interactions of several one-dimensional particles [CDR08].

Theorem 10.1 is taken from [EI01], a similar claim for interactions supported by surfaces was demonstrated in [EK03]. The proof of *Proposition* 10.1.1 is based on an abstract analogue to *Theorem* 7.7 due to [Pos01]. It again employs traces of the free resolvent, which is now given by $G_k(x-y) = \frac{e^{ik|x-y|}}{4\pi|x-y|}$, however, one has to interpret the embedding operators not as maps between L^2 spaces, but rather to consider $R_\Gamma^k : H^2(\mathbb{R}^3) \to L^2(\mathbb{R})$ and its counterpart $\breve{R}_\Gamma^k$ as the Banach space dual to $\overline{R_\Gamma^k}$. Then

$$R^k = R_0^k - \breve{R}_\Gamma^k (Q^k - \alpha)^{-1} R_\Gamma^k ,$$

where the operator Q^k is the counterpart to $R_{m,m}^k$ of *Theorem* 6.7 which now cannot be written simply as an integral operator and requires a renormalization analogous to the replacement of boundary values by the generalized ones given by (5.1), cf. [EK02] for more details. The number α can be, of course, replaced by a function which produces a wider class of singular Schrödinger operators.

The approximation described in *Theorem* 10.1 elucidates the meaning of the singular Schrödinger operators considered here but it does not help much when we want to find spectral properties of a particular operator $H_{\alpha,\Gamma}$. This can be done using another approximation using a family of point-interaction Hamiltonians; recall that if the number of the δ-potentials is finite the spectral analysis reduces by means of Krein's formula to an algebraic problem. One replaces the curve Γ by arrays of two-dimensional point interactions in such a way that their distances tend to zero and the coupling parameters are inversely proportional to the distances. This may look strange, but one has to keep in mind that the coupling described by the boundary conditions (5.2) becomes *weaker* as α increases. In this way one can construct operators approximating $H_{\alpha,\Gamma}$ in the strong resolvent sense [EN03b, Ož06] and similar approximations to δ-interaction supported by surfaces [BFT98]; moreover, a norm resolvent convergence can be achieved if a Δ^2 type perturbation is added [BO07].

Section 10.2 *Theorem* 10.2 comes from [EI01], the analogous result for curves in $\mathbb{R}^3$ is due to [EK02]. Note that in both cases assumptions *(i)* and *(ii)* are sufficient to the establish the existence of a geometrically induced spectrum. Its discreteness requires an asymptotic straightness of Γ; in case of *Theorem* 10.3 a local smoothness is also needed due to the more strongly singular character of the interaction. The discussion of leaky star graphs in Example 10.2.1 comes from [EN01], see also [EN03b]. *Theorem* 10.4 describing scattering for a locally deformed Γ is taken from [EK05].

Spectral properties of operators $H_{\alpha,\Gamma}$ in $L^2(\mathbb{R}^d)$ for a compact Γ of codimension one are discussed in [BEKŠ94, BLL13]. The fact that $H_{\alpha,\Gamma_\varepsilon}$ has for small ε the same number of eigenvalues as $H_{\alpha,\Gamma}$ and that $\lambda_j(\varepsilon) \to \lambda_j(0)$ holds as $\varepsilon \to 0$ follows from the convergence of the corresponding quadratic forms on $H^1(\mathbb{R}^d)$, cf. [Ka, Theorem VIII.3.15]. To prove *Theorem* 10.5, however, one cannot employ Kato's perturbation theory of quadratic forms due to the strongly singular character of the perturbation. Indeed, we have $t_{-\alpha m_{\Gamma_\varepsilon}}[\psi] = t_{-\alpha m_\Gamma}[\psi] + \alpha\, m_\Gamma(\mathcal{P}_\varepsilon)|\psi(0)|^2 + \mathcal{O}(\varepsilon^d)$ as $\varepsilon \to 0$ for $\psi \in C_0^\infty(\mathbb{R}^d)$ and the quadratic form $\psi \mapsto |\psi(0)|^2$ does not extend from $C_0^\infty(\mathbb{R}^d)$ to a bounded form on $H^1(\mathbb{R}^d)$, because the set of $\psi \in C_0^\infty(\mathbb{R}^d)$ vanishing at the origin is dense in $H^1(\mathbb{R}^d)$. A way to eliminate this difficulty is to use the compactness of the map $H^1(\mathbb{R}^d) \ni f \mapsto f|_\Gamma \in L^2(\Gamma)$; the argument is worked out in [EY03]. Note that there are other ways to derive such asymptotic expansions, for instance, the technique of matching of asymptotic expansions [Il] where, however, additional assumption are required such as self-similarity properties of the family of shrinking sets $\mathcal{P}_\varepsilon$. Numerical results illustrating such geometric perturbations for curves in the plane were worked out in [ET04]. The asymptotic expansion for a curve with a hiatus in $\mathbb{R}^3$ mentioned at the end of Sect. 10.2.2 was derived in [EK08].

Theorem 10.6 comes from [EHL06]. The geometric result of *Lemma* 10.2.1 on which the proof is based can be demonstrated in alternative ways [Lü66, ACF03]; a proof of local validity of the isoperimetric inequality can be found in [Ex05]. Similarly as in Problem 8, there is a discrete analogue concerning the ground state of a Hamiltonian with N point interactions in $\mathbb{R}^d$, $d = 2, 3$, arranged like beads on a necklace with a uniform bound of the neighbor distances: a sharp maximum is now achieved if they are placed at vertices of a regular planar polygon [Ex06]. Another partial analogue of *Theorem* 10.6 concerns δ-interactions supported by a surface $\Gamma \subset \mathbb{R}^3$. If the latter is a sphere and the coupling constant α is such that the corresponding operator $H_{\alpha,\Gamma}$ is critical, then any small area-preserving deformation of Γ gives rise to a non-void discrete spectrum [EFr09]. This result holds only locally, however, there are large deformations which do not produce any eigenvalues.

Section 10.3 Most of *Theorem* 10.7 comes from [EY02a], the claim concerning finite arcs has been demonstrated in [EPa14]. One can also conjecture that in the other cases discussed here the strong-coupling asymptotical behavior for interaction supported by manifolds with a boundary will be expressed in terms of the appropriate operator with Dirichlet boundary conditions. *Theorem* 10.8 and *Corollary* 10.3.1 were demonstrated in [EK04], *Theorems* 10.9 and 10.10 come from [EK03]. The strong coupling behavior in the two-dimensional situation when the coupling strength α in (10.1) is non-constant along the curve has been discussed in [Kon13].

Theorem 10.11 comes from [EY01], its three-dimensional analogue in Problem 16b was demonstrated in [EK04]. Note that the hypotheses about matching the period cells of the curve and the space can be weakened; what we really need is a complete "tiling" of $\mathbb{R}^d$ by domains with piecewise smooth boundaries. In the case $d = 3$ such "bricks" need not even be simply connected: remember what your grandmother was doing with her crochet to get an example of a curve which is topologically inequivalent to a line, or in other words, such that you cannot disentangle it by any local deformation—you can only unwind it by "pulling the ends".

Corollary 10.3.2 was also proved in [EY01] and [EK04], respectively. *Corollary* 10.3.3 comes from [BDE03]. To appreciate this result recall that if the period cell of $H_{\alpha,\Gamma}$ is compact there is a way to establish the (global) absolute continuity of such operators [BSŠ00, SŠ01]. The case $n < d$ is more difficult; the global absolute continuity has been so far demonstrated only in the related case of a straight Γ supporting a singular interaction with the strength α periodically modulated [EF07].

Theorem 10.12 was proved in [Ex03], for *Theorem* 10.13 see [EY02b]. Note that the last named result does not require the magnetic field to be homogeneous. The important quantity is the magnetic flux through the loop, the field shape influences the error term but not the first two terms of the expansion. In particular, the case of an Aharonov-Bohm flux line was worked out in [HH04]. Numerical results for curves Γ in the plane illustrating the strong coupling limit in both the nonmagnetic and magnetic cases can be found in [ET04].

Interpretation of the strong coupling results in this section can also be viewed from another angle. The spectral properties of $H_{\alpha,\Gamma}$ can be related to quantum tunneling, hence they should be sensitive to the Planck's constant if we reintroduce it into the picture. In the case of codim $\Gamma = 1$, for instance, the operator $-h^2\Delta - v\delta(x-\Gamma)$ is the h^2 multiple of (10.1) with the coupling constant $\alpha := vh^{-2}$; in this sense therefore one can regard the obtained asymptotic formulæ as a semiclassical approximation, in particular, the leading term in $\sharp\sigma_{\mathrm{disc}}(H_{\alpha,\Gamma})$ is the usual Weyl expression. With *Theorem* 10.1 in mind, we can regard the asymptotic behavior discussed here also as a counterpart to the shrinking potential channels mentioned in the notes to Sect. 1.1.

10.5 Problems

1. Check that the measure (10.2) belongs to the generalized Kato class.
Hint: Cf. [BEKŠ94, Theorem IV.1].

2. (a) The closure of $\dot{H}_{\alpha,\Gamma}$ defined by the conditions (10.3) coincides with $H_{\alpha,\Gamma}$.
(b) Extend this identification to the case where α is a bounded Borel function.
(c) Do the same in the situation when the curve Γ is replaced by a smooth hypersurface $\Sigma \subset \mathbb{R}^d$, $d > 2$, of codimension one.
Hint: Use Green's formula and a suitable core, e.g. $C_0^\infty(\mathbb{R}^d)$.

3. Using the notation introduced in Sect. 6.3.1 prove that an eigenfunction of $H_{\alpha,\Gamma}$ associated with an eigenvalue k^2 can be expressed as $\psi(x) = \int_0^L R^k_{\mathrm{dx},m}(x,s)\phi(s)\,\mathrm{d}s$, where ϕ is the corresponding eigenfunction of $\alpha R^k_{m,m}$ with eigenvalue one.

4. (a) Fill in the details of the proof of *Theorem* 10.1. Extend the claim to operators (10.1) with a non-constant coupling strength α using a family of bounded potentials $W : \Sigma_0^1 \to \mathbb{R}$ such that $\alpha(x) = \int_{-1}^1 W(x,y)\,\mathrm{d}y$.
(b) Prove an analogous claim for the operator $H_{\alpha,\Sigma}$ with the δ-interaction supported by a smooth surface $\Sigma \subset \mathbb{R}^3$.
Hint: Cf. [EI01, Appendix A] and [EK03].

5. (a) Fill in the details of the proof of *Theorem* 10.2.

(b) Suppose that $\Gamma \in C^2$, then condition *(iii)* of Sect. 10.2 with $\tau > \frac{1}{2}$ is valid provided the curvature of Γ satisfies $\gamma(s) = \mathcal{O}(|s|^{-\beta})$ with $\beta > \frac{5}{4}$ as $|s| \to \infty$.
(c) Let $\Gamma \subset \mathbb{R}^2$ be a graph which outside a compact consists of a finite number of straight halflines separated by wedges of nonzero angles, then $\sigma_{\mathrm{ess}}(H_{\alpha,\Gamma}) = [-\frac{1}{4}\alpha^2, \infty)$.
Hint: Cf. [EI01, Sect. 5].

6. Prove that the bound $\inf \sigma(H_2(\beta)) \geq -\alpha^2 \left(1 + \sin(\frac{\beta}{2})\right)^{-2}$ holds for the operator $H_2(\beta)$ of Example 10.2.1. Similarly, for $H_4(\tilde{\beta})$ describing a leaky star of two crossing lines, $\tilde{\beta} = \{\beta, \pi - \beta, \beta\}$, we have $\inf \sigma(H_4(\tilde{\beta})) \geq -\alpha^2(1 + \sin\beta)^{-1}$.
Hint: Use properties of Sobolev spaces on wedges, cf. [LP08] and [Lo13].

7. Prove *Theorem* 10.3.
Hint: Use the generalized BS principle, cf. notes to Sect. 10.1 and [EK02].

8. Let $H_{\alpha,Y}$ be the Hamiltonian of a particle interacting with a chain of point potentials in $\mathbb{R}^d$, $d = 2, 3$, all of the same strength $\alpha \in \mathbb{R}$, defined by the conditions of the type (5.2). The interaction support $Y = \{y_j\}_{j\in\mathbb{Z}}$ is such that $|y_j - y_{j+1}| = \ell$ for a fixed $\ell > 0$. Furthermore, there is a $c \in (0, 1)$ such that $|y_j - y_{j'}| \geq c\,\ell\,|j - j'|$, and $d_0 > 0$, $\tau > \frac{1}{2}$, and $\omega \in (0, 1)$ such that

$$1 - \frac{|y_j - y_{j'}|}{|j - j'|} \leq d_0 \left[1 + |j + j'|^{2\tau}\right]^{-1/2}.$$

holds in the sector $S_\omega := \{(j, j') : j, j' \neq 0' \omega < j/j' < \omega^{-1}\} \in \mathbb{Z}^2$. The essential spectrum of $H_{\alpha,Y}$ consists of two absolutely continuous bands, possibly overlapping. If the chain is not straight, $|y_j - y_{j'}| < \ell\,|j - j'|$ for some $j, j' \in \mathbb{Z}$, the operator has at least one isolated eigenvalue below $\inf \sigma_{\mathrm{ess}}(H_{\alpha,Y})$.
Hint: For the straight-chain spectrum see [AGHH]; using Krein's formula, mimick then the proof of *Theorem* 10.2, cf. [Ex01].

9. Prove *Proposition* 10.2.1. and show that $\sigma_{\mathrm{ess}}(H_{\alpha,\Gamma}) = \sigma_{\mathrm{ess}}(H_{\alpha,\Gamma_0}) = [-\frac{1}{4}\alpha^2, \infty)$.
Hint: To prove the last claim, check that $R^k_{\Gamma_0,\nu}$ is Hilbert-Schmidt under the given assumptions and the other two factors are bounded.

10. Fill in the details of the proof of *Theorem* 10.4.
Hint: Use Kuroda-Birman theorem to prove (a). To check that $B^{i\kappa}$ is trace class, find a two-sided estimate of $\Theta^{i\kappa}_\alpha$ by suitable sign-definite integral operators, cf. [BT92]. For the asymptotics computation see [EK02, EK05].

11. Given a smooth, piecewise C^2 loop $\Gamma : [0, L] \to \mathbb{R}^d$, $d \geq 2$, with $\dot{\Gamma}(s) = 1$, consider the following inequalities labeled as $C^p_L(u)$ with $p \in \mathbb{R}$ and $u > 0$,

$$\operatorname{sgn} p \int_0^L |\Gamma(s+u) - \Gamma(s)|^p \, ds \leq \operatorname{sgn} p \, \frac{L^{1+p}}{\pi^p} \sin^p \frac{\pi u}{L}.$$

Prove the following claims:
(a) $C^p_L(u)$ implies $C^{p'}_L(u)$ if $p \geq p' > 0$.

(b) $C_L^p(u)$ implies $C_L^{-p}(u)$ for any $p > 0$.
(c) $C_L^2(u)$ holds for any $u \in (0, \frac{1}{2}L]$ with strict inequality unless Γ is a planar circle.
Hint: Use convexity to prove the claims (a) and (b). As for (c), put $L = 2\pi$ and write $\Gamma(s) = \sum_{0\neq n\in\mathbb{Z}} c_n e^{ins}$ with $c_{-n} = \bar{c}_n$. Use $\dot{\Gamma}(s) = 1$ and Parseval's identity to reduce the problem to checking the inequality $|\sin nx| \leq n \sin x$ for any positive integer n and all $x \in (0, \frac{1}{2}\pi]$, cf. [EHL06].

12. Prove the inequalities (10.15).

13. Fill in the details of the proofs of *Theorem* 10.8 and *Corollary* 10.2.1.
Hint: To estimate the 3D transverse part, replace (10.15) by ground-state bounds for a point interaction in the center of the disc with Dirichlet and Neumann boundary.

14. Fill in the details of the proof of *Theorem* 10.9.
Hint: For the spectral analysis of $-\Delta_\Sigma$ see [Ch99].

15. Prove *Theorem* 10.10. Show that the assumptions (i) and (iii) are satisfied if the normal vector to Σ satisfies $n \to n_0$ as the geodesic radius $r \to \infty$, where n_0 is a fixed vector.

16. (a) Prove the three-dimensional analogue of *Theorem* 10.11: Let Γ be a periodic curve, without self-intersections and with the global Frenet frame, given by a C^4 smooth function $\Gamma : \mathbb{R} \to \mathbb{R}^3$. Suppose, in addition, that the period cells Γ_p of Γ and Λ referring to the corresponding operator $H_{\alpha,\Gamma}$ match in the sense that $\Gamma_\mathrm{p} = \Gamma \cap \Lambda$. Then $\sigma_\mathrm{disc}(H_{\alpha,\Gamma}(\theta))$ is as in *Theorem* 10.11, with the jth eigenvalue of $H_{\alpha,\Gamma}(\theta)$ having the asymptotic expansion

$$\lambda_j(\alpha, \theta) = \zeta_\alpha + \mu_j(\theta) + \mathcal{O}(e^{\pi\alpha}) \quad \text{as} \quad \alpha \to -\infty\,,$$

where $\mu_j(\theta)$ is the jth eigenvalue of $S_\Gamma(\theta)$ and the error is uniform w.r.t. $\theta \in \mathcal{B}$.
(b) Prove *Theorem* 10.12.
Hint: (b) Cf. [Ex03].

Appendix A
Coda

The road goes ever on and on, down from the door where it began.

J.R.R. Tolkien, The Hobbit

Time came to stop. As good Dyson frogs we have explored the pool discovered in the late eighties and collected various beautiful water lilies we found on the way. Some pools are big, though. It is not only that there are numerous ways in which the results of this treatise can be improved technically, more important is the existence of wide areas and deep questions inviting to be explored.

The most prominent among them is the many-body theory of quantum waveguides. We have briefly touched the subject in Sect. 3.3, but this was just a small foray into a mostly unchartered territory. Needles to say, the problem has a strong physical motivation because most effects investigated in waveguide structures have a many-body character. The challenge includes both a true quantum mechanical description of interacting particle ensembles confined to tubular and more complicated regions as well as effective theories, and one should not also forget about quantum-field aspects of this problem.

The focus of this book was on nonrelativistic quantum dynamics governed by Schrödinger equation. Leaving aside the much older and well studied subject of electromagnetic waveguides, one may wonder whether other equations are of interest. The recent discovery of graphene, in which electron behavior is described by a massless Dirac equation, shows that one's mind should be kept open. Graphene ribbons as waveguides with boundary conditions depending on the way they are cut from a sheet offer various open questions.

A lot remains to be learned about time evolution of such systems. We have discussed here mostly bound and scattering states, simple from that point of view. It is known, however, that presence of Dirichlet boundaries can force wave packets to behave in unusual ways. Moreover, classical hard-wall tubes with local geometric perturbations can exhibit mixed phase space picture which should be reflected in the behavior of their quantum counterparts.

P. Exner and H. Kovařík, *Quantum Waveguides*,
Theoretical and Mathematical Physics, DOI 10.1007/978-3-319-18576-7

Another subject we touched only lightly upon through the example discussed in Sect. 9.3 was the behavior of waveguide systems under random perturbations, especially those of geometric nature. A particularly deep aspect of the problem is associated with the "mixed dimensionality" of such objects. It is known that randomness leads to localization at all energies in one-dimensional systems, while existence of a mobility edge is expected (albeit not proven in general) in higher dimensions; it is not easy to predict what will happen with the motion infinitely extended in one direction and confined in the others.

The mixed dimensionality of waveguides also plays a role in another open problem, namely the validity of Bethe-Sommerfeld conjecture which we have mentioned in the notes to Sect. 9.1. One expects that spectra of waveguides with periodic perturbations of the classes considered here will exhibit only finite numbers of open gaps, but this fact remains to be demonstrated.

Our discussion of electric and magnetic effects is quantum waveguides presented in Chap. 7 leaves various questions open. Let us mention two of them. Once a homogeneous electric field is not perpendicular to the waveguide outside a compact a variety of Stark effect situations can occur which are worth exploration. Another problem concerns magnetic waveguides: one can ask whether a strong enough magnetic field can destroy the geometrically induced bound states.

While for squeezed Neumann networks we have a prescription which allows us to approximate any self-adjoint vertex coupling, the Dirichlet case is less well understood. We know that some nontrivial couplings can be obtained using threshold resonances, however, it is not clear how far one can extend the approximation procedure in this case. And nothing is known about the related problem of strong-coupling limit for leaky graphs with nontrivial branchings.

And this, of course, is not all ...

Bibliography

Today, we are neglecting the theory of solids, in which a student has to study perhaps six hundred papers before he reaches the frontier and can do research on his own, we concentrate instead on quantum electrodynamics, in which he has to study six papers.

Eugene P. Wigner, 1950

[AS] Abramowitz, M.S., Stegun, I.A. (eds.): Handbook of Mathematical Functions. Dover, New York (1965)

[ACF03] Abrams, A., Cantarella, J., Fu, J.G., Ghomi, M., Howard, R.: Circles minimize most knot energies. Topology **42**, 381–394 (2003)

[Ad] Adams, R.A.: Sobolev Spaces. Academic Press, New York (1975)

[Ag82] Agmon, S.: Lectures on Exponential Decay of Solutions of Second-Order Elliptic Equations. Princeton University Press, Princeton (1982)

[AC71] Aguilar, J., Combes, J.-M.: A class of analytic perturbation for one-body Schrödinger Hamiltonians. Commun. Math. Phys. **22**, 262–269 (1971)

[AS00] Aizenman, M., Schenker, J.H.: The creation of spectral gaps by graph decorations. Lett. Math. Phys. **53**, 253–279 (2000)

[ASW06] Aizenman, M., Sims, R., Warzel, S.: Absolutely continuous spectra of quantum tree graphs with weak disorder. Commun. Math. Phys. **264**, 371–389 (2006)

[AG] Akhiezer, N.I., Glazman, I.M.: Theory of Linear Operators in Hilbert Space, 3rd edn. Viša Škola, Kharkov (1978). (in Russian. English translation of the first edition: F. Ungar Co., New York 1961, 1963)

[AVD95] Akis, R., Vasilopoulos, P., Debray, P.: Ballistic transport in electron stub tuners: shape and temperature dependence, tuning of the conductance output, and resonant tunneling. Phys. Rev. B **52**, 2805–2813 (1995)

[AEu90] del Alamo, J.A., Eugster, C.C.: Quantum field-effect directional coupler. Appl. Phys. Lett. **56**, 78–80 (1990)

[ACF07] Albeverio, S.A., Cacciapuoti, C., Finco, D.: Coupling in the singular limit of thin quantum waveguides. J. Math. Phys. **48**, 032103 (2007)

P. Exner and H. Kovařík, *Quantum Waveguides*,
Theoretical and Mathematical Physics, DOI 10.1007/978-3-319-18576-7

[AGHH] Albeverio, S., Gesztesy, F., Høegh-Krohn, R., Holden, H.: Solvable Models in Quantum Mechanics, 2nd edn with an Appendix by P. Exner. AMS Chelsea Publishing, Providence, R.I. (2005)

[AK] Albeverio, S., Kurasov, P.: Singular Perturbations of Differential Operators. Cambridge University Press, Cambridge (2000)

[AN00] Albeverio, S., Nizhnik, L.: Approximation of general zero-range potentials. Ukrainian Math. J. **52**, 582–589 (2000)

[Al83] Alexander, S.: Superconductivity of networks. A percolation approach to the effects of disorder. Phys. Rev. B **27**, 1541–1557 (1983)

[ART12] Amore, P., Rodriguez, M., Terrero-Escalante, C.A.: Bound states in open-coupled asymmetrical waveguides and quantum wires. J. Phys. A: Math. Theor. **45**, 105303 (2012)

[ABG] Amrein, W., Boutet de Monvel, A., Georgescu, V.: C_0-Groups. Commutator Methods and Spectral Theory of N-Body Hamiltonians. Birkhäuser, Basel (1996)

[AGS87] Antoine, J.-P., Gesztesy, F., Shabani, J.: Exactly solvable models of sphere interactions in quantum mechanics. J. Phys. A: Math. Gen. **20**, 3687–3712 (1987)

[AB92a] Ashbaugh, M.S., Benguria, R.D.: A sharp bound for the ratio of the first two eigenvalues of Dirichlet Laplacians and extensions. Ann. Math. **135**, 601–628 (1992)

[AB92b] Ashbaugh, M.S., Benguria, R.D.: A second proof of the Payne-Pólya-Weinberger conjecture. Commun. Math. Phys. **147**, 181–190 (1992)

[AE90] Ashbaugh, M.S., Exner, P.: Lower bounds to bound state energies in bent tubes. Phys. Lett. A **150**, 183–186 (1990)

[APV00] Aslanyan, A., Parnovski, L., Vassiliev, D.: Complex resonances in acoustic waveguides. Quart. J. Mech. Appl. Math. **53**, 429–447 (2000)

[AJ09] Avila, A., Jitomirskaya, S.: The Ten Martini problem. Ann. Math. **170**, 303–342 (2009)

[ABGM91] Avishai, Y., Bessis, D., Giraud, B.M., Mantica, G.: Quantum bound states in open geometries. Phys. Rev. B **44**, 8028–8034 (1991)

[Ba90] Bagwell, P.F.: Evanescent modes and scattering in quasi-one-dimensional wires. Phys. Rev. B **41**, 10354–10371 (1990)

[BPB09] Band, R., Parzanchevski, O., Ben-Shach, G.: The isospectral fruits of representation theory: quantum graphs and drums. J. Phys. A: Math. Theor. **42**, 175202 (2008)

[Barg52] Bargmann, V.: On the number of bound states in a central field of force. Proc. Nat. Acad. Sci. U.S.A. **38**, 961–966 (1952)

[BD79] Baumgärtel, H., Demuth, M.: Decoupling by a projection. Rep. Math. Phys. **15**, 173–186 (1979)

[BLL13] Behrndt, J., Langer, M., Lotoreichik, V.: Schrödinger operators with δ and δ'-potentials supported on hypersurfaces. Ann. H. Poincaré **14**, 385–423 (2013)

[BCEZ99] Bentosela, F., Cavalcanti, R.M., Exner, P., Zagrebnov, V.A.: Anomalous electron trapping by localized magnetic fields. J. Phys. A **32**, 3029–3039 (1999)

[BDE03] Bentosela, F., Duclos, P., Exner, P.: Absolute continuity in periodic thin tubes and strongly coupled leaky wires. Lett. Math. Phys. **65**, 75–82 (2003)

[BF61] Berezin, F.A., Faddeev, L.D.: A remark on Schrödinger equation with a singular potential. Sov. Acad. Sci. Doklady **137**, 1011–1014 (1961)

[BSh91] Berezin, F., Shubin, M.: The Schrödinger equation, Mathematics and its Applications (Soviet Series), vol. 66. Kluwer Academic Publishers Group, Dordrecht (1991)

[BCFK] Berkolaiko, G., Carlson, R., Fulling, S., Kuchment, P. (eds.): Quantum Graphs and Their Applications, Contemporary Mathematics, vol. 415. American Mathematical Society, Providence (2006)

[BK] Berkolaiko, G., Kuchment, P.: Introduction to Quantum Graphs. American Mathematical Society, Providence (2013)

[Bi61] Birman, M.S.: The spectrum of singular boundary problems. Math. Sbornik **55**, 124–174 (1961)

[BiSo] Birman, M.S., Solomyak, M.Z.: Spectral Theory of Self-Adjoint Operators in Hilbert-Spaces. Reidel Publishing Company, Dordrecht (1987)

[BiSu97] Birman, M.S., Suslina, T.: Two-dimensional periodic magnetic Hamiltonian is absolutely continuous. Algebra i Analiz **9**, 32–48 (1997) (in Russian) (Engl. Trans. St Petersburg Math. J. **9**, 21–32 (1998))

[BSŠ00] Birman, M.S., Suslina, T.A., Shterenberg, R.G.: Absolute continuity of the two-dimensional Schrödinger operator with delta potential concentrated on a periodic system of curves, Algebra i Analiz **12**, 140–177 (2000); English translation in St. Petersburg Math. J. **12**, 535–567 (2001)

[BDM13] Bittner, S., Dietz, B., Miski-Oglu, M., Richter, A., Ripp, C., Sadurní, E., Schleich, W.P.: Bound states in sharply bent waveguides: analytical and experimental approach. Phys. Rev. E **87**, 042912 (2013)

[BEH] Blank, J., Exner, P., Havlíček, M.: Hilbert Space Operators in Quantum Physics. American Institute of Physics, New York (1994) [2nd extended edn. Springer, Dordrecht (2008)]

[BGS77] Blanckenbecler, R., Goldberger, M.L., Simon, B.: The bound states of weakly coupled long-range one-dimensional quantum Hamiltonians. Ann. Phys. **108**, 89–78 (1977)

[BGW85] Bollé, D., Gesztesy, F., Wilk, S.F.J.: A complete treatment of low energy scattering in one dimension. J. Oper. Theory **13**, 3–32 (1985)

[BE09] Bolte, J., Endres, S.: The trace formula for quantum graphs with general self-adjoint boundary conditions. Ann. Henri Poincaré **10**, 189–223 (2009)

[BH03] Bolte, J., Harrison, J.: Spectral statistics for the Dirac operator on graphs. J. Phys. A: Math. Gen. **36**, 2747–2769 (2003)

[BK05] Boman, J., Kurasov, P.: Symmetries of quantum graphs and the inverse scattering problem. J. Adv. Appl. Math. **35**, 58–70 (2005)

[Bo46] Borg, G.: Eine Umkehrung der Sturm-Liouvillschen Eigenwertaufgabe. Bestimmung der Differentialgleichung durch die Eigenwerte. Acta. Math. **78**, 1–96 (1946)

[Bor07] Borisov, D.: The spectrum of two quantum layers coupled by a window. J. Phys. A: Math. Theor. **40**, 5045–5066 (2007)

[BBC10] Borisov, D., Bunoiu, R., Cardone, G.: On a waveguide with frequently alternating boundary conditions: homogenized Neumann condition. Ann. Henri Poincaré **11**, 1591–1627 (2010)

[BBC11a] Borisov, D., Bunoiu, R., Cardone, G.: On a waveguide with an infinite number of small windows. C. R. Math. Acad. Sci. Paris **349**, 53–56 (2011)

[BBC11b] Borisov, D., Bunoiu, R., Cardone, G.: Homogenization and asymptotics for a waveguide with an infinite number of closely located small windows. J. Math. Sci. **176**, 774–785 (2011)

[BC11] Borisov, D., Cardone, G.: Planar waveguide with "twisted" boundary conditions: discrete spectrum. J. Math. Phys. **52**, 123513 (2011)

[BC12] Borisov, D., Cardone, G.: Planar waveguide with "twisted" boundary conditions: small width. J. Math. Phys. **5**, 023503 (2012)

[BEK05] Borisov, D., Ekholm, T., Kovařík, H.: Spectrum of the magnetic Schrödinger operator in a waveguide with combined boundary conditions. Ann. Henri Poincaré **6**, 327–342 (2005)

[BE04] Borisov, D., Exner, P.: Exponential splitting of bound states in a waveguide with a pair of distant windows. J. Phys. A: Math. Gen. **37**, 3411–3428 (2004)

[BE07] Borisov, D., Exner, P.: Distant perturbation asymptotics in window-coupled waveguides. I. The non-threshold case. J. Math. Phys. **47**, 113502 (2007)

[BEGK01] Borisov, D., Exner, P., Gadyl'shin, R., Krejčiřík, D.: Bound states in weakly deformed strips and layers. Ann. Henri Poincaré **2**, 553–572 (2001)

[BEG13] Borisov, D., Exner, P., Golovina, A.: Tunneling resonances in systems without a classical trapping. J. Math. Phys. **54**, 012102 (2013)

[BP13] Borisov, D., Pankrashkin, K.: Gap opening and split band edges for waveguides coupled by a periodic system of small windows. Math. Notes **93**, 660–675 (2013)

[BMT07] Bouchitté, G., Mascarenhas, M.L., Trabucho, L.: On the curvature and torsion effects in one dimensional waveguides. ESAIM: Control Optim. Calc. Var. 13, 793–808 (2007)

[BMT12] Bouchitté, G., Mascarenhas, M.L., Trabucho, L.: Thin waveguides with Robin boundary conditions. J. Math. Phys. **53**, 123517 (2012)

[BEKŠ94] Brasche, J.F., Exner, P., Kuperin, Y.A., Šeba, P.: Schrödinger operators with singular interactions. J. Math. Anal. Appl. **184**, 112–139 (1994)

[BFT98] Brasche, J.F., Figari, R., Teta, A.: Singular Schrödinger operators as limits of point interaction Hamiltonians. Potential Anal. **8**, 163–178 (1998)

[BO07] Brasche, J.F., Ožanová, K.: Convergence of Schrödinger operators. SIAM J. Math. Anal. **39**, 281–297 (2007)

[BT92] Brasche, J.F., Teta, A.: Spectral analysis and scattering for Schrödinger operators with an interaction supported by a regular curve. In: Albeverio, S., Fenstadt, J.E., Holden, H., Lindstrøm, T. (eds.) Ideas and Methods in Quantum and Statistical Physics, pp. 197–211. Cambridge University Press, Cambridge (1992)

[BF09] Breuer, J., Frank, R.L.: Singular spectrum for radial trees. Rev. Math. Phys. **21**, 1–17 (2009)

[BKR14] Briet, Ph, Kovařík, H., Raikov, G.: Scattering in twisted waveguides. J. Funct. Anal. **266**, 1–35 (2014)

[BKRS09] Briet, Ph, Kovařík, H., Raikov, G., Soccorsi, E.: Eigenvalue asymptotics in a twisted waveguide. Commun. PDE **34**, 818–836 (2009)

[BRS08] Briet, Ph, Raikov, G., Soccorsi, E.: Spectral properties of a magnetic quantum Hamiltonian on a strip. Asymptot. Anal. **58**, 127–155 (2008)

[BG06] Buchendorfer, Ch., Graf, G.M.: Scattering of magnetic edge states. Ann. Henri Poincaré **7**, 303–333 (2006)

[BEPS02] Bulgakov, E.N., Exner, P., Pichugin, K.N., Sadreev, A.F.: Multiple bound states in scissor-shaped waveguides. Phys. Rev. B **66**, 155109 (2002)

[BG85] Bulla, W., Gesztesy, F.: Deficiency indices and singular boundary conditions in quantum mechanics. J. Math. Phys. **26**, 2520–2528 (1985)

[BGRS97] Bulla, W., Gesztesy, F., Renger, W., Simon, B.: Weakly coupled bound states in quantum waveguides. Proc. Am. Math. Soc. **127**, 1487–1495 (1997)

[BT90] Bulla, W., Trenckler, T.: The free Dirac operator on compact and non-compact graphs. J. Math. Phys. **31**, 1157–1163 (1990)

[Bü88] Büttiker, M.: Absence of backscattering in the quantum Hall effect in multiprobe conductors. Phys. Rev. B **38**, 9375–9389 (1988)

[CaE07] Cacciapuoti, C., Exner, P.: Nontrivial edge coupling from a Dirichlet network squeezing: the case of a bent waveguide. J. Phys. A: Math. Theor. **40**, F511–F523 (2007)

[CDN10] Cardone, G., Durante, T., Nazarov, S.A.: The localization effect for eigenfunctions of the mixed boundary value problem in a thin cylinder with distorted ends. SIAM J. Math. Anal. **42**, 2581–2609 (2010)

[CNR12] Cardone, G., Nazarov, S.A., Ruotsalainen, K.M.: Asymptotics of an eigenvalue in the continuous spectrum of a narrowing waveguide. Mat. Sbornik **203**, 3–32 (2012); English translation in Sbornik Math. **203**, 153–182 (2012)

[CNR13] Cardone, G., Nazarov, S.A., Ruotsalainen, K.M.: Bound states of a converging quantum waveguide. ESAIM Math. Model. Numer. Anal. **47**, 305–315 (2013)

[CLM92] Carini, J.P., Londergan, J.T., Mullen, K., Murdock, D.P.: Bound states and resonances in quantum wires. Phys. Rev. B **46**, 15538–15541 (1992)

[CLM93] Carini, J.P., Londergan, J.T., Mullen, K., Murdock, D.P.: Multiple bound states in sharply bent waveguides. Phys. Rev. B **48**, 4503–4514 (1993)

[CLMT97] Carini, J.P., Londergan, J.T., Murdock, D.P., Trinkle, D., Yung, C.S.: Bound states in waveguides and bent quantum wires, I. Applications to waveguide systems. Phys. Rev. B **55**, 9842–9851 (1997)

[CL] Carmona, R., Lacroix, J.: Spectral Theory of Random Schrödinger Operators. Probability and Its Applications, Birkhäuser (1990)

[CEK04] Carron, G., Exner, P., Krejčiřík, D.: Topologically non-trivial quantum layers. J. Math. Phys. **45**, 774–784 (2004)

[Ca97] Cattaneo, C.: The spectrum of the continuous Laplacian on a graph. Monatsh. Math. **124**, 215–235 (1997)

[Ch99] Chavel, I.: The Laplacian on Riemannian manifolds. In: Proceedings of ICMS Instructional Conference Spectral Theory and Geometry, pp. 30–75. Cambridge University Press, Cambridge (1999)

[CBC92] Chaudhury, S., Bandyopadhyay, S., Cahay, M.: Spatial distribution of current and Fermi carriers around localized elastic scatterers in quantum transport. Phys. Rev. B **45**, 11126–11135 (1992)

[CMV06] Chen, K., Molchanov, S., Vainberg, B.: Localization on Avron-Exner-Last graphs, I. Local perturbations, in [BCFK], pp. 81–91

[CDFK05] Chenaud, B., Duclos, P., Freitas, P., Krejčiřík, D.: Geometrically induced discrete spectrum in curved tubes. Diff. Geom. Appl. **22**, 95–105 (2005)

[CE04] Cheon, T., Exner, P.: An approximation to δ' couplings on graphs. J. Phys. A: Math. Gen. **37**, L329–L335 (2004)

[CET10] Cheon, T., Exner, P., Turek, O.: Approximation of a general singular vertex coupling in quantum graphs. Ann. Phys. **325**, 548–578 (2010)

[CS98] Cheon, T., Shigehara, T.: Realizing discontinuous wave functions with renormalized short-range potentials. Phys. Lett. A **243**, 111–116 (1998)

[CBr96] Clark, I.J., Bracken, A.J.: Effective potentials of quantum strip waveguide and their dependence upon torsion. J. Phys. A: Math. Gen. **29**, 339–348 (1996)

[CVo35] Cohn-Vossen, S.: Kürzeste Wege und Totalkrümmung auf Flächen. Compos. Math. **2**, 69–133 (1935)

[CDR08] Cornean, H.D., Duclos, P., Ricaud, B.: On the skeleton method and an application to a quantum scissor. In: [EKKST], pp. 657–672

[CJM05] Cornean, H., Jensen, A., Moldoveanu, V.: A rigorous proof for the Landauer-Büttiker formula. J. Math. Phys. **46**, 042106 (2005)

[dC81] da Costa, R.C.T.: Quantum mechanics of a constrained particle. Phys. Rev. A **23**, 1982–1987 (1981)

[dC82] da Costa, R.C.T.: Constraints in quantum mechanics. Phys. Rev. A **25**, 2893–2900 (1982)

[dC83] da Costa, R.C.T.: Constrained particles in quantum mechanics. Lett. N. Cim. **36**, 393–396 (1983)

[CFKS] Cycon, H.L., Froese, R.G., Kirsch, W., Simon, B.: Schrödinger Operators, with Applications to Quantum Mechanics and Global Geometry. Springer, Berlin (1987)

[DR12] Dauge, M., Raymond, N.: Plane waveguides with corners in the small angle limit. J. Math. Phys. **53**, 123529 (2012)

[Da] Davies, E.B.: Heat Kernels and Spectral Theory. Cambridge University Press, Cambridge (1989)

[DEL10] Davies, E.B., Exner, P., Lipovský, J.: Non-Weyl asymptotics for quantum graphs with general coupling conditions. J. Phys. A: Math. Theor. **43**, 474013 (2010)

[DP98] Davies, E.B., Parnovski, L.: Trapped modes in acoustic waveguides. Quart. J. Mech. Appl. Math. **51**, 477–492 (1998)

[DP11] Davies, E.B., Pushnitski, A.: Non-Weyl resonance asymptotics for quantum graphs. Anal. PDE **4**, 729–756 (2011)

[dBP99] de Biévre, S., Pulé, J.V.: Propagating edge states for a magnetic Hamiltonian. Math. Phys. El. J. **5**, 3 (1999)

[DAC10] Dell'Antonio, G.F., Costa, E.: Effective Schrödinger dynamics on ε-thin Dirichlet waveguides via quantum graphs: I. Star-shaped graphs. J. Phys. A: Math. Theor. **43**, 474014 (2010)

[DH10] Demirel, S., Harrell, E.: On semiclassical and universal inequalities for eigenvalues of quantum graphs. Rev. Math. Phys. **22**, 305–329 (2010)

[DO] Demkov, Y.N., Ostrovskii, V.N.: The Zero-Range Potential Method in Atomic Physics. Leningrad University Press, Leningrad (1975). (In Russian. Plenum, New York, English translation 1988)

[DvC] Demuth, M., van Casteren, J.: Stochastic Spectral Theory for Selfadjoint Feller Operators: A Functional Integration Approach. Birkhäuser, Basel (2000)

[DDI98] Dermenjian, Y., Durand, M., Iftimie, V.: Spectral analysis of an acoustic multistratified cylinder. Commun. PDE **23**, 141–169 (1998)

[DKř02a] Dittrich, J., Kříž, J.: Curved planar quantum wires with Dirichlet and Neumann boundary conditions. J. Phys. A: Math. Gen. **A35**, L269–275 (2002)

[DKř02b] Dittrich, J., Kříž, J.: Bound states in straight quantum waveguides with combined boundary conditions. J. Math. Phys. **43**, 3892–3915 (2002)

[DGR11] Dombrowski, N., Germinet, F., Raikov, G.: Quantization of edge currents along magnetic barriers and magnetic guides. Ann. Henri Poincaré **12**, 1169–1197 (2011)

[Do81] Donnelly, H.: On the essential spectrum of a complete Riemannian manifold. Topology **20**, 1–14 (1981)

[DMP99] Dorlas, T.C., Macris, N., Pulé, J.: Characterization of the spectrum of the Landau Hamiltonian with point impurities. Commun. Math. Phys. **204**, 367–396 (1999)

[DE95] Duclos, P., Exner, P.: Curvature-induced bound states in quantum waveguides in two and three dimensions. Rev. Math. Phys. **7**, 73–102 (1995)

[DEK00] Duclos, P., Exner, P., Krejčiřík, D.: Locally curved quantum layers. Ukrainian J. Phys. **45**, 595–601 (2000)

[DEK01] Duclos, P., Exner, P., Krejčiřík, D.: Bound states in curved quantum layers. Commun. Math. Phys. **223**, 13–28 (2001)

[DEM98] Duclos, P., Exner, P., Meller, B.: Exponential bounds on curvature-induced resonances in a two-dimensional Dirichlet tube. Helv. Phys. Acta **71**, 133–162 (1998)

[DEM01] Duclos, P., Exner, P., Meller, B.: Open quantum dots: resonances from perturbed symmetry and bound states in strong magnetic fields. Rep. Math. Phys. **47**, 253–267 (2001)

[DEŠ95] Duclos, P., Exner, P., Šťovíček, P.: Curvature-induced resonances in a two-dimensional Dirichlet tube. Ann. Inst. H. Poincaré: Phys. Théor. **62**, 81–101 (1995)

[DET08] Duclos, P., Exner, P., Turek, O.: On the spectrum of a bent chain graph. J. Phys. A: Math. Theor. **41**, 415206 (2008)

[DJ93] Dunne, G., Jaffe, R.L.: Bound states in twisted Aharonov-Bohm tubes. Ann. Phys. **223**, 180–196 (1993)

[EEK10] Ekholm, T., Enblom, A., Kovařík, H.: Schrödinger operators on regular metric trees with long range potentials: weak coupling behavior. J. Diff. Eqs. **248**, 850–865 (2010)

[EFK11] Ekholm, T., Frank, R.L., Kovařík, H.: Eigenvalue estimates for Schrödinger operators on metric trees. Adv. Math. **226**, 5165–5197 (2011)

[EkKo05] Ekholm, T., Kovařík, H.: Stability of the magnetic Schrödinger operator in a waveguide. Commun. PDE **30**, 539–565 (2005)

[EKK08] Ekholm, T., Kovařík, H., Krejčiřík, D.: A Hardy inequality in twisted waveguides. Arch. Rat. Mech. Anal. **188**(2), 245–264 (2008)

[ELU93] Evans, D.V., Linton, C.M., Ursell, F.: Trapped mode frequencies embedded in the continuous spectrum. Q. J. Mech. Appl. Math. **46**, 253–274 (1993)

[Ex] Exner, P.: Open Quantum Systems and Feynman Integrals; Fundamental Theories of Physics, vol. 6. D. Reidel, Dordrecht (1985)

[Ex90a] Exner, P.: A model of resonance scattering on curved quantum wires. Ann. Phys. **47**, 123–138 (1990)

[Ex90b] Exner, P.: Bound states and resonances in quantum wires. Order. Disorder and Chaos in Quantum Systems, Integral Equations and Operator Theory, vol. 46, pp. 65–84. Birkhauser, Basel (1990)

[Ex93] Exner, P.: A twisted Landau gauge. Phys. Lett. A **178**, 236–238 (1993)

[Ex95] Exner, P.: A quantum pipette. J. Phys. A: Math. Gen. **28**, 5323–5330 (1995)

[Ex96a] Exner, P.: Contact interactions on graph superlattices. J. Phys. A: Math. Gen. **29**, 87–102 (1996)

[Ex96b] Exner, P.: Weakly coupled states on branching graphs. Lett. Math. Phys. **38**, 313–320 (1996)

[Ex97a] Exner, P.: A duality between Schrödinger operators on graphs and certain Jacobi matrices. Ann. Inst. H. Poincaré: Phys. Théor. **66**, 359–371 (1997)

[Ex00] Exner, P.: Point interactions in a tube, in Stochastic Processes: Physics and Geometry II (In honor of Albeverio, S., Gesztesy, F., et al., (eds.)) In: CMS Conference Proceedings, vol. 29, pp. 165–174. Providence (2000)

[Ex01] Exner, P.: Bound states of infinite curved polymer chains. Lett. Math. Phys. **57**, 87–96 (2001)

[Ex03] Exner, P.: Spectral properties of Schrödinger operators with astrongly attractive δ interaction supported by a surface. InProceedings of the NSF Summer Research Conference (Mt. Holyoke 2002); AMS "Contemporary Mathematics" Series, vol. 339, pp. 25–36. Providence (2003)

[Ex05] Exner, P.: An isoperimetric problem for leaky loops and related mean-chord inequalities. J. Math. Phys. **46**, 062105 (2005)

[Ex06] Exner, P.: Necklaces with interacting beads: isoperimetric problems. In: Proceedings of the International Conference on Differential Equations and Mathematical Physics. AMS Contemporary Mathematics Series, Providence, R.I., 2006, vol. 412, pp. 141–149. Birmingham (2006)

[Ex08] Exner, P.: Leaky quantum graphs: a review. In: EKKST, pp. 523–564

[EB13] Exner, P., Barseghyan, D.: Spectral estimates for Dirichlet Laplacians and Schrödinger operators on geometrically nontrivial cusps. J. Spect. Theory **3**, 465–484 (2013)

[EFr07] Exner, P., Fraas, M.: A remark on helical waveguides. Phys. Lett. A **369**, 393–399 (2007)

[EFr09] Exner, P., Fraas, M.: On geometric perturbations of critical Schrödinger operators with a surface interaction. J. Math. Phys. **50**, 112101 (2009)

[EF07] Exner, P., Frank, R.: Absolute continuity of the spectrum for periodically modulated leaky wires in $\mathbb{R}^3$. Ann. H. Poincaré **8**, 241–263 (2007)

[EFK04] Exner, P., Freitas, P., Krejčiřík, D.: A lower bound to the spectral threshold in curved tubes. Proc. R. Soc. **A469**, 3457–3467 (2004)

[EGŠT96] Exner, P., Gawlista, R., Šeba, P., Tater, M.: Point interactions in a strip. Ann. Phys. **252**, 133–179 (1996)

[EHL99] Exner, P., Harrell, E.M., Loss, M.: Optimal eigenvalues for some Laplacians and Schrödinger operators depending on curvature. Mathematical Results in Quantum Mechanics. Operator Theory: Advances and Applications, vol. 108, pp. 47–53. Birkhäuser, Basel (1999)

[EHL06] Exner, P., Harrell, E.M., Loss, M.: Inequalities for means of chords, with application to isoperimetric problems. Lett. Math. Phys. 75:225–233 (2006) (addendum 77:219 (2006))

[EHS07] Exner, P., Helm, M., Stollmann, P.: Localization on a quantum graph with a random potential on the edges. Rev. Math. Phys. **19**, 923–939 (2007)

[EI01] Exner, P., Ichinose, T.: Geometrically induced spectrum in curved leaky wires. J. Phys. A: Math. Gen. **34**, 1439–1450 (2001)

[EJK99] Exner, P., Joye, A., Kovařík, H.: Edge currents in the absence of edges. Phys. Lett. A **264**, 124–130 (1999)

[EKKST] Exner, P., Keating, J.P., Kuchment, P., Sunada, T., Teplayaev, A. (eds.): Analysis on graphs and its applications. In: Proceedings of the Symposium Pure Mathematics, vol. 77. American Mathematical Society, Providence (2008)

[EK02] Exner, P., Kondej, S.: Curvature-induced bound states for a δ interaction supported by a curve in $\mathbb{R}^3$. Ann. H. Poincaré **3**, 967–981 (2002)

[EK03] Exner, P., Kondej, S.: Bound states due to a strong δ interaction supported by a curved surface. J. Phys. A: Math. Gen. **36**, 443–457 (2003)

[EK04] Exner, P., Kondej, S.: Strong-coupling asymptotic expansion for Schrödinger operators with a singular interaction supported by a curve in $\mathbb{R}^3$. Rev. Math. Phys. **16**, 559–582 (2004)

[EK05] Exner, P., Kondej, S.: Scattering by local deformations of a straight leaky wire. J. Phys. A: Math. Gen. **38**, 4865–4874 (2005)

[EK08] Exner, P., Kondej, S.: Hiatus perturbation for a singular Schrödinger operator with an interaction supported by a curve in $\mathbb{R}^3$. J. Math. Phys. **49**, 032111 (2008)

[EKo00] Exner, P., Kovařík, H.: Magnetic strip waveguides. J. Phys. A: Math. Gen. **33**, 3297–3311 (2000)

[EKo05] Exner, P., Kovařík, H.: Spectrum of the Schrödinger operator in a perturbed periodically twisted tube. Lett. Math. Phys. **73**, 183–192 (2005)

[EKr99] Exner, P., Krejčiřík, D.: Quantum waveguide with a lateral semitransparent barrier: spectral and scattering properties. J. Phys. A: Math. Gen. **32**, 4475–4494 (1999)

[EKr01a] Exner, P., Krejčiřík, D.: Bound states in mildly curved layers. J. Phys. A: Math. Gen. **34**, 5969–5985 (2001)

[EKr01b] Exner, P., Krejčiřík, D.: Waveguides coupled through a semitransparent barrier: a Birman-Schwinger analysis. Rev. Math. Phys. **13**, 307–334 (2001)

[ELW04] Exner, P., Linde, H., Weidl, T.: Lieb-Thirring inequalities for geometrically induced bound states. Lett. Math. Phys. **70**, 83–95 (2004)

[EL10a] Exner, P., Lipovský, J.: On the absence of absolutely continuous spectra for Schrödinger operators on radial tree graphs. J. Math. Phys. **51**, 122107 (2010)

[EL10b] Exner, P., Lipovský, J.: Resonances from perturbations of quantum graphs with rationally related edges. J. Phys. A: Math. Theor. **43**, 105301 (2010)

[EL11] Exner, P., Lipovský, J.: Non-Weyl resonance asymptotics for quantum graphs in a magnetic field. Phys. Lett. A **375**, 805–807 (2011)

[ENZ01] Exner, P., Neidhardt, H., Zagrebnov, V.A.: Potential approximations to δ': an inverse Klauder phenomenon with norm-resolvent convergence. Commun. Math. Phys. **224**, 593–612 (2001)

[EN01] Exner, P., Němcová, K.: Bound states in point interaction star graphs. J. Phys. A: Math. Gen. **34**, 7783–7794 (2001)

[EN02] Exner, P., Němcová, K.: Quantum mechanics of layers with a finite number of point perturbations. J. Math. Phys. **43**, 1152–1184 (2002)

[EN03a] Exner, P., Němcová, K.: Magnetic layers with periodic point perturbations. Rep. Math. Phys. **52**, 255–280 (2003)

[EN03b] Exner, P., Němcová, K.: Leaky quantum graphs: approximations by point interaction Hamiltonian. J. Phys. A: Math Gen. **36**, 10173–10193 (2003)

[EPa14] Exner, P., Pankrashkin, K.: Strong coupling asymptotics for a singular Schrödinger operator with an interaction supported by an open arc. Comm. PDE **39**, 193–212 (2014)

[EP05] Exner, P., Post, O.: Convergence of spectra of graph-like thin manifolds. J. Geom. Phys. **54**, 77–115 (2005)

[EP07] Exner, P., Post, O.: Convergence of resonances on thin branched quantum wave guides. J. Math. Phys. **48**, 092104 (2007)

[EP09] Exner, P., Post, O.: Approximation of quantum graph vertex couplings by scaled Schrödinger operators on thin branched manifolds. J. Phys. A: Math. Theor. **42**, 415305 (2009)

[EP13] Exner, P., Post, O.: A general approximation of quantum graph vertex couplings by scaled Schrödinger operators on thin branched manifolds. Commun. Math. Phys. **322**, 207–227 (2013)

[ESŠF98] Exner, P., Sadreev, A.F., Šeba, P., Středa, P., Feher, P.: Strength of topologically induced magnetic moments in a quantum device. Phys. Rev. Lett. **80**, 1710–1713 (1998)

[EŠ89a] Exner, P., Šeba, P.: Bound states in curved quantum waveguides. J. Math. Phys. **30**, 2574–2580 (1989)

[EŠ89b] Exner, P., Šeba, P.: Free quantum motion on a branching graph. Rep. Math. Phys. **28**, 7–26 (1989)

[EŠ90] Exner, P., Šeba, P.: Trapping modes in a curved electromagnetic waveguide with perfectly conducting walls. Phys. Lett. A **144**, 347–350 (1990)

[EŠ96] Exner, P., Šeba, P.: Point interactions in dimension two and three as models of small scatterers. Phys. Lett. A **222**, 1–4 (1996)

[EŠŠ89] Exner, P., Šeba, P., Šťovíček, P.: On existence of a bound state in an L-shaped waveguide. Czech. J. Phys. **B39**, 1181–1191 (1989)

[EŠŠ90] Exner, P., Šeba, P.: P. Šťovíček: semiconductor edges can bind electrons. Phys. Lett. **A150**, 179–182 (1990)

[EŠTV96] Exner, P., Šeba, P., Tater, M., Vaněk, D.: Bound states and scattering in quantum waveguides coupled laterally through a boundary window. J. Math. Phys. **37**, 4867–4887 (1996)

[EŠe94] Exner, P., Šerešová, E.: Appendix resonances on a simple graph. J. Phys. A: Math. Gen. **27**, 8269–8278 (1994)

[ET04] Exner, P., Tater, M.: Spectra of soft ring graphs. Waves Random Media **14**, S47–60 (2004)

[ET10] Exner, P., Tater, M.: Spectrum of Dirichlet Laplacian in a conical layer. J. Phys. A: Math. Theor. **43**, 474023 (2010)

[ET07] Exner, P., Turek, O.: Approximations of singular vertex couplings in quantum graphs. Rev. Math. Phys. **19**, 571–606 (2007)

[EV96] Exner, P., Vugalter, S.A.: Asymptotic estimates for bound states in quantum waveguides coupled laterally through a narrow window. Ann. Inst. H. Poincaré: Phys. Théor. **65**, 109–123 (1996)

[EV97a] Exner, P., Vugalter, S.A.: Bounds states in a locally deformed waveguide: the critical case. Lett. Math. Phys. **39**, 59–68 (1997)

[EV97b] Exner, P., Vugalter, S.A.: Bound-state asymptotic estimates for window-coupled Dirichlet strips and layers. J. Phys. A: Math. Gen. **30**, 7863–7878 (1997)

[EV99] Exner, P., Vugalter, S.A.: On the number of particles that a curved quantum waveguide can bind. J. Math. Phys. **40**, 4630–4638 (1999)

[EW02] Exner, P., Weidl, T.: Lieb-Thirring inequalities on trapped modes in quantum wires. In: Proceedings of the XIII International Congress on Mathematical Physics (London 2000) (pp. 437–443). International Press, Boston (2001)

[EY01] Exner, P., Yoshitomi, K.: Band gap of the Schrödinger operator with a strong δ-interaction on a periodic curve. Ann. Henri Poincaré **2**, 1139–1158 (2001)

[EY02a] Exner, P., Yoshitomi, K.: Asymptotics of eigenvalues of the Schrödinger operator with a strong δ-interaction on a loop. J. Geom. Phys. **41**, 344–358 (2002)

[EY02b] Exner, P., Yoshitomi, K.: Persistent currents for 2D Schrödinger operator with a strong δ-interaction on a loop. J. Phys. A: Math. Gen. **35**, 3479–3487 (2002)

[EY03] Exner, P., Yoshitomi, K.: Eigenvalue asymptotics for the Schrödinger operator with a δ-interaction on a punctured surface. Lett. Math. Phys. **65**, 19–26 (2003) (erratum 67, 81–82 (2004))

[Fa23] Faber, G.: Beweiss das unter allen homogenen Membranen von Gleicher Fläche und gleicher Spannung die kreisförmige den Tiefsten Grundton gibt, Sitzungber. der math.-phys. Klasse der Bayerische Akad. der Wiss. zu München 169–172 (1923)

[Fe22] Fermi, E.: Sopra i fenomeni che avvengono in vicinanza di una linea oraria. Atti R. Accad. Lincei Rend. Cl. Sci. Fis. Mat. Nat. **31**, 21–23 (1922) [English translation in E. Fermi Collected Papers, vol. I, Italy, 1921–1938, University of Chicago Press 1962; pp. 17–19]

[Fe36] Fermi, E.: Sul moto dei neutroni nelle sostanze idrogenate, Ricerca Scientifica **7**, 13–52 (1936); English transl. ibidem 980–1016

[FK96] Figotin, A., Kuchment, P.: Band-gap structure of spectra of periodic dielectric and acoustic media I. II. SIAM J. Appl. Math. **56**(68–88), 1561–1620 (1996)

[FK11] Filonov, N., Kachkovskiy, I.: Absolute continuity of the spectrum of the periodic Schrödinger operator in a layer and in a smooth cylinder. J. Math. Sci. (N.Y.) **178**, 274–281 (2011)

[FLS10] Frank, R.L., Lieb, E.H., Seiringer, R.: Equivalence of Sobolev inequalities and Lieb-Thirring inequalities. In: Proceedings of the XVI International Congress on Mathematical Physics (Prague 2009), World Scientific, Singaprore (2010), pp. 523–535

[FW93] Freidlin, M.I., Wentzell, A.D.: Diffusion processes on graphs and the averaging principle. Ann. Probab. **21**, 2215–2245 (1993)

[FKr06] Freitas, P., Krejčiřík, D.: Waveguides with combined Dirichlet and Robin boundary conditions. Math. Phys. Anal. Geom. **9**, 335–352 (2006)

[FH01] Froese, R., Herbst, I.: Realizing holonomic constraints in classical and quantum mechanics. Commun. Math. Phys. **220**, 489–535 (2001)

[FGW00] Fröhlich, J., Graf, G.M., Walcher, J.: On the extended nature of edge states of quantum Hall Hamiltonians. Ann. Henri Poincaré **1**, 405–442 (2000)

[Fu] Fukushima, M.: Dirichlet Forms and Markov Processes. North-Holland, Amsterdam (1980)

[FT00] Fülöp, T., Tsutsui, I.: A free particle on a circle with point interaction. Phys. Lett. A **264**, 366–374 (2007)

[FKW07] Fulling, S., Kuchment, P., Wilson, J.: Index theorems for quantum graphs. J. Phys. A: Math. Theor. **40**, 14165 (2007)

[Ga] Garabedian, P.R.: Partial Differential Equations. AMS Chelsea Publishing, Providence, R.I. (1998)

[GB13] Gaulter, S.N., Biggs, R.T.: Acoustic trapped modes in a three-dimensional waveguide of slowly varying cross section. Proc. R. Soc. **A469**, 20120384 (2013)

[GW11] Geisinger, L., Weidl, T.: Sharp spectral estimates in domains of infinite volume. Rev. Math. Phys. **23**, 615–641 (2011)

[GHKSV88] Gesztesy, F., Gurarie, D., Holden, H., Klaus, M., Sadun, L., Simon, B., Vogl, P.: Trapping and cascading in the large coupling constant limit. Commun. Math. Phys. **118**, 597–634 (1988)

[GZ94] Gesztesy, F., Zhao, Z.: Domain perturbations, Brownian motion, capacities, and ground states of Dirichlet Schrödinger operators. Math. Z. **215**, 143–150 (1994)

[G92] Geyler, V.: The two-dimensional Schrödinger operator with a homogeneous magnetic field and its perturbations by periodic zero-range potentials. St. Petersburg Math. J. **3**, 489–532 (1992)

[GS97] Geyler, V., Senatorov, M.: The structure of the spectrum of the Schrödinger operator with a magnetic field in a strip, and finite-gap potentials, Mat. Sbornik 188: 21–32. Engl. Trans. Sbornik Math. **188**(1997), 657–669 (1997)

[GJ92] Goldstone, J., Jaffe, R.L.: Bound states in twisting tubes. Phys. Rev. B **45**, 14100–14107 (1992)

[GG] Gorbachuk, V.I., Gorbachuk, M.L.: Boundary Value Problems for Operator Differential Equations. Kluwer, Dordrecht (1991)

[GR] Gradshtein, I.S., Ryzhik, I.M.: Tables of Integral, Series, and Products. Academic Press, New York (1980)

[Gra] Gray, A.: Tubes. Addison-Wesley, New York (1990)

[Gr02] Grieser, D.: Uniform bounds for eigenfunctions of the Laplacian on manifolds with boundary. Comm. PDE **27**, 1283–1299 (2002)

[Gr08] Grieser, D.: Spectra of graph neighborhoods and scattering. Proc. Lond. Math. Soc. **97**, 718–752 (2008)

[GS09] Grigor'yan, A., Saloff-Coste, L.: Heat kernel on manifolds with ends. Ann. Inst. Fourier **59**, 825–890 (2009)

[GKP14] Grillo, G., Kovařík, H., Pinchover, Y.: Sharp two-sided heat kernel estimates of twisted tubes and applications. Arch. Rat. Mech. Anal. **213**, 215–243 (2014)

[GM69] Gromoll, D., Meyer, W.: On complete open manifolds of positive curvature. Ann. Math. **90**, 75–90 (1969)

[Gr04] Grushin, V.V.: On the eigenvalues of finitely perturbed Laplace operators in infinite cylindrical domains. Math. Notes 75(3), 331–340 (2004) [translation from Mat. Zametki 75(3), 360–371 (2004)]

[Gr05] Grushin, V.V.: Asymptotic behavior of the eigenvalues of the Schrödinger operator with transversal potential in a weakly curved infinite cylinder. Math. Notes 77(5), 606–613 (2005) [translation from Mat. Zametki 77(5), 656–664 (2005)]

[Gr09] Grushin, V.V.: Asymptotic behavior of the eigenvalues of the Laplace operator in thin infinite tubes. Math. Notes **85**(5), 661–673 (2009) [translation from Mat. Zametki **85**(5), 687–701 (2009)]

[GS01] Gutkin, B., Smilansky, U.: Can one hear the shape of a graph? J. Phys. A.: Math. Gen. **34**, 6061–6068 (2001)

[Ha08] Hadamard, J.: Mémoire sur le problème d'analyse relatif à l'équilibre des plaques élastiques encastrées. Méms. prés. par div. savants a l'Acad. des Sci. **33**, 1–128 (1908)

[Ha82] Halperin, B.I.: Quantized Hall conductance, current carrying edge states, and the existence of extended states in two-dimensional disordered potential. Phys. Rev. B **25**, 2185–2190 (1982)

[Ha00] Harmer, M.: Hermitian symplectic geometry and extension theory. J. Phys. A: Math. Gen. **33**, 9193–9203 (2000)

[HKK01] Harrell, E., Kröger, P., Kurata, K.: On the placement of an obstacle or a well so as to optimize the fundamental eigenvalue. SIAM J. Math. Anal. **33**, 240–259 (2001)

[HSS91] Hempel, R., Seco, L.A., Simon, B.: The essential spectrum of Neumann Laplacians on some bounded singular domains. J. Funct. Anal. **102**, 448–483 (1991)

[He89] Herczynski, J.: On Schrödinger operators with a distributional potential. J. Oper. Theory **21**, 273–295 (1989)

[He65] Hersch, J.: Erweiterte Symmetrieeigenschaften von Lösungen gewisser linearer Rand- und Eigenwertprobleme. J. Reine Angew. Math. **218**, 143–158 (1965)

[HS08a] Hislop, P., Soccorsi, E.: Edge currents for quantum Hall systems. I. One-edge, unbounded geometries. Rev. Math. Phys. **20**, 71–115 (2008)

[HS08b] Hislop, P., Soccorsi, E.: Edge currents for quantum Hall systems. II. Two-edge, bounded and unbounded geometries, Ann. Henri Poincaré **9**, 1141–1175 (2008)

[HTW93] Hirayama, Y., Tokura, Y., Wieck, A.D., Koch, S., Haug, R.J., von Klitzing, K., Ploog, K.: Transport characteristics of a window-coupled in-plane-gated wire system. Phys. Rev. B **48**, 7991–7998 (1993)

[HP09] Hislop, P., Post, O.: Anderson localization for radial tree-like random quantum graph. Waves Random Media **19**, 216–261 (2009)

[Ho65] Hochstadt, H.: On the determination of a Hill's equation from its spectrum. Arch. Rat. Mech. Anal. **19**, 353–362 (1965)

[HH04] Honnouvo, G., Hounkonnou, M.N.: Asymptotics of eigenvalues of the Aharonov-Bohm operator with a strong δ-interaction on a loop. J. Phys. A: Math. Gen. **37**, 693–700 (2004)

[Hu57] Huber, A.: On subharmonic functions and differential geometry in the large. Comment. Math. Helv. **32**, 13–72 (1957)

[HBPSŻ04] Hul, O., Bauch, S., Pakon'ski, P., Savytskyy, N., Życzkowski, K., Sirko, L.: Experimental simulation of quantum graphs by microwave networks. Phys. Rev. E **69**, 056205 (2004)

[HLW00] Hundertmark, D., Laptev, A., Weidl, T.: New bounds on the Lieb-Thirring constants. Invent. Math. **140**, 693–704 (2000)

[HLT98] Hundertmark, D., Lieb, E.H., Thomas, L.E.: A sharp bound for an eigenvalue moment of the one-dimensional Schrödinger operator. Adv. Theor. Math. Phys. **2**, 719–731 (1998)

[Il] Il'in, A.: Matching of Asymptotic Expansions of Solutions of Boundary Value Problems, Translations of Mathematical Monographs, vol. 102. AMS, Providence, R.I. (1992)

[Iw85] Iwatsuka, A.: Examples of absolutely continuous Schrödinger operators in magnetic fields. Publ. RIMS **21**, 385–401 (1985)

[JMS92] Jakšić, V., Molčanov, S., Simon, B.: Eigenvalue asymptotics of the Neumann Laplacian of regions and manifolds with cusps. J. Funct. Anal. **106**, 59–79 (1992)

[Je78] Jensen, A.: Local decay in time of solutions to Schrödinger equation with a dilation-analytic interaction. Manuscr. Math. **25**, 61–77 (1978)

[JK71] Jensen, H., Koppe, H.: Quantum mechanics with constraints. Ann. Phys. **63**, 586–591 (1971)

[Jí07] M. Jílek: Straight quantum waveguide with Robin boundary conditions, SIGMA 3, Paper 108 (2007)

[JLP05] Johnson, E.R., Levitin, M., Parnovski, L.: Existence of eigenvalues of a linear operator pencil in a curved waveguide-localized shelf waves on a curved coast. SIAM J. Math. Anal. **37**, 1465–1481 (2006)

[Ka66] Kato, T.: Wave operators and similarity for some non-self-adjoint operators. Math. Ann. **162**, 258–279 (1966)

[Ka73] Kato, T.: Schrödinger operators with singular potentials. Israel J. Math. **13**, 135–148 (1973)

[Ka] Kato, T.: Perturbation Theory for Linear Operators, 2nd edn. Springer, Berlin (1976)

[KPV00] Khallaf, N.S.A., Parnovski, L., Vassiliev, D.: Trapped modes in a waveguide with a long obstacle. J. Fluid Mech. **403**, 251–261 (2000)

[KSH94] Kirczenow, G., et al.: Artificial impurities in quantum wires: From classical to quantum behavior. Phys. Rev. Lett. **72**, 2069–2072 (1994)

[KS87] Kirsch, W., Simon, B.: Comparison theorems for the gap of Schrödinger operators. J. Funct. Anal. **75**, 396–410 (1987)

[Kl77] Klaus, M.: On the bound state of Schrödinger operators in one dimension. Ann. Phys. **108**, 288–300 (1977)

[KS00] Kleespies, F., Stollmann, P.: Lifshitz asymptotics and localization for random quantum waveguides. Rev. Math. Phys. **12**, 1345–1365 (2000)

[Kl94] Klein, A.: Absolutely continuous spectrum in the Anderson model on the Bethe lattice. Math. Res. Lett. **1**, 399–407 (1994)

[Kli] Klingenberg, W.: A Course in Differential Geometry. Springer, New York (1978)

[KP09] Klopp, F., Pankrashkin, K.: Localization on quantum graphs with random edge lengths. Lett. Math. Phys. **87**, 99–114 (2009)

[KR10] Klopp, F., Raikov, G.: The fate of Landau levels under perturbations of constant sign. Int. Math. Res. Notices **2009**, 4726–4734 (2009)

[Kon13] Kondej, S.: Schrödinger operator with a strong varying interaction on a curve in $\mathbb{R}^2$. J. Math. Phys **54**, 093511 (2013)

[KonK13] Kondej, S., Krejčiřík, D.: Spectral analysis of a quantum system with a double line singular interaction. Publ. RIMS **49**, 831–859 (2013)

[Kon67] Kondratev, V.A.: Boundary-value problems for elliptic equations in domains with conical or angular points. Trans. Mosc. Math. Soc. **16**, 227–313 (1967)

[KJ71] Koppe, H., Jensen, H.: Das Prinzip von d'Alembert in der klassischen Mechanik und in der Quantentheorie. Sitzungber. der Heidelberger Akad. der Wiss., Math.-Naturwiss. Klasse **5**, 127–140 (1971)

[KoS99] Kostrykin, V., Schrader, R.: Kirchhoff's rule for quantum wires. J. Phys. A: Math. Gen. **32**, 595–630 (1999)

[KoS00] Kostrykin, V., Schrader, R.: Kirchhoff's rule for quantum wires, II. The inverse problem with possible applications to quantum computers. Fortschr. Phys. **48**, 703–716 (2000)

[KSm99] Kottos, T., Smilansky, U.: Periodic orbit theory and spectral statistics for quantum graphs. Ann. Phys. **274**, 76–124 (1999)

[Ko07] Kovařík, H.: Weakly coupled Schrödinger operators on regular metric trees. SIAM J. Math. Anal. **39**, 1135–1149 (2007)

[Ko11] Kovařík, H.: Eigenvalue asymptotic of Robin Laplace operators on two-dimensional domains with cusps. J. Lond. Math. Soc. **83**, 256–271 (2011)

[KS07] Kovařík, H., Sacchetti, A.: Resonances in twisted quantum waveguides. J. Phys. A: Math. Theor. **40**, 8371–8384 (2007)

[KoKr08] Kovařík, H., Krejčiřík, D.: A Hardy inequality in a twisted Dirichlet-Neumann waveguide. Math. Nachr. **281**, 1159–1168 (2008)

[KV08] Kovařík, H., Vugalter, S.: Estimates on trapped modes in deformed quantum layers. J. Math. Anal. Appl. **345**, 566–572 (2008)

[Kra25] Krahn, E.: Über eine von Rayleigh formulierte minimal Eigenschaft des Kreises. Ann. Math. **94**, 97–100 (1925)

[KL71] Krein, M.G., Langer, H.: On defect subspaces and generalized resolvents of an Hermitian operator in the space Π_κ. Funct. Anal. Appl. **5**(2), 59–71 (1971)

[Kr01] D. Krejčiřík: Quantum layers. PhD thesis, Marseille (2001)

[Kr03] Krejčiřík, D.: Quantum strips on surfaces. J. Geom. Phys. **45**, 203–217 (2003)

[Kr08] Krejčiřík, D.: Twisting versus bending in quantum waveguides, in [EKKST], pp. 617–636

[Kr09] Krejčiřík, D.: Spectrum of the Laplacian in a narrow curved strip with combined Dirichlet and Neumann boundary conditions, ESAIM: Control Optim. Calc. Var. **15**, 555–568 (2009)

[KŠ12] Krejčiřík, D., Šediváková, H.: The effective Hamiltonian in curved quantum waveguides under mild regularity assumptions. Rev. Math. Phys. **24**, 1250018 (2012)

[KrT04] Krejčiřík, D., Tiedra de Aldecoa, R.: The nature of the essential spectrum in curved quantum waveguides. J. Phys. A: Math. Gen. **37**, 5449–5466 (2004)

[KrZu10] Krejčiřík, D., Zuazua, E.: The Hardy inequality and the heat equation in twisted tubes. J. Math. Pures Appl. **94**, 277–303 (2010)

[KP31] de Kronig, R.L., Penney, W.G.: Quantum mechanics of electrons in crystal lattices. Proc. Roy. Soc. Lon. **130A**, 499–513 (1931)

[Ku01] Kuchment, P.: The mathematics of photonic crystals, Chap. 7. In: Bao, G., Cowsar, L., Masters, W. (eds). Mathematical Modeling in Optical Science. Frontiers in Applied Mathematics, vol. 22, pp. 207–272, SIAM 2001

[Ku04] Kuchment, P.: Quantum graphs: I. Some basic structures. Waves Random Media **14**, S107–S128 (2004)

[Ku05] Kuchment, P.: Quantum graphs: II. Some spectral properties of quantum and combinatorial graphs, J. Phys. A.: Math. Gen. **38**, 4887–4900 (2005)

[KZ01] Kuchment, P., Zeng, H.: Convergence of spectra of mesoscopic systems collapsing onto a graph. J. Math. Anal. Appl. **258**, 671–700 (2001)

[Ku93] Kunze, Ch.: Leaky and mutually coupled quantum wires. Phys. Rev. B **48**, 14338–14346 (1993)

[La70] Landauer, R.: Electrical resistance of disordered one-dimensional lattices. Phil. Mag. **21**, 863–867 (2000)

[LL] Landau, L.D., Lifshitz, E.M.: Quantum Mechanics. Nonrelativistic Theory. Nauka, Moscow (1974)

[LW00] Laptev, A., Weidl, T.: Sharp Lieb-Thirring inequalities in high dimensions. Acta Math. **184**, 87–111 (2000)

[LLM86] Lenz, F., Londergan, J.T., Moniz, R.J., Rosenfelder, R., Stingl, M., Yazaki, K.: Quark confinement and hadronic interactions. Ann. Phys. **170**, 65–254 (1986)

[LP08] Levitin, M., Parnovski, L.: On the principal eigenvalue of a Robin problem with a large parameter. Math. Nachr. **281**, 272–281 (2008)

[LPX00] Li, J., Papadopoulos, C., Xu, J.M.: Growing Y-junction carbon nanotubes. Nature **402**, 253–254 (2000)

[Li76] Lieb, E.: Bounds on the eigenvalues of the Laplace and Schrödinger operator. Bull. Am. Math. Soc. **82**, 751–753 (1976)

[LT76] Lieb, E.H., Thirring, W.: Inequalities for the moments of the eigenvalues of the Schrödinger Hamiltonian and their relation to Sobolev inequalities. Studies in Mathematical Physics. Essays in Honor of Valentine Bargmann (pp. 269–303). Princeton University Press, Princeton (1976)

[LL07] Lin, Ch., Lu, Z.: Existence of bound states for layers built over hypersurfaces in $\mathbb{R}^{n+1}$. J. Funct. Anal. **244**, 1–25 (2007)

[LCM] Londergan, J.T., Carini, J.P., Murdock, D.P.: Binding and Scattering in Two-Dimensional Systems. Applications to Quantum Wires, Waveguides and Photonic Crystals. Springer LNP m60, Berlin (1999)

[Lo13] Lotoreichik, V.: Note on $2D$ Schrödinger operators with δ-interactions on angles and crossing lines. Nanosyst. Phys. Chem. Math. **3**, 166–172 (2013)

[LR12] Lu, Z., Rowlett, J.: On the discrete spectrum of quantum layers. J. Math. Phys. **53**, 073519 (2012)

[Lü66] Lükő, G.: On the mean lengths of the chords of a closed curve. Israel J. Math. **4**, 23–32 (1966)

[MMP99] Macris, N., Martin, PhA, Pulé, J.: On edge states in semi-infinite quantum Hall systems. J. Phys. A: Math. Gen. **32**, 1985–1996 (1999)

[Mad] Maddox, I.J.: Elements of Functional Analysis. Cambridge University Press, Cambridge (1970)

[MP97] Mantoiu, M., Purice, R.: Some propagation properties of the Iwatsuka model. Commun. Math. Phys. **188**, 691–708 (1997)

[MKSW91] Martorell, J., Klarsfeld, S., Sprung, D.W.L., Wu, H.: Analytical treatment of electron wave propagation in two-dimensional structures. Solid State Commun. **78**, 13–18 (1991)

[Mo98] Moiseyev, N.: Quantum theory of resonances: calculating energies, widths and cross-sections by complex scaling. Phys. Rep. **302**, 212–293 (1998)

[MV07] Molchanov, S., Vainberg, B.: Scattering solutions in networks of thin fibers. Commun. Math. Phys. **273**, 533–559 (2007)

[Mor98] Morame, A.: Absence of singular spectrum for a perturbation of two-dimensional Laplace-Beltrami operator with periodic electromagnetic potential. J. Phys. A.: Math. Gen. **31**, 7593–7601 (1998)

[Mou81] Mourre, E.: Absence of singular continuous spectrum for certain selfadjoint operators. Comm. Math. Phys. **78**, 519–567 (1981)

[NS00] Naimark, K., Solomyak, M.: Eigenvalue estimates for the weighted Laplacian on metric trees. Proc. Lond. Math. Soc. **80**, 690–724 (2000)

[NO11] Najar, H., Olendski, O.: Spectral and localization properties of the Dirichlet wave guide with two concentric Neumann discs. J. Phys. A: Math. Theor. **44**, 305304 (2011)

[Na11a] Nazarov, S.A.: Eigenvalues of the Laplace operator with the Neumann conditions at regular perturbed walls of a waveguide. J. Math. Sci. (N.Y.) **172**, 555–588 (2011)

[Na11b] Nazarov, S.A.: The discrete spectrum of cranked, branching, and periodic waveguides. Algebra i Analiz **23**, 206–247 (2011) [English translation in St. Petersburg Math. J. **23**, 351–379 (2012)]

[Ne97] Nedelec, L.: Sur les résonances de l'opérateur de Dirichlet dans un tube. Comm. PDE **22**, 143–163 (1997)

[vN] von Neumann, J.: Mathematische Grundlagen der Quantenmechanik, Springer, Berlin (1932). Princeton University Press, English translation (1955)

[Ne83] Newton, R.G.: Bounds for the number of bound states for Schrödinger equation in one and two dimensions. J. Oper. Theory **10**, 119–125 (1983)

[Nö92] Nöckel, J.U.: Resonances in quantum-dot transport. Phys. Rev. B **46**, 15348–15356 (1992)

[OM03] Olendski, O., Mikhailovska, L.: Localized-mode evolution in a curved planar waveguide with combined Dirichlet and Neumann boundary conditions. Phys. Rev. E **67**, 056625 (2003)

[OM10] Olendski, O., Mikhailovska, L.: Theory of a curved planar waveguide with Robin boundary conditions. Phys. Rev. E **81**, 036606 (2010)

[Ol] Olver, F.W.J.: Asymptotics and Special Functions. Academic Press, New York (1974)

[OT69] Opechowski, W., Tam, W.G.: Invariance groups of the Schrödinger equation for the case of uniform magnetic field I. Physica **42**, 529–556 (1969)

[Ož06] Ožanová, K.: Approximation by point potentials in a magnetic field. J. Phys. A: Math. Gen. **39**, 3071–3083 (2006)

[Pa12] Pankrashkin, K.: Unitary dimension reduction for a class of self-adjoint extensions with applications to graph-like structures. J. Math. Anal. Appl. **396**, 640–655 (2012)

[Pa08] Parnovski, L.: Bethe-Sommerfeld conjecture. Ann. Henri Poincaré **9**, 457–508 (2008)

[PF] Pastur, L., Figotin, A.: Spectra of Random and Almost-Periodic Operators. Springer, Berlin (1992)

[PPW55] Payne, L.E., Pólya, G., Weinberger, H.F.: Sur le quotient de deux fréquences propres consecutives. Comp. Rend. **241**, 917–919 (1955)

[Pi82] Picca, D.: Constraints in quantum mechanics. Lett. N. Cim. **34**, 449–452 (1982)

[Pop86] Popov, A.N.: On the existence of eigenoscillations of a resonator open to a waveguide. Sov. J. Tech. Phys. **56**, 1916–1922 (1986)

[Pos01] Posilicano, A.: A Krein-like formula for singular perturbations of self-adjoint operators and applications. J. Funct. Anal. **183**, 109–147 (2001)

[Po05] Post, O.: Branched quantum wave guides with Dirichlet boundary conditions: the decoupling case. J. Phys. A.: Math. Gen. **38**, 4917–4931 (2005)

[Po06] Post, O.: Spectral convergence of quasi-one-dimensional spaces. Ann. Henri Poincaré **7**(5), 933–973 (2006)

[Po11] Post, O.: Spectral Analysis on Graph-Like Spaces. Lecture Notes in Mathematics, vol. 2039. Springer, Berlin (2011)

[Pr81] Prange, R.E.: Quantized Hall resistance and the measurement of the fine structure constant. Phys. Rev. B **23**, 4802–4805 (1981)

[PBM] Prudnikov, A.P., Brychkov, Y.O., Marichev, O.I.: Integrals and Series, I. Elementary Functions, II. Special Functions. Nauka, Moscow (1981–1983)

[Rao] Rao, M.M.: Measure Theory and Integration. Wiley, New York (1987)

[RT75] Rauch, J., Taylor, M.: Potential and scattering theory on wildly perturbed domains. J. Funct. Anal. **18**, 27–59 (1975)

[RS] Reed, M., Simon, B.: Methods of Modern Mathematical Physics, including I. Functional Analysis, II. Fourier Analysis. Self-adjointness, III. Scattering Theory, IV. Analysis of Operators. Academic Press, New York (1972–1979)

[RP00] Reijniers, J., Peeters, F.M.: Snake orbits and related magnetic edge states. J. Phys.: Cond. Matt. **12**, 9771–9786 (2000)

[RB95] Renger, W., Bulla, W.: Existence of bound states in quantum waveguides under weak conditions. Lett. Math. Phys. **35**, 1–12 (1995)

[Ro83] Roth, J.-P.: Spectre du laplacien sur un graphe, C. R. Acad. Sci. Paris Sér. I Math. **296**, 793–795 (1983)

[RS98] Rozenblum, G., Solomyak, M.: CLR-estimate for the generators of positivity preserving and positively dominated semigroups. St. Petersburg. Math. J. **9**, 1195–1211 (1998)

[RuS01] Rubinstein, J., Schatzman, M.: Variational problems on multiply connected thin strips, I. Basic estimates and convergence of the Laplacian spectrum. Arch. Rat. Mech. Anal. **160**, 271–308 (2001)

[RSch53] Ruedenberg, K., Scherr, C.W.: Free-electron network model for conjugated systems. I. Theory. J. Chem. Phys. **21**, 1565–1581 (1953)

[Ru83] Ruijsenaars, S.N.M.: The Aharonov-Bohm effect and scattering theory. Ann. Phys. **146**, 1–34 (1983)

[Sa00] Saito, Y.: The limiting equation for Neumann Laplacians on shrinking domains, Electron. J. Differ. Eqs. 31, 25 p. (2000)

[SP] Sánchez-Palencia, E.: Non-Homogeneous Media and Vibration Theory. Springer, New York (1980)

[Sch] Schechter, M.: Operator Methods in Quantum Mechanics. North Holland, New York (1981)

[SRW89] Schult, R.L., Ravenhall, D.G., Wyld, H.W.: Quantum bound states in a classically unbounded system of crossed wires. Phys. Rev. B **39**, 5476–5479 (1989)

[Sch61] Schwinger, J.: On the bound states of a given potential. Proc. Nat. Acad. Sci. U.S.A. **47**, 122–129 (1961)

[SeS89] Seeger, A., Sogge, C.D.: Bounds for eigenfunctions of differential operators. Indiana Univ. Math. J. **38**, 669–682 (1989)

[Se74] Seto, N.: Bargmann's inequalities in spaces of arbitrary dimension. Publ. RIMS **9**, 429–461 (1974)

[Sha88] Shabani, J.: Finitely many delta interactions with supports on concentric spheres. J. Math. Phys. **29**, 660–664 (1988)

[Si76] Simon, B.: The bound state of weakly coupled Schrödinger operators in one and two dimensions. Ann. Phys. **97**, 279–288 (1976)

[Si92] Simon, B.: The Neumann Laplacians of a jelly roll. Proc. AMS **114**, 783–785 (1992)

[SoS02] Sobolev, A., Solomyak, M.: Schrödinger operator on homogeneous metric trees: spectrum in gaps. Rev. Math. Phys. **14**, 421–467 (2002)

[SW02] Sobolev, A., Walthoe, J.: Absolute continuity in periodic waveguides. Proc. London Math. Soc. **85**, 717–741 (2002)

[So09] Solomyak, M.: Remarks on counting negative eigenvalues of Schrödinger operators on regular metric trees. Am. Math. Soc. Transl. **226**(2), 165–178 (2009)

[SB33] A. Sommerfeld, H. Bethe: Elektronentheorie der Metalle, in Handbuch der Physik. In: Geiger, H., Scheel, K. (eds.), vol. 24/2, pp. 333–622. Springer (1933)

[SM90] Sols, F., Macucci, M.: Circular bends in electron waveguides. Phys. Rev. B **41**, 11887–11891 (1990)

[St] Stollmann, P.: Caught by Disorder. Bound States in Random Media, Birkhäuser (2001)

[SV96] Stollmann, P., Voigt, J.: Perturbations of Dirichlet forms by measures. Potential Anal. **5**, 109–138 (1996)

[SŠ01] Suslina, T.A., Shterenberg, R.G.: Absolute continuity of the spectrum of the Schrödinger operator with the potential concentrated on a periodic system of hypersurfaces, Algebra i Analiz **13**, 197–240 (2001). English Transl. St. Petersburg Math. J. **13**, 859–891 (2002)

[T95] Talagrand, M.: Concentration of measure and isoperimetric inequalities in product spaces. Publ. Math. IHES **81**, 73–205 (1995)

[Th73] Thomas, L.E.: Time dependent approach to scattering from impurities in a crystal. Commun. Math. Phys. **33**, 335–343 (1973)

[Ti] Titchmarch, E.C.: Eigenfunction Expansions Associated with Second-Order Differential Equations. Clarendon Press, Oxford (1958)

[To88] Tolar, J.: On a Quantum Mechanical d'Alembert Principle, in Group Theoretical Methods in Physics. Lecture Notes in Physics, vol. 313. Springer, Berlin (1988)

[TG89] Tsao, C.Y., Gambling, W.A.: Curvilinear optical fibre waveguide: characterization of bound modes and radiative field. Proc. R. Soc. Lond. **A425**, 1–16 (1989)

[U61] Ungar, P.: Stable Hill equations. Comm. Pure Appl. Math. **14**, 707–710 (1961)

[WT10] Wachsmuth, J., Teufel, S.: Constrained quantum systems as an adiabatic problem. Phys. Rev. A **82**, 022112 (2010)

[We] Weidmann, J.: Linear Operators in Hilbert Space. Springer, New York (1980)

[WSM92] Wu, H., Sprung, D.W., Martorell, J.: Electronic properties of a quantum wire with arbitrary bending angle. J. Appl. Phys. **72**, 151–154 (1992)

[Y99] Yngvason, J.: Quantum dots. A survey of rigorous results. Mathematical Results in Quantum Mechanics. Operator Theory: Advances and Applications, vol. 108, pp. 161–180. Birkhäuser, Basel (1999)

[Yo98] Yoshitomi, K.: Band gap of the spectrum in periodically curved quantum waveguides. J. Diff. Eqs **142**, 123–166 (1998)

[Za64] Zak, J.: Magnetic translation group. Phys. Rev **134**, A1602–A1606 (1964)

[Zo80] Zorbas, J.: Perturbation of self-adjoint operators by Dirac distributions. J. Math. Phys. **21**, 840–847 (1980)

Index

P. Exner and H. Kovařík, *Quantum Waveguides*,
Theoretical and Mathematical Physics, DOI 10.1007/978-3-319-18576-7

MIX
Papier aus verantwortungsvollen Quellen
Paper from responsible sources
FSC® C105338

If you have any concerns about our products,
you can contact us on
ProductSafety@springernature.com

In case Publisher is established outside the EU,
the EU authorized representative is:
Springer Nature Customer Service Center GmbH
Europaplatz 3, 69115 Heidelberg, Germany

Printed by Libri Plureos GmbH
in Hamburg, Germany